Philosophy in the Flesh

Philosophy in the Flesh

THE EMBODIED MIND AND ITS CHALLENGE TO WESTERN THOUGHT

George Lakoff and Mark Johnson

BASIC BOOKS

A Member of the Perseus Books Group

Published by Basic Books, A Member of the Perseus Books Group

Library of Congress Cataloging-in-Publication Data
Lakoff, George.
    Philosophy in the flesh : the embodied mind and its challenge to western thought / by George Lakoff and Mark Johnson.
        p.   cm.
    Includes index.
    ISBN-10 0-465-05673-3 (cloth); 0-465-05674-1 (pbk.)
    ISBN-13 978-0-465-05673-6 (cloth); 978-0-465-05674-3 (pbk.)
    1. Philosophy of mind.   2. Cognitive science.   I. Johnson, Mark,
1949–   . II. Title.
BD418.3.L35   1999
128—dc21   98-37113
                                                                  CIP

LSC-C

Printing 17, 2021

For
Three generations of Lakoffs
Herman, Sandy, and Andy

and for
Sandra McMorris Johnson

Contents

Part III

THE COGNITIVE SCIENCE OF PHILOSOPHY

Part IV

EMBODIED PHILOSOPHY

Acknowledgments

An undertaking of this scope would not be possible without a great deal of help. Our students and colleagues at Berkeley and at Oregon have contributed immeasurably to this work, and we would like to thank them, as well as other friends who have gone out of their way to help in this enterprise.

Claudia Baracchi, Thomas Alexander, David Barton, and Robert Hahn helped us greatly to appreciate the subtleties of Greek philosophy and provided valuable assistance with Chapters 16, 17, and 18.

Rick Grush, Michele Emanatian, and Tim Rohrer read and gave extensive comments on various drafts of the manuscript as a whole.

Eve Sweetser was involved in the project from the beginning and has made innumerable suggestions and contributions over the years. Her groundbreaking research on the metaphors for mind laid the foundation for the study presented in Chapter 12.

Jerome Feldman's definitive contributions to the neural theory of language helped lay the foundations for the approach to the embodiment of mind that is central to this work. In addition, we have benefited immensely from his detailed suggestions on various drafts of the manuscript.

Robert Powell went above and beyond the call of collegiality and friendship in helping to work out the mathematical details of rational action in Chapter 23.

Joseph Grady's investigations into the experiential basis of metaphor played a major role in the theory presented in this book, as did Srini Narayanan's neural theory of metaphor, which meshed perfectly with Grady's results.

Jane Espenson spent more than a year researching the metaphorical structure of causation before leaving Berkeley for a writing career in Hollywood.

None of this work would have been possible without the development of cognitive linguistics as a discipline. Our debt to those who have defined the discipline is both immense and evident throughout the entire project. In particular, our gratitude goes to Ron Langacker, Leonard Talmy, Gilles Fauconnier, Eve Sweetser, Charles Fillmore, Mark Turner, Claudia Brugman, Adele Goldberg, and Alan Cienki.

We owe a special debt to Evan Thompson, Francisco Varela, and Eleanor Rosch, whose extensive work on embodied cognition has been inspirational and has informed our thinking throughout.

In addition, many students and colleagues have made analytic suggestions and contributed unpublished research that has greatly enriched this work. We would like to recognize their contributions, chapter by chapter.

Chapter 3, The Embodied Mind and Appendix: Jerome Feldman, Terry Regier, David Bailey, and Srini Narayanan.

Chapter 10, Time: Rafael Núñez, Kevin Moore, Jeong-Woon Park, Mark Turner, and John Robert Ross.

Chapter 11, Events and Causes: Jane Espenson, Karin Myhre, Sharon Fischler, Claudia Brugman, Adele Goldberg, Sarah Taub, and Tim Rohrer.

Chapter 12, The Mind: Eve Sweetser, Alan Schwartz, Michele Emanatian, and György László.

Chapter 13, The Self: Miles Becker, Andrew Lakoff, Yukio Hirose, and Bart Wisialowski.

Chapter 14, Morality: Bruce Buchanan, Sarah Taub, Chris Klingebiel, and Tim Adamson.

We also extend our deepest thanks to others who have helped in various ways: Robert Adcock, David Abram, Michael Barzelay, David Collier, Owen Flanagan, David Galin, Christopher Johnson, Dan Jurafsky, Jean-Pierre Koenig, Tony Leiserowitz, Robert McCauley, James D. McCawley, Laura Michaelis, Pamela Morgan, Charlene Spretnak, and Lionel Wee. We would also like to thank all the participants in seminars we have held on philosophy and cognitive science at Berkeley and at Oregon, as well as the participants in our seminar at the 1996 Berkeley Summer Institute. We are indebted to several anonymous readers whose comments made this a much better book.

We are especially grateful for the diligent copyediting of Ann Moru and the editorial guidance of William Frucht.

Our wives, Kathleen Frumkin and Sandi Johnson, have not only provided support above and beyond the call of duty with their love and abundant patience extending over many years, but have also given us regular and invaluable feedback on our ideas.

We would especially like to express our gratitude for the privilege of working in the extraordinary intellectual communities at the University of California at Berkeley and the University of Oregon at Eugene. We owe a particular debt to the Institute for Cognitive Studies and the International Computer Science Institute at Berkeley.

Finally, we want to honor the two greatest philosophers of the embodied mind. Any book with the words "philosophy" and "flesh" in the title must express its obvious debt to Maurice Merleau-Ponty. He used the word "flesh" for our primordial embodied experience and sought to focus the attention of philosophy on what he called "the flesh of the world," the world as we feel it by living in it. John Dewey, no less than Merleau-Ponty, saw that our bodily experience is the primal basis for everything we can mean, think, know, and communicate. He understood the full richness, complexity, and philosophical importance of bodily experience. For their day, Dewey and Merleau-Ponty were models of what we will refer to as "empirically responsible philosophers." They drew upon the best available empirical psychology, physiology, and neuroscience to shape their philosophical thinking.

A Note on the References

In citing the many sources we used in preparing this book, we have departed somewhat from the familiar author-date reference system by arranging the sources into subject categories. Each citation begins with a capitalized letter and a number (or sometimes just a letter) keyed to its location in the reference list. For example, our book *Metaphors We Live By* is cited in the text as (A1, Lakoff and Johnson 1980). This tells readers that the full listing can be found in section A1 of the references, "Metaphor Theory." We have done this to make it easier to find specific references and also to provide a helpful start for readers who want to dig deeper into the literature on particular topics.

Here is a complete list of the subject categories:

A. Cognitive Science and Cognitive Linguistics
   A1. Metaphor Theory
   A2. Experimental Studies in Metaphor
   A3. Metaphor in Gesture and American Sign Language
   A4. Categorization
   A5. Color
   A6. Framing
   A7. Mental Spaces and Conceptual Blending
   A8. Cognitive Grammar and Image Schemas
   A9. Discourse and Pragmatics
   A10. Decision Theory: The Heuristics and Biases Approach

B. Neuroscience and Neural Modeling
   B1. Basic Neuroscience
   B2. Structured Connectionist Modeling

Part I

How the Embodied Mind Challenges the Western Philosophical Tradition

1

Introduction: Who Are We?

How Cognitive Science Reopens
Central Philosophical Questions

The mind is inherently embodied.

Thought is mostly unconscious.

Abstract concepts are largely metaphorical.

These are three major findings of cognitive science. More than two millennia of a priori philosophical speculation about these aspects of reason are over. Because of these discoveries, philosophy can never be the same again.

When taken together and considered in detail, these three findings from the science of the mind are inconsistent with central parts of Western philosophy. They require a thorough rethinking of the most popular current approaches, namely, Anglo-American analytic philosophy and postmodernist philosophy.

This book asks: What would happen if we started with these empirical discoveries about the nature of mind and constructed philosophy anew? The answer is that an empirically responsible philosophy would require our culture to abandon some of its deepest philosophical assumptions. This book is an extensive study of what many of those changes would be in detail.

Our understanding of what the mind is matters deeply. Our most basic philosophical beliefs are tied inextricably to our view of reason. Reason has been taken for over two millennia as the defining characteristic of human beings. Reason includes not only our capacity for logical inference, but also our ability to conduct inquiry, to solve problems, to evaluate, to criticize, to deliberate about how we should act, and to reach an understanding of ourselves, other people, and the world. A radical change in our understanding of reason is therefore a radical change in our understanding of ourselves. It is surprising to discover, on the basis of empirical research, that human rationality is not at all what the Western philosophical tradition has held it to be. But it is shocking to discover that we are very different from what our philosophical tradition has told us we are.

Let us start with the changes in our understanding of reason:

- Reason is not disembodied, as the tradition has largely held, but arises from the nature of our brains, bodies, and bodily experience. This is not just the innocuous and obvious claim that we need a body to reason; rather, it is the striking claim that the very structure of reason itself comes from the details of our embodiment. The same neural and cognitive mechanisms that allow us to perceive and move around also create our conceptual systems and modes of reason. Thus, to understand reason we must understand the details of our visual system, our motor system, and the general mechanisms of neural binding. In summary, reason is not, in any way, a transcendent feature of the universe or of disembodied mind. Instead, it is shaped crucially by the peculiarities of our human bodies, by the remarkable details of the neural structure of our brains, and by the specifics of our everyday functioning in the world.

- Reason is evolutionary, in that abstract reason builds on and makes use of forms of perceptual and motor inference present in "lower" animals. The result is a Darwinism of reason, a rational Darwinism: Reason, even in its most abstract form, makes use of, rather than transcends, our animal nature. The discovery that reason is evolutionary utterly changes our relation to other animals and changes our conception of human beings as uniquely rational. Reason is thus not an essence that separates us from other animals; rather, it places us on a continuum with them.

- Reason is not "universal" in the transcendent sense; that is, it is not part of the structure of the universe. It is universal, however, in that it is a capacity shared universally by all human beings. What allows it to be shared are the commonalities that exist in the way our minds are embodied.

- Reason is not completely conscious, but mostly unconscious.

- Reason is not purely literal, but largely metaphorical and imaginative.

- Reason is not dispassionate, but emotionally engaged.

This shift in our understanding of reason is of vast proportions, and it entails a corresponding shift in our understanding of what we are as human beings. What we now know about the mind is radically at odds with the major classical philosophical views of what a person is.

For example, there is no Cartesian dualistic person, with a mind separate from and independent of the body, sharing exactly the same disembodied transcendent reason with everyone else, and capable of knowing everything about his or her mind simply by self-reflection. Rather, the mind is inherently embodied, reason is shaped by the body, and since most thought is unconscious, the mind cannot be known simply by self-reflection. Empirical study is necessary.

There exists no Kantian radically autonomous person, with absolute freedom and a transcendent reason that correctly dictates what is and isn't moral. Reason, arising from the body, doesn't transcend the body. What universal aspects of reason there are arise from the commonalities of our bodies and brains and the environments we inhabit. The existence of these universals does not imply that reason transcends the body. Moreover, since conceptual systems vary significantly, reason is not entirely universal.

Since reason is shaped by the body, it is not radically free, because the possible human conceptual systems and the possible forms of reason are limited. In addition, once we have learned a conceptual system, it is neurally instantiated in our brains and we are not free to think just anything. Hence, we have no absolute freedom in Kant's sense, no full autonomy. There is no a priori, purely philosophical basis for a universal concept of morality and no transcendent, universal pure reason that could give rise to universal moral laws.

The utilitarian person, for whom rationality is economic rationality—the maximization of utility—does not exist. Real human beings are not, for the most part, in conscious control of—or even consciously aware of—their reasoning. Most of their reason, besides, is based on various kinds of prototypes, framings, and metaphors. People seldom engage in a form of economic reason that could maximize utility.

The phenomenological person, who through phenomenological introspection alone can discover everything there is to know about the mind and the nature of experience, is a fiction. Although we can have a theory of a vast, rapidly and automatically operating cognitive unconscious, we have no direct conscious access to its operation and therefore to most of our thought. Phenomenological reflection, though valuable in revealing the structure of experience, must be supplemented by empirical research into the cognitive unconscious.

There is no poststructuralist person—no completely decentered subject for whom all meaning is arbitrary, totally relative, and purely historically contingent, unconstrained by body and brain. The mind is not merely embodied, but embodied in such a way that our conceptual systems draw largely upon the commonalities of our bodies and of the environments we live in. The result is that much of a person's conceptual system is either universal or widespread across languages and cultures. Our conceptual systems are not totally relative and not *merely* a matter of historical contingency, even though a degree of conceptual relativity does exist and even though historical contingency does matter a great deal. The grounding of our conceptual systems in shared embodiment and bodily experience creates a largely centered self, but not a monolithic self.

There exists no Fregean person—as posed by analytic philosophy—for whom thought has been extruded from the body. That is, there is no real person whose embodiment plays no role in meaning, whose meaning is purely objective and defined by the external world, and whose language can fit the external world with no significant role played by mind, brain, or body. Because our conceptual systems grow out of our bodies, meaning is grounded in and through our bodies. Because a vast range of our concepts are metaphorical, meaning is not entirely literal and the classical correspondence theory of truth is false. The correspondence theory holds that statements are true or false objectively, depending on how they map directly onto the world—independent of any human understanding of either the statement or the world. On the contrary, truth is mediated by embodied understanding and imagination. That does not mean that truth is purely subjective or that there is no stable truth. Rather, our common embodiment allows for common, stable truths.

There is no such thing as a computational person, whose mind is like computer software, able to work on any suitable computer or neural hardware—whose mind somehow derives meaning from taking meaningless symbols as input, manipulating them by rule, and giving meaningless symbols as output. Real people have embodied minds whose conceptual systems arise from, are shaped by, and are given meaning through living human bodies. The neural structures of our brains produce conceptual systems and linguistic structures that cannot be adequately accounted for by formal systems that only manipulate symbols.

Finally, there is no Chomskyan person, for whom language is pure syntax, pure form insulated from and independent of all meaning, context, perception, emotion, memory, attention, action, and the dynamic nature of communication. Moreover, human language is not a totally genetic innovation. Rather, central aspects of language arise evolutionarily from sensory, motor, and other neural systems that are present in "lower" animals.

Classical philosophical conceptions of the person have stirred our imaginations and taught us a great deal. But once we understand the importance of the cognitive unconscious, the embodiment of mind, and metaphorical thought, we can never go back to a priori philosophizing about mind and language or to philosophical ideas of what a person is that are inconsistent with what we are learning about the mind.

Given our new understanding of the mind, the question of what a human being is arises for us anew in the most urgent way.

Asking Philosophical Questions Requires Using Human Reason

If we are going to ask philosophical questions, we have to remember that we are human. As human beings, we have no special access to any form of purely objective or transcendent reason. We must necessarily use common human cognitive and neural mechanisms. Because most of our thought is unconscious, a priori philosophizing provides no privileged direct access to knowledge of our own mind and how our experience is constituted.

In asking philosophical questions, we use a reason shaped by the body, a cognitive unconscious to which we have no direct access, and metaphorical thought of which we are largely unaware. The fact that abstract thought is mostly metaphorical means that answers to philosophical questions have always been, and always will be, mostly metaphorical. In itself, that is neither good nor bad. It is simply a fact about the capacities of the human mind. But it has major consequences for every aspect of philosophy. Metaphorical thought is the principal tool that makes philosophical insight possible and that constrains the forms that philosophy can take.

Philosophical reflection, uninformed by cognitive science, did not discover, establish, and investigate the details of the fundamental aspects of mind we will be discussing. Some insightful philosophers did notice some of these phenomena, but lacked the empirical methodology to establish the validity of these results and to study them in fine detail. Without empirical confirmation, these facts about the mind did not find their way into the philosophical mainstream.

Jointly, the cognitive unconscious, the embodiment of mind, and metaphorical thought require not only a new way of understanding reason and the nature of a person. They also require a new understanding of one of the most common and natural of human activities—asking philosophical questions.

What Goes into Asking and Answering Philosophical Questions?

If you're going to reopen basic philosophical issues, here's the minimum you have to do. First, you need a method of investigation. Second, you have to use that method to understand basic philosophical concepts. Third, you have to apply that method to previous philosophies to understand what they are about and what makes them hang together. And fourth, you have to use that method to ask the big questions: What it is to be a person? What is morality? How do we understand the causal structure of the universe? And so on.

This book takes a small first step in each of these areas, with the intent of giving an overview of the enterprise of rethinking what philosophy can become. The methods we use come from cognitive science and cognitive linguistics. We discuss these methods in Part I of the book.

In Part II, we study the cognitive science of basic philosophical ideas. That is, we use these methods to analyze certain basic concepts that any approach to philosophy must address, such as time, events, causation, the mind, the self, and morality.

In Part III, we begin the study of philosophy itself from the perspective of cognitive science. We apply these analytic methods to important moments in the history of philosophy: Greek metaphysics, including the pre-Socratics, Plato, and Aristotle; Descartes's theory of mind and Enlightenment faculty psychology; Kant's moral theory; and analytic philosophy. These methods, we argue, lead to new and deep insights into these great intellectual edifices. They help us understand those philosophies and explain why, despite their fundamental differences, they have each seemed intuitive to many people over the centuries. We also take up issues in contemporary philosophy, linguistics, and the social sciences, in particular, Anglo-American analytic philosophy, Chomskyan linguistics, and the rational-actor model used in economics and foreign policy.

Finally, in Part IV, we summarize what we have learned in the course of this inquiry about what human beings are and about the human condition.

What emerges is a philosophy close to the bone. A philosophical perspective based on our empirical understanding of the embodiment of mind is a philosophy in the flesh, a philosophy that takes account of what we most basically are and can be.

2

The Cognitive Unconscious

Living a human life is a philosophical endeavor. Every thought we have, every decision we make, and every act we perform is based upon philosophical assumptions so numerous we couldn't possibly list them all. We go around armed with a host of presuppositions about what is real, what counts as knowledge, how the mind works, who we are, and how we should act. Such questions, which arise out of our daily concerns, form the basic subject matter of philosophy: metaphysics, epistemology, philosophy of mind, ethics, and so on.

Metaphysics, for example, is a fancy name for our concern with what is real. Traditional metaphysics asks questions that sound esoteric: What is essence? What is causation? What is time? What is the self? But in everyday terms there is nothing esoteric about such questions.

Take our concern with morality. Does morality consist of a set of absolute moral laws that come from universal reason? Or is it a cultural construct? Or neither? Are there unchanging universal moral values? Where does morality come from? Is it part of the essence of what it is to be a human being? *Is* there an essence of what it is to be a human being? And what, exactly, is an essence anyway?

Causation might appear to be another esoteric topic that only a philosopher could care about. But our moral and political commitments and actions presuppose implicit views on whether there are social causes and, if so, what they might be. Whenever we attribute moral or social responsibility, we are implicitly assuming the possibility of causation, as well as very specific notions of what a cause is.

9

Or take the self. Asking about the nature of the self might seem to be the ultimate in esoteric metaphysical speculation. But we cannot get through a day without relying on unconscious conceptions of the internal structure of the self. Have you taken a good look at yourself recently? Are you trying to find your "true self"? Are you in control of yourself? Do you have a hidden self that you are trying to protect or that is so awful you don't want anyone to know about it? If you have ever considered any matters of this sort, you have been relying on unconscious models of what a self is, and you could hardly live a life of any introspection at all without doing so.

Though we are only occasionally aware of it, we are all metaphysicians—not in some ivory-tower sense but as part of our everyday capacity to make sense of our experience. It is through our conceptual systems that we are able to make sense of everyday life, and our everyday metaphysics is embodied in those conceptual systems.

The Cognitive Unconscious

Cognitive science is the scientific discipline that studies conceptual systems. It is a relatively new discipline, having been founded in the 1970s. Yet in a short time it has made startling discoveries. It has discovered, first of all, that most of our thought is unconscious, not in the Freudian sense of being repressed, but in the sense that it operates beneath the level of cognitive awareness, inaccessible to consciousness and operating too quickly to be focused on.

Consider, for example, all that is going on below the level of conscious awareness when you are in a conversation. Here is only a small part of what you are doing, second by second:

Accessing memories relevant to what is being said
Comprehending a stream of sound as being language, dividing it into distinctive phonetic features and segments, identifying phonemes, and grouping them into morphemes
Assigning a structure to the sentence in accord with the vast number of grammatical constructions in your native language
Picking out words and giving them meanings appropriate to context
Making semantic and pragmatic sense of the sentences as a whole
Framing what is said in terms relevant to the discussion
Performing inferences relevant to what is being discussed

Constructing mental images where relevant and inspecting them
Filling in gaps in the discourse
Noticing and interpreting your interlocutor's body language
Anticipating where the conversation is going
Planning what to say in response

Cognitive scientists have shown experimentally that to understand even the simplest utterance, we must perform these and other incredibly complex forms of thought automatically and without noticeable effort below the level of consciousness. It is not merely that we occasionally do not notice these processes; rather, they are inaccessible to conscious awareness and control.

When we understand all that constitutes the cognitive unconscious, our understanding of the nature of consciousness is vastly enlarged. Consciousness goes way beyond mere awareness of something, beyond the mere experience of qualia (the qualitative senses of, for example, pain or color), beyond the awareness that you are aware, and beyond the multiple takes on immediate experience provided by various centers of the brain. Consciousness certainly involves all of the above plus the immeasurably vaster constitutive framework provided by the cognitive unconscious, which must be operating for us to be aware of anything at all.

Why "Cognitive" Unconscious?

The term *cognitive* has two very different meanings, which can sometimes create confusion. In cognitive science, the term *cognitive* is used for any kind of mental operation or structure that can be studied in precise terms. Most of these structures and operations have been found to be unconscious. Thus, visual processing falls under the *cognitive,* as does auditory processing. Obviously, neither of these is conscious, since we are not and could not possibly be aware of each of the neural processes involved in the vastly complicated total process that gives rise to conscious visual and auditory experience. Memory and attention fall under the *cognitive.* All aspects of thought and language, conscious or unconscious, are thus *cognitive.* This includes phonology, grammar, conceptual systems, the mental lexicon, and all unconscious inferences of any sort. Mental imagery, emotions, and the conception of motor operations have also been studied from such a cognitive perspective. And neural modeling of any cognitive operation is also part of cognitive science.

Confusion sometimes arises because the term *cognitive* is often used in a very different way in certain philosophical traditions. For philosophers in these traditions, *cognitive* means *only* conceptual or propositional structure. It also includes rule-governed operations on such conceptual and propositional structures. Moreover, *cognitive* meaning is seen as truth-conditional meaning, that is, meaning defined not internally in the mind or body, but by reference to things in the external world. Most of what we will be calling the *cognitive unconscious* is thus for many philosophers not considered *cognitive* at all.

As is the practice in cognitive science, we will use the term *cognitive* in the richest possible sense, to describe any mental operations and structures that are involved in language, meaning, perception, conceptual systems, and reason. Because our conceptual systems and our reason arise from our bodies, we will also use the term *cognitive* for aspects of our sensorimotor system that contribute to our abilities to conceptualize and to reason. Since cognitive operations are largely unconscious, the term *cognitive unconscious* accurately describes all unconscious mental operations concerned with conceptual systems, meaning, inference, and language.

The Hidden Hand That Shapes Conscious Thought

The very existence of the cognitive unconscious, a fact fundamental to all conceptions of cognitive science, has important implications for the practice of philosophy. It means that we can have no direct conscious awareness of most of what goes on in our minds. The idea that pure philosophical reflection can plumb the depths of human understanding is an illusion. Traditional methods of philosophical analysis alone, even phenomenological introspection, cannot come close to allowing us to know our own minds.

There is much to be said for traditional philosophical reflection and phenomenological analysis. They can make us aware of many aspects of consciousness and, to a limited extent, can enlarge our capacities for conscious awareness. Phenomenological reflection even allows us to examine many of the background prereflective structures that lie beneath our conscious experience. But neither method can adequately explore the cognitive unconscious—the realm of thought that is completely and irrevocably inaccessible to direct conscious introspection. It is this realm that is the primary focus of cognitive science, which allows us to theorize about the cognitive unconscious on the basis of evidence. Cognitive science, however, does not allow us direct access to what the cognitive unconscious is doing as it is doing it.

Conscious thought is the tip of an enormous iceberg. It is the rule of thumb among cognitive scientists that unconscious thought is 95 percent of all thought—and that may be a serious underestimate. Moreover, the 95 percent below the surface of conscious awareness shapes and structures all conscious thought. If the cognitive unconscious were not there doing this shaping, there could be no conscious thought.

The cognitive unconscious is vast and intricately structured. It includes not only all our automatic cognitive operations, but also all our implicit knowledge. All of our knowledge and beliefs are framed in terms of a conceptual system that resides mostly in the cognitive unconscious.

Our unconscious conceptual system functions like a "hidden hand" that shapes how we conceptualize all aspects of our experience. This hidden hand gives form to the metaphysics that is built into our ordinary conceptual systems. It creates the entities that inhabit the cognitive unconscious—abstract entities like friendships, bargains, failures, and lies—that we use in ordinary unconscious reasoning. It thus shapes how we automatically and unconsciously comprehend what we experience. It constitutes our unreflective common sense.

For example, let us return to our commonsense understanding of the self. Consider the common experience of struggling to gain control over ourselves. We not only feel this struggle within us, but conceptualize the "struggle" as being between two distinct parts of our self, each with different values. Sometimes we think of our "higher" (moral and rational) self struggling to get control over our "lower" (irrational and amoral) self.

Our conception of the self, in such cases, is fundamentally metaphoric. We conceptualize ourselves as split into two distinct entities that can be at war, locked in a struggle for control over our bodily behavior. This metaphoric conception is rooted deep in our unconscious conceptual systems, so much so that it takes considerable effort and insight to see how it functions as the basis for reasoning about ourselves.

Similarly, when you try to find your "true self," you are using another, usually unconscious metaphorical conceptualization. There are more than a dozen such metaphorical conceptions of the self, and we will discuss them below. When we consciously reason about how to gain mastery over ourselves, or how to protect our vulnerable "inner self," or how to find our "true self," it is the hidden hand of the unconscious conceptual system that makes such reasoning "common sense."

Metaphysics as Metaphor

A large part of this book will be devoted to exploring in detail what the hidden hand of our unconscious conceptual system looks like and how it shapes not only everyday commonsense reasoning but also philosophy itself. We will discuss some of the most basic of philosophical concepts, not only the self but also time, events, causation, essence, the mind, and morality. What is startling is that, even for these most basic of concepts, the hidden hand of the unconscious mind uses metaphor to define our unconscious metaphysics—the metaphysics used not just by ordinary people, but also by philosophers to make sense of these concepts. As we will see, what counts as an "intuitive" philosophical theory is one that draws upon these unconscious metaphors. In short, philosophical theories are largely the product of the hidden hand of the cognitive unconscious.

Throughout history it has been virtually impossible for philosophers to do metaphysics without such metaphors. For the most part, philosophers engaged in making metaphysical claims are choosing from the cognitive unconscious a set of existing metaphors that have a consistent ontology. That is, using unconscious everyday metaphors, philosophers seek to make a noncontradictory choice of conceptual entities defined by those metaphors; they then take those entities to be real and systematically draw out the implications of that choice in an attempt to account for our experience using that metaphysics.

Metaphysics in philosophy is, of course, supposed to characterize what is real—literally real. The irony is that such a conception of the real depends upon unconscious metaphors.

Empirically Responsible Philosophy: Beyond Naturalized Epistemology

For more than two thousand years, philosophy has defined metaphysics as the study of what is literally real. The weight of that tradition is so great that it is hardly likely to change in the face of empirical evidence against the tradition itself. Nevertheless that evidence, which comes from cognitive science, exists and raises deep questions not only about the project of philosophical metaphysics but also about the nature of philosophy itself.

Throughout most of our history, philosophy has seen itself as being independent of empirical investigation. It is that aspect of philosophy that is called into question by results in cognitive science. Through the study of the cognitive unconscious, cognitive science has given us a radically new view of how we conceptualize our experience and how we think.

Cognitive science—the empirical study of the mind—calls upon us to create a new, empirically responsible philosophy, a philosophy consistent with empirical discoveries about the nature of mind. This is not just old-fashioned philosophy "naturalized"—making minor adjustments, but basically keeping the old philosophical superstructure.

A serious appreciation of cognitive science requires us to rethink philosophy from the beginning, in a way that would put it more in touch with the reality of how we think. It would be based on a detailed understanding of the cognitive unconscious, the hidden hand that shapes our conscious thought, our moral values, our plans, and our actions.

Unless we know our cognitive unconscious fully and intimately, we can neither know ourselves nor truly understand the basis of our moral judgments, our conscious deliberations, and our philosophy.

3
The Embodied Mind

What does it mean to say that concepts and reason are embodied? This chapter takes a first step toward answering that question. It takes up the role that the perceptual and motor systems play in shaping particular kinds of concepts: color concepts, basic-level concepts, spatial-relations concepts, and aspectual (event-structuring) concepts.

Any reasoning you do using a concept requires that the neural structures of the brain carry out that reasoning. Accordingly, the architecture of your brain's neural networks determines what concepts you have and hence the kind of reasoning you can do. Neural modeling is the field that studies which configurations of neurons carry out the neural computations that we experience as particular forms of rational thought. It also studies how such neural configurations are learned.

Neural modeling can show in detail one aspect of what it means for the mind to be embodied: how particular configurations of neurons, operating according to principles of neural computation, compute what we experience as rational inferences. At this point the vague question "Can reason make use of the sensorimotor system?" becomes the technically answerable question "Can rational inferences be computed by the same neural architecture used in perception or bodily movement?" We now know that, in some cases, the answer to this question is yes. Those cases will be discussed in this chapter.

How the Body and Brain Shape Reason

We have inherited from the Western philosophical tradition a theory of faculty psychology, in which we have a "faculty" of reason that is separate from and independent of what we do with our bodies. In particular, reason is seen as independent of perception and bodily movement. In the Western tradition, this autonomous capacity of reason is regarded as what makes us essentially human, distinguishing us from all other animals. If reason were not autonomous, that is, not independent of perception, motion, emotion, and other bodily capacities, then the philosophical demarcation between us and all other animals would be less clearly drawn. This view was formulated prior to the emergence of evolutionary theory, which shows that human capacities grow out of animal capacities.

The evidence from cognitive science shows that classical faculty psychology is wrong. There is no such fully autonomous faculty of reason separate from and independent of bodily capacities such as perception and movement. The evidence supports, instead, an evolutionary view, in which reason uses and grows out of such bodily capacities. The result is a radically different view of what reason is and therefore of what a human being is. This chapter surveys some of the evidence for the view that reason is fundamentally embodied.

These findings of cognitive science are profoundly disquieting in two respects. First, they tell us that human reason is a form of animal reason, a reason inextricably tied to our bodies and the peculiarities of our brains. Second, these results tell us that our bodies, brains, and interactions with our environment provide the mostly unconscious basis for our everyday metaphysics, that is, our sense of what is real.

Cognitive science provides a new and important take on an age-old philosophical problem, the problem of what is real and how we can know it, if we can know it. Our sense of what is real begins with and depends crucially upon our bodies, especially our sensorimotor apparatus, which enables us to perceive, move, and manipulate, and the detailed structures of our brains, which have been shaped by both evolution and experience.

Neural Beings Must Categorize

Every living being categorizes. Even the amoeba categorizes the things it encounters into food or nonfood, what it moves toward or moves away from. The amoeba cannot choose whether to categorize; it just does. The same is true at every level of the animal world. Animals categorize food, predators, possible mates, members of their own species, and so on. How animals categorize depends upon their sensing apparatus and their ability to move themselves and to manipulate objects.

Categorization is therefore a consequence of how we are embodied. We have evolved to categorize; if we hadn't, we would not have survived. Categorization is, for the most part, not a product of conscious reasoning. We categorize as we do because we have the brains and bodies we have and because we interact in the world the way we do.

The first and most important thing to realize about categorization is that it is an inescapable consequence of our biological makeup. We are neural beings. Our brains each have 100 billion neurons and 100 trillion synaptic connections. It is common in the brain for information to be passed from one dense ensemble of neurons to another via a relatively sparse set of connections. Whenever this happens, the pattern of activation distributed over the first set of neurons is too great to be represented in a one-to-one manner in the sparse set of connections. Therefore, the sparse set of connections necessarily groups together certain input patterns in mapping them across to the output ensemble. Whenever a neural ensemble provides the same output with different inputs, there is neural categorization.

To take a concrete example, each human eye has 100 million light-sensing cells, but only about 1 million fibers leading to the brain. Each incoming image must therefore be reduced in complexity by a factor of 100. That is, information in each fiber constitutes a "categorization" of the information from about 100 cells. Neural categorization of this sort exists throughout the brain, up through the highest levels of categories that we can be aware of. When we see trees, we see them as trees, not just as individual objects distinct from one another. The same with rocks, houses, windows, doors, and so on.

A small percentage of our categories have been formed by conscious acts of categorization, but most are formed automatically and unconsciously as a result of functioning in the world. Though we learn new categories regularly, we cannot make massive changes in our category systems through conscious acts of recategorization (though, through experience in the world, our categories are subject to unconscious reshaping and partial change). We do not, and cannot, have full conscious control over how we categorize. Even when we think we are deliberately forming new categories, our unconscious categories enter into our choice of possible conscious categories.

Most important, it is not just that our bodies and brains determine *that* we will categorize; they also determine what kinds of categories we will have and what their structure will be. Think of the properties of the human body that contribute to the peculiarities of our conceptual system. We have eyes and ears, arms and legs that work in certain very definite ways and not in others. We have a visual system, with topographic maps and orientation-sensitive cells, that provides structure for our ability to conceptualize spatial relations. Our abilities to move in the ways we do and to track the motion of other things give motion a major role in our conceptual system. The fact that we have muscles and use them to apply force in certain ways leads to the structure of our system of causal concepts. What is important is not just that we have bodies and that thought is somehow embodied. What is important is that the peculiar nature of our bodies shapes our very possibilities for conceptualization and categorization.

The Inseparability of Categories, Concepts, and Experience

Living systems must categorize. Since we are neural beings, our categories are formed through our embodiment. What that means is that the categories we form are *part of our experience!* They are the structures that differentiate aspects of our experience into discernible kinds. Categorization is thus not a purely intellectual matter, occurring after the fact of experience. Rather, the formation and use of categories is the stuff of experience. It is part of what our bodies and brains are constantly engaged in. We cannot, as some meditative traditions suggest, "get beyond" our categories and have a purely uncategorized and unconceptualized experience. Neural beings cannot do that.

What we call *concepts* are neural structures that allow us to mentally characterize our categories and reason about them. Human categories are typically conceptualized in more than one way, in terms of what are called *prototypes*. Each prototype is a neural structure that permits us to do some sort of inferential or imaginative task relative to a category. Typical-case prototypes are used in drawing inferences about category members in the absence of any special contextual information. Ideal-case prototypes allow us to evaluate category members relative to some conceptual standard. (To see the difference, compare the prototypes for the ideal husband and the typical husband.) Social stereotypes are used to make snap judgments, usually about people. Salient exemplars (well-known examples) are used for making probability judgments. (For a survey of kinds of conceptual prototypes, see A4, Lakoff 1987.) In short, prototype-based reasoning constitutes a large proportion of the actual reasoning that we do. Reasoning with prototypes is, indeed, so common that it is inconceivable that we could function for long without it.

Since most categories are matters of degree (e.g., tall people), we also have graded concepts characterizing degrees along some scale with norms of various kinds for extreme cases, normal cases, not quite normal cases, and so on. Such graded norms are described by what are called *linguistic hedges* (A4, Lakoff 1972), for example, *very, pretty, kind of, barely,* and so on. For the sake of imposing sharp distinctions, we develop what might be called *essence prototypes,* which conceptualize categories as if they were sharply defined and minimally distinguished from one another.

When we conceptualize categories in this way, we often envision them using a spatial metaphor, as if they were containers, with an interior, an exterior, and a boundary. When we conceptualize categories as containers, we also impose complex hierarchical systems on them, with some category-containers inside other category-containers. Conceptualizing categories as containers hides a great deal of category structure. It hides conceptual prototypes, the graded structures of categories, and the fuzziness of category boundaries.

In short, we form extraordinarily rich conceptual structures for our categories and reason about them in many ways that are crucial for our everyday functioning. All of these conceptual structures are, of course, neural structures in our brains. This makes them embodied in the trivial sense that any mental construct is realized neurally. But there is a deeper and more important sense in which our concepts are embodied. What makes concepts concepts is their inferential capacity, their ability to be bound together in ways that yield inferences. *An embodied concept is a neural structure that is actually part of, or makes use of, the sensorimotor system of our brains. Much of conceptual inference is, therefore, sensorimotor inference.*

If concepts are, as we believe, embodied in this strong sense, the philosophical consequences are enormous. The locus of reason (conceptual inference) would be the same as the locus of perception and motor control, which are bodily functions. If this seems like a radical claim, it is radical only from the perspective of faculty psychology, a philosophy that posits a radical separation between rational abilities and the sensorimotor system. It is not at all radical from the point of view of the brain, which is the joint locus of reason, perception, and movement. The question from the viewpoint of the brain is whether conceptual inference makes use of the same brain structures as perceptual motor inference. In other words, does reason piggyback on perception and motor control? From the perspective of the brain, the locus of all three functions, it would be quite natural if it did.

Realism, Inference, and Embodiment

The question of what we take to be real and the question of how we reason are inextricably linked. Our categories of things in the world determine what we take to be real: trees, rocks, animals, people, buildings, and so on. Our concepts determine how we reason about those categories. In order to function *realistically* in the world, our categories and our forms of reason must "work" very well together; our concepts must characterize the structure of our categories sufficiently well enough for us to function.

Mainstream Western philosophy adds to this picture certain claims that we will argue are false. Not trivially false, but so false as to drastically distort our understanding of what human beings are, what the mind and reason are, what causation and morality are, and what our place is in the universe. Here are those claims:

1. Reality comes divided up into categories that exist independent of the specific properties of human minds, brains, or bodies.

2. The world has a rational structure: The relationships among categories in the world are characterized by a *transcendent* or *universal* reason, which is independent of any peculiarities of human minds, brains, and bodies.

3. The concepts used by mind-, brain-, and body-free reason correctly characterize the mind-, brain-, and body-free categories of reality.

4. Human reason is the capacity of the human mind to use transcendent reason, or at least a portion of it. Human reason may be performed by the human brain, but the structure of human reason is defined by transcendent reason, independent of human bodies or brains. Thus, the structure of human reason is disembodied.

5. Human concepts are the concepts of transcendent reason. They are therefore defined independent of human brains or bodies, and so they too are disembodied.

6. Human concepts therefore characterize the objective categories of mind-, brain, and body-free reality. That is, the world has a unique, fixed category structure, and we all know it and use it when we are reasoning correctly.

7. What makes us essentially human is our capacity for disembodied reason.

8. Since transcendent reason is culture-free, what makes us essentially human is not our capacity for culture or for interpersonal relations.
9. Since reason is disembodied, what makes us essentially human is not our relation to the material world. Our essential humanness has nothing to do with our connection to nature or to art or to music or to anything of the senses.

Much of the history of mainstream Western philosophy consists of exploring variations on these themes and drawing out the consequences of these claims. A given philosopher may not hold all of these tenets in the strong form that we have stated them; however, together these claims form a picture of concepts, reason, and the world that any student of philosophy will be familiar with. If they are false, then large parts of the Western philosophical tradition and many of our most common beliefs have to be rethought.

These tenets were not adopted on the basis of empirical evidence. They arose instead out of a priori philosophy. Contemporary cognitive science calls this entire philosophical worldview into serious question on empirical grounds. Here is the reason why cognitive science has a crucial bearing on these issues.

At the heart of this worldview are tenets 4, 5, and 6—that human reason and human concepts are mind-, brain-, and body-free and characterize objective, external reality. If these tenets are false, the whole worldview collapses. Suppose human concepts and human reason are body- and brain-dependent. Suppose they are shaped as much by the body and brain as by reality. Then the body and brain are essential to our humanity. Moreover, our notion of what reality is changes. There is no reason whatever to believe that there is a disembodied reason or that the world comes neatly carved up into categories or that the categories of our mind are the categories of the world. If tenets 4, 5, and 6 are empirically incorrect, then we have a lot of rethinking to do about who we are and what our place is in the universe.

Embodied Concepts

In this chapter and the next, we will review some of the results of cognitive science research that bear on these issues. We will suggest, first, that human concepts are not just reflections of an external reality, but that they are crucially shaped by our bodies and brains, especially by our sensorimotor system. We will do so by looking at three kinds of concepts: color concepts, basic-level concepts, and spatial-relations concepts. After that, we will use studies of neural modeling to argue that certain human concepts and forms of conceptual reasoning make use of the sensorimotor system.

The philosophical stakes here are high. As we shall see in later chapters, these arguments have far-reaching implications for who we are and what our role in the world is.

Color Concepts

What could be simpler or more obvious than colors? The sky is blue. Fresh grass is green. Blood is red. The sun and moon are yellow. We see colors as inhering in things. Blue is in the sky, green in the grass, red in the blood, yellow in the sun. We see color, and yet it is false, as false as another thing we see, the moving sun rising past the edge of the stationary earth. Just as astronomy tells us that the earth moves around the sun, not the sun around a stationary earth, so cognitive science tells us that colors do not exist in the external world. Given the world, our bodies and brains have evolved to create color.

Our experience of color is created by a combination of four factors: wavelengths of reflected light, lighting conditions, and two aspects of our bodies: (1) the three kinds of color cones in our retinas, which absorb light of long, medium, and short wavelengths, and (2) the complex neural circuitry connected to those cones.

Here are some crucial things to bear in mind. One physical property of the surface of an object matters for color: its reflectance, that is, the relative percentages of high-, medium-, and low-frequency light that it reflects. That is a constant. But the actual wavelengths of light reflected by an object are not a constant. Take a banana. The wavelengths of light coming from the banana depend on the nature of the light illuminating it: tungsten or fluorescent, daylight on a sunny or a cloudy day, the light of dawn or dusk. Under different conditions the wavelengths of light coming from the banana will differ considerably, yet the color of the banana will be relatively constant; it will look pretty much the same. Color, then, is not just the perception of wavelength; color constancy depends on the brain's ability to compensate for variations in the light source. Moreover, there is not a one-to-one correspondence between reflectance and color; two different reflectances can both be perceived as the same red.

Another crucial thing to bear in mind is that light is not colored. Visible light is electromagnetic radiation, like radio waves, vibrating within a certain frequency range. It is not the kind of thing that could be colored. Only when this electromagnetic radiation impinges on our retinas are we able to see. We see a particular color when the surrounding lighting conditions are right, when radiation in a certain range impinges on our retina, and when our color cones absorb the radiation, producing an electrical signal that is appropriately processed by the neural circuitry of our brains. The qualitative experience that this produces in us is what we call "color."

One might suppose that color is an internal representation of the external reality of the reflectance properties of the surface of an object. If this were true, then the properties of colors and color categories would be representations of reflectances and categories of reflectances. But it is not true. Color concepts have internal structure, with certain colors being "focal." The category *red*, for instance, contains central red as well as noncentral, peripheral hues such as purplish red, pinkish red, and orangish red. The center-periphery structure of categories is a result of the neural response curves for color in our brains. Focal hues correspond to frequencies of maximal neural response. The internal structure of color categories is not out there in the surface reflectances. The same is true of the relationships among colors. The opposition between red and green or blue and yellow is a fact about our neural circuitry, not about the reflectance properties of surfaces. Color is not just the internal representation of external reflectance. And it is not a thing or a substance out there in the world.

To summarize, our color concepts, their internal structures, and the relationships between them are inextricably tied to our embodiment. They are a consequence of four interacting factors: lighting conditions, wavelengths of electromagnetic radiation, color cones, and neural processing. Colors as we see them, say, the red of blood or the blue of the sky, are not out there in the blood or the sky. Indeed, the sky is not even an object. It has no surface for the color to be in. And without a physical surface, the sky does not even have a surface reflectance to be detected as color. The sky is blue because the atmosphere transmits only a certain range of wavelengths of incoming light from the sun, and of the wavelengths it does transmit, it scatters some more than others. The effect is like a colored lightbulb that only lets certain wavelengths of light through the glass. Thus, the sky is blue for a very different reason than a painting of the sky is blue. What we perceive as blue does not characterize a single "thing" in the world, neither "blueness" nor wavelength reflectance.

Color concepts are "interactional"; they arise from the interactions of our bodies, our brains, the reflective properties of objects, and electromagnetic radiation. Colors are not objective; there is in the grass or the sky no greenness or blueness independent of retinas, color cones, neural circuitry, and brains.

Nor are colors purely subjective; they are neither a figment of our imaginations nor spontaneous creations of our brains.

The philosophical consequences are immediate. Since colors are not things or substances in the world, metaphysical realism fails. The meaning of the word *red* cannot be just the relation between the word and something in the world (say, a collection of wavelengths of light or a surface reflectance). An adequate theory of the conceptual structure of *red,* including an account of why it has the structure it has (with focal red, purplish red, orangish red, and so on) cannot be constructed solely from the spectral properties of surfaces. It must make reference to color cones and neural circuitry. Since the cones and neural circuitry are embodied, the internal conceptual properties of *red* are correspondingly embodied.

Subjectivism in its various forms—radical relativism and social constructionism—also fails to explain color, since color is created jointly by our biology and the world, not by our culture. This is not to say that color does not differ in its significance from culture to culture. It clearly does. Rather, color is a function of the world and our biology interacting.

Philosophically, color and color concepts make sense only in something like an embodied realism, a form of interactionism that is neither purely objective nor purely subjective. Color is also important for the "realism" of embodied realism. Evolution has worked with physical limitations: only certain wavelengths of light get through the atmosphere, only certain chemicals react to short, medium, and long wavelengths, and so on. We have evolved within these limitations to have the color systems we have, and they allow us to function well in the world. Plant life has been important to our evolution, and so the ability to place in one category the things that are green has apparent value for survival and flourishing. The same goes for blood and the color red, water and the sky and the color blue, and the sun and the moon and the color yellow. We have the color concepts we do because the physical limitations constraining evolution gave evolutionary advantages to beings with a color system that enabled them to function well in crucial respects.

Color, of course, does more than just help us recognize things in the world. It is an evolved aspect of the brain that plays many roles in our lives, cultural, aesthetic, and emotional. Thinking of color as merely the internal representation of the external reality of surface reflectance is not merely inaccurate; it misses most of the function of color in our lives.

At least since John Locke, philosophers have known that color is an interactional property of objects, what Locke called a "secondary quality" that does not exist in the object itself. Locke contrasted secondary qualities with "primary qualities," which were assumed to exist objectively in things independent of any perceiver. Primary qualities were seen as having metaphysical import, as determining what is real, while secondary qualities were seen as perceiver-dependent and therefore not constitutive of objective reality.

But giving up on color as a metaphysically real "primary quality" has profound philosophical consequences. It means abandoning the correspondence theory of truth, the idea that truth lies in the relationship between words and the metaphysically and objectively real world external to any perceiver. Since there is no color in the world in itself, a sentence like "Blood is red," which we all take to be true, would not be true according to the correspondence theory.

Since the correspondence theory of truth is the one thing many philosophers are not willing to give up, they go to extraordinary lengths to salvage it. Some attempt to see color as the internal representation of external reflectance of surfaces, and to say that "Blood is red" is true if and only if blood has such and such a surface reflectance. As we have seen, the same reasoning cannot work for "The sky is blue," since the sky cannot have a surface reflectance. Some philosophers have even been willing on these grounds to say that "The sky is blue" is false, granting that the sky has no surface reflectance but trying to keep the correspondence theory nonetheless. They claim that those of us who think that it is true that the sky is blue are simply being fooled by an optical illusion! Getting philosophers to give up on the correspondence theory of truth will not be easy. (For a thorough discussion of the details of the color debate in philosophy, see Thompson [A5, 1995]. For an account of the general philosophical implications of color research, see Varela, Thompson, and Rosch [C2, 1991], who argue, as we do, that color is interactional in nature and hence neither objective nor subjective. Defenses of objectivism and subjectivism can be found in Hilbert [A5, 1987, 1992] and Hardin [A5, 1988].)

As we are about to see, color is the tip of the iceberg. What Locke recognized as perceiver-dependence is a fully general phenomenon. Cognitive science and neuroscience suggest that the world as we *know* it contains *no* primary qualities in Locke's sense, because the qualities of things as we can experience and comprehend them depend crucially on our neural makeup, our bodily interactions with them, and our purposes and interests. For real human beings, the only realism is an embodied realism.

Basic-Level Categories

Why has metaphysical realism been so popular over the centuries? Why is it so common to feel that our concepts reflect the world as it is—that our categories of mind fit the categories of the world? One reason is that we have evolved to form at least one important class of categories that optimally fit our bodily experiences of entities and certain extremely important differences in the natural environment—what are called *basic-level categories.*

Our perceptual systems have no problem distinguishing cows from horses, goats from cats, or elephants from giraffes. In the natural world, the categories we distinguish among most readily are the folk versions of biological genera, namely, those that have evolved significantly distinct shapes so as to take advantage of different features of their environments. Go one level down in the biological hierarchy and it is a lot harder to distinguish one species of elephant from another (A4, Berlin et al. 1974). It's the same for physical objects. It's easy to tell cars from boats or trains, but a lot less easy to tell one kind of car from another.

Consider the categories *chair* and *car,* which are "in the middle" of the category hierarchies *furniture–chair–rocking chair* and *vehicle–car–sports car.* In the mid-1970s, Brent Berlin, Eleanor Rosch, Carolyn Mervis, and their coworkers discovered that such mid-level categories are cognitively "basic"— that is, they have a kind of cognitive priority, as contrasted with "superordinate" categories like *furniture* and *vehicle* and with "subordinate" categories like *rocking chair* and *sports car* (A4, Berlin et al. 1974; Mervis and Rosch 1981).

The Body-Based Properties of Basic-Level Categories

Basic-level categories are distinguished from superordinate categories by aspects of our bodies, brains, and minds: mental images, gestalt perception, motor programs, and knowledge structure. The basic level, as Berlin and Rosch found, is characterized by at least four conditions.

Condition 1: It is the highest level at which a single mental image can represent the entire category. For example, you can get a mental image of a chair. You can get mental images of other categories at the basic level such as tables and beds. But you cannot get a mental image of a general piece of furniture—a thing that is not a chair, table, or bed, but something more general. Similarly, you can get a mental image of a car. You can also get mental images of opposing categories at this level such as trains, boats, and planes. But you cannot get a mental image of a generalized vehicle—a thing that is not a car, train, boat, or plane, but a vehicle in general. The basic level is the highest level at which we have mental images that stand for the entire category.

Condition 2: It is the highest level at which category members have similarly perceived overall shapes. You can recognize a chair or a car by its overall shape. There is no overall shape that you can assign to a generalized piece of furniture or a vehicle so that you could recognize the category from that shape. The basic level is the highest level at which category members are recognized by gestalt perception (perception of overall shape).

Condition 3: It is the highest level at which a person uses similar motor actions for interacting with category members. You have motor programs for interacting with objects at the basic level—for interacting with chairs, tables, and beds. But you have no motor programs for interacting with generalized pieces of furniture.

Condition 4: It is the level at which most of our knowledge is organized. You have a lot of knowledge at the basic level. Think for a moment of all that you know about cars versus what you know about vehicles. You know a handful of things about vehicles in general, but a great many things about cars. You know much less about lower-level categories, unless you are an expert.

As a result of these characteristics, the basic level has other priorities over the superordinate and subordinate levels: It is named and understood earlier by children, enters a language earlier in its history, has the shortest primary lexemes, and is identified faster by subjects. The basic level also tends to be used in neutral contexts, that is, contexts in which there is no explicit indication of which level is most appropriate. From the perspective of an overall theory of the human mind, these are important properties of concepts and cannot be ignored.

The Philosophical Significance of the Basic Level

The philosophical significance of these results follows directly. First, the division between basic-level and nonbasic-level categories is body-based, that is, based on gestalt perception, motor programs, and mental images. Because of this, classical metaphysical realism cannot be right, since the properties of categories are mediated by the body rather than determined directly by a mind-independent reality.

Second, the basic level is that level at which people interact optimally with their environments, given the kinds of bodies and brains they have and the kinds of environments they inhabit. How is this possible? The best answer we know, suggested by Tversky and Hemenway (A4, 1984), is that the properties that make for basic-level categories are responses to the part-whole structure of physical beings and objects. Gestalt perception is about overall part-whole structure, as is mental imagery. The use of motor schemas to interact with objects depends significantly on their overall part-whole structure. Moreover, the functions something can perform, and hence what we know about it, likewise depend to a significant degree on part-whole structure. That is why there is a basic-level category structure with respect to which we can function optimally.

Third, basic-level categorization tells us why metaphysical realism makes sense for so many people, where it seems to work, and where it goes wrong. Metaphysical realism seems to work primarily at the basic level. If you look only at examples of basic-level categories, at the level of category where we interact optimally with the world, then it appears as if our conceptual categories fit the categories of the world. If you look at categories at other levels, it does not (A4, Berlin et al. 1974). It is not surprising, therefore, that philosophical discussions about the relationship between our categories and things in the world tend to use basic-level examples. Philosophical examples like "The cat is on the mat" or "The boy hit the ball" typically use basic-level categories like *cat, mat, boy,* and *ball* or basic-level substances like *water* and *gold.* It is no accident that philosophers do not try to make their argument with things farther down on the biological taxonomy: brown-capped chickadees, brown-headed nuthatches, Bewick's wrens, bushtits, and so on.

The basic level, of course, is not just about objects. There are basic-level actions, actions for which we have conventional mental images and motor programs, like swimming, walking, and grasping. We also have basic-level social concepts, like families, clubs, and baseball teams, as well as basic-level social actions, like arguing. And there are basic emotions, like happiness, anger, and sadness.

Fourth, the properties of the basic level explain an important aspect of the stability of scientific knowledge. For basic-level physical objects and basic-level actions or relations, the link between human categories and divisions of things in the world is quite accurate. We can think of scientific instruments as extending these basic-level abilities to perceive, image, and intervene. Telescopes, microscopes, cameras, and delicate probing instruments of all sorts extend our capacity for basic-level perception, imaging, and intervention. Such instruments allow us to greatly extend the range of our categories of mind to fit important distinctions in the world.

For basic-level categories, the idea that our categories of mind fit the categories of the world is not that far off. When our basic-level capacities are extended by scientific instrumentation, our ability to select useful real-world divisions is improved. Basic-level categories are the source of our most stable knowledge, and the technological capacity to extend them allows us to extend our stable knowledge.

In summary, our categories arise from the fact that we are neural beings, from the nature of our bodily capacities, from our experience interacting in the world, and from our evolved capacity for basic-level categorization—a level at which we optimally interact with the world. Evolution has not required us to be as accurate above and below the basic level as at the basic level, and so we are not.

There is a reason why our basic-level categorization and evolution match up. In the natural world, basic-level categories of organisms are genera. That means that they are for the most part determined by their overall part-whole structure. The part-whole structure of a class of organisms is, significantly, what determines whether it will survive and function well in a given environment. Thus, part-whole structure determines the natural categories of existing genera. And it is what our perceptual and motor systems have evolved to recognize at the basic level. That is why we have tended over our evolutionary history to function optimally in our basic-level interactions.

Though the facts of basic-level categorization do not fit metaphysical realism, they do provide us with the basis for embodied realism, which is an improvement over metaphysical realism in that it provides a link between our ideas and the world, at least at the level that matters most for our survival. The facts of basic-level categorization also remind us that our bodies contribute to our sense of what is real.

We turn next to spatial-relations concepts. These too are embodied. They have to be, because they allow us to negotiate space, to function in it as well as to conceptualize it and talk about it.

Spatial-Relations Concepts

Spatial-relations concepts are at the heart of our conceptual system. They are what make sense of space for us. They characterize what spatial form is and define spatial inference. But they do not exist as entities in the external world. We do not see spatial relations the way we see physical objects.

We do not see nearness and farness. We see objects where they are and we attribute to them nearness and farness from some landmark. The relations *in front of* and *in back of* are imposed by us on space in a complex way. When you go *in the front* of a church, you find yourself *in the back* of it. Or take the concept *across*. Suppose you are to row across a round pond. If you row "straight across" it (at a 90-degree angle from the shore), you have certainly rowed across it. If you row at a 45-degree angle, it is not as clear. If you row at a 15-degree angle, certainly not. Here, what counts as *across* varies with the shape of the area crossed and the angle of crossing and is also a matter of degree. Spatial-relations concepts are not simple or straightforward, and they vary considerably from language to language.

We use spatial-relations concepts unconsciously, and we impose them via our perceptual and conceptual systems. We just automatically and unconsciously "perceive" one entity as *in, on,* or *across from* another entity. However, such perception depends on an enormous amount of automatic unconscious mental activity on our part. For example, to see a butterfly as *in* the garden, we have to project a nontrivial amount of imagistic structure onto a scene. We have to conceptualize the boundaries of the garden as a three-dimensional container with an interior that extends into the air. We also have to locate the butterfly as a figure (or *trajector*) relative to that conceptual container, which serves as a ground (or *landmark*). We perform such complex, though mundane, acts of imaginative perception during every moment of our waking lives.

Most spatial relations are complexes made up of elementary spatial relations. English *into* is a composite of the English elementary spatial relations *in* and *to*. English *on* in its central sense is a composite of *above, in contact with,* and *supported by*. Each of these is an elementary spatial relation. Elementary spatial relations have a further internal structure consisting of an *image schema, a profile*, and a *trajector-landmark structure*.

To see what these terms mean, let us take a simple example.

The Container Schema

English *in* is made up of a container schema (a bounded region in space), a profile that highlights the interior of the schema, and a structure that identifies the boundary of the interior as the landmark (LM) and the object overlapping with the interior as a trajector (TR). In "Sam is in the house," the house is the landmark (LM) relative to which Sam, the trajector (TR), is located.

Spatial relations also have built-in spatial "logics" by virtue of their image-schematic structures. Figure 3.1 illustrates the spatial logic built into the container schema:

- Given two containers, *A* and *B*, and an object, *X*, if *A* is *in B* and *X* is *in A*, then *X* is *in B*.

We don't have to perform a deductive operation to compute this. It is self-evident simply from the image in Figure 3.1.

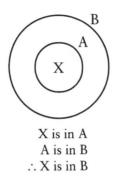

X is in A
A is in B
∴ X is in B

FIGURE 3.1   (Container Schema Logic)

A container schema has the following structure: an inside, a boundary, and an outside. This is a gestalt structure, in the sense that the parts make no sense without the whole. There is no inside without a boundary and an outside, no outside without a boundary and an inside, and no boundary without sides. The structure is topological in the sense that the boundary can be made larger, smaller, or distorted and still remain the boundary of a container schema.

A container schema, like any other image schema, is conceptual. Such a container schema can, however, be physically instantiated, either as a concrete object, like a room or a cup, or as bounded region in space, like a basketball court or a football field.

Suppose the boundary of a container schema is physically instantiated in a concrete object, say, a box. A physical boundary can impose forceful and visual constraints: It can protect the container's contents, restrict their motion, and render them inaccessible to vision. It is important to distinguish a purely conceptual schema from a physically instantiated one; they have different properties.

Container schemas, like other image schemas, are cross-modal. We can impose a conceptual container schema on a visual scene. We can impose a container schema on something we hear, as when we conceptually separate out one part of a piece of music from another. We can also impose container schemas on our motor movements, as when a baseball coach breaks down a batter's swing into component parts and discusses what goes on "inside" each part.

The Source-Path-Goal Schema

As with a container schema, there is a spatial logic built into the source-path-goal schema (Figure 3.2). The source-path-goal schema has the following elements (or "roles"):

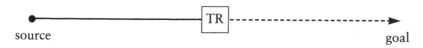

FIGURE 3.2   (Source-Path-Goal Schema)

A trajector that moves
A source location (the starting point)
A goal, that is, an intended destination of the trajector
A route from the source to the goal
The actual trajectory of motion
The position of the trajector at a given time
The direction of the trajector at that time
The actual final location of the trajector, which may or may not be the intended destination

Extensions of this schema are possible: a vehicle, the speed of motion, obstacles to motion, forces that move one along a trajectory, additional trajectors, and so on.

This schema is topological in the sense that a path can be expanded, shrunk, or deformed and still remain a path. Trajectories are imaginative insofar as they are not entities in the world; they are conceptualized as a linelike "trail" left by an object as it moves and projected forward in the direction of motion.

As with the container schema, we can form spatial relations from this schema by the addition of profiling (also called *highlighting*) and a trajector-landmark relation. The concept expressed by *to* profiles the goal and identifies it as the landmark relative to which the motion takes place. The concept expressed by *from* profiles the source, taking the source as the landmark relative to which the motion takes place.

The source-path-goal schema also has an internal spatial "logic" and built-in inferences:

- If you have traversed a route to a current location, you have been at all previous locations on that route.
- If you travel from *A* to *B* and from *B* to *C*, then you have traveled from *A* to *C*.

- If there is a direct route from *A* to *B* and you are moving along that route toward *B*, then you will keep getting closer to *B*.
- If *X* and *Y* are traveling along a direct route from *A* to *B* and *X* passes *Y*, then *X* is further from *A* and closer to *B* than *Y* is.
- If *X* and *Y* start from *A* at the same time moving along the same route toward *B* and if *X* moves faster than *Y*, then *X* will arrive at *B* before *Y*.

Our most fundamental knowledge of motion is characterized by the source-path-goal schema, and this logic is implicit in its structure. Many spatial-relations concepts are defined using this schema and depend for their meaning on its inherent spatial logic, for example, *toward, away, through,* and *along.*

Bodily Projections

Bodily projections are especially clear instances of the way our bodies shape conceptual structure. Consider examples such as *in front of* and *in back of.* The most central senses of these terms have to do with the body. We have inherent fronts and backs. We see from the front, normally move in the direction the front faces, and interact with objects and other people at our fronts. Our backs are opposite our fronts; we don't directly perceive our own backs, we normally don't move backwards, and we don't typically interact with objects and people at our backs.

We project fronts and backs onto objects. What we understand as the front of a stationary artifact, like a TV or a computer or a stove, is the side we normally interact with using our fronts. What we take to be the front of a moving object like a car is that part of the object that "faces" the direction in which it normally moves. We project fronts onto stationary objects without inherent fronts such as trees or rocks. English speakers project fronts onto such objects so the front faces the speaker. In other languages (e.g., Hausa), speakers project fronts onto such objects in the opposite direction, facing away from the speaker.

The concepts *front* and *back* are body-based. They make sense only for beings with fronts and backs. If all beings on this planet were uniform stationary spheres floating in some medium and perceiving equally in all directions, they would have no concepts of *front* or *back.* But we are not like this at all. Our bodies are symmetric in some ways and not in others. We have faces and move in the direction in which we see. Our bodies define a set of fundamental spatial orientations that we use not only in orienting ourselves, but in perceiving the relationship of one object to another.

When we perceive a cat as being in front of a car or behind a tree, the spatial relationships *in front of* and *behind*, between cat and car or between cat and tree, are not objectively there in the world. The spatial relation is not an entity in our visual field. The cat is behind the tree or in front of the car only relative to our capacity to project fronts and backs onto cars and trees and to impose relations onto visual scenes relative to such projections. In this way, perceiving the cat as being behind the tree requires an imaginative projection based on our embodied nature.

Compared to certain other languages, English is relatively impoverished in its use of bodily projections to conceptualize spatial relations. By contrast, languages of the Otomonguean family, such as Mixtec, use bodily projections as their primary means of characterizing spatial relations (A1, Brugman 1985).

For example, in Mixtec, there is no unitary concept or word corresponding to English *on*. The range of cases covered by English *on* is instead described by using body-part projections. Suppose you want to say "He is on top of the hill." You say the equivalent of "He is located head hill." If you want to say "I was on the roof of the house," you say the Mixtec equivalent of "I was located animal-back house," in which an animal back, being canonically oriented horizontally, is projected onto the house. If you want to say "I am sitting on the branch of the tree," you say the equivalent of "I am sitting arm tree."

One way in which languages differ is that, while some have mainly body-centered relations like *in front of*, others have mainly externally based relations, like *to the north of*, and still others have mixed systems (A8, Levinson 1992–present).

Other Image Schemas and Elements of Spatial Relations

The study of spatial-relations concepts within cognitive linguistics has revealed that there is a relatively small collection of primitive image schemas that structure systems of spatial relations in the world's languages. Here are some examples, without the full detail given above: part-whole, center-periphery, link, cycle, iteration, contact, adjacency, forced motion (e.g., pushing, pulling, propelling), support, balance, straight-curved, and near-far. Orientations also used in the spatial-relations systems of the world's languages include vertical orientation, horizontal orientation, and front-back orientation. (For a fuller discussion see A4, Lakoff 1987, case study 2; A1, Johnson 1987; A8, Talmy 1983; and B2, Regier 1996.)

One of the important discoveries of cognitive science is that the conceptual systems used in the world's languages make use of a relatively small number of basic image schemas, though the range of complex spatial relations that can be built out of these schemas is very large. As we shall see when we get to the discussion of conceptual metaphor, the spatial logics of these body-based image schemas are among the sources of the forms of logic used in abstract reason.

The Embodied Nature of Spatial-Relations Concepts

Spatial-relations concepts are embodied in various ways. Bodily projections are obviously based on the human body. Concepts like *front* and *back* and those in Mixtec arise from the body, depend on the body, and would not exist if we did not have the kinds of bodies we have. The same is true of fundamental force-dynamic schemas: pushing, pulling, propelling, supporting, and balance. We comprehend these through the use of our body parts and our ability to move them, especially our arms, hands, and legs.

Other image schemas are also comprehended through the body. Our bodies are containers that take in air and nutrients and emit wastes. We constantly orient our bodies with respect to containers—rooms, beds, buildings. We spend an inordinate amount of time putting things in and taking things out of containers. We also project abstract containers onto areas in space, as when we understand a swarm of bees as being *in* the garden. Similarly, every time we see something move, or move ourselves, we comprehend that movement in terms of a source-path-goal schema and reason accordingly.

These forms of embodiment arise from the way we schematize our own bodies and things we interact with daily (C2, Gallagher 1995). We will refer to this as *phenomenological embodiment*. But there is also *neural embodiment,* as we saw in the case of color. Neural embodiment characterizes the neural mechanisms that give rise to concepts—for example, the neural circuitry connected to the color cones that brings color into existence and characterizes the structure of color categories. These neural mechanisms explain why color categories have many of the phenomenological properties they have.

We do not yet know the exact neural mechanisms that give rise to spatial-relations concepts, but a beginning has been made. A computational neural model has been constructed that characterizes certain image schemas neurally, explains why they should exist, and accounts for their topological and orientational properties. Let us now turn to this research.

The Neural Modeling of Spatial and Motor Concepts

As we mentioned above, much of the Western philosophical tradition assumes a form of faculty psychology, according to which we have a faculty of reason separate from our faculties of perception and bodily movement. Concepts and the forms of reason based on them are assumed to be purely part of the faculty of reason. Perception may inform reason, and movement may be a consequence of reason, but in the tradition no aspect of perception or movement is *part* of reason.

Consequently, there is assumed to be an absolute dichotomy between *perception* and *conception*. While *perception* has always been accepted as bodily in nature, just as movement is, *conception*—the formation and use of concepts—has traditionally been seen as purely mental and wholly separate from and independent of our abilities to perceive and move.

We have already begun to get intimations that this picture is false. We have seen that basic-level concepts depend on motor movement, gestalt perception, and mental imagery, which is carried out in the visual system of the brain. We have seen that color is anything but purely mental, that our color concepts are intimately shaped not merely by perception as a faculty of mind but by such physical parts of our bodies as color cones and neural circuitry. And we have seen that spatial-relations concepts like *front* and *back* are not characterized by some abstract, disembodied mental capacity but rather in terms of bodily orientation. In these cases, the body is not merely somehow involved in conceptualization but is shaping its very nature.

Embodiment Not as Realization but as Shaping

What is the view that the mind is disembodied? It is the view that the contents of mind, the actual concepts, are not crucially shaped or given any significant inferential content by the body. It is the view that concepts are formal in nature and arise from the mind's capacity to generate formal structure in such a way as to derive further, inferred, formal structures. Advocates of the disembodied mind will, of course, say that conceptual structure must have a neural *realization* in the brain, which just *happens* to reside in a body. But they deny that anything about the body is essential for characterizing what concepts are.

The claim that the mind is embodied is, therefore, far more than the simpleminded claim that the body is needed if we are to think. Advocates of the disembodied-mind position agree with *that*. Our claim is, rather, that the very properties of concepts are created as a result of the way the brain and body are structured and the way they function in interpersonal relations and in the physical world.

The embodied-mind hypothesis therefore radically undercuts the *perception/conception* distinction. In an embodied mind, it is conceivable that the same neural system engaged in *perception* (or in bodily movement) plays a central role in *conception*. That is, the very mechanisms responsible for perception, movements, and object manipulation could be responsible for conceptualization and reasoning. Indeed, in recent neural modeling research, models of perceptual mechanisms and motor schemas can actually do *conceptual* work in language learning and in reasoning. This is a startling result. It flies in the face of time-honored philosophical theories of faculty psychology and their recent reincarnation in strong modularity theories of mind and language, each of which insists on a separation of the mechanisms for perception and conception.

Neural Modeling as an Existence Proof
for the Embodiment of Mind

As yet, we do not have any strong neurophysiological evidence, say from PET scan or functional MRI results, that the same neural mechanisms used in perception and movement are also used in abstract reasoning. What we do have is an existence proof that this is possible and good reasons to believe that it is plausible. The existence proof comes from the field of neural modeling, and it comes in the following form. A neural model of a perceptual or motor mechanism is constructed, and that very same mechanism is used for conceptual tasks as well. The conceptual tasks are of two sorts: (1) learning the structure of a semantic field of lexical items so as to get the relationships among the lexical items correct and (2) performing abstract inferences.

These models are existence proofs in the sense that they show that neural structures that can carry out sensorimotor functions in the brain can in principle do both jobs at once—the job of perception or motor control, on the one hand, and the job of conceptualizing, categorizing, and reasoning, on the other.

What is particularly impressive about these models is that they are computational. The field of computational neuroscience is concerned not merely with *where* the neural computations are done but with *how*, that is, with precise neural computational mechanisms that perform sensorimotor operations and that carry out conceptualizing, categorizing, reasoning, and language learning. Each of the models we will discuss does such jobs in detail.

Models have been constructed for three kinds of concepts:

1. Spatial-relations concepts, for example, those named by English words like *in, on, over, through,* and *under.*

2. Concepts of bodily movement, represented by verbs like *grasp, pull, lift, tap,* and *punch.*

3. Concepts indicating the structure of actions or events (what linguists call *aspectual concepts*) like *starting, stopping, resuming, continuing, finishing,* including those indicated grammatically as in process (in English *is/are* plus the verb stem plus *-ing: is running*) or completed (*has/have* plus the verb stem plus *-ed: has lifted*).

Since these concepts are about what the body does, namely, perceive and move, one would expect that what the body actually does should shape these concepts. In particular:

- Since spatial-relations concepts are about space, it should not be surprising if our capacities for vision and negotiating space are used in constituting spatial-relations concepts and their logics.
- Since concepts of bodily movement are about motor actions, it should not be surprising if our motor schemas and parameters of bodily movement structure those concepts and their logics.
- Since moving the body is our most common form of action, it should not be surprising if the general structure of control schemas for bodily movements should be used to characterize aspectual structure, the structure we find in actions and events in general.

These models suggest some things that make eminently good sense: The visual systems of our brains are used in characterizing spatial-relations concepts. Our actual motor schemas and motor synergies are involved in what verbs of motor movement mean. And the general form of motor control gives general form to all our actions and the events we perceive. The point is this: In such models, there is no absolute perceptual/conceptual distinction, that is, the conceptual system makes use of important parts of sensorimotor system that impose crucial conceptual structure.

The Three Models

The three models we are about to discuss are highly complex, and we can give only a very brief overview of them here. A more detailed discussion is found in the Appendix. (For full discussions of all the technical details, see B2, Regier 1996; Bailey 1997; and Narayanan 1997a, b.)

Regier's Model for Learning Spatial-Relations Terms

Terry Regier (B2, 1996) constructed a neural model for learning spatial-relations terms in the world's languages. Given a model of retinal input with geometric figures in various spatial configurations together with a linguistic description correctly describing the configuration in a given language, the neural model was to learn the system of spatial-relations concepts and terms so that it could correctly categorize and label novel configurations. It was to do this both in cases of static spatial configurations (e.g., *on*) and in cases involving motion (e.g., *onto*). The model learned using no negative evidence, that is, no incorrectly labeled cases, only correctly labeled ones.

Here is the idea behind the model: Though spatial-relations terms differ wildly across the world's languages, they have to categorize using structures found in the visual system of the brain. Spatial-relations concepts should therefore depend on neural structures found in the brain's visual system. Consequently, Regier's model is designed to make maximal use of the types of structures known to exist in the human visual system. Regier's major insights were, first, that topographic maps of the visual field should be instrumental in the computation of image schemas that have topological properties (e.g., the container schema); second, that orientation-sensitive cell assemblies should be able compute the orientational aspects of spatial concepts that rely on bodily orientation (e.g., *above*); third, that center-surround receptive fields should be crucial to characterizing concepts like *contact*; and finally that the "filling-in" architecture discovered by Ramachandran and Gregory (B1, 1991) should play a central role in characterizing the notion of *containment*.

The Regier model is simultaneously both perceptual and conceptual. By virtue of the way the perceptual mechanisms work, it accomplishes the conceptual task of categorizing spatial configurations adequately to fit the conceptual distinctions and contrasts among spatial-relations terms in natural languages. It thereby gives us some insight into how the neural structures in the brain that do perceptual work might be recruited to do conceptual work as well.

Bailey's Model for Learning Verbs of Hand Motion

David Bailey's model (B2, 1997) learns not only how to categorize and name hand motions in the world's languages but also how to use those verbs correctly to give orders to produce the corresponding hand motion in a computer model of the body. At the heart of Bailey's model are models of high-level motor-control schemas that operate dynamically in time to control motor synergies—subcortical neural circuits that act automatically to produce small, low-level movements. These synergies provide the parameters used by the motor-control schemas, called *X-schemas* (for *executing schemas*).

The idea behind the model is this: Verbs of hand action differ considerably around the world, categorizing actual hand actions in markedly different ways from language to language. Yet the categorization should depend on the actual motor schemas used in moving things with the hand and on parameters given by actual motor synergies. Thus, the actual motor mechanisms should also be doing the conceptual work of categorizing actions for the purpose of naming them. The success of the Bailey model suggests how neural circuitry used for motor control can be recruited for conceptual purposes.

Narayanan's Model of Motor Schemas, Linguistic Aspect, and Metaphor
Srini Narayanan (B2, 1997a, b), working with Bailey on modeling motor schemas, discovered that all motor schemas have the same high-level control structure:

Getting into a state of readiness
The initial state
The starting process
The main process (either instantaneous or prolonged)
An option to stop
An option to resume
An option to iterate or continue the main process
A check to see if a goal has been met
The finishing process
The final state .

This should come as no surprise. Any high-level motor activity you undertake, from scratching your head to turning on a light switch to sipping a cup of tea, will have this structure. (It is actually more complex; for the sake of a brief presentation, we have simplified it a bit.) Narayanan then constructed a model of this control structure so that it could be structured separately from the individual special cases (e.g., lifting a cup). That permitted a great simplification in characterizing neural control structures.

Linguists should recognize this model immediately. It characterizes the semantic structure of events in general, what linguists call *aspect*. Any action one undertakes, whether a bodily movement or a more abstract undertaking like planning what to have for dinner, has such a structure. And each language has a linguistic means of highlighting aspects of such a structure. In English, for example, the present imperfect form of the verb (*is/are* plus the present stem of the verb plus *-ing*, as in *is walking*) focuses on the main process as it is happening.

Aspect—the general structure of events—has a conceptual structure and a logic. What Narayanan discovered was that exactly the same neural structure that can perform motor control also characterizes the conceptual structure of linguistic aspect, and the same neural mechanism that can control bodily movements can perform logical inferences about the structure of actions in general.

Narayanan devised an ingenious way to test whether his model of general high-level motor control could handle purely abstract inferences, inferences having nothing to do with bodily movement. He constructed a neural model of conceptual metaphor and then found cases in which body-based metaphors were used in an abstract domain, in this case, international economics. Prominent newspapers and journals use such metaphors every day in economic news reports; for example, "India loosened its stranglehold on business," "France fell into a recession and Germany pulled it out." Narayanan then showed that models of the motor schemas for physical actions can—under metaphoric projection—perform the appropriate abstract inferences about international economics.

The Body in the Mind

Each of these neural modeling studies constitutes an existence proof. Spatial-relations concepts *can* be represented and spatial-relations terms learned on the basis of neural perceptual apparatus in the brain's visual system (topographic maps of the visual field, orientation-sensitive cells, and so on). Concepts for hand motions *can* be represented and hand-motion terms learned on the basis of detailed models of high-level motor control and motor synergies. Aspectual concepts that characterize the structure of events *can* be adequately represented in terms of general motor-control schemas, and abstract reasoning using those schemas can be carried out using neural motor-control simulations. None of this proves that people actually use those parts of the brain involved in perception and motor control to do such reasoning, but it is in principle possible. At present, these systems that use neural models of motor-control schemas are the only ones capable of carrying out the given tasks.

Now that we know that there can be such a direct embodiment of reason, the question becomes an empirical one, to be settled in experimental neuro-science, not in the arena of philosophical argumentation. The evidence so far favors embodied cognition, and there are general reasons for believing that something like the embodied cognition theory will turn out to be true.

Brains tend to optimize on the basis of what they already have, to add only what is necessary. Over the course of evolution, newer parts of the brain have built on, taken input from, and used older parts of the brain. Is it really plausi-ble that, if the sensorimotor system can be put to work in the service of reason, the brain would build a whole new system to duplicate what it could do al-ready?

Regier has shown that the topological properties of spatial relations can be explained on the basis of the topological properties arising from applying cen-ter-surround receptive fields and Ramachandran's filling-in process to topo-graphic maps of the visual field. Is it really plausible that the brain would develop another, nonvisual system with the same topological properties to rea-son about space, when we obviously already use vision to get around in space?

Narayanan has shown that the neural structure of motor control must al-ready have all the capacities necessary to characterize aspect (the structure of events) and its logic. If the brain can reason about actions using the structure already present to perform actions, is it plausible that the brain would build another system to do the same thing? And if it did, is it plausible that it would take a significantly different neural form?

From a biological perspective, it is eminently plausible that reason has grown out of the sensory and motor systems and that it still uses those systems or structures developed from them. This explains why we have the kinds of con-cepts we have and why our concepts have the properties they have. It explains why our spatial-relations concepts should be topological and orientational. And it explains why our system for structuring and reasoning about events of all kinds should have the structure of a motor-control system.

It is only from a conservative philosophical position that one would want to believe in the old faculty psychology—in the idea that the human mind has nothing about it that animals share, that reason has nothing about it that smells of the body.

Philosophically, the embodiment of reason via the sensorimotor system is of great importance. It is a crucial part of the explanation of why it is possible for our concepts to fit so well with the way we function in the world. They fit so well because they have evolved from our sensorimotor systems, which have in turn evolved to allow us to function well in our physical environment. The embodiment of mind thus leads us to a philosophy of embodied realism. Our concepts cannot be a direct reflection of external, objective, mind-free reality because our sensorimotor system plays a crucial role in shaping them. On the other hand, it is the involvement of the sensorimotor system in the conceptual system that keeps the conceptual system very much in touch with the world.

4

Primary Metaphor and Subjective Experience

Our subjective mental life is enormous in scope and richness. We make subjective judgments about such abstract things as importance, similarity, difficulty, and morality, and we have subjective experiences of desire, affection, intimacy, and achievement. Yet, as rich as these experiences are, much of the way we conceptualize them, reason about them, and visualize them comes from other domains of experience. These other domains are mostly sensorimotor domains (A1, Lakoff and Johnson 1980; Lakoff 1993), as when we conceptualize understanding an idea (subjective experience) in terms of grasping an object (sensorimotor experience) and failing to understand an idea as having it go right by us or over our heads. The cognitive mechanism for such conceptualizations is conceptual metaphor, which allows us to use the physical logic of grasping to reason about understanding.

Metaphor allows conventional mental imagery from sensorimotor domains to be used for domains of subjective experience. For example, we may form an image of something going by us or over our heads (sensorimotor experience) when we fail to understand (subjective experience). A gesture tracing the path of something going past us or over our heads can indicate vividly a failure to understand.

Conceptual metaphor is pervasive in both thought and language. It is hard to think of a common subjective experience that is not conventionally conceptualized in terms of metaphor. But why does such a huge range of conventional conceptual metaphor exist? How is it learned and what are the precise details?

What is the mechanism by which we reason metaphorically? And which metaphors are universal (or at least widespread) and why?

We now have preliminary answers to such questions. They come from separate strands of investigation by Christopher Johnson (A1), Joe Grady (A1), Srini Narayanan (B2), and Mark Turner and Gilles Fauconnier (A7). This chapter weaves those strands into an integrated account of how we conceptualize and describe subjective experience.

The Integrated Theory of Primary Metaphor

The overall theory of primary metaphor has four parts. We will look at each part in more detail below, but let us begin with a brief outline of each of the parts and how they fit together.

Part 1: Johnson's *theory of conflation* in the course of learning. For young children, subjective (nonsensorimotor) experiences and judgments, on the one hand, and sensorimotor experiences, on the other, are so regularly conflated—undifferentiated in experience—that for a time children do not distinguish between the two when they occur together. For example, for an infant, the subjective experience of affection is typically correlated with the sensory experience of warmth, the warmth of being held. During the period of conflation, associations are automatically built up between the two domains. Later, during a period of *differentiation*, children are then able to separate out the domains, but the cross-domain associations persist. These persisting associations are the mappings of conceptual metaphor that will lead the same infant, later in life, to speak of "a *warm* smile," "a *big* problem," and "a *close* friend."

Part 2: Grady's *theory of primary metaphor.* All complex metaphors are "molecular," made up of "atomic" metaphorical parts called *primary metaphors.* Each primary metaphor has a minimal structure and arises naturally, automatically, and unconsciously through everyday experience by means of conflation, during which cross-domain associations are formed. Complex metaphors are formed by conceptual blending. Universal early experiences lead to universal conflations, which then develop into universal (or widespread) conventional conceptual metaphors.

Part 3: Narayanan's *neural theory of metaphor.* The "associations" made during the period of conflation are realized neurally in simultaneous activations that result in permanent neural connections being made across the neural networks that define conceptual domains. These connections form the anatomical basis of source-to-target activations that constitute metaphorical entailments.

Briefly, an entailment at the neural level in Narayanan's theory occurs when some sequence of neural activations, *A*, results in a further neural activation, *B*. If *B* is connected to a neuronal cluster, *C*, in the network that characterizes another conceptual domain, then *B* can activate *C*. In the theory, this constitutes a metaphorical entailment: The activation of *B* is a literal entailment; *C* is "metaphorically" linked to *B*, since it is in another conceptual domain; therefore the activation of *C* is a metaphorical entailment.

Part 4: Fauconnier and Turner's *theory of conceptual blending.* Distinct conceptual domains can be coactivated, and under certain conditions connections across the domains can be formed, leading to new inferences. Such "conceptual blends" may be either conventional or wholly original. Grady suggests that conventional blends are the mechanism by which two or more primary metaphors can be brought together to form larger complex metaphors.

The integrated theory—the four parts together—has an overwhelming implication: We acquire a large system of primary metaphors automatically and unconsciously simply by functioning in the most ordinary of ways in the everyday world from our earliest years. We have no choice in this. Because of the way neural connections are formed during the period of conflation, we all naturally think using hundreds of primary metaphors.

Let us flesh out this theory with some examples.

The Sensorimotor Structuring of Subjective Experience

In *Metaphors We Live By*, we gave evidence that conceptual metaphors are mappings across conceptual domains that structure our reasoning, our experience, and our everyday language. We pointed to the existence of *experientially grounded mappings,* for example, More Is Up, as in "Prices rose" and "Stocks plummeted." In More Is Up, a subjective judgment of quantity is conceptualized in terms of the sensorimotor experience of verticality.

This correspondence between quantity and verticality arises from a correlation in our normal everyday experiences, like pouring more water into the glass and seeing the level go up. Early in development, Johnson hypothesizes, such correlations are "conflations" in which quantity and verticality are not seen as separate, and associations between them are formed. After the conflation period, according to Grady, the associations between More and Up and between Less and Down constitute a cross-domain mapping between the sensorimotor concept of verticality (the source domain) and the subjective judgment of quantity. Conventional linguistic metaphors like "Prices fell" are secondary manifestations of the primary cross-domain mapping.

Conflation

Let us look a little more closely at Christopher Johnson's work on conflation. In research on metaphor acquisition in children, Johnson (A1, 1997b, c) studied the Shem corpus in detail. This is a well-known collection of the utterances of a child named Shem, recorded over the course of his language development (D, MacWhinney 1995). In an attempt to discover the age at which Shem acquired a commonplace metaphor, Johnson looked at Shem's use of the verb *see*. His objective was to discover the mechanism involved in the acquisition of metaphor. He had hypothesized conflation as a possible mechanism, and he wanted to find out whether there is indeed a stage of conflation prior to the use of the metaphor. His test case was Knowing Is Seeing, as in sentences like "I see what you're saying." In such metaphorical examples, knowing is the subject matter. Seeing is the metaphorical source domain used to conceptualize knowledge, but it is not used literally.

Johnson discovered that, prior to using metaphor, Shem went through a stage in which the knowing and seeing domains were conflated. Since we normally get most of our knowledge from seeing, a conflation of these domains would have been expected. In such conflations, the domains of knowing and seeing are coactive and the grammar of *know* is used with the verb *see* in a context in which seeing and knowing occur together—for instance, "Let's see what's in the box." Here, seeing what's in the box correlates with knowing what's in the box.

Metaphorical cases such as "I see what you mean," which do not involve literal seeing, are absent at this stage. Such metaphorical cases develop later according to Johnson's hypothesis. The conflations provide the basis for the learning of primary conceptual metaphors. Subsequent to the conflation experience, the child is able to differentiate the two conceptual domains. Only then does conceptual metaphor emerge. In the neural theory, the conflations are instances of coactivation of both domains, during which permanent neural connections between the domains develop.

In short, Johnson hypothesizes that conceptual metaphor emerges in two stages: (1) the conflation stage, during which connections between coactive domains are established and the domains are not experienced as separate, and (2) the differentiation stage, during which domains that were previously coactive are differentiated into metaphorical sources and targets.

This does not, of course, imply that all *linguistic* metaphorical expressions are learned the way primary metaphors are. For example, *illuminate,* an extended instance of the general Knowing Is Seeing metaphor, is learned well after the conceptual primary metaphor Knowing Is Seeing is learned.

Grady's Theory of Primary Metaphor

Johnson's theory of conflation is the basis for Grady's theory of primary metaphor. Early conflations in everyday experience should lead to the automatic formation of hundreds of primary metaphors that pair subjective experience and judgment with sensorimotor experience. Each primary metaphor, Grady hypothesizes, is simple, an atomic component of the molecular structure of complex metaphors.

Complex metaphors are formed from primary ones through conventional conceptual blending, that is, the fitting together of small metaphorical "pieces" into larger wholes. In the process, long-term connections are learned that coactivate a number of primary metaphorical mappings. Each such coactive structure of primary metaphors constitutes a complex metaphorical mapping. We will give examples of this process in Chapter 5, but first, let us look at a range of primary metaphors to get a feel for what they are like.

Table 4.1 shows a short, representative list of primary metaphors. In each case, we state the primary metaphorical mapping, distinguish its sensorimotor component from its subjective component, and describe the primary experiences of domain conflation that give rise to it. The examples are derived from (A1, Grady 1997).

Primary Metaphor Within Narayanan's Neural Theory

In Chapter 3, we briefly described Srini Narayanan's neural theory of metaphor. Though Narayanan's model did not learn metaphors, the recruitment learning mechanism in Bailey's model ought, with suitable modification,

TABLE 4.1   Representative Primary Metaphors

Affection Is Warmth

Subjective Judgment: Affection
Sensorimotor Domain: Temperature
Example: "They greeted me *warmly*."
Primary Experience: Feeling warm while being held affectionately

Important Is Big

Subjective Judgment: Importance
Sensorimotor Domain: Size
Example: "Tomorrow is a *big* day."
Primary Experience: As a child, finding that big things, e.g., parents, are
    important and can exert major forces on you and dominate your visual
    experience

Happy Is Up

Subjective Judgment: Happiness
Sensorimotor Domain: Bodily orientation
Example: "I'm feeling *up* today."
Primary Experience: Feeling happy and energetic and having an upright
    posture (correlation between affective state and posture)

Intimacy Is Closeness

Subjective Experience: Intimacy
Sensorimotor Experience: Being physically close
Example: "We've been *close* for years, but we're beginning to *drift
    apart*."
Primary Experience: Being physically close to people you are intimate with

Bad Is Stinky

Subjective Judgment: Evaluation
Sensorimotor Domain: Smell
Example: "This movie *stinks*."
Primary Experience: Being repelled by foul-smelling objects (correlation
    between evaluative and olfactory experience)

Difficulties Are Burdens

Subjective Judgment: Difficulty
Sensorimotor Domain: Muscular exertion
Example: "She's *weighed down* by responsibilities."
Primary Experience: The discomfort or disabling effect of lifting or carry-
    ing heavy objects

More Is Up

Subjective Judgment: Quantity
Sensorimotor Domain: Vertical orientation
Example: "Prices are *high*."
Primary Experience: Observing rise and fall of levels of piles and fluids as more is added or subtracted

Categories Are Containers

Subjective Judgment: Perception of kinds
Sensorimotor Domain: Space
Example: "Are tomatoes *in* the fruit or vegetable category?"
Primary Experience: Observing that things that go together tend to be in the same bounded region (correlation between common location and common properties, functions, or origins)

Similarity Is Closeness

Subjective Judgment: Similarity
Sensorimotor Domain: Proximity in space
Example: "These colors aren't quite the same, but they're *close*."
Primary Experience: Observing similar objects clustered together (flowers, trees, rocks, buildings, dishes)

Linear Scales Are Paths

Subjective Judgment: Degree
Sensorimotor Domain: Motion
Example: "John's intelligence *goes way beyond* Bill's."
Primary Experience: Observing the amount of progress made by an object in motion (correlation between motion and scalar notion of degree)

Organization Is Physical Structure

Subjective Judgment: Abstract unifying relationships
Sensorimotor Domain: Experience of physical objects
Example: "How do the *pieces* of this theory *fit together*?"
Primary Experience: Interacting with complex objects and attending to their structure (correlation between observing part-whole structure and forming cognitive representations of logical relationships)

(continues)

TABLE 4.1 *(continued)*

Help Is Support

Subjective Judgment: Assistance

Sensorimotor Domain: Physical supportExample: "*Support* your local charities."

Primary Experience: Observing that some entities and people require physical support in order to continue functioning

Time Is Motion

Subjective Judgment: The passage of time

Sensorimotor Domain: Motion

Example: "Time *flies*."

Primary Experience: Experiencing the passage of time as one moves or observes motion

States Are Locations

Subjective Judgment: A subjective state

Sensorimotor Experience: Being in a bounded region of space

Example: "I'm *close to* being *in* a depression and the next thing that goes wrong will *send me over the edge*."

Primary Experience: Experiencing a certain state as correlated with a certain location (e.g., being cool under a tree, feeling secure in bed)

Change Is Motion

Subjective Judgment: Experiencing a change of state

Sensorimotor Domain: Moving

Example: "My car has *gone from* bad *to* worse lately."

Primary Experience: Experiencing the change of state that goes with the change of location as you move

Actions Are Self-Propelled Motions

Subjective Experience: Action

Sensorimotor Experience: Moving your body through space

Example: "I'm *moving* right along on the project."

Primary Experience: The common action of moving yourself through space, especially in the early years of life

Purposes Are Destinations

Subjective Judgment: Achieving a purpose

Sensorimotor Experience: Reaching a destination

Example: "He'll ultimately be successful, but he isn't *there* yet."

Primary Experience: Reaching destinations throughout everyday life and thereby achieving purposes (e.g., if you want a drink, you have to go to the water cooler)

Purposes Are Desired Objects

Subjective Judgment: Achieving a purpose

Sensorimotor Domain: Object manipulation

Example: "I saw an opportunity for success and *grabbed* it."

Primary Experience: Grasping a desired object (correlation between satisfaction and holding a desired physical object)

Causes Are Physical Forces

Subjective Judgment: Achieving results

Sensorimotor Domain: Exertion of force

Example: "They *pushed* the bill *through* Congress."

Primary Experience: Achieving results by exerting forces on physical objects to move or change them

Relationships Are Enclosures

Subjective Experience: An interpersonal relationship

Sensorimotor Experience: Being in an enclosure

Example: "We've been *in* a *close* relationship for years, but it's beginning to seem *confining*."

Primary Experience: Living in the same enclosed physical space with the people you are most closely related to

Control Is Up

Subjective Judgment: Being in control

Sensorimotor Domain: Vertical orientation

Example: "Don't worry! I'm *on top of* the situation."

Primary Experience: Finding that it is easier to control another person or exert force on an object from above, where you have gravity working with you

Knowing Is Seeing

Subjective Judgment: Knowledge

Sensorimotor Domain: Vision

(continues)

TABLE 4.1   *(continued)*

Example: "I *see* what you mean."
Primary Experience: Getting information through vision

Understanding Is Grasping

Subjective Judgment: Comprehension
Sensorimotor Domain: Object manipulation
Example: "I've never been able to *grasp* transfinite numbers."
Primary Experience: Getting information about an object by grasping and
    manipulating it

Seeing Is Touching

Subjective Judgment: Visual perception
Sensorimotor Domain: Touch
Example: "She *picked* my face *out of* the crowd."
Primary Experience: Correlation between the visual and tactile explo-
    ration of objects

---

to be able to learn metaphorical connections across domains. Let us consider
how such a model might work in the case of More Is Up.

Experiencing the More Is Up correlation over and over should lead to the establishment of connections between those neural networks in the brain characterizing More in the domain of quantity and those networks characterizing Up
in the domain of verticality. In the model, such neural connections would carry
out the function of a conceptual mapping between More and Up and make it
possible (though not necessary) for the words for verticality (such as *rise, fall,
skyrocket, plummet, high, low, dip,* and *peak*) to be used conventionally to indicate quantity as well.

Such a metaphor is embodied in three important ways. First, the correlation
arises out of our embodied functioning in the world, where we regularly encounter cases in which More correlates with Up. Second, the source domain of
the metaphor comes from the body's sensorimotor system. Finally, the correlation is instantiated in the body via neural connections.

Here are the characteristics of primary metaphor from a neural modeling
perspective:

- A primary metaphor like More Is Up arises via a neurally instantiated
  correlation between (1) a sensorimotor operation (such as a determina-

tion of a degree or change of verticality) and (2) a subjective experience or judgment (such as a judgment of degree or change of quantity). The conflation of these two is the simultaneous activation of their respective neural networks.

- Neural connections are established in early childhood during such a period of conflation, when the networks characterizing the domains are coactivated in everyday experience, as when we pile more books on the desk and their height goes up. The sensorimotor networks perform complex inferences; for example, if something *shoots up,* it moves upward rapidly and in a short time is much higher than before. Via the neural connections, the results of these inferences are "projected" from the sensorimotor *source network* (verticality) to the subjective judgment *target network* (quantity).

Here's how that projection might work. In the Narayanan model, activation flows both ways between the source and target networks. For example, a Decrease in the quantity-domain network is connected with Motion Downward in the verticality-domain network. In an example like "Prices hit bottom," *prices* activates the quantity-domain network, which sends activation to the corresponding elements in the source-domain verticality network. *Hit bottom* activates the source-domain inference mechanism that computes that the entity *hit bottom,* went as far down as it can go. Activation then flows back to the quantity-domain network indicating Maximum Negative Change. Narayanan (B2, 1997a, b) has other examples.

Via this mechanism, reasoning about vertical motion in the spatial domain is thus used to reason about quantity. But the reverse is not true. We do not reason about verticality in terms of quantity. If activation flows both ways, why are inferences and language about quantity not mapped onto verticality? Why, for example, does *too much* not mean *too high*? Within Narayanan's theory, the explanation would go as follows:

- The theory assumes that a sensorimotor neural system has more inferential connections, and therefore a greater inferential capacity, than a neural system characterizing subjective experience in itself. This is the source of the asymmetry of primary conceptual metaphor. The asymmetry arises because results of inferences flow in one direction only, from the sensorimotor domain to the domain of subjective judgment. Because of the one-way flow of activation during the conflation period, long-term one-way connections are established via recruitment learning. It is the direction of inference that determines what is source and what is target. Sensorimotor inferences are performed in the sensorimotor domain (e.g., where inferences about verticality are computed). The results of those inferences flow from the sensorimotor domain to the domain of abstract subjective experience via the neural connections.

- Conventional language connected to a concept in the sensorimotor source network may develop a connection as well to the corresponding target-domain network. For example, the phonological form *rise,* which names a motion upward in physical space, may also name, by virtue of the metaphor, an increase in quantity as well. This process may also apply to imagery. Mental images associated with source-domain entities can be activated and thereby associated with target-domain entities.

- The neural connections between the domains, which constitute the metaphorical mapping, may or may not be activated. Indeed they may be inhibited, perhaps by the choice of another metaphor. The results of source-domain inferences flow to the target domain only when the connections are activated.

- When both domains are active, imagery associated with source-domain entities can be activated and thereby associated with the target-domain entities neurally connected to them.

The Embodiment of Primary Metaphor

The neural perspective provided by Feldman's NTL paradigm (B2, Bailey et al. 1997) and by Narayanan's and Bailey's models together gives a clear idea of what it means for metaphor to be embodied. It provides a neural learning mechanism and a precise neural computational mechanism for acquiring the metaphors and carrying out metaphorical inferences.

Primary metaphors are part of the cognitive unconscious. We acquire them automatically and unconsciously via the normal process of neural learning and may be unaware that we have them. We have no choice in this process. When the embodied experiences in the world are universal, then the corresponding primary metaphors are universally acquired. This explains the widespread occurrence around the world of a great many primary metaphors. Copious examples will be provided throughout this book.

Universal conceptual metaphors are learned; they are universals that are not innate. These conceptual universals contribute to linguistic universals, for example, how time is expressed in languages around the world (see Chapter 10). There appear to be at least several hundred such widespread, and perhaps universal, metaphors.

It is also important to stress that not all conceptual metaphors are manifested in the words of a language. Some are manifested in grammar, others in gesture, art, or ritual. These nonlinguistic metaphors may, however, be secondarily expressed through language and other symbolic means.

Contrary to long-standing opinion about metaphor, primary metaphor is *not* the result of a conscious multistage process of interpretation. Rather it is a matter of immediate conceptual mapping via neural connections.

The Inevitability of Primary Metaphor

If you are a normal human being, you inevitably acquire an enormous range of primary metaphors just by going about the world constantly moving and perceiving. Whenever a domain of subjective experience or judgment is coactivated regularly with a sensorimotor domain, permanent neural connections are established via synaptic weight changes. Those connections, which you have unconsciously formed by the thousands, provide inferential structure and qualitative experience activated in the sensorimotor system to the subjective domains they are associated with.

Our enormous metaphoric conceptual system is thus built up by a process of neural selection. Certain neural connections between the activated source- and target-domain networks are randomly established at first and then have their synaptic weights increased through their recurrent firing. The more times those connections are activated, the more the weights are increased, until permanent connections are forged.

Metaphor as Cross-Domain Conceptual Mapping

Primary metaphors, from a neural perspective, are neural connections learned by coactivation. They extend across parts of the brain between areas dedicated to sensorimotor experience and areas dedicated to subjective experience. The greater inferential complexity of the sensory and motor domains gives the metaphors an asymmetric character, with inferences flowing in one direction only.

From a conceptual point of view, primary metaphors are cross-domain mappings, from a *source domain* (the sensorimotor domain) to a *target domain* (the domain of subjective experience), preserving inference and sometimes preserving lexical representation. Indeed, the preservation of inference is the most salient property of conceptual metaphors.

We will be using two conventional notations for conceptual metaphors interchangeably throughout the remainder of this book. The first is the one we have used in this chapter, for example, Similarity Is Proximity, with the target domain in subject position (Similarity), the source domain in predicate nominal position (Proximity), and the mapping represented by the capitalized copula (Is). This takes the superficial form of a an English sentence just to make it easier to read. But technically, it is intended not as a sentence in English, but as a name for a metaphorical mapping across conceptual domains.

When we want to stress the structure of the mapping, we will use an alternative notation, for example, Proximity → Similarity, where the source domain (Proximity) is to the left of the arrow, the target domain (Similarity) is to the right of the arrow, and the arrow indicates the cross-domain mapping. In both cases, the notation is just a name for a mapping, that is, a name for a reality at either the neural or conceptual level.

Can We Think Without Metaphor?

The pervasiveness of primary conceptual metaphor in no way denies the existence of nonmetaphorical concepts. Quite the contrary. As we have seen, there is a vast system of literal concepts, for example, the basic-level concepts and the spatial-relations concepts. All basic sensorimotor concepts are literal. *Cup* (the object you drink from) is literal. *Grasp* (the action of holding) is literal. *In* (in its spatial sense) is literal.

Concepts of subjective experience and judgment, when not structured metaphorically, are literal; for example, "These colors are similar" is literal, while "These colors are close" uses the metaphor Similarity Is Proximity. "He achieved his purpose" is literal, while "He got what he wanted most" can be metaphorical. Without metaphor, such concepts are relatively impoverished and have only a minimal, "skeletal" structure. A primary metaphor adds sensorimotor inferential structure. As we shall see in the next chapter, such sensorimotor inferential capacity is considerably multiplied when two or more primary metaphors are combined to create complex conceptual metaphors. For example, A Purposeful Life Is A Journey lets us use our rich knowledge of journeys to derive rich inferences about purposeful lives.

Can we think about subjective experience and judgment without metaphor? Hardly. If we consciously make the enormous effort to separate out metaphorical from nonmetaphorical thought, we probably can do some very minimal and unsophisticated nonmetaphorical reasoning. But almost no one ever does this, and such reasoning would never capture the full inferential capacity of complex metaphorical thought.

Consider the Similarity Is Proximity metaphor, in which Similarity Is Spatial Closeness and Difference Is Spatial Distance. It is very hard for us to imagine thinking about similarity without this metaphor. Mathematical accounts of similarity typically set up a metaphorical "similarity space" in which similar things are close in that space and dissimilar things are at a distance. Similarity metrics use the same metaphor. Without such metaphors, abstract thought is virtually impossible.

But even if nonmetaphorical thought about subjective experience and judgment is occasionally possible, it almost never happens. We do not have a choice as to whether to acquire and use primary metaphor. Just by functioning normally in the world, we automatically and unconsciously acquire and use a vast number of such metaphors. Those metaphors are realized in our brains *physically* and are mostly beyond our control. They are a consequence of the nature of our brains, our bodies, and the world we inhabit.

Summary

There are hundreds of primary metaphors. Together these metaphors provide subjective experience with extremely rich inferential structure, imagery, and qualitative "feel," when the networks for subjective experience and the sensorimotor networks neurally connected to them are coactivated. They also allow a great many of the words of sensorimotor experience to be used to name aspects of metaphorically conceptualized subjective experience.

Narayanan's neural theory of metaphor gives us an account of how primary metaphors are learned, an explanation of why we have the ones we have, and a neural mechanism for metaphorical inference. We have a system of primary metaphors simply because we have the bodies and brains we have and because we live in the world we live in, where intimacy does tend to correlate significantly with proximity, affection with warmth, and achieving purposes with reaching destinations.

5

The Anatomy of Complex Metaphor

The Construction of Complex Metaphors

Primary metaphors are like atoms that can be put together to form molecules. A great many of these complex molecular metaphors are stable—conventionalized, entrenched, fixed for long periods of time. They form a huge part of our conceptual system and affect how we think and what we care about almost every waking moment. Beyond that, they structure our dreams (A1, Lakoff 1997) and form the bases of new metaphorical combinations, both poetic and ordinary (A1, Lakoff and Turner 1989; A7, Turner 1995).

This chapter is about how complex, everyday metaphors are built out of primary metaphors plus forms of commonplace knowledge: cultural models, folk theories, or simply knowledge or beliefs that are widely accepted in a culture. Let us begin with a common complex metaphor that affects most people in Western culture in order to see how it is built up from some of the primary metaphors and image schemas we have examined earlier.

A Purposeful Life Is a Journey

In our culture, there is a profoundly influential folk model according to which people are supposed to have a purpose in life, and there is something wrong

60

with you if you don't. If you are purposeless, you are seen as "lost," "without direction" in your life, as "not knowing which way to turn." Having purpose in your life gives you "goals to reach" and forces you to map out a way to reach those goals, to see what other intermediate goals you would have to reach to get there, to contemplate what might be standing in your way, how to get around obstacles, and so on.

The result is a complex metaphor that affects us all, the metaphor A Purposeful Life Is A Journey, which is built up out of primary metaphors in the following way. Start with the cultural belief:

People are supposed to have purposes in life, and they are supposed to act so as to achieve those purposes.

The primary metaphors are:

Purposes Are Destinations
Actions Are Motions

Turn this into a metaphorical version of that cultural belief:

People are supposed to have destinations in life, and they are supposed to move so as to reach those destinations.

These are then combined with a simple fact, namely,

A long trip to a series of destinations is a journey.

When these are taken together, they entail a complex metaphorical mapping:

A Purposeful Life Is A Journey Metaphor

A Purposeful Life Is A Journey
A Person Living A Life Is A Traveler
Life Goals Are Destinations
A Life Plan Is An Itinerary

Using the equivalent arrow notation, this can be expressed alternatively in the form:

Journey   →   Purposeful Life
Traveler   →   Person Living A Life
Destinations   →   Life Goals
Itinerary   →   Life Plan

This mapping defines a complex metaphor made up of four submetaphors. It is a consequence of (a) the cultural belief that everyone is supposed to have a purpose in life, (b) the primary metaphors Purposes Are Destinations and Action Is Motion, and (c) the fact that a long trip to a series of destinations is a journey.

The full import of this metaphor for our lives arises through its entailments. Those entailments are consequences of our commonplace cultural knowledge about journeys, especially:

A journey requires planning a route to your destinations.
Journeys may have obstacles, and you should try to anticipate them.
You should provide yourself with what you need for your journey.
As a prudent traveler you should have an itinerary indicating where you are supposed to be at what times and where to go next. You should always know where you are and where you are going next.

The three submappings of the A Purposeful Life Is A Journey metaphor turn this knowledge about travel into guidelines for life:

A purposeful life requires planning a means for achieving your purposes.
Purposeful lives may have difficulties, and you should try to anticipate them.
You should provide yourself with what you need to pursue a purposeful life
As a prudent person with life goals you should have an overall life plan indicating what goals you are supposed to achieve at what times and what goals to set out to achieve next. You should always know what you have achieved so far and what you are going to do next.

We have presented the logic of these mappings and their entailments in a linear sequential fashion. Though this is necessary for explication, it can be misleading. From a neural perspective, what we have discussed in a linear fashion arises from parallel connections and the passing of neural activations in parallel. The internal logic of the metaphor, rather than operating sequentially, is activated and computed in parallel.

It is important to bear in mind that conceptual metaphors go beyond the conceptual; they have consequences for material culture. For example, the metaphor A Purposeful Life Is A Journey defines the meaning of an extremely important cultural document, the Curriculum Vitae (from the Latin, "the course of life"). The CV indicates where we have been on the journey and whether we are on schedule. We are supposed to be impressed with people who have come very far very fast and less impressed with people who are "behind schedule." People who have not "found a direction in life" are seen as being in need of help. We are supposed to feel bad for people who have "missed the boat," who have waited too long to start on the journey. And we are supposed to envy those who have gotten much farther than we have much faster.

If you have any doubt that you think metaphorically or that a culture's metaphors affect your life, take a good look at the details of this metaphor and at how your life and the lives of those around you are affected by it every day. As you do so, recall that there are cultures around the world in which this metaphor does not exist; in those cultures people just live their lives, and the very idea of being without direction or missing the boat, of being held back or getting bogged down in life, would make no sense.

The Grounding of the Whole
Is the Grounding of Its Parts

The complex metaphor we have just examined, A Purposeful Life Is A Journey, does not have an experiential grounding of its own. There is no correlation between purposeful lives and journeys in our everyday experience. Does this mean that this metaphor has no grounding of any kind?

Not at all. It is composed of primary metaphors, as we have seen. Those primary metaphors are grounded. For example, Purposes Are Destinations and Action Are Motions each have their own experiential grounding. That grounding is preserved when the primary metaphors are combined into the larger complex metaphor. The grounding of A Purposeful Life Is A Journey is given by the individual groundings of each component primary metaphor.

Love Is a Journey

Complex metaphors can be used as the basis for even more complex metaphors. There is not only structure within a single complex metaphor. There is also structure in the metaphorical conceptual system as a whole. The neural connectivity of the brain makes it natural for complex metaphorical mappings to be built out of preexisting mappings, starting with primary metaphors. Let us consider one more example, a metaphor that builds on A Purposeful Life Is A Journey.

In our culture, people in a long-term love relationship are expected not only to have individual purposes in life, but to have a joint purpose in life. Not only is each individual life a journey, but a couple's life together is also supposed to be a journey to common goals. Each individual life journey is difficult enough, but the task of choosing common goals and of pursuing them together in spite of differences is that much more difficult. The result is a complex metaphor that concerns the difficulties faced in setting and pursuing common goals by people in a long-term love relationship.

In this Love Is A Journey metaphor, the lovers' common goals in life are destinations, the lovers are travelers, and their difficulties are impediments to motion. But what about the love relationship? Recall the primary metaphors A Relationship Is An Enclosure and Intimacy Is Closeness. When joined together, these form the complex metaphor An Intimate Relationship Is A Close Enclosure. Given that the lovers are travelers in this metaphor, the most natural close enclosure is a vehicle of some sort. The complex metaphor that results from putting together all these parts and deriving entailments is:

THE LOVE IS A JOURNEY METAPHOR

Love Is A Journey
The Lovers Are Travelers
Their Common Life Goals Are Destinations
The Relationship Is A Vehicle
Difficulties Are Impediments To Motion

In our culture, this is a well-entrenched, stable, conventionalized understanding of a love relationship and the difficulties involved in setting and achieving joint purposes. This conceptual metaphor is reflected in conventional expressions:

Look how *far* we've *come*. It's been a *long, bumpy road*. We can't *turn back* now. We're at a *crossroads*. We're heading *in different directions*. We may have to *go our separate ways*. The relationship is *not going anywhere*. We're *spinning our wheels*. The marriage is *out of gas*. Our relationship is *off the track*. The marriage is *on the rocks*. We're trying to keep the relationship *afloat*. We may have to *bail out* of this relationship.

The Love Is A Journey metaphor systematically links the literal the meanings of these expressions about travel to corresponding meanings in the domain of love.

Metaphors Are Used to Reason With

Perhaps the most important thing to understand about conceptual metaphors is that they are used to reason with. The Love Is a Journey mapping does not just permit the use of travel *words* to speak of love. That mapping allows forms of *reasoning* about travel to be used in reasoning about love. It functions so as to map inferences about travel into inferences about love, enriching the concept of love and extending it to love-as-journey.

Consider, for example, four of the things you know about dead-end streets:

1. A dead-end street leads nowhere.
2. Suppose two *travelers* have common *destinations* they are trying to reach. A dead-end street will not allow them to keep making continuous progress toward those *destinations*.
3. The dead-end street constitutes an *impediment* to the *motion* of the *vehicle* and continuing the present course of the *vehicle* is impossible.
4. *Traveling* in a *vehicle* toward given *destinations* takes effort, and if the *travelers* have been on a dead-end street, then their effort has been wasted.

Now take the Love Is A Journey mapping, repeated here for convenience:

Love Is A Journey
The Lovers Are Travelers
Their Common Life Goals Are Destinations
The Love Relationship Is A Vehicle
Difficulties Are Impediments To Motion

and apply it to the italicized expressions in the travel knowledge given in 1 through 4. You then get 1' through 4', which are about love relationships:

1'. A "dead-end street" doesn't allow the *pursuit of common life goals*.
2'. Suppose two *lovers* have common *life goals* they are trying to achieve. A "dead-end street" will not allow them to keep making continuous progress toward those *life goals*.

3'. The "dead-end street" constitutes a *difficulty* for the *love relationship,* and continuing the present course of the *love relationship* is impossible.

4'. *Functioning* in a *love relationship* toward given *life goals* takes effort, and if the lovers have been on a "dead-end street," then effort has been wasted.

Of course, love does not have to be conceptualized as a journey. Indeed, in many cultures, there is no such conventional conceptualization of love. But in America, it is common to conceptualize love this way automatically, typically without conscious choice or reflection. The Love Is A Journey metaphor imposes the inferential structure of travel on a love relationship. And when one reasons about love in terms of travel, one talks about it in those terms.

The Love Is A Journey mapping states a generalization over both inference patterns and language. It maps inference patterns about travel like those in 1–4 onto inference patterns about love like those in 1'-4'. It also maps expressions like *dead-end street, stuck, spinning one's wheels,* and *bail out,* with meanings in the travel domain, onto occurrences of those expressions with meanings in the domain of love. In short, the same mapping states a generalization over two kinds of data—inferential data and linguistic data.

Is this mapping cognitively real? That is, is it a live correspondence in the conceptual systems of speakers or just an after-the-fact analysis of something that may have been alive in the past but is not now, something that is merely a linguistic remnant of a now-dead conceptual mapping? One type of evidence that conventionalized everyday conceptual metaphors are alive is that we can use them in a systematic way to understand new extended metaphors automatically and without conscious reflection.

Novel Metaphor

Shortly after the Love Is A Journey mapping was discovered, there appeared a song lyric that goes, "We're driving in the fast lane on the freeway of love." Most people have no trouble in grasping immediately what this means. Indeed, they may not even notice that it required a process of interpretation. How is this possible?

If we are right that there is, in our conceptual system, a cognitively real conceptual mapping of the sort discussed above, then this novel expression would make sense as a systematic extension of that mapping. Love here is also being conceptualized as a journey. Here too, there are inferences from the domain of travel to the domain of love. And here too the language reflects that love is being conceptualized in terms of travel.

The question arises as to whether this novel metaphor is really an instance of the same mapping. It is easy to show that it is. The same mapping applies to inference patterns about driving in fast lanes on freeways and yields inference patterns about love relationships. Consider the following inference pattern about driving in the fast lane.

FL: Travelers in a vehicle driving in the fast lane make a lot of progress in a short time. But there is sometimes a danger that the vehicle will be wrecked and the travelers hurt. Yet the travelers find both the speed of the vehicle and the danger exciting.

Apply the following parts of the Love Is A Journey mapping to FL:

The Lovers Are Travelers
The Love Relationship Is A Vehicle

The result is FL':

FL': Lovers in a love relationship "driving in the fast lane" make a lot of progress in a short time. But there is sometimes a danger that the relationship will be wrecked and the lovers hurt. Yet the lovers find both the speed of the relationship and the danger exciting.

It is not just that terms for travel are being used to talk about love, as they are in everyday use of the Love Is A Journey metaphor. What is significant is that *the same mapping* is used to map the new inference patterns about travel onto inference patterns about love. That's what it means for this metaphorical expression to be a novel instance of the same Love Is A Journey metaphorical mapping.

Metaphorical Idioms and Mental Imagery

A significant portion of the linguistic expressions of the Love Is A Journey metaphor are idioms: *spinning one's wheels, off the track, on the rocks*. In traditional linguistics, idioms were seen as arbitrary—sequences of words that can mean anything at all. But these idioms are not arbitrary. Their meaning is motivated by the metaphorical mapping and certain conventional mental images. For example, consider the sentence "We're spinning our wheels in this relationship." There is a rich conventional mental image associated with the idiom *spinning one's wheels,* and we have a lot of knowledge about this image:

> The wheels are the wheels of a car. The wheels are spinning, but the car is not moving. The car is stuck (either on ice, or in mud, sand, or snow). The travelers want the car to be moving so that they can make progress on their journey. They are not happy that it is stuck. They are putting a lot of energy into getting the car unstuck, and they feel frustrated.

The Love Is A Journey metaphor maps this knowledge about the conventional image onto knowledge about the love relationship. Since the car is a vehicle, the submapping A Love Relationship Is A Vehicle applies to the car. But since the Love Is A Journey mapping does not mention wheels, knowledge about the wheels themselves and their spinning is not mapped. The medium in which the car is stuck (the ice, mud, etc.) does not get mapped either. Here is the knowledge that gets mapped:

> The relationship is stuck. The lovers want the relationship to be functioning so that they can continue making progress toward common life goals. They are not happy that the relationship is stuck. They are putting a lot of energy into getting the relationship unstuck, and they feel frustrated.

This is what it means to be spinning one's wheels in a relationship.

We will refer to such idioms as "metaphorical idioms." Each metaphorical idiom comes with a conventional mental image and knowledge about that image. A conventional metaphorical mapping maps that source-domain knowledge onto target-domain knowledge.

It has often been observed that in idioms, the meaning of the whole is not simply a function of the meaning of the parts. That is true in the case of metaphorical idioms. But that does not mean that the meaning of the parts of the idiom plays no cognitive role in the meaning of the whole idiom. In the above example, the meanings of *spinning* and *wheels* play an important cognitive role. They jointly evoke the conventional image and knowledge about it.

In the image, *wheels* designate the wheels of the car and *spinning* designates what the wheels are doing. But the cognitive function of the meanings of these parts of the idiom ends there. The Love Is A Journey mapping maps a portion of the knowledge evoked, but not the portion about the wheels and their spinning.

The cognitive reality of such images and such knowledge mappings has been established in a series of experiments by Gibbs and his coworkers (A2, Gibbs 1994).

Metaphorical idioms are philosophically important in a number of ways. First, they show something important about meaning, namely, that words can designate portions of conventional mental images.

Second, they show that mental images do not necessarily vary wildly from person to person. Instead, there are conventional mental images that are shared across a large proportion of the speakers of a language.

Third, they show that a significant part of cultural knowledge takes the form of conventional images and knowledge about those images. Each of us appears to have thousands of conventional images as part our long-term memory.

Fourth, they open the possibility that a significant part of the lexical differences across languages may have to do with differences in conventional imagery. The same metaphorical mappings applied to different images will give rise to different linguistic expressions of those mappings.

Fifth, they show dramatically that the meaning of the whole is not a simple function of the meanings of the parts. Instead, the relationship between the meaning of the parts and the meaning of the whole is complex. The words evoke an image; the image comes with knowledge; conventional metaphors map appropriate parts of that knowledge onto the target domain; the result is the meaning of the idiom. Thus, a metaphorical idiom is not just a linguistic expression of a metaphorical mapping. It is the linguistic expression of an image plus knowledge about the image plus one or more metaphorical mappings. It is important to separate that aspect of the meaning that has to do with the general metaphorical mapping from that portion that has to do with the image and knowledge of the image.

Why the Term *Metaphor*?

We can now see why it is appropriate to use the term *metaphor* for both everyday and novel cases. The reason is the same mappings cover both kinds of cases. Traditionally, only novel cases were called *metaphors*. But as Lakoff and Turner (A1, 1989) show in great detail in their study of poetic metaphor, the theory of the novel cases is the same as the theory of the conventional cases. Thus, the theory of conceptual cross-domain mapping is exactly the theory needed to account for traditional cases of novel metaphorical expressions. It is thus best called a *theory of metaphor.*

Metaphorical Pluralism:
Multiple Metaphors for a Single Concept

So far, we have only discussed cases of a single conceptual metaphor for a single concept, for example, the journey metaphor for love. But abstract concepts are typically structured by more than one conventional metaphor. Let us look at how the concept of love is structured by multiple metaphors.

Gibbs (A2, 1994) gives a protocol taken from his research on the conceptualization of love. Here a young woman describes, first, her definition of love and, second, her description of her first love experience:

> The overall concern for another person. Sharing of yourself but not giving yourself away. Feeling like you are both one, willing to compromise, knowing the other person well with excitement and electrical sparks to keep you going.
>
> It kicked me in the head when I first realized it. My body was filled with a current of energy. Sparks filled through my pores when I thought of him or things we'd done together. Though I could not keep a grin off my face, I was scared of what this really meant. I was afraid of giving myself to someone else. I got that feeling in my stomach that you get the first time you make eye contact with someone you like. I enjoyed being with him, never got tired of him. I felt really overwhelmed, excited, comfortable, and anxious. I felt warm when I heard his voice, the movements of his body, his smell. When we were together, we fit like a puzzle, sharing, doing things for each other, knowing each other, feeling each other breathe.

Our experience of love is basic—as basic as our experience of motion or physical force or objects. But as an experience, it is not highly structured on its own terms. There is some literal (i.e., nonmetaphorical) inherent structure to love in itself: a lover, a beloved, feelings of love, and a relationship, which has an onset and often an end point.

But that is not very much inherent structure. The metaphor system gives us much more. When we comprehend the experience of love, when we think and talk about love, we have no choice but to conceptualize mostly in terms of our conventional metaphors—to conceptualize it not on its own terms, but in terms of other concepts such as journeys and physical forces. When we reason and talk about love, we import inferential structure and language from those other conceptual domains. The cognitive mechanism we use is cross-domain conceptual mapping. The neural mechanism, so far as we can estimate at present, is one like that in Narayanan's neural theory.

Each mapping is rather limited: a small conceptual structure in a source domain mapped onto an equally small conceptual structure in the target domain. For a rich and important domain of experience like love, a single conceptual mapping does not do the job of allowing us to reason and talk about the experience of love as a whole. More than one metaphorical mapping is needed.

We (A1, Lakoff and Johnson 1980) and Kövecses (A1, 1986, 1988, 1990) have written extensively on the conventional system of metaphors for love. Love is conventionally conceptualized, for example, in terms of a journey, physical force, illness, magic, madness, union, closeness, nurturance, giving of oneself, complementary parts of single object and heat. The young woman's definition and description above reflect all these conceptual metaphors, which are conventional in our culture.

In philosophy, metaphorical pluralism is the norm. Our most important abstract philosophical concepts, including time, causation, morality, and the mind, are all conceptualized by multiple metaphors, sometimes as many as two dozen. What each philosophical theory typically does is to choose one of those metaphors as "right," as the true literal meaning of the concept. One reason there is so much argumentation across philosophical theories is that different philosophers have chosen different metaphors as the "right" one, ignoring or taking as misleading all other commonplace metaphorical structurings of the concept. Philosophers have done this because they assume that a concept must have one and only one logic. But the cognitive reality is that our concepts have multiple metaphorical structurings. A common philosophical response is that no metaphorical structure enters into the concept at all, that concepts are literal and independent of all metaphor.

Is the Concept Independent of the Metaphors for That Concept?

Is the concept of love independent of the metaphors for love? The answer is a loud "No!" The metaphors for love are significantly constitutive of our concept of love. Imagine a concept of love without physical force—that is, without attraction, electricity, magnetism—and without union, madness, illness, magic, nurturance, journeys, closeness, heat, or giving of oneself. Take away all those metaphorical ways of conceptualizing love, and there's not a whole lot left. What's left is the mere literal skeleton: a lover, a beloved, feelings of love, and a relationship, which has an onset and an end point. Without the conventional conceptual metaphors for love, we are left with only the skeleton, bereft of the richness of the concept. If somehow everyone had been forced to speak and think about love using only the little that is literal about it, most of what has been thought and said about love over the ages would not exist. Without those conventional metaphors, it would be virtually impossible to reason or talk about love. Most of the love poetry in our tradition simply elaborates those conceptual metaphors.

The Aptness of Metaphor

What does it mean for a metaphor to be apt? First, a metaphor may play some significant role in structuring one's experience. For example, take the metaphor Emotional Experiences Are Physical Forces, in which one can be *overcome* by emotion, or in which emotional experiences can be *jarring* or *painful*. We may very well experience emotions in the same way we experience certain physical forces. An emotional experience can be *painful* or *disruptive*. In short, there are certain metaphorical entailments based on the logic of the source domain, that can be true because the metaphor structures experience itself. Thus, when our emotional experiences are the subject matter we are thinking and talking about, the Emotional Experiences Are Physical Forces metaphor can be apt.

Another way a metaphor can be apt is if it has nonmetaphorical entailments. Take, for example, the Love Is A Journey metaphor. Consider the expression "We're going in different directions" as said of a marital relationship. Given that Common Life Goals Are Destinations in this mapping, the metaphorical idea of going in different directions entails that the spouses have different life goals that are incompatible with the marriage. This is a metaphorical entailment that can be literally true or false. In situations where the metaphorical entailments are nonmetaphorical and true, the metaphor can be said to be apt.

Does this mean that we can simply replace the metaphor by literal truth conditions? Not at all. The metaphor is, in most cases, used for reasoning, it may impose a nonliteral ontology that is crucial to this reasoning, and there may be no nonmetaphorical conceptualization that is adequate for reasoning with the concept. Moreover, not all of its entailments may be literally true. In other words, a metaphorical mapping may be apt in some respects, but not in others.

The point here is that one cannot ignore conceptual metaphors. They must be studied carefully. One must learn where metaphor is useful to thought, where it is crucial to thought, and where it is misleading. Conceptual metaphor can be all three.

The very notion of the aptness of a metaphorical concept requires an embodied realism. Aptness depends on basic-level experience and upon a realistic body-based understanding of our environment.

Summary

Our most important abstract concepts, from love to causation to morality, are conceptualized via multiple complex metaphors. Such metaphors are an essential part of those concepts, and without them the concepts are skeletal and bereft of nearly all conceptual and inferential structure.

Each complex metaphor is in turn built up out of primary metaphors, and each primary metaphor is embodied in three ways: (1) It is embodied through bodily experience in the world, which pairs sensorimotor experience with subjective experience. (2) The source-domain logic arises from the inferential structure of the sensorimotor system. And (3) it is instantiated neurally in the synaptic weights associated with neural connections.

In addition, our system of primary and complex metaphors is part of the cognitive unconscious, and most of the time we have no direct access to it or control over its use.

Thus, abstract concepts structured by multiple complex metaphors exemplify the three aspects of mind that are the central themes of this book: the cognitive unconscious, the embodiment of mind, and metaphorical thought.

6

Embodied Realism: Cognitive Science Versus A Priori Philosophy

We have outlined the three results from cognitive science research that we have taken as themes of this book. At this point two objections naturally arise.

First, not every cognitive scientist accepts all these as "results." Many cognitive scientists were raised in the tradition of analytic philosophy, which asserts that concepts are literal and disembodied. Those who were educated to assume such a view tend to reject out of hand either (1) the existence of metaphorical concepts or (2) the imposition and definition of rational structure by the body and brain, or both.

Second, many postmodern philosophers and other post-Kuhnian philosophers of science deny that cognitive science can have "results" that could provide a basis for criticizing a particular philosophical view. They argue instead that all it can do is make claims on the basis of culturally constructed narratives. From the radical postmodern perspective, no science, including cognitive science, can be free of crucial philosophical assumptions that determine the so-called results. Therefore, they argue, cognitive science can neither function as the basis for a critique of existing philosophy nor provide the basis for an alternative philosophical theory.

Both of these objections raise the question of whether cognitive science can be free of a priori philosophical assumptions that determine its "results." In the first case, the question is whether cognitive science can, or should, be free of the assumptions of analytic philosophy. In the second, the question is whether scientific inquiry in the study of mind in general can ever produce results not determined by some philosophy or other.

Let us consider these two objections in order.

Two Conceptions of Cognitive Science

Philosophy is so much an implicit, though not always recognized, part of all intellectual disciplines that it has determined, for many investigators, the conception of what cognitive science is. There are at least two approaches to cognitive science defined by different philosophical commitments: a first-generation cognitive science that assumed most of the fundamental tenets of traditional Anglo-American philosophy and a second generation that called most of those tenets into question on empirical grounds.

These two versions of cognitive science entail two very different conceptions of the nature of philosophy, and so it is crucial to examine their philosophical assumptions in detail.

The First Generation:
The Cognitive Science of the Disembodied Mind

Cognitive science got its start within a context defined by traditional Anglo-American philosophical assumptions (see Chapters 12 and 21). First-generation cognitive science evolved in the 1950s and 1960s, centering on ideas about symbolic computation (A, Gardner 1985). It accepted without question the prevailing view that reason was disembodied and literal—as in formal logic or the manipulation of a system of signs. In those years, Anglo-American philosophy fit very well with certain dominant paradigms of that era: early artificial intelligence, information-processing psychology, formal logic, generative linguistics, and early cognitive anthropology, all of which played a role in first-generation cognitive science. This was no accident. Many of the practitioners in these paradigms had been trained using the assumptions of Anglo-American philosophy.

Accordingly, it seemed natural to assume that the mind could be studied in terms of its cognitive functions, ignoring any ways in which those functions arise from the body and brain. The mind, from this "functionalist" perspective, was seen metaphorically as a kind of abstract computer program that could be run on any appropriate hardware. A consequence of the metaphor was that the hardware—or rather "wetware"—was seen as determining nothing at all about the nature of the program. That is, the peculiarities of the body and brain contributed nothing to the nature of human concepts and reason. This was philosophy without flesh. There was no body in this conception of mind.

Early cognitive science thus assumed a strict dualism in which the mind was characterized in terms of its formal functions, independent of the body (C2, Haugeland 1985). What was added from artificial intelligence, formal logic, and generative linguistics was that thought could be represented using formal symbol systems. As in a computer language, these symbols were meaningless in themselves, and thought was seen as the manipulation of such symbols according to formal rules that do not look at any meanings that might be attributed to the symbols.

There were two attitudes about meanings. In the first, meanings are what the symbols compute. Meanings are defined entirely in terms of the internal relationships among symbols. On the second view, the symbols characterizing thought were taken as internal *representations* of an external reality. In other words, the symbols were to be given meaning through reference to that external reality, that is, to things in the world—objects, their properties, relations between them, and classical categories of objects.

Thus, the term *mental representation* had two different meanings in these traditions. In the first, a *representation* was seen as a representation of a *concept*, which in turn was taken to be defined solely in terms of its relationships to other concepts within a formal system. Thus, on this account, a representation was a symbolic expression that was purely internal to a given formal system. In the second, a *representation* was taken to be a symbol representation of something outside the formal system.

Mind, on both conceptions, happened to be embodied in the brain in the trivial sense in which software needs hardware to run on: the brain was the hardware on which the mind's software happened to be running, but the brain-hardware was seen as being capable of running any appropriate software, and so was assumed to play no essential or even important role in characterizing the mind-software. Functionally, mind was disembodied. Moreover, thought was seen as literal; imaginative capacities did not enter the picture at all. This was a modern version of the Cartesian view that reason is transcendental, universal, disembodied, and literal. This view of mind is sometimes referred to as *philosophical cognitivism*.

The Second Generation:
The Cognitive Science of the Embodied Mind

By the mid- to late 1970s, a body of empirical research began to emerge that called into question each of these fundamental tenets of Anglo-American "cognitivism." Responding to this research, a competing view of cognitive science developed in which all the above assumptions had to be abandoned in the face of two kinds of evidence: (1) a strong dependence of concepts and reason upon the body and (2) the centrality to conceptualization and reason of imaginative processes, especially metaphor, imagery, metonymy, prototypes, frames, mental spaces, and radial categories.

These empirical results directly contradicted the assumptions of Anglo-American philosophy. The key points of the second-generation embodied view of mind are the following:

- Conceptual structure arises from our sensorimotor experience and the neural structures that give rise to it. The very notion of "structure" in our conceptual system is characterized by such things as image schemas and motor schemas.
- Mental structures are intrinsically meaningful by virtue of their connection to our bodies and our embodied experience. They cannot be characterized adequately by meaningless symbols.
- There is a "basic level" of concepts that arises in part from our motor schemas and our capacities for gestalt perception and image formation.
- Our brains are structured so as to project activation patterns from sensorimotor areas to higher cortical areas. These constitute what we have called *primary metaphors.* Projections of this kind allow us to conceptualize abstract concepts on the basis of inferential patterns used in sensorimotor processes that are directly tied to the body.
- The structure of concepts includes prototypes of various sorts: typical cases, ideal cases, social stereotypes, salient exemplars, cognitive reference points, end points of graded scales, nightmare cases, and so on. Each type of prototype uses a distinct form of reasoning. Most concepts are not characterized by necessary and sufficient conditions.
- Reason is embodied in that our fundamental forms of inference arise from sensorimotor and other body-based forms of inference.
- Reason is imaginative in that bodily inference forms are mapped onto abstract modes of inference by metaphor.

- Conceptual systems are pluralistic, not monolithic. Typically, abstract concepts are defined by multiple conceptual metaphors, which are often inconsistent with each other.

In short, second-generation cognitive science is in every respect a cognitive science of the embodied mind (C2, Varela et al. 1991). Its findings reveal the central role of our embodied understanding in all aspects of meaning and in the structure and content of our thought. Meaning has to do with the ways in which we function meaningfully in the world and make sense of it via bodily and imaginative structures. This stands in contrast with the first-generation view that meaning is only an abstract relation among symbols (in one view) or between symbols and states of affairs in the world (in another view), having nothing to do with how our understanding is tied to the body.

What we are calling "first-generation" versus "second-generation" cognitive science has nothing to do with the age of any individual or when one happened to enter the field. The distinction could just as well be called "disembodied" versus "embodied" or "assuming tenets of formalist analytic philosophy" versus "not assuming tenets of formalist analytic philosophy." The distinction is one of philosophical and methodological assumptions.

The Issue of Initial Philosophical Commitments

First-generation cognitive science, as we have just seen, is based on very specific a priori commitments about what concepts, reason, and meaning are:

- *Functionalism:* The mind is essentially disembodied; it can be studied fully independently of any knowledge of the body and brain, simply by looking at functional relations among concepts represented symbolically.
- *Symbol manipulation:* Cognitive operations, including all forms of thought, are formal operations on symbols without regard to what those symbols mean.
- *Representational theory of meaning:* Mental representations are symbolic; they get their meaning either by relations to other symbols or by relations to external reality.
- *Classical categories:* Categories are defined by necessary and sufficient conditions.

- *Literal meaning:* All meaning is literal; no meaning is fundamentally metaphorical or imagistic.

These views about the nature of mind were not based on empirical results. They came from an a priori philosophy. This is the opposite of the situation in second-generation cognitive science, where views about the nature of mind have come from empirical evidence, rather than a priori philosophical assumptions.

First-generation cognitive science is based on analytic philosophy and, for that reason, denies many of the "results" that we are reporting on. How has second-generation cognitive science managed to free itself from the dominating influence of analytic philosophy?

General Methodological Assumptions Versus Specific Philosophical Assumptions

What needs to be avoided in science are assumptions that predetermine the results of the inquiry before any data is looked at. We also need to avoid all assumptions that circumscribe what is to count as data in such a way as to predetermine the outcome. To keep the data from being artificially circumscribed, we need assumptions that will guarantee an appropriately wide range of data. To make sense of the data—to see the structure in it—we need to require that maximal generalizations be stated wherever possible.

What we have just described are methodological assumptions. In applying a method, we need to be as sure as we can that the method itself does not either determine the outcome in advance of the empirical inquiry or artificially skew it. A common method for achieving this, especially in the studies we will be discussing, is to seek converging evidence using the broadest available range of differing methodologies. Ideally, the skewing effects of any one method will be canceled out by the other methods. The more sources of evidence we have, the more likely this is to happen. Where one has five to ten sources of converging evidence, the probability of any particular methodological assumption skewing the results falls considerably.

Thus, certain commitments are required for an empirically responsible inquiry. They include:

The Cognitive Reality Commitment: An adequate theory of concepts and reason must provide an account of mind that is cognitively and neurally realistic.

The Convergent Evidence Commitment: An adequate theory of concepts and reason must be committed to the search for converging evidence from as many sources as possible.

The Generalization and Comprehensiveness Commitment: An adequate theory must provide empirical generalizations over the widest possible range of phenomena.

Where second-generation cognitive science differs most strongly from the first-generation theories is that it has steadfastly resisted putting a priori philosophical assumptions from analytic philosophy, generative linguistics, and so forth, ahead of these basic methodological commitments.

We need assumptions like this to minimize the possibility that the results of the inquiry will be predetermined. The assumption that we should seek generalizations over the widest possible range of data does not guarantee that we will find any generalizations at all; nor does it determine the content of any generalizations found.

Assumptions That Do Not Determine Results

Assumptions such as these do not tell you what answers you will come up with—if any—when you look at empirical data. For example, they do not determine, prior to looking at empirical data, the three results that we are discussing in this book. They do not predetermine that there is (or is not) unconscious thought, that thought can (or cannot) be metaphorical, or how (if at all) the body shapes how we think. It is only when such assumptions are applied to a broad range of data of many sorts using many different convergent methodologies that these "results" appear.

For instance, in the emergence of second-generation cognitive science, there were no a priori commitments to the existence of prototypes, conceptual metaphors, image schemas, radial categories, embodiment, and so on. There was, however, a commitment to make sense of a vast range of phenomena that included polysemy (systematically related linguistic forms), inference, historical change, psychological experiments, poetic extensions of everyday language, gesture, language acquisition, grammar, and iconicity in signed languages. The evidence from these diverse empirical domains converges: It is all made sense of by conceptual metaphors, image schemas, and radial categories—and by no other theory of concepts yet proposed (A2, Gibbs 1994; A1, Lakoff 1993). The concrete results about conceptual and inferential structure were empirical discoveries not anticipated in advance. Indeed, they were quite surprising.

In the move from first- to second-generation cognitive science then, the relationship between philosophy and cognitive science has reversed. In the first generation, philosophy was in control of much of cognitive science; basic tenets of Anglo-American cognitivist philosophy were taken as true prior to empirical research, and cognitive science was expected to conform to its assumptions. Second-generation cognitive science argues that philosophy must begin with an empirically responsible cognitive science based on the above methodological assumptions, especially the assumption of convergent evidence.

This stance is what sanctions our use of the word *results* in connection with second-generation cognitive science. It is what allows us to assess the empirical adequacy of philosophical claims about concepts, mind, and language. And it is what allows us to begin philosophical inquiry anew on the basis of these results.

As we have seen, the phenomenon of complex metaphor involves all of the three major types of results we are considering: the cognitive unconscious, the embodiment of mind, and metaphorical thought. Complex conceptual metaphor therefore provides an extended example of the methodology of convergent evidence.

Convergent Evidence for the Existence of Conceptual Metaphor

If conceptual metaphor is part of the cognitive unconscious, if we have no conscious direct access to it, how do we know it exists at all? What kind of evidence is there?

The first three types of evidence are generalization evidence, in which a conceptual structure is taken as existing if it is required to explain generalizations over the data. Let us begin with the Love Is A Journey metaphor discussed above.

The Love Is A Journey mapping is needed to account for generalizations of at least three kinds: inference generalizations, polysemy generalizations, and novel-case generalizations.

Inference Generalizations

In Chapter 5, we saw that the Love Is A Journey mapping states the generalization governing the use of conventional travel-domain inferences to reason about love.

The main function of conceptual metaphor is to project inference patterns from one conceptual domain onto another. The result is that conceptual metaphor allows us to reason about the target domain in a way that we otherwise would not, as when we use inference patterns for travel to draw conclusions about love. What makes the mapping a *generalization* is that it covers multiple cases in which ways of reasoning about travel systematically correspond to ways of reasoning about love.

Polysemy Generalizations

Words like *crossroads, stuck,* and *dead end* are words whose primary meaning is in the domain of travel. As we have seen, these words can also be used to speak about love, and when they are, they have a meaning in the love domain that is related systematically to the meaning in the travel domain. Such cases of systematically related meanings for a single word are referred to as instances of *polysemy.* The Love Is A Journey mapping states the generalization linking various travel-domain words and the meanings of those words to the corresponding uses of those words in the love domain.

In the case of inferential generalizations, the generalizations were about concepts, not about *words*, that is, not about the *sound sequences* used to express ideas. In the case of polysemy, the generalizations do concern the words, that is, the sound sequences. And not just individual words, but whole systems of words.

Novel-Case Generalizations

The very same mappings that state the polysemy and inferential generalizations for conventional metaphorical expressions also cover novel cases. Novel-case generalizations are extremely important for showing that the metaphorical mapping is alive, not "dead."

To date, nine major kinds of convergent evidence have contributed to the conclusion that conceptual metaphor is cognitively real. We just looked at three types of generalization evidence: Generalizations over (1) inference patterns, (2) polysemy, and (3) novel extensions. This is often sufficient to establish the case.

But cognitive scientists, like other scientists, are happier when the evidence is absolutely overwhelming. Moreover, cognitive scientists tend to be a bit chauvinistic when it comes to evidence, preferring evidence from their home field. Thus, linguists tend to prefer linguistic evidence, such as generalizations over inference, polysemy, and novel cases. Historical linguists prefer that the evidence be historical, either from etymology or grammaticalization. Cognitive psychologists tend to prefer evidence using paradigms they are familiar with—priming, problem solving, and so on. Developmental psychologists prefer language acquisition data. Gesture analysts prefer gestural evidence, and prefer it from signed languages over evidence from spoken languages. And discourse analysts prefer evidence from real discourses, either live ones or written texts. That is why we need to be able to draw upon several additional types of evidence.

Psychological Experiments

The experimental techniques (for a survey, see A2, Gibbs 1994, 161–167, 252–257) include the following seven types: *priming, problem solving, inferential reasoning, image analysis, classification, verbal protocol analysis,* and *discourse comprehension.* Because the range of convergent methodologies within such experiments is quite wide, we could technically consider each kind of experiment as presenting a different type of convergent evidence. If we counted each experimental methodology as a distinct source of convergent evidence, we would have fifteen sources. Here is a brief account of one such experiment, just to get the flavor of the kind of evidence that comes from experiments in cognitive psychology.

Albritton (A2, 1992) devised an experiment to test the cognitive reality of the Love Is A Physical Force metaphor (A1, Lakoff and Johnson 1980). The Love Is A Physical Force mapping goes as follows:

Physical Force	→	Love
Objects Affected By Force	→	Lovers

The conventional metaphorical expressions that express this mapping include:

She *knocked me out.* I was *bowled over* by him. We were immediately *attracted* to each other. There was a *magnetism* between us. We were *drawn to* each other. He *swept her off her feet.*

Albritton gave participants in the experiment little stories like the following:

> (1) John and Martha met a party about a month ago. (2) Since then they have hardly ever been seen apart from each other. (3) The attraction between John and Martha was overwhelming. (4) Sparks flew the moment they first saw one another. (5) It was a classic case of love at first sight.

In this story, the third and fourth sentences are verbal expressions of the Love Is A Physical Force metaphor. Afterwards, the participants were presented with one of two prime sentences taken from the story. One was a verbal expression of Love Is A Physical Force (e.g., "The attraction between John and Martha was overwhelming"); the other was neutral (e.g., "John and Martha met at a party about a month ago"). At this point Albritton presented a different sentence from the story that was a verbal expression of the Love Is A Physical Force metaphor (e.g., "Sparks flew the moment they first saw one another"). The participants' task was to decide whether or not they had read the test sentence earlier.

The experiment was designed to find out whether the metaphor was "live," that is, cognitively real and active in the minds of speakers, or "dead," that is, nonexistent in the minds of speakers now and merely a historical remnant of a live metaphor from an earlier time. The dead-metaphor hypothesis says there is no live Love Is A Physical Force metaphor. Sentences 3 and 4 are just dead metaphors, which means they now have only a literal meaning about love. Thus, all the sentences in the story are literal. Since both sentences 1 and 3 are literal in this hypothesis, and since both occur earlier in the story, their effect on the recognition of sentence 4 should be identical.

The live-metaphor hypothesis says there is a live Love Is A Physical Force metaphor. Sentences 3 and 4 in the story are expressions of it, while sentence 1 is not. Since sentence 3 is an expression of the Love Is A Physical Force metaphor, it should activate that metaphor, while sentence 1 should not. Thus, sentence 3 should make the recognition of sentence 4, which is also an expression of that metaphor, occur faster than sentence 1, which does not activate the Love Is A Physical Force metaphor.

The result was that the participants who had been primed by reading sentence 3, the other verbal expression of the Love Is A Physical Force metaphor, were significantly faster at making this recognition judgment than those who had been primed by sentence 1. Thus, the experiment confirmed the hypothesis that there is a live Love Is A Physical Force metaphor, and the results contradicted the dead-metaphor hypothesis.

This was only one in a series of experiments by Albritton (A2, 1992), and other experiments have also confirmed that conceptual metaphors are cognitively real and alive (A2, Kemper 1989; Gibbs and O'Brien 1990; Nayak and Gibbs 1990; Gentner and Gentner 1982).

Historical Semantic Change

Sweetser (A1, 1990) demonstrated that conceptual metaphor provides "routes" for possible changes of word meaning over the course of history. For example, she provides extensive evidence for the existence of the Knowing Is Seeing metaphor over the whole range of the Indo-European languages, going as far back into antiquity as is possible. Her data includes a myriad of cases from various branches and times in which words from the domain of vision change to acquire additional meanings in the domain of knowledge. For example, consider the Indo-European root *weid-,* whose reconstructed meaning is "see." This develops in Greek into both *eidon,* "see," and *oida,* "know" (from which we get English "idea"). In English, it becomes both the vision word "witness" and the knowledge words "wit" and "wise." In Latin, it shows up as *video,* "see," while in Irish it becomes *fios,* "knowledge." Other roots meaning "see" have similar histories: roots originally meaning "see" come to mean "know" throughout the Indo-European language family at various times in various branches.

These are all independent developments, occurring at different times in different places with different roots. They cannot be random changes. Sweetser's argument is that they can all be explained if one assumes that the Knowing Is Seeing metaphor developed early in Indo-European and has been naturally learned by generation after generation of Indo-European language speakers. The existence of this conceptual metaphor in the minds of speakers made these independent changes natural.

Spontaneous Gesture Studies

McNeill (A3, 1992) has shown that spontaneous unconsciously performed gestures accompanying speech often trace out images from the source domains of conceptual metaphors. For example, in one of his early studies, a speaker said that he could not decide whether to stay home or go to the movies. The gesture he made as he said this consisted of his holding his hands in front of him palm up, with the hands alternately going up and down, as if his palms were the balance pans on a scale. He was talking about choosing and was using the metaphor Choosing Is Weighing; his hands were the scale doing the weighing of the two choices. McNeill analyzes a large number of such examples.

Language Acquisition Studies

As we noted above, Christopher Johnson (A1, 1997b, c), in studies of the acquisition of the Knowing Is Seeing metaphor by children, has found that the acquisition of such conceptual metaphors goes through two stages: conflation and differentiation. In the conflation stage, seeing is correlated with gaining knowledge, as when the child sees that Daddy is home or sees what he spilled. At this stage, a child might say "See what I spilled?" but not "See what I mean?" At the first stage, the use of the word *see* is conventionalized to mean both *see* and *know* together, but there is no metaphor at this stage. At a later stage, *see* can be used metaphorically to mean *know* when no actual seeing takes place, as in "See what I mean?"

This means that two senses of *see* cannot be mere homonyms, words with the same spelling that happen to be used for unrelated concepts. The reason is that *see* cannot be learned as meaning *know* independent of seeing contexts during the first stage.

Sign Language Metaphor Studies

The lexicon of American Sign Language (ASL) is replete with metaphorical signs that reflect common cross-cultural conceptual metaphors (A3, Taub 1997). For example, the sign meaning *know* has the dominant forefinger moving to the forehead, tracing the path of a piece of knowledge coming into the head. The sign meaning *past* gestures toward the region behind the signer, tracing out the metaphor that the past is behind us. It is used in expressions like "Let's put that *behind* us" or "Let's not look *back* to the past." Such cases have been found in ASL by the hundreds.

Discourse Coherence Studies

As we noted in Chapter 3, Srini Narayanan, in a study of examples of the uses of metaphor in news stories about international economics (B2, 1997a), has observed that conceptual metaphor is necessary to make coherent sense of such examples of written discourse.

What the Evidence Shows

Different forms of evidence show different things. Let us look, case by case, at what follows from each type of evidence.

The generalizations over inferences and polysemy show the following: There are systematic correlations across conceptual domains in the use of source-domain inference patterns for target-domain inferences. The exact *same* systematic correlation accounts for the use of source-domain words to name target-domain concepts.

The generalizations over novel cases show that the conventionalized conceptual mapping is in fact productive for new cases. This indicates that it is presently psychologically real and not just a remnant of some earlier stage in history or development.

The psychological experiments also show that the conventional mappings are not dead, but alive. They are psychologically real, they can be activated, and we think using them.

The historical data show that such conceptual metaphors must have been real in the minds of speakers of Indo-European languages in all branches over thousands of years.

The gesture evidence indicates that all of us who gesture spontaneously during speech, and most of us do, unconsciously use the conceptual metaphors in shaping our gestures.

The language acquisition evidence shows that there is a natural mechanism by which these conceptual metaphors are acquired in a developmental sequence. It also shows that the use of *see* to mean *know* in the second stage is indeed metaphorical rather than just a case of nonmetaphorical homonymy. The contrasting homonymy hypothesis would predict that *see* should be able to mean *know* or anything else in the first stage.

The American Sign Language evidence shows that a huge number of iconic signs systematically fit patterns defined by conventional metaphorical mappings. In addition to such conventional signs, novel metaphorical signs that are instances of existing conceptual metaphors can be made up on the spot, indicating that the conceptual metaphor is alive for signers of ASL.

The discourse coherence evidence shows that there are conventional metaphors for thinking about specialized subject matters such as economics. These metaphors are very much alive, so much so that they are used to draw inferences that make discourse coherent.

How Convergent Evidence Can Help Free Cognitive Science from A Priori Philosophizing

According to mainstream analytic philosophy, all concepts are literal and there are no such things as metaphorical concepts. Because tenets from analytic philosophy placed constraints on the views of first-generation cognitive scientists, many of them have simply not accepted the existence of conceptual metaphors in spite of the array of kinds of evidence listed above. This is a remarkable contemporary case of philosophy placing limits on science. Many cognitive scientists have, in the course of their education, internalized tenets of analytic philosophy, consciously or not. Among these is the tenet that concepts are necessarily *defined* as being literal. If this tenet is assumed, the evidence for conceptual metaphor will not matter, because analytic philosophy rules it out a priori. When such a *definition* of what a concept is is accepted a priori, no evidence contradicting that philosophical definition *could* matter.

How did second-generation cognitive scientists escape such conclusions? They put convergent evidence ahead of a priori philosophical views about concepts, meaning, and language. Second-generation cognitive scientists were aware of the constraints placed on cognitive science by analytic philosophy, but refused to accept philosophy as the ultimate arbiter of scientific arguments.

Philosophical awareness matters here. Many first-generation cognitive scientists assent to the idea that empirical evidence should take precedence over a priori philosophy. But if they are not aware of how a priori philosophy shapes their scientific worldviews, they will simply not notice its effects. They may assume, for example, that it is *somehow* given in advance that concepts cannot be metaphorical, without being aware of the source of the prejudice. For this reason, cognitive scientists need to be as aware of philosophy as philosophers should be of cognitive science.

A Response to the Postmodern Critique of Science

Any student of twentieth-century history and philosophy of science will be aware that there can be no science without at least *some* assumptions. As we saw, even second-generation cognitive science makes methodological assumptions, which, however spare, are nonetheless philosophical assumptions. This raises the postmodern challenge: No "scientific results" can be used to criticize a philosophical position, since those "results" themselves are based on a competing philosophical position. Science, the charge goes, is just one more philosophical narrative with no privileged status relative to any other philosophical narrative.

We are well aware of this argument and, as post-Kuhnians, we are also well aware of the fallacies in classical logical empiricism: There are no pure observation sentences from which a scientific theory can be arrived at through induction. There can be no assumption-free scientific observations. And there is no correct logic of induction that will yield correct laws directly from observational data.

Science, as Kuhn rightly observed, does not always proceed by the linear accretion of objective knowledge. Science is a social, cultural, and historical practice, knowledge is always situated, and what counts as knowledge may depend on matters of power and influence. Accordingly, we reject the simpleminded ideas that all science is purely objective, that issues of power and politics never enter into science, that science progresses linearly, and that it can always be trusted. Moreover, we strongly reject the myths that science provides the ultimate means of understanding everything and that humanistic knowledge has no standing relative to anything that calls itself science.

But this does not mean that there is no reliable or stable science at all and that there can be no lasting scientific results. Now that we have photographs of the earth from the moon, any lingering doubts that the earth is round have been removed. We are not likely to discover that there are no such things as cells or that DNA does not have a double-helix structure. Many scientific results are stable. Indeed, we believe that we have some insight into what makes scientific results stable, and we will discuss those insights below in our treatment of embodied scientific realism.

This is also true of the science of the mind. We are not likely to discover that there are no neurons or neurotransmitters. Nor are we likely to discover that there is no distinction between short-term and long-term memory. We know from neuroscience that our brains contain topographic maps and that our visual systems contain orientation-sensitive cells. Much of what we have learned about the brain and the mind is now stable knowledge.

We believe that the three results from cognitive science research on which this book rests are also stable. We maintain that they deserve to be called "results" because of all the converging evidence supporting them. The existence of so many forms of convergent evidence demonstrates that what we take as specific results are not merely the consequences of assumptions underlying a particular method of inquiry. The methodology of convergent evidence and the masses of different types of evidence minimize the probability that the results will be an artifact of any one specific methodology.

Embodied Scientific Realism

We will be using the results of second-generation cognitive science to rethink philosophy. In doing so, we are committed to at least some form of scientific realism. We are basing our argument on the existence of at least three stable scientific findings—the embodied mind, the cognitive unconscious, and metaphorical thought. Just as the ideas of cells and DNA in biology are stable and not likely to be found to be mistakes, so we believe that there is more than enough converging evidence to establish at least these three results.

Ironically, these scientific results challenge the classical philosophical view of scientific realism, a disembodied objective scientific realism that can be characterized by the following three claims:

1. There is a world independent of our understanding of it.
2. We can have stable knowledge of it.
3. Our very concepts and forms of reason are characterized not by our bodies and brains, but by the external world in itself. It follows that scientific truths are not merely truths as we understand them, but absolute truths.

Obviously, we accept (1) and (2) and we believe that (2) applies to the three findings of cognitive science we are discussing on the basis of converging evidence. But those findings themselves contradict (3). The doctrine of disembodied reason has, unfortunately, been applied to yield an untenable version of scientific realism: disembodied scientific realism. The evidence we will be looking at concerns the embodiment of mind and, as we shall see, allows us to keep a scientific realism in an embodied form, one that is cognitively and neurally realistic: an embodied scientific realism.

At the heart of embodied realism is our physical engagement with an environment in an ongoing series of interactions. There is a level of physical interaction in the world at which we have evolved to function very successfully, and an important part of our conceptual system is attuned to such functioning. The existence of such "basic-level concepts"—characterized in terms of gestalt perception, mental imagery, and motor interaction—is one of the central discoveries of embodied cognitive science.

For example, Berlin, Breedlove, and Raven (A4, 1974) and Hunn (A4, 1977), in detailed studies of Tzeltal plant and animal categories, note that at the basic level (they call it the "folk-generic" level) Tzeltal speakers are extremely accurate (in the 90–95 percent range) at identifying plants and animals relative to scientific biological classification. At lower levels—the species and variety—their accuracy drops off precipitously to around 50 percent and below. In short, we are better equipped to recognize plants and animals at the level of the genus, that is, at the basic level, than at lower biological levels. (For further discussion, see A4, Lakoff 1987.)

Our embodied system of basic-level concepts has evolved to "fit" the ways in which our bodies, over the course of evolution, have been coupled to our environment, partly for the sake of survival, partly for the sake of human flourishing beyond mere survival, and partly by chance. It is not that every basic-level concept exists because of its survival value, but without such an embodied system coupled to our environment, we would not have survived. The basic level of conceptualization is the cornerstone of embodied realism.

One thing that science has done successfully in many (but by no means all) cases has been to extend our basic-level capacities for perception and manipulation via technology. Instruments like telescopes, microscopes, and spectroscopes have extended our basic-level perception, and other technologies have expanded our capacities for manipulation. In addition, computers have enlarged our basic capacity for calculation. Such enhancements of basic bodily capacities extend the basic level for us, the level that is at the heart of embodied realism.

What fills out embodied realism, permitting us to move far beyond mere observation and manipulation, are several crucial findings about our embodied concepts and imaginative capacities. The first important finding is that there are perceptual and motor "inferences" and that there is a neurally instantiated logic of perception and motor movements. The second crucial finding is the existence of conceptual metaphor, which allows us to conceptualize one domain of experience in terms of another, preserving in the target domain the inferential structure of the source domain. Mathematics allows us to model metaphorical theories and to calculate precisely inferences about literal basic-level categories. Such inferences can then be projected onto scientific subject matters to give explanatory accounts for existing data and to make predictions. What permits this is that metaphoric theories can have literal, basic-level entailments.

Interesting scientific theories have inferences about multiple subject matters, for instance, language acquisition, historical semantics, gesture studies, and priming experiments. Each such subject matter is thus a test bed for such a theory. We speak of evidence for a scientific theory as being "convergent" when the results all support the same explanatory hypothesis.

Such convergent evidence tests inferences that are different for different subject matters and yet confirm the same theory. What makes converging evidence convincing is that the theory cannot follow from any *one* set of methodological assumptions. Rather, our confidence in it increases as converging evidence from various methodologies mounts up. The degree of confirmation of a theory thus goes up exponentially with the number of distinct subject matters having distinct methodologies for testing inferences of the theory.

Embodied scientific realism is thus compatible both with the success of science and with what we have learned in the Kuhnian tradition: that theories change over time, that new theories often don't cover previously known phenomena, that theories can be incommensurable, and that politics, culture, and personal issues enter into science. Successful sciences are those for which there is broad and deep converging evidence. But not all areas of science are on a par with respect to evidential criteria; some have more converging evidence over more subject matters than others.

Moreover, it may be the case that the limitations of human conceptual systems will make it impossible for there to be fully general, global scientific theories. For example, general relativity and quantum mechanics are incompatible theories, each with an enormous range of converging evidence. String theorists seek a unified theory of physics, but we do not yet know—and we may never know—whether that is possible. All that may be possible are partial theories, theories that are what we will call "locally optimal"—incompatible, but widely comprehensive (though not fully comprehensive) theories supported by considerable converging evidence. Perhaps locally optimal theories are the best we can do using human minds. We don't know. Quantum mechanics and general relativity may be locally optimal, globally incommensurable theories.

As Kuhn saw, the history of science yields cases of scientific revolutions. For us, these are cases in which new metaphors replace old ones, in which the new metaphor is incommensurable with the old metaphor, and hence an entire discipline is reconceptualized. This can even happen in sciences in which old theories have a great deal of convergent evidence. In such cases, the new theory must also have a range of convergent evidence at least approaching that of the old theory, though the details may be quite different. But broadly convergent evidence remains a crucial standard for the wide acceptance of a new scientific theory, whatever other factors may enter in. In short, embodied scientific realism makes sense of both stable scientific knowledge and scientific revolutions.

Replacing disembodied scientific realism with embodied scientific realism is a gain for realism, not a loss, since it brings our understanding of what science is in line with the best neuroscience and cognitive science of our age. It allows us to understand science better. And it allows us to appreciate Kuhn's contributions while recognizing, as he did, the success of science.

Beyond Subject and Object

Embodied realism can work for science in part because it rejects a strict subject-object dichotomy. Disembodied scientific realism creates an unbridgeable ontological chasm between "objects," which are "out there," and subjectivity, which is "in here." Once the separation is made, there are then only two possible, and equally erroneous, conceptions of objectivity: Objectivity is either given by the "things themselves" (the objects) or by the intersubjective structures of consciousness shared by all people (the subjects).

The first is erroneous because the subject-object split is a mistake; there are no objects-with-descriptions-and-categorizations existing in themselves. The second is erroneous because mere intersubjectivity, if it is nothing more than social or communal agreement, leaves out our contact with the world. The alternative we propose, embodied realism, relies on the fact that we are coupled to the world through our embodied interactions. Our directly embodied concepts (e.g., basic-level concepts, aspectual concepts, and spatial-relations concepts) can reliably fit those embodied interactions and the understandings of the world that arise from them.

The problem with classical disembodied scientific realism is that it takes two intertwined and inseparable dimensions of all experience—the awareness of the experiencing organism and the stable entities and structures it encounters—and erects them as separate and distinct entities called subjects and objects. What disembodied realism (what is sometimes called "metaphysical" or "external" realism) misses is that, as embodied, imaginative creatures, *we never were separated or divorced from reality in the first place.* What has always made science possible is our embodiment, not our transcendence of it, and our imagination, not our avoidance of it.

7

Realism and Truth

Direct, Representational, and Embodied Realism

Perhaps the oldest of philosophical problems is the problem of what is real and how we can know it, if we can know it. Greek philosophy began with that question. The issue for the Greeks was whether, as Greek religion assumed, our fates were ruled by the whims of the gods or, as Greek philosophy asserted, our capacity for reason gave us a sufficient understanding of the world to survive and flourish.

The Greek philosophers asked how we could know. Their answer was that we could know directly. For them, knowledge that worked was knowledge of Being. Aristotle, for example, saw an identity between ideas in the mind and the essences of things in the world. That identity answered the problem of knowledge. Aristotle concluded that we could know because our minds could directly grasp the essences of things in the world. This was ultimate *metaphysical realism.* There was no split between ontology (what there is) and epistemology (what you could know), because the mind was in direct touch with the world.

With Descartes, philosophy opened a gap between the mind and the world. If the mind and the world were not one, then they had to be different kinds of things. The body was flesh and of the world; the mind was not. The mind, separate from the body and the world, could not be directly in touch with the world. Ideas (other than those that were assumed to be innate) became internal "representations" of external reality, forever distant from the world but some-

94

how "corresponding" to it. This split metaphysics from epistemology, and that split still plagues philosophy today. Once the mind is taken to be disembodied, the gap between mind and world becomes unbridgeable. For this reason, there has been notoriously little progress to this day in philosophical attempts to characterize the "representation-to-reality correspondence" (C2, Putnam 1981).

Indeed, in the most popular current version, the representations have shrunk to their minimum—mere symbolic representations, with symbols being abstract entities having no properties other than being distinct from one another (e.g., C2, Fodor 1975, 1981, 1987). This "symbol-system realism" maximizes the chasm between mind and world, since the abstract entity of the symbol shares nothing with anything in the world, not even physical reality. Nor is there any natural correlation between symbols and things in the world. In symbol-system realism, the mind-world gap is not only maximal, but maximally arbitrary, since there is nothing—not correlation, not similarity, not even common physicality—to make the correspondence nonarbitrary.

The embodiment of reason, as revealed by cognitive science, provides a new understanding of the fit between the mind and reality, a view that we will call *embodied realism*. It is closer to the direct realism of the Greeks than it is to the disembodied representational realism of Cartesian and analytic philosophy, which is fundamentally separated from the world. Embodied realism, rejecting the Cartesian separation, is, rather, a realism grounded in our capacity to function successfully in our physical environments. It is therefore an evolution-based realism. Evolution has provided us with adapted bodies and brains that allow us to accommodate to, and even transform, our surroundings.

Realism is fundamentally about our success in functioning in the world. Someone who is "not realistic" is someone who is ill-adapted, someone who is out of touch and out of harmony with the world. Realism is about being in touch with the world in ways that allow us to survive, to flourish, and to achieve our ends. But being in touch requires something that touches—a body.

The embodiment of mind thus brings us far closer to the direct realism that the Greeks assumed than does the disembodied and mere symbol-system realism of present-day analytic philosophy. It gives up on being able to know things-in-themselves, but, through embodiment, explains how we can have knowledge that, although it is not absolute, is nonetheless sufficient to allow us to function and flourish.

The direct realism of the Greeks can thus be characterized as having three aspects:

1. The Realist Aspect: The assumption that the material world exists and an account of how we can function successfully in it.
2. The Directness Aspect: The lack of any mind-body gap.
3. The Absoluteness Aspect: The view of the world as a unique, absolutely objective structure of which we can have absolutely correct, objective knowledge.

Symbol-system realism of the sort found in analytic philosophy accepts (3), denies (2), and claims that (1) follows from (3), given a scientifically unexplicated notion of "correspondence."

Embodied realism accepts (1) and (2), but denies that we have any access to (3).

All three of these views are "realist" by virtue of their acceptance of (1). Embodied realism is close to the direct realism of the Greeks in its denial of a mind-body gap. It differs from direct and symbol-system realism in its epistemology, since it denies that we can have objective and absolute knowledge of the world-in-itself.

Since embodied realism denies, on empirical grounds, that there exists one and only one correct description of the world, it may appear to some to be a form of relativism. However, while it does treat knowledge as relative—relative to the nature of our bodies, brains, and interactions with our environment—it is not a form of extreme relativism, because it has an account of how real, stable knowledge, both in science and the everyday world, is possible. That account has two aspects. First, there are the directly embodied concepts, such as basic-level concepts, spatial-relations concepts, and event-structure concepts. These concepts have an evolutionary origin and enable us to function extremely successfully in our everyday interactions in the world. They also form the basis of our stable scientific knowledge.

Second, primary metaphors make possible the extension of these embodied concepts into abstract theoretical domains. The primary metaphors are anything but arbitrary social constructs, since they are highly constrained both by the nature of our bodies and brains and by the reality of our daily interactions.

Embodied realism, however, does recognize a central insight of relativist thought, namely, that in many important cases, concepts *do* change over time, vary across cultures, have multiple inconsistent structures, and reflect social conditions. Embodied realism also provides mechanisms for characterizing these changes, variations, multiplicities, and instances of "social construction."

The formation of complex metaphors and other conceptual blends appears to be the major mechanism for them (A7, Fauconnier and Turner 1994).

Philosophical Precursors of Embodied Realism

The embodied realism we are developing here is not created out of nothing. It is anticipated by two of our greatest philosophers of the embodied mind, John Dewey and Maurice Merleau-Ponty. Despite their wide differences of temperament and style, both Dewey and Merleau-Ponty believed that philosophy must be informed by the best scientific understanding available, and they each made extensive use of the empirical psychology, neuroscience, and physiology of their day. They both argued that mind and body are not separate metaphysical entities, that experience is embodied, not ethereal, and that when we use the words *mind* and *body* we are imposing bounded conceptual structures artificially on the ongoing integrated process that constitutes our experience.

Dewey (C2, 1922, 1925) focused on the whole complex circuit of organism-environment interactions that makes up our experience, and he showed how experience is at once bodily, social, intellectual, and emotional. Merleau-Ponty (C2, 1962) argued that "subjects" and "objects" are not independent entities, but instead arise from a background, or "horizon," of fluid, integrated experience on which we impose the concepts "subjective" and "objective."

More recently, Varela, Thompson, and Rosch (C2, 1991) have drawn on embodied cognitive science, phenomenology, and Buddhist awareness practices to explain their "enactive" notion of experience. They explain the two basic tenets of their own version of embodied realism: "First, that cognition depends upon the kinds of experience that come from having a body with various sensorimotor capacities, and second, that these individual sensorimotor capacities are themselves embedded in a more encompassing biological, psychological, and cultural context" (C2, 1991, 173).

Other philosophical thinkers, such as Alfred North Whitehead, Terry Winograd and Fernando Flores, Drew Leder, and Eugene Gendlin, have explored dimensions of some form of embodied realism. What distinguishes the view of embodied realism we are proposing is the use we make of empirical evidence from recent cognitive neuroscience and embodied cognitive science. This empirical research makes it possible for us to explore in a suitably detailed way the workings of the embodied mind in its structuring of experience via neural cognition. It gives us ways to explain why we have the categories we do, why we have the concepts we have, and how our embodiment shapes our reasoning and the structure of understanding that forms the basis for what we take to be true.

Realism and Truth

In contemporary analytic philosophy, the theories of reference and truth are central because the analytic program depends on them to fill the gap between symbols and the world. Analytic philosophy's symbol-system realism makes the weight placed on problems of reference and truth all the heavier, since symbol-system realism maximizes the distance and arbitrariness between the symbolic representations and the world.

What analytic philosophy must resort to is a "correspondence theory of truth," in which the chasm between abstract symbols and what they refer to in the world is to be bridged by a notion of "correspondence," a notion that has very little content. Analytic philosophy parades its realism and downplays the chasm it has introduced between symbols and the world, yet the enterprise depends crucially upon the "correspondence" that is to bridge the symbol-world gap.

The Correspondence Theory of
Truth in Its Simple, Intuitive Version

In its simplest form, the correspondence theory can be stated as follows:

> A statement is true when it fits the way things are in the world. It is false when it fails to fit the way things are in the world.

Put this way, how could one object? It seems as American as apple pie, as British as the Union Jack. But when spelled out in greater detail, the problems begin to appear. What does "fit" mean when there is a mind-world chasm and a need to say what the correspondence between symbols and the world consists in. It is not enough to say, "Poof! Let there be a correspondence." The correspondence is left up to the theory of reference, a theory that is supposed to bridge the symbol-world gap. But theories of reference have been of little help for what we will see is a deep reason.

There are two major types of theories of reference in analytic philosophy. The first says the meaning of an expression determines what it refers to. The second says reference is determined causally, that is, by acts of referring by particular people. Theories of the first type stem from the work of Gottlob Frege. The Fregean theory is that senses, which are abstract entities independent of both mind and body, somehow pick out referents correctly, but there is no scientific account of how. In Richard Montague's version, Fregean senses, or "intensions," are mathematical functions that pick out reference. This leaves us in the dark as to how people are supposed to do it. John Searle would have the mind-brain determining Fregean reference to an objective reality. But that objective reality is external to and independent of the mind-brain, and Searle offers no scientific explanation as to how that gap is to be bridged.

One of the second type of theories of reference is the Kripke-Putnam causal theory, which has two parts: (1) Historical individuals, by an act of pointing, indicate the fixed referent of a specific word and (2) somehow both the identity of the object and the relation of that object to the name remain fixed through history. No account is given of either part. It is not explained *how* mere pointing and naming can establish a "rigid" symbol-to-world designation or how that "rigid designation" is able to remain in place over millennia.

What is inadequate in all of these accounts is the statement of the problem, the assumption that truth is a matter of correspondence between symbols and a mind-, brain-, and body-independent world. To see the inadequacies, we need to consider the more technical current versions of the correspondence theory.

The first level of increased technicality comes from observing that *statements,* which are either spoken or written sentences, express *propositions.* Propositions, in turn, are structures made up of symbols, and it is these symbolic propositional structures that are taken as corresponding, or failing to correspond, to reality. The internal structure of propositions is seen, variously, as having one of a number of structures—a subject-predicate structure, predicate-argument structure, and so on. The claim is that, by virtue of this structuring of symbols, the proposition can be made to correspond to structure in the world and thereby make truth claims about the world.

Propositions are introduced to neutralize the differences among languages. The typical examples given are quite simple: "Snow is white" and *"Schnee ist weiss"* are both supposed to name *the same* proposition, namely, the assertion that concepts for what we call *snow* and what we call *being white* go together in a way that corresponds to the way snow and whiteness go together in the world.

The introduction of propositions thus turns the gap between words and the world into two gaps:

Gap 1: The gap between sentences of natural languages and propositions, which are language-neutral structures consisting of abstract symbols.

Gap 2: The gap between the symbol structures and the world.

However, this technical device of language-neutral propositions in the form of symbolic structures doesn't solve the problem of how propositions can fit the world any more than the simple intuitive account in terms of sentences does. One must still show how sentences can correspond to the world. Only now one must show two things: how sentences of a specific language can correspond to symbolic propositions and how the symbolic propositions can correspond to the world.

Formal analytic philosophy introduces an additional level of complexity—the level of models of situations—into the correspondence theory. The correspondence between a statement and the way the world is is now broken down into *three* correspondences, and the gap has now become three gaps:

Gap 1: The gap between the natural language and the symbols in a "formal language" that are used to represent aspects of the natural language.

Gap 2: The gap between the symbols of the formal language and the sets of arbitrary abstract entities in the set-theoretical model of the language.

Gap 3: The gap between the set-theoretical models of the world and the world itself.

The correspondence theory is in serious trouble on all fronts. The first gap, the gap between natural languages and formal symbol systems, was supposed to be filled by formal linguistics. That promise has not been kept, and it appears that it cannot be. The first full-blown attempt was Lakoff and McCawley's generative semantics, which sought to combine formal logic and generative linguistics. By the mid-1970s, results like those we cited above concerning the embodiment of spatial-relations concepts (image schemas), basic-level categories, various types of prototypes, radial categories, color concepts, aspectual concepts, and conceptual metaphors undermined the possibility that a cognitively and neurally real linguistics could be accommodated within the limited resources of formal syntax and model theory.

In the intervening years, the development of the field of cognitive linguistics has turned up ever more phenomena that cannot be accounted for by the formal-syntax-and-semantics paradigm (see Chapter 22). Though the formal linguistic project is still ongoing in existing varieties of formal syntax and semantics, the vast range of cognitively real linguistic phenomena beyond its scope appears to be directly at odds with any prospect of success. Indeed, research in the formal linguistics paradigm does not even address most of these phenomena.

The second gap, between the formal language and the set-theoretical models is in a different kind of trouble. As Quine and Putnam have repeatedly emphasized, statements in a formal language (a symbol system) vastly underdetermine the models that the symbols can map onto (for details, see Chapter 21). As Putnam has pointed out (C2, Putnam 1981), this indeterminacy of reference dooms the project of characterizing symbol-to-model correspondences in a way that is needed to satisfy the requirements of the correspondence theory.

Formal model theory cannot fill the second gap for an empirical reason as well: The meanings of the words and grammatical constructions in real natural languages cannot be given in terms of set-theoretical models (see A4, Lakoff 1987). This is clear when you look at the details. Spatial-relations concepts (image schemas), which fit visual scenes, are not characterizable in terms of set-theoretical structures. Motor concepts (verbs of bodily movement), which fit the body's motor schemas, cannot be characterized by set-theoretical models. Set-theoretical models simply do not have the kind of structure needed to fit visual scenes or motor schemas, since all they have in them are abstract entities, sets of those entities, and sets of those sets. These models have no structure appropriate to embodied meaning—no motor schemas, no visual or imagistic mechanisms, and no metaphor.

The third gap, which may be the most difficult of all to bridge, has barely even been discussed. That is the gap between the set-theoretical models and the real world. Most formal philosophers don't see the problem, because they have adopted a metaphysics that appears to make the problem go away. That metaphysics goes like this: The world is made up of distinct objects having determinate properties and standing in definite relations at any given time. These entities form categories called natural kinds, which are defined by necessary and sufficient conditions.

If you assume this objectivist metaphysics, then it follows that certain set-theoretical models should be able to map onto the world: abstract entities onto real-world individuals, sets onto properties, sets of n-tuples onto relations, and so on. But such a mapping must bridge the gap between the model and the world. No progress whatever has been made in demonstrating that the world is the way objectivist metaphysics claims it is. Nor has anyone even tried to fit such a set-theoretical model to the world. The problem is rarely discussed in any real detail.

Once one looks seriously at the problems in bridging all three gaps, one can see why the correspondence theory is in trouble even on its own terms. But the situation gets much more desperate if one looks at the question of truth from a cognitive and neural perspective. The correspondence theory simply does not fit cognitive and neural facts of the sort we have been discussing. Let us now turn to a discussion of why.

Embodiment and Truth

What we understand the world to be like is determined by many things: our sensory organs, our ability to move and to manipulate objects, the detailed structure of our brain, our culture, and our interactions in our environment, at the very least. *What we take to be true in a situation depends on our embodied understanding of the situation,* which is in turn shaped by all these factors. Truth for us, any truth that we can have access to, depends on such embodied understanding.

The classical correspondence theory of truth is disembodied. The sensorimotor system plays no role in it. Bodily functioning in the world plays no role in it. The brain plays no role in it. There is no body at all in the correspondence theory of truth.

But truth is not simply a relation between words and the world, as if there were no being with a brain and a body interposed. Indeed, the very idea that beings embodied in all these concept-shaping ways could arrive at a disembodied truth based on disembodied concepts is not merely arrogant, but utterly unrealistic.

To see why embodiment presents insuperable problems for the correspondence theory of truth, we have to look at embodiment in more detail.

Levels of Embodiment

There are at least three levels to what we are calling the embodiment of concepts: the neural level, phenomenological conscious experience, and the cognitive unconscious.

Neural embodiment concerns structures that characterize concepts and cognitive operations at the neural level. The neural circuitry characterizing color concepts, Regier-style models of spatial-relations concepts, and Narayanan's neural models of aspectual concepts are examples.

It is important to recall that the neural level is, of course, arrived at through scientific investigation, including sophisticated experimental techniques and a heavy dose of theoretical abstraction. Even PET scans, which present us with pictures, require a lot of theorizing in order for us to get some sense of what the pictures are pictures *of*. When we speak of "neural circuitry," we are, of course, using an important metaphor to conceptualize neural structure in electronic terms. The circuitry metaphor is used by the neuroscience community at large and is taken as providing crucially important insights into the behavior of the brain. "Truths" about the neural level are commonly stated in terms of this metaphor. We mention this because the neural level is seen quite properly as a "physical" level, and yet much of what we take as true about it is stated in terms of the metaphor of neural circuitry, which abstracts away from ion channels and glial cells.

The *phenomenological level* is conscious or accessible to consciousness. It consists of everything we can be aware of, especially our own mental states, our bodies, our environment, and our physical and social interactions. This is the level at which we speak of the "feel" of experience, of the way things appear to us, and of qualia, that is, the distinctive qualities of experiences such as a toothache, the taste of dark chocolate, the sound of a violin, or the redness of a ripe bing cherry. Most of what is known as phenomenological reflection is about this level. However, phenomenology also hypothesizes nonconscious structures that underlie and make possible the structure of our conscious experience.

The *cognitive unconscious* is the massive portion of the iceberg that lies below the surface, below the visible tip that is consciousness. It consists of all those mental operations that structure and make possible all conscious experience, including the understanding and use of language. The cognitive unconscious makes use of and guides the perceptual and motor aspects of our bodies, especially those that enter into basic-level and spatial-relations concepts. It includes all our unconscious knowledge and thought processes. Therefore it includes all aspects of linguistic processing—phonetics, phonology, morphology, syntax, semantics, pragmatics, and discourse.

The cognitive unconscious is posited in order to explain conscious experience and behavior that cannot be directly understood on its own terms. That is, the cognitive unconscious is what has to be hypothesized to account for generalizations governing conscious behavior as well as a wide range of unconscious behavior. The details of these unconscious structures and processes are arrived at through convergent evidence, gathered from various methodologies used in studying the mind. What has been concluded on the basis of such studies is that there exists a highly structured level of mental organization and processing that functions unconsciously and is inaccessible to conscious awareness.

To say that the cognitive unconscious is real is very much like saying that neural "computation" is real. The Neural Computation metaphor uses numbers and numerical calculations in representing "activations," "inhibitions," "thresholds," "synaptic weights," and so on. It is hypothesized to make sense of what happens in the vast complex of neurochemistry in the brain. Similarly, the detailed processes and structures of the cognitive unconscious (e.g., basic-level categories, prototypes, image schemas, nouns, verbs, and vowels) are hypothesized to make sense of conscious behavior.

These three levels are obviously not independent of one another. The details of the character of the cognitive unconscious and of conscious experience arise from the details of neural structure. We would not have the spatial-relations concepts we have without topographic maps or orientation-sensitive cells. We would not have the color concepts we have without the specific kind of neural circuitry that creates color categories. The neural level is not merely some hardware that happens to be able to run an independently existing software. The neural level significantly determines (together with experience of the external world) what concepts can be and what language can be.

A full understanding of the mind requires descriptions and explanations at all three levels. Descriptions at the neural level alone—at least given our current understanding of it—are not sufficient to explain *all* aspects of the mind. Many aspects of mind are about the feel of experience and the level at which our bodies function in the world, what we have called the phenomenological level. Other aspects of mind depend on the effects of causally efficacious higher-level patterns of neural connectivity, which constitute the cognitive unconscious.

People are not just brains, not just neural circuits. Neither are they mere bundles of qualitative experiences and patterns of bodily interaction. Nor are they just structures and operations of the cognitive unconscious. All three are present—and much more that we are not discussing here. Explanations at all three levels are necessary (though certainly not sufficient) for an adequate account of the human mind.

The Levels-of-Truth Dilemma

Here is the problem that levels of embodiment present for classical correspondence theory of truth: Truth claims at one level may be inconsistent with those at another. Color provides an obvious example. At the phenomenological level of conscious experience, we perceive colors as being "in" the objects that "are" colored. At this level, there are common truths: Grass is green, the sky is blue, blood is red. *Green, blue,* and *red* are one-place predicates holding of grass, the sky, and blood.

Here is what the correspondence theory would say about sentences like "Grass is green." The word *grass* names things (or stuff) in the world. The word *green* names a property that inheres in things in the world. If the green-property inheres in the grass-things, then the sentence "Grass is green" is true.

This is a phenomenology-first account of truth, because it implicitly privileges that level over scientific truth claims. The science of color is irrelevant here. The word *green* has a meaning that reflects our conscious (phenomenological) experience of colors as properties inhering in objects themselves; that is, the meaning of *green* is a one-place predicate denoting a physical property in the world. If grass is green, then there is greenness in the grass.

In much of the Western philosophical tradition, truth is taken to be absolute and scientific truth claims take priority over nonscientific truth claims. We know from the neurophysiology of color vision that colors do not inhere in objects themselves. They are created by our color cones and neural circuitry together with the wavelength reflectances of objects and local light conditions. *At the neural level, green is a multiplace interactional property, while at the phenomenological level, green is a one-place predicate characterizing a property that inheres in an object.* Here is the dilemma: A scientific truth claim based on knowledge about the neural level is contradicting a truth claim at the phenomenological level.

The dilemma arises because the philosophical theory of truth as correspondence does not distinguish such levels and assumes that all truths can be stated at once from a neutral perspective. Yet there are distinct "truths" at different levels; and there is no perspective that is neutral between these levels. To state both the phenomenological and neural truths at once requires looking at both levels at once. The problem is that the truth-as-correspondence theory requires one consistent, level-independent truth. This raises the question as to which is to be given priority, the phenomenological experience or science.

We could adopt a science-first strategy and say that the sentence "Grass is green" is always false, since the meaning of the word *green* is a one-place inherent property, which is not the way color really works. On this strategy, no color terms could ever produce true predications. The problem with this strategy, of course, is that it does violence to what we mean by "truth" and requires us to deny an entire range of truths about our phenomenological experience.

Another strategy is possible while keeping science first: Redefine *green* as a multiplace predicate by bringing into the meaning of *green* our color cones, neural circuitry, local lighting conditions, and so on. The problem here is that this strategy does violence to what most people normally mean by *green*. Moreover, it would get the truth conditions wrong for a range of sentences in which *green* must have its phenomenological meaning. We have in mind sentences like "Most people see grass as being inherently green." If *green* in this sentence is given the multiplace meaning, with color cones and neural circuitry included, then no objects appear or could appear "inherently green." In this sentence *green* must be assigned its normal meaning at the phenomenological level. Scientific truths about color, truths depending on knowledge of the neural level, are not what people are normally concerned with when they use everyday words.

Both the phenomenology-first and science-first strategies are inadequate in one way or other. If we take the phenomenology-first strategy, we miss what we know scientifically is true about color. We get the scientific metaphysics of color wrong. Our "truth conditions" do not reflect what we know to be true. If we take the science-first strategy, we do violence to the normal meanings of the word and to what ordinary people mean by "truth."

Truth Depends on Understanding

In our earlier writings (A1, Lakoff and Johnson 1980; A4, Lakoff 1987), we recognized this dilemma and saw that it could be avoided by taking into account the role of embodied understanding. There is no truth for us without understanding. Any truth must be in a humanly conceptualized and understandable form if it is to be a truth for us. If it's not a truth for us, how can we make sense of its being a truth at all?

Embodied Truth

A person takes a sentence as "true" of a situation if what he or she understands the sentence as expressing accords with what he or she understands the situation to be.

The phenomenological and neural levels provide different modes of understanding, the first in terms of everyday experience and the second in scientific terms. In the first case, we experience colors as inhering in objects and that experience defines a mode of understanding in which "Grass is green" is true. In the second case, the neurophysiology of color vision defines a very different mode of understanding in which "Grass is green" would be false if green were taken as inhering in the grass. There is no contradiction here. Nor is there a single, perspective-neutral truth. That is just as it should be.

Embodied truth is not, of course, absolute objective truth. It accords with how people use the word *true*, namely, relative to understanding.

Embodied truth is also not purely subjective truth. Embodiment keeps it from being purely subjective. Because we all have pretty much the same embodied basic-level and spatial-relations concepts, there will be an enormous range of shared "truths," as in such clear cases as when the cat is or isn't on the mat.

Social truths also make much better sense on this account than on the correspondence theory. They are based on enormously wide understandings and experiences of culture, institutions, and interpersonal relations, including social practices. So-called institutional facts are facts only relative to massive social understanding (C2, Searle 1969, 1995). For example, who the World Heavyweight Champion of boxing is depends on which of the two major boxing associations you think has the right to award the championship.

Social truth can only be embodied truth since it makes no sense without understanding. Moreover, there are conflicting social truths based on conflicting understandings. Different understandings of the nature of such contested concepts as justice, rights, democracy, and freedom lead us to have considerably different competing understandings of what society is about and therefore of what constitutes social truth. Take the sentence "Affirmative action is just." You will take this as true or false depending on your understanding of what constitutes social justice.

Embodied scientific realism gives rise to a corresponding notion of embodied truth for science. As we saw above, statements like "There are cells" are stable scientific truths, embodied truths depending on the capacity for scientific instrumentation to extend our basic-level abilities to perceive and manipulate. Or take the theoretical metaphor of "neural computation." This is a central scientific metaphor in neuroscience with entailments so robust and stable that it has gained the status of embodied truth. Being a metaphor, it could not, of course, have the standing of a literal truth in the correspondence theory. However, the metaphor provides a form of understanding absolutely crucial to neuroscience.

Such an embodied account of truth will, of course, not satisfy the advocate of the traditional correspondence theory, for that theory was predicated on the existence of a single unified mind-independent metaphysics for both language and the world—in all contexts. What second-generation cognitive science has found is that there is no single unified metaphysics; nor is there any that is mind- and body-independent.

The question of what we take truth to be is therefore a matter for cognitive science because it depends on the nature of human understanding: what conceptual systems are, what metaphors are and how we use them, how we frame situations, and what grounds our concepts. Truth is, for this reason, not something subject to definition by an a priori philosophy.

Phenomenology, Functionalism, and Materialism: The Issue of Privileging the Metaphysics of Only One Level

The existence of these three levels of description and explanation required by second-generation cognitive science confronts us with a classic problem: Should one make a metaphysical commitment to only one of these levels to the exclusion of the others, or should one advocate a metaphysical pluralism? That is, do we want to say that only one of these levels is relevant to explanation?

There have been phenomenologists, for example, Edmund Husserl and Hubert Dreyfus, who insist that, with respect to truths about human experience, the ultimate level of explanation is the one that relies on the phenomenological analysis of lived experience. This privileges the metaphysics only of the level of experience that is conscious or accessible to consciousness. There are functionalists, for example, Noam Chomsky and Jerry Fodor, who privilege the level of the cognitive unconscious (as they conceive of it) as legitimate for explanation of the use of human language and thought. Finally, there are eliminative materialists, for example, Patricia and Paul Churchland, who privilege the neural level as the only and ultimate source of explanation for all aspects of cognition.

In addition, there are mixed cases. John Searle, for example, holds that explanation is relevant at the level of the phenomenological and the neural, but rejects the reality of the cognitive unconscious (C2, 1995). There are functionalist developmental psychologists who study the child's acquisition of language and concepts and who privilege both the phenomenological level and the cognitive unconscious as being relevant to explanation of the acquisition of concepts and language.

We, the authors, recognize the validity of all three levels, because we see all three as relevant to a complete description and explanation of thought, language, and other cognitive phenomena such as memory and attention. Some generalizations can be stated with full generality only by using the metaphysics of the neural level, for example, the explanation for the structure of color categories or the role of topographic maps of the visual field in explaining the topological properties of spatial-relations concepts. Other generalizations require the metaphysics of the cognitive unconscious, for example, the statement of metaphorical mappings or grammatical constructions. Still others require the metaphysics of the phenomenological level, for example, the fact that basic color terms have the lexical semantics of a one-place predicate indicating a property inherent in the object. We accept all these as significant modes of understanding relative to which "truths" can be characterized.

In sum, embodied truth requires us to give up the illusion that there exists a unique correct description of any situation. Because of the multiple levels of our embodiment, there is no one level at which one can express all the truths we can know about a given subject matter. But even if there is no *one* correct description, there can still be many correct descriptions, depending on our embodied understandings at different levels or from different perspectives.

Each different understanding of a situation provides a commitment to what is real about that situation. Each such reality commitment is a version of a commitment to truth.

What we mean by "real" is what we need to posit conceptually in order to be realistic, that is, in order to function successfully to survive, to achieve ends, and to arrive at workable understandings of the situations we are in.

When, for example, we say that a construct of cognitive science such as "verb" or "concept" or "image schema" is "real," we mean the same thing as any scientist means: It is an ontological commitment of a scientific theory and therefore can be used to make predictions and can function in explanations. It is like the physicist's ontological commitment to "energy" and "charge" as being real. Neither can be directly observed, but both play a crucial role in explanation and prediction. The same can be said of neural computation, conceptual metaphors, prototypes, phonemes, morphemes, verbs, and so on.

The Embodied Mind Without Eliminativism

Phonemes and verbs are real. They are entities in language. Of course, no one who takes these as real thinks that they are physical entities. We take them as real because they are required if we are to make sense of the nature of language. Any explanatorily adequate theory of language will have to posit them. That is, they are real relative to an understanding, in this case, a scientific understanding of language.

Eliminativism is a philosophical position that says that the only things that are real are physically existing entities. Obviously, as proponents of embodied realism we are not eliminativists. But, at the same time, we are *physicalists*, in the sense that we believe that there is an ultimate material basis for what we take, from a scientific perspective, as being real. That is, we, along with others in the cognitive science community, argue that there is a physical basis in the body and brain for such entities in scientific theories of mind and language as phonemes, verbs, and metaphors.

What it means to be a physicalist, but not an eliminativist, can best be illustrated by a discussion of the neural modeling research discussed in Chapter 3 and in the Appendix. The research paradigm of the Neural Theory of Language Group at Berkeley (the NTL paradigm) is a multilevel paradigm, in which each level contributes something necessary to explanation in cognitive science. That is, such a model implicitly claims that in an explanatorily adequate cognitive science there are truths at each level that cannot be stated adequately at some other level.

The NTL paradigm is an instance of a common paradigm that most cognitive scientists share, at least in principle.

The Common Paradigm

The common paradigm is:

Top Level: Cognitive
Middle Level: Neurocomputational
Bottom Level: Neurobiological

In this paradigm, the top level is a description of cognitive structures and mechanisms in functional terms. It includes such notions as phonemes, verbs, and concepts. The bottom level is a description of the neural system of the brain in biological terms. It includes notions of ion channels, axons, dendrites, synapses, and so on. The role of the neurocomputational level is to link the two—to model the neural structure of the brain or some aspect of it, while using that model to account for aspects of thought, language, and other cognitive functions.

Most cognitive scientists have a theoretical commitment to the reality of such things as phonemes, verbs, and concepts, not in the same sense as a commitment to the reality of chairs and rocks, but nonetheless a commitment at some appropriate level of understanding. Similarly, cognitive neuroscientists engaged in neural computation have a theoretical commitment to the reality of neural gates, synaptic weights, thresholds, and mathematical operations "performed by neurons" (addition, subtraction, multiplication, differentiation, integration, vector addition, and so on). Of course, *the numbers used in such calculations are not literally there in the cell bodies.* The mathematics used in the computations is part of a critically important scientific metaphor for understanding how neurons function: the Neural Computation metaphor. This is the central metaphor of computational neuroscience, a metaphor that appears to be apt, that is, to accurately characterize how biological neural networks function. In computational neuroscience, where the metaphor is taken for granted, synaptic weights (which are numbers) are seen as properties of neural networks, and learning is understood as the changing of synaptic weights, which in the technical models is the changing of numbers. Embodied realism makes sense of commitments to the scientific reality of such metaphorical entities as the numbers in a computational model referred to as "synaptic weights."

Finally, it is important to notice that neurobiologists and computational neuroscientists commonly use the Neural Computation metaphor without noticing that it is a metaphor. Indeed, it is extremely common for computational neurobiologists to form what linguists call a "conceptual blend" of the source and target domains of the metaphor (A7, Fauconnier and Turner 1994, 1996). In the blend, the target-domain biological structures (containing cell bodies, dendrites, synapses, and ion channels) are brought together with the source-domain neural circuitry (containing circuit ideas like connections and gates, as well as numbers indicating synaptic weights, activation, and inhibition). In such a blended discourse, biological structures are conceptualized as if they "changed (the numbers indicating) synaptic weights," "sent inhibition," "formed gates," and so on. Conceptual blends of this sort are the norm in scientific discourse.

Noneliminative Physicalism

We have seen that reality and truth occur relative to our understanding at many levels and from many perspectives. This is inconsistent with the classical eliminativist program in the philosophy of science, which asserts that the only realities and the only truths are at the "lowest level," here the neural level, that is, the level of neurochemistry and cellular physiology.

Virtually no neuroscientists hold this position, as can be seen by the ubiquity of the Neural Computation metaphor. Strictly speaking, "neural circuits" with their neural computational numerical algorithms are not a direct part of neurochemistry and cellular physiology, which talks about such things as ion channels, neurotransmitters, and cell permeability. But the Neural Computation metaphor, which defines the field of computational neuroscience (linking the middle level to the bottom level in the common paradigm), is absolutely necessary to an adequate understanding of how the brain and body function. No serious neuroscience could "eliminate" these higher, metaphorically constituted levels of scientific understanding at which computations using numbers are taken as real.

The same is true of the models of linguistic and cognitive behavior constructed by cognitive linguists and other cognitive scientists, for example, structures like conceptual systems and theoretical constructs such as basic-level categories, conceptual metaphors, image schemas, and prototypes. When there is sufficient convergent evidence, such theoretical constructs are taken as "real." Since we are not, and could not be, aware of them, they are postulated as part of the cognitive unconscious. What we call "the cognitive unconscious" is the totality of those theoretical cognitive mechanisms above the neural level that we have sufficient evidence for, but that we do not have conscious access to. Like each of the cognitive mechanisms that constitute it, the cognitive unconscious as a whole, as a general phenomenon, is taken to be real.

This is significant for the philosophy of science in the following way. Consider the NTL paradigm in its simplified version, with cognitive, neurocomputational, and neurobiological levels:

Top Level: Cognitive
Middle Level: Neurocomputational
Bottom Level: Neurobiological

In an eliminative physicalist theory, explanation would flow in one direction only, bottom to top, with only the neurobiology taken as real and the other levels taken as epiphenomena.

This is not the case in the NTL paradigm. The paradigm is physicalist in that it does not claim that any mystical nonphysical entities such as soul, or spirit, or a disembodied Cartesian mind exist. Ultimately, the brain is all neurochemistry and neurophysiology. But it is noneliminative in two ways. First, each level is taken as real, as having a theoretical ontology necessary to explain phenomena. Second, explanation and motivation flow in both directions. To explain how the neurochemistry and neurophysiology function in networks of neurons, we need a theoretical level of neural computation. Explanation of what the physical neurons are doing flows from the middle level to the bottom level. This is a noneliminative, top-to-bottom form of explanation.

Next, consider the architecture of the circuitry at the level of neural computation. It is motivated bottom to top by physical structures at the neurobiological level: topographic maps, center-surround receptive fields, and so on. It is also motivated top to bottom by cognitive phenomena such as priming, presupposition projection, and other "spreading activation" effects at the cognitive level.

The structures at the cognitive level require bottom-to-top explanation. The structure of color categories is explained by the color cones and neural circuitry of the brain's system for color vision. Image-schema structure is explained in part by topographic maps and neural gating (B2, Regier 1995). The nature of event structure and linguistic aspect is explained in part by the structure of neural motor control systems (B2, Narayanan 1997a, b).

Furthermore, the entire paradigm, involving all three levels, makes sense only relative to the fact that we are organisms functioning in a physical and social environment and that we have evolved to survive in such an environment. The very existence of our color systems can only be explained relative to the fact that surfaces have reflectances, that certain chemicals in our retinal cones are altered by certain wavelengths of light, and so on. Similarly, the existence of image schemas makes sense only because the spatial relations they characterize are useful to us: containment, trajectories of motion, centrality, force-dynamics, balance, contact, and so on. The same is true of conceptual metaphor, as we saw in our discussion of primary metaphor. Primary metaphors can be explained partly in the neural terms described by Narayanan (B2, 1997a, b), but in order for neural connections to form, there must be physical and social correlations in the experience of organisms with our size, physical characteristics and capacities, and social needs. This ecological and evolutionary aspect of the NTL paradigm is also noneliminative. That is, it is more than neurobiology, and it is absolutely essential to an explanation of why we have the conceptual and linguistic systems we have.

In short, the NTL paradigm is an instance of a noneliminative physicalism for three reasons: First, all levels are taken as real. Second, explanation and motivation spread top to bottom as well as bottom to top. And third, explanation and motivation can only be adequate relative to larger ecological and evolutionary considerations.

Embodied Realism Enters

We have seen that the basic empirical results of second-generation cognitive science require an embodied realism. Embodied realism is correspondingly necessary to make sense of cognitive science. To see why it is needed, consider the following objection: How can you be a physicalist when you believe in the reality of such nonphysical things as neural computation and the cognitive unconscious?

In a philosophy in which metaphysics is independent of, and prior to, epistemology, a physicalist is a philosopher who claims that the only things that exist "objectively"—that is, that exist independent of any understanding by any beings—are physical. However, in order to conceptualize anything physical, one must use one's understanding. In embodied realism, where truth depends on understanding, there is no such metaphysics-epistemology split. Hence the term *physicalist* takes on a very different meaning, one concerning the nature of scientific explanation and motivation and what one takes as real for the purpose of scientific explanation. A *physicalist* is someone who believes that there is a material basis for all entities taken as real within any scientific theory.

Embodied scientific realism thus makes for sensible science. Once levels of understanding are distinguished, one can speak sensibly of what is real or unreal, true or false relative to those levels of scientific explanation. One can discuss whether neural network structures of a certain architecture are real or whether it is true that neural computation works according to one theory or another. One does not have to believe that the numbers used in the computation are individually physically real, but rather that the relative values of those numbers accord in some appropriate way with physical reality. At the level of the cognitive unconscious, one can discuss whether conceptual metaphors or phonemes are real, and if they are, what their properties are. Accordingly, one can give a physicalist neural theory of how metaphors and phonemes—as scientifically understood entities—arise and function in the brain. It makes perfect sense in embodied realism to be a physicalist, yet to speak of such nonphysical things as real relative to forms of scientific theorizing. The only kinds of nonphysical entities and structures taken as "real" are those that are hypothesized on the basis of convergent evidence and that are required for scientific explanation.

The Efficacious Cognitive Unconscious

To say that the cognitive unconscious is efficacious is to say that the theoretically postulated cognitive mechanisms that compose it do real cognitive work, that they play a central role in conceptualization and reasoning and therefore are intentional, representational, and truth characterizing. Recently, John Searle (C2, 1995, 127–148) has claimed that what we and other cognitive scientists have called the cognitive unconscious is merely a "background" with none of the properties we attribute to it. Indeed, Searle does not believe in the reality of a cognitive unconscious at all. "To put the point crudely, I believe that in most appeals to the unconscious in Cognitive Science we really have no clear idea of what we are talking about" (C2, Searle 1995, 128).

Searle characterizes the background as "a certain category of neurophysiological causation." It is neurophysiology without any real structure. Searle prefers the parallel distributed processing version of connectionism, "neuronal net modeling, where there is a meaningful input and a meaningful output, but in between there are no symbol-processing steps; rather there is just a series of nodes with different connection strengths between them, and signals pass from one node to another, and eventually changes in connection strengths give the right match of inputs and outputs." (C2, Searle 1995, 140–141).

Such a background, Searle claims, could constitute capacities that could enable consciousness without any of the structure or characteristics of rational thought. Unstructured neural networks are safely nonrational; they are nonconceptual and nonpropositional. Neurons operate below the level of consciousness, but need not constitute an unconscious level of conceptual structure. And neurons are not about anything. So if the background has a reality only at the neural level, where that level has no discernible propositional or other conceptual structure, then the background would not amount to a cognitive unconscious—a true level of conceptual structure with the efficacious properties of the mind.

On the contrary, we can see that the cognitive unconscious is efficacious and quite real by looking at Searle's own criteria for conceptual meaning and rational structure. If we look carefully, we can see that the cognitive unconscious has all the following properties: It is intentional, representational, propositional, and hence truth characterizing and causal. To demonstrate this, we will take four examples of structures from the cognitive unconscious: basic-level concepts, conceptual frames, spatial-relations concepts, and conceptual metaphors.

First, consider basic-level concepts (e.g., *chair*), which are characterized by mental imagery, motor movement, and gestalt perception. Because we are embodied beings functioning in the world, a basic-level concept like *chair* is intentional and representational. What makes it *intentional* is that the concept picks out the things that fit our mental image of a chair, fit our motor program for sitting in chair, and fit our gestalt perception of chairs. The mental image, the motor program, and the perceptual gestalt together form an embodied representation of category members. This is not merely a symbolic representation, that is, not merely an internal symbolization of an external reality. Rather it *is* the embodied structure that is constitutive of the experience of a chair.

Second, take semantic frames (A6, Fillmore 1982b), which provide an overall conceptual structure defining the semantic relationships among whole "fields" of related concepts and the words that express them. Our restaurant frame, which characterizes our general knowledge of restaurants, is not only intentional and representational, but also *propositional*. The frame characterizes the structured background knowledge relative to which concepts like *restaurants, waiters, maître d's, menus,* and *checks* make sense. It contains propositional information: A waiter brings you a menu, takes your order, brings you your food, and so on. The *propositional* information is *intentional*: It is *about* waiters, menus, food, and so on. The frame *represents* the structure of the experience of restaurants.

In addition, the conceptual frames that inhabit our cognitive unconscious *contribute semantically* to the meanings of words and sentences. Thus, a word like *waiter* is defined relative to the restaurant frame. Consider the sentence "Harry has functioned as a waiter for twenty years, but he has never taken an order for food, never delivered food, never written up or delivered a check, and never even worked in a restaurant or had any other type of food service job." One would be hard pressed to understand how Harry could have functioned as a waiter and how such a sentence could be true given the ordinary meaning of *waiter*. The problem is that key elements of the frame with respect to which *waiter* is defined are being denied. Thus, frames in the cognitive unconscious must certainly be parts of the meanings of sentences when words in those sentences are defined relative to the frame. In such cases, frames in the cognitive unconscious definitely contribute to the semantic content of words and to the meanings of sentences.

Moreover, even when a frame is used purely as a background with no lexical items in the sentence defined relative to it, the frame will typically enter *causally* into inferences made on the basis of what *is* in the sentence. Take the sentence "After we ate, we got up and left." Said in the context of a restaurant frame, one would normally infer that we had gotten the check for the meal and that we paid. In short, frames used as a background are *inference generating*. And inference generation is both *causal* and part of what we take semantics to be about.

Next, consider spatial-relations concepts and image schemas. Such elements of the cognitive unconscious can be *causal* of understandings. When we understand a bee as being *in* the garden, we are imposing an imaginative container structure on the garden, with the bee *inside* the container. The cognitive structure imposed on the garden is called *the container image schema*. That cognitive structure plays a *causal role* in bringing about an understanding—a conceptualization of the bee as being *in* something. Similarly, when we understand a cat as being *behind* a tree, we are imposing a front and a back on the tree. This spatial-relations concept is *causal* in that it imposes on the scene something that it not externally there: the front and back of a tree!

Finally, let us turn to conceptual metaphor. As we shall see below, when we conceptualize Christmas as *coming a week ahead of* New Year's Day, we are imposing a metaphorical front on New Year's Day that allows Christmas to be *ahead* of it in a conceptualization of times as objects moving in a metaphorical space. We are also characterizing the conditions for metaphorical truths: New Year's Day does not come a week ahead of Christmas; the reverse is true. Conceptual metaphor thus plays a causal role in something real that we do, namely, conceptualize time in terms of motion in space.

As we shall see in Part III, philosophical theories are structured by conceptual metaphors that constrain what inferences can be drawn within that philosophical theory. The (typically unconscious) conceptual metaphors that are constitutive of a philosophical theory have the *causal effect* of constraining how you can reason within that philosophical framework.

In short, the cognitive unconscious is thoroughly efficacious: intentional, representational, propositional, truth characterizing, inference generating, imaginative, and causal. The fact that it is efficacious indicates that it is real. The mode of its efficaciousness indicates that it has real conceptual structure—structure at the level of intentionality, representation, propositions, truth, and inference—and not just structure at the neural level.

8

Metaphor and Truth

Whether reason is embodied and whether metaphor is conceptual may sound like obscure pedantic issues. They aren't. They cut to the deepest questions of what we as human beings are and how we understand our everyday world. If you hold the traditional views about metaphor, then you inherit views about what reality is, what truth is, how language is connected to the world, whether we can have objective knowledge, and even what morality is.

We have seen that the traditional view of metaphor is empirically false, because metaphor is conceptual and everyday thought is largely metaphorical. Therefore, the views of reality, truth, language, knowledge, and morality that are tied to the traditional theory of metaphor must also be false. This is disturbing, because it calls into question many of our most basic commonsense views of the world as well as the philosophical theories that elaborate those views.

But how could a challenge to the traditional theory of metaphor possibly have anything to do with undermining the general view of reality and truth and the possibilities for knowledge inherent in most of Western philosophy? Metaphor is usually seen as irrelevant to such topics—and that is just the point! The traditional theory *has to* treat metaphor as irrelevant to fundamental questions about the nature of the world and our knowledge of it, or else a basic part of our commonsense worldview will be placed at risk.

To see just why the philosophical stakes are so high, let us look once more at the traditional theory of metaphor.

The Traditional Theory

The traditional theory of metaphor has persisted for twenty-five hundred years in the philosophical and literary traditions, and the weight of all that tradition cannot easily be overcome by empirical evidence for the existence of conceptual metaphor. The traditional theory has fostered a number of empirically false beliefs about metaphor that have become so deeply entrenched that they have been taken as necessary truths, just as the traditional theory has been taken as definitional. Here are the central tenets of the traditional theory:

1. Metaphor is a matter of words, not thought. Metaphor occurs when a word is applied not to what it normally designates, but to something else.
2. Metaphorical language is not part of ordinary conventional language. Instead, it is novel and typically arises in poetry, rhetorical attempts at persuasion, and scientific discovery.
3. Metaphorical language is deviant. In metaphor, words are not used in their proper senses.
4. Conventional metaphorical expressions in ordinary everyday language are "dead metaphors," that is, expressions that once were metaphorical, but have become frozen into literal expressions.
5. Metaphors express similarities. That is, there are preexisting similarities between what words normally designate and what they designate when they are used metaphorically.

This is not an accidental list. This theory is deeply rooted in the Western philosophical tradition and makes intuitive sense to many people, because it fits an extremely common folk theory about language and truth.

The Commonsense Theory of Language and Truth

The world consists of objects and living beings that have certain properties and, at any given time, stand in certain relations to one another. There is only one way the world is. Our language consists of words expressing ideas that literally fit the world. The primary role of language is to express and communicate such basic truths about the world.

This folk theory is by no means all wrong. For basic-level concepts, the commonsense theory is fundamentally right. Given the centrality of basic-level concepts in our embodied understanding and their prevalence in our mundane experience, the commonsense theory does make sense. This is an embodied interpretation of the commonsense theory. It is not the usual interpretation. The commonsense theory is usually interpreted as an objectivist theory, in which the body and embodied understanding do not enter in at all.

Much of Western philosophy has turned the commonsense folk theory into an expert, objectivist theory. What makes sense for basic-level concepts has been turned into a theory that is supposed to be true for all thought and language. It is a theory that has held sway for well over two millennia, and its dominance has for all that time hidden from view one of the most powerful influences over our daily lives—conventional metaphorical thought.

Why, exactly, does the commonsense theory hide the true nature of metaphorical thought? Why should it have given rise to the false traditional theory of metaphor? The answer, briefly, is this:

If the commonsense theory were true, metaphor would not serve the central function of language, which is supposedly to express and communicate literal truths about the world. Because of this, metaphor has been traditionally relegated to a theory of tropes, which is intended to handle uses of language in which truth is not thought to be at issue: poetry, rhetorical flourish, fictional discourse, and so on. The banishment of metaphor from the realm of truth explains why metaphor has traditionally been left to rhetoric and literary analysis, rather than being taken seriously by science, mathematics, and philosophy, which are seen as truth-seeking enterprises.

We can now see why the traditional theory of metaphor goes hand in hand with the objectivist interpretation of the commonsense theory of language and truth. Consider those parts of the traditional theory of metaphor that are implied by that interpretation of the commonsense theory.

First, because ideas have to be literal if they are to fit the world, they cannot be metaphorical. Therefore, metaphor must be a matter of words, not thoughts. That is why the very idea of *conceptual* metaphor is at odds with this interpretation of the commonsense theory.

Second, if ordinary everyday words are used in their "proper" senses, then they would be literal. Metaphorical language must therefore be "deviant," a means for saying one thing while meaning some other thing.

Third, assuming that the function of ordinary everyday language is to communicate about the world as it really, objectively *is*, language that does not behave in this way is not ordinary everyday language. It must be either poetic, or especially rhetorical, or fictional. That is why the idea that metaphor can be *conventional* and part of normal thought is at odds with the commonsense theory.

Fourth, if it is admitted that an everyday conventional expression is metaphorical at all, it cannot be a "live" metaphor, it can only be "dead," that is, it can no longer be a real metaphor. The dead metaphor is really literal and its apparent metaphorical character must be attributed to an earlier historical stage of the language.

The first four theses of the traditional theory of metaphor thus follow from the objectivist interpretation of the commonsense theory of language and truth, though not from the embodied realist interpretation. The fifth thesis, that metaphor is based on similarity, is consistent with the objectivist interpretation of the commonsense theory and implies certain aspects of it.

In the objectivist theory, since all meaning is held to be literal, a metaphor does not have a capacity to express truth claims. It can only make truth claims if it does so indirectly by expressing some other literal meaning. This, in turn, requires a systematic relationship between a metaphorical use of language and what it indirectly, but literally, expresses. The theory that metaphor is based on similarity provides such a relationship and is consistent with the objectivist interpretation of the commonsense theory of truth and language.

Moreover, the similarity condition has an important implication. The metaphorical expression, typically a word or phrase in a sentence, has a normal meaning that does not fit the subject matter of the rest of the sentence. It therefore seems anomalous, and hence must mean something other than what it appears to mean if it is to make any sense. What else could it mean? The similarity theory has an answer: It must mean something similar to its normal meaning.

For metaphor to be based on similarity in this way, two parts of the objectivist interpretation of the commonsense theory must hold:

1. All concepts must be literal and must name objectively existing things and objectively existing categories in the world.
2. Similarity must be defined by shared properties that really exist objectively in the world.

Though the similarity condition neither entails nor is entailed by the objectivist interpretation of the commonsense theory, it fits that interpretation very well and entails certain important parts of it.

We can now see why there have been within philosophy two long-standing views about the nature of metaphor. Since concepts must be able to accurately fit the world as it really is in itself, there can be no such thing as metaphorical concepts. All there can be are metaphorical uses of language. Those uses can be either (1) *indirectly literal,* in that their meaning must be reducible to literal concepts, or else (2) *meaninglessly fanciful,* in that they do not express literal ideas at all and thus have no meaning, but are only flights of the imagination. Theories of the first sort reduce all metaphor to "proper," but indirect, literal language, while theories of the second sort treat it either as irrelevant to meaning and rational thought or as an imaginative disruption of rational thought, one that, for better or worse, destabilizes meaning.

Aristotle, the father of the traditional theory, was a literalist, as is John Searle, whose version of speech-act theory requires all propositions, that is, anything that can be true, to be literal. Donald Davidson and Richard Rorty are antimeaning theorists, who deny that metaphor has anything to do with meaning or truth (C2, Davidson 1978; Rorty 1989). In general, views of metaphor within romanticism and postmodernism fall under antimeaning theories. Thus, when a romantic like Nietzsche or a postmodernist like Derrida analyzes someone's metaphors, he sees the use of metaphor in formulating a position as invalidating any absolute truth claims that the author was making. Davidson is an unusual case of an objectivist philosopher who is antimeaning about metaphor. He denies metaphor any serious role in truth so that he can preserve the traditional notion that all truth is literal.

These philosophers are either objectivists who want to preserve the commonsense folk theory or radical relativists who want to abandon it. The objectivists can be either literalist (like Aristotle or Searle) or antiliteralist, while the radical relativists, of course, can only be antiliteralist. What these seemingly wildly disparate traditions share is the assumption that, ultimately, metaphor can have nothing directly to do with objective truth. Either it is simply another way of stating literal truth or else it undermines any claims to objective truth.

Why the Traditional Theory Fails

All the evidence for conceptual metaphor that we have discussed in previous chapters is evidence against the traditional theory of metaphor. Since the objectivist interpretation of the commonsense theory of language and truth entails the first four tenets of the traditional theory, the fact that the traditional theory is false entails that objectivist interpretation of the commonsense theory is false, too. We will discuss some of the implications of this below. But for now, it is important to see, point by point, exactly why each tenet of the traditional theory of metaphor is false.

> *Tenet 1:* Metaphor is a matter of words, not thought. Metaphor occurs when a word is applied not to what it normally designates, but to something else.

The Love Is A Journey example reveals the fallacy in tenet 1 clearly. If metaphor were just a matter of words, then each different linguistic expression should be a different metaphor. Thus, each of the example sentences should be entirely different metaphors, with nothing in common among them. "Our relationship has hit a dead-end street" should be distinct from and unrelated to "Our relationship is spinning its wheels," which in turn should be different from and unrelated to "We're going in different directions" and "Our relationship is at a crossroads," and so on. But these are not simply distinct, different, and unrelated metaphorical expressions. They are all instances of a single conceptual metaphor, namely, Love Is A Journey, which is characterized by the conceptual cross-domain mapping stated in Chapter 5. There is one conceptual metaphor here, not dozens of unrelated linguistic expressions that happen to be used metaphorically. Metaphor is centrally a matter of thought, not just words. Metaphorical language is a reflection of metaphorical thought. Metaphorical thought, in the form of cross-domain mappings is primary; metaphorical language is secondary.

> *Tenet 2:* Metaphorical language is not part of ordinary conventional language. Instead, it is novel and typically arises in poetry, rhetorical attempts at persuasion, and scientific discovery.

Aristotle was also mistaken about metaphorical language being only poetic and rhetorical in nature and not part of ordinary everyday language. Expressions like "This relationship isn't going anywhere" or "We're at a crossroads in our relationship" are ordinary everyday expressions, not novel poetic or rhetorical expressions. Such expressions can be part of our everyday language, because the Love Is A Journey mapping is part of our ordinary everyday way of conceptualizing love and reasoning about it.

Tenet 3: Metaphorical language is deviant. In metaphor, words are not used in their proper senses.

Metaphorical thought is normal, not deviant. Conceptualizing love as a journey is one of our normal ways of conceptualizing love. "We're at a crossroads in our relationship" is a normal, not a deviant, expression.

The empirical incorrectness of Aristotle's theory is especially striking, because the theory was taken for granted for so long that it came to be thought of as a definition rather than a theory. For many people, the term *metaphor* was defined by those conditions. But the nature of metaphor is a matter for empirical study, not for a priori definition.

Tenet 4: Conventional metaphorical expressions in ordinary everyday language are "dead metaphors," that is, expressions that once were metaphorical, but have become frozen into literal expressions.

Conventional metaphorical expressions like "at a crossroads" in "The relationship is at a crossroads" are sometimes mistakenly confused with "dead metaphors." As we have seen above, such cases are very much alive and cognitively real. There are real cases of dead metaphors, but conventional conceptual metaphors are not among them.

A dead metaphor is a linguistic expression that came into the language long ago as a product of a live conceptual metaphor. The conceptual mapping has long since ceased to exist, and the expression now has only its original target-domain meaning. A good example is the word *pedigree,* which came from the French *ped de gris,* which means "foot of a grouse." It was based on an image metaphor in which the image of a grouse's foot was mapped onto a family-tree diagram, which had the same general shape. The family-tree diagram was thereafter called a *ped de gris*—"a grouse's foot"—which came to be spelled "pedigree." Nowadays, the image mapping from a grouse's foot to a family-tree diagram has ceased to exist as a living part of our conceptual system. Moreover, English speakers no longer call a grouse's foot a *ped de gris.* Both the conceptual and linguistic aspects of the mapping are dead.

Interestingly enough, it is possible for a conceptual metaphor to remain alive, while a word initially expressing that metaphor may come to lose its metaphorical meaning. To see this, consider the words *grasp* and *comprehend.* *Grasp* can either mean to hold an object tightly or to understand an idea. As we shall see below, there is a general Ideas Are Objects metaphor, with the submapping Understanding Is Grasping (that is, Grasping→Understanding). This submapping maps the sense of the word *grasp* meaning *hold tightly* onto the corresponding sense of the same word meaning *understand*. This mapping produces entailments like:

- If an object has gone over one's head, then one hasn't grasped it.
- If an idea has "gone over one's head," then one hasn't understood it.

This metaphorical mapping is very much alive and is used not only for expressions like "I don't grasp what you're saying" and "That went over my head" but also for novel metaphorical expressions like "Now throw me one I can catch," which you might say after someone says something that goes over your head.

Now consider *comprehend,* which in Latin used to mean both *hold tightly* and *understand,* but which now only means *understand.* In Latin, *comprehend* worked like *grasp* now works in English. In other words, Latin had the same conceptual metaphor of Understanding Is Grasping, and it applied to *comprehend.* But, in being borrowed into English, the word *comprehend* lost its source-domain sense of *hold tightly.* Thus, though English has the live Understanding Is Grasping metaphor in its conceptual system, that metaphor applies to the word *grasp,* but not to the word *comprehend.*

Is *comprehend* an instance of a "dead metaphor"? No, since the Understanding Is Grasping mapping is still very much alive. *Comprehend* is simply a word that changed its meaning by losing its old source-domain sense of *holding tightly.*

The point is that the term *dead metaphor* applies to only a very narrow range of cases, like *pedigree.* Indeed, it takes some effort to come up with such cases. Cases like *pedigree* work differently from cases like *comprehend,* in which the conceptual metaphorical mapping is still alive, but the term has ceased to be a linguistic expression of that mapping. *Pedigree* is also quite different from conventional metaphorical expressions like *not going anywhere,* which instantiate the live Love Is A Journey mapping.

Cases of dead metaphors like *pedigree* are the very opposite of cases like the Love Is A Journey mapping, which is so alive that it keeps producing more examples of new metaphorical expressions in song lyrics, poems, self-help books, and marriage ceremonies. Like principles of phonology and grammar, conventional metaphors are relatively fixed, unconscious, automatic, and so alive that they are used regularly without awareness or noticeable effort.

One easy way of telling whether a metaphorical mapping is alive is to see if a novel metaphorical expression can be an instance of that mapping, which is what we saw in the "freeway of love" example. If a metaphorical mapping can give rise to new metaphoric expressions in poetry, rhetoric, and songs, then that metaphor is alive.

> *Tenet 5:* Metaphors express similarities. That is, there are preexisting similarities between what words normally designate and what they designate when they are used metaphorically.

There are at least four arguments against the hypothesis that metaphor expresses literal similarity instead of being a cross-domain mapping. The first is the simplest. In many cases, there is just no preexisting similarity there. For example, there is no preexisting similarity between the inherent (skeletal) concept of love and the concept of a journey. However, the conventional Love Is A Journey metaphor creates a fleshed out Love Is A Journey concept, which of course has similarities to journeys—exactly the similarities expressed in the mapping, since the mapping creates the similarities.

The second case is one in which the source and target domains do happen to share something, but the metaphor does not just express a similarity. Consider the Knowing Is Seeing metaphor, as shown in examples like "I see what you mean," "That's a murky argument," and "Let's shed some light on the subject." In the source domain of seeing, it is the case that seeing typically results in knowing. Thus, both the source domain of vision and the target domain of knowledge involve knowing. Yet these sentences do not express a similarity between source and target. Since one cannot literally see what someone else means, there can be no literal similarity between knowing what someone else means and seeing what someone else means. Simple sharing of concepts between source and target does not guarantee that a metaphor expresses a similarity.

Third, similarity is a symmetric notion. If metaphors just expressed similarities, they should be symmetric. There should be no target-source distinction. The source should be just as expressible in terms of the target as the target is in terms of the source. But that is not what occurs in the vast number of cases. The words that describe aspects of journeys can be used to characterize love, but the reverse in not the case. Love expressions do not characterize the corresponding journey concepts. Cars, for example, are not referred to as "relationships." This is true both for conventional and novel cases.

Semantic change also works this way. Words that mean *see* can come to mean *know* throughout the Indo-European family for different roots at different times in different languages. But the converse is not true. Similarly, when we gesture spontaneously, we trace images from the source domain in discussing the target domain, but not conversely. And the same asymmetry holds in the case of reasoning. We import inferential structure about journeys to conceptualize love, but we don't use our forms of reasoning about love to conceptualize and reason about journeys.

A fourth argument comes from the fact that concepts can be metaphorically conceptualized in inconsistent ways through different conceptual metaphors. For example, marriage is often conceptualized either as a business partnership or a parent-child relationship. Both are possible in our culture. If it is conceptualized as a business partnership, then the relationship is seen as an equal one. By contrast, if the marriage is being conceptualized as a parent-child relationship, then the relationship is seen as an unequal one. Marriage in itself is a sufficiently fluid concept in this culture that it can permit both metaphorical conceptualizations—though not both at once!

Suppose metaphor necessarily expressed a preexisting similarity. Then the Marriage As Business Partnership metaphor would express a preexisting equal relationship. That is, marriage would inherently have to involve equality of the spouses. But the Marriage As Parent-Child Relationship metaphor also exists. It posits an unequal relationship. If metaphor expressed a preexisting similarity, then marriage would have to be inherently an unequal relationship. But the marriage relationship cannot be both inherently equal and inherently unequal. Since both metaphors exist, the similarity theory would require a contradiction! The mapping theory does not, since both mappings need not be simultaneously activated.

For all these reasons, the similarity hypothesis is false.

Some Philosophical Implications of Metaphorical Thought

Even the few simple examples we have looked at so far have radical implications for philosophy. It is no small matter to say that ordinary, everyday reason can be metaphorical. Even at this preliminary stage, before we go on to analyze philosophically important concepts, we can see a great many implications for philosophy:

- Correlations in our everyday experience inevitably lead us to acquire primary metaphors, which link our subjective experiences and judgments to our sensorimotor experience. These primary metaphors supply the logic, the imagery, and the qualitative feel of sensorimotor experience to abstract concepts. We all acquire these metaphorical modes of thought automatically and unconsciously and have no choice as to whether to use them.

- Many, if not all, of our abstract concepts are defined in significant part by conceptual metaphor. Abstract concepts have two parts: (1) an inherent, literal, nonmetaphorical skeleton, which is simply not rich enough to serve as a full-fledged concept; and (2) a collection of stable, conventional metaphorical extensions that flesh out the conceptual skeleton in a variety of ways (often inconsistently with one another).

- The fundamental role of metaphor is to project inference patterns from the source domain to the target domain. Much of our reasoning is therefore metaphorical.

- Metaphorical thought is what makes abstract scientific theorizing possible.

- Metaphorical concepts are inconsistent with the classical correspondence theory of truth. Instead, what is required is embodied truth.

- Formal logic has no resources for characterizing any of the aspects of human concepts and human reason discussed so far in this book. The reason is that formal logic is disembodied, literal, nonimagistic, and nonmetaphorical.

- Reason and conceptual structure are shaped by our bodies, brains, and modes of functioning in the world. Reason and concepts are therefore not transcendent, that is, not utterly independent of the body.

- Much of everyday metaphysics arises from metaphor.

The analyses to follow in the rest of the book will give more substance to these claims. They will also allow us to explore more fully the far-ranging implications for philosophy of unconscious metaphorical thought.

As we will see, our most fundamental concepts—time, events, causation, the mind, the self, and morality—are multiply metaphorical. So much of the ontology and inferential structure of these concepts is metaphorical that, if one somehow managed to eliminate metaphorical thought, the remaining skeletal concepts would be so impoverished that none of us could do any substantial everyday reasoning.

Eliminating metaphor would eliminate philosophy. Without a very large range of conceptual metaphors, philosophy could not get off the ground.

The metaphoric character of philosophy is not unique to philosophic thought. It is true of all abstract human thought, especially science. Conceptual metaphor is what makes most abstract thought possible. Not only can it not be avoided, but it is not something to be lamented. On the contrary, it is the very means by which we are able to make sense of our experience. Conceptual metaphor is one of the greatest of our intellectual gifts.

Part II

The Cognitive Science of Basic Philosophical Ideas

*Events, Causation, Time, the Self,
the Mind, and Morality*

9

The Cognitive Science of Philosophical Ideas

Certain ideas are so basic to philosophical inquiry that any systematic approach to philosophy must discuss them. Since experientialism is a philosophy that begins from the findings of second-generation cognitive science, we will begin by asking what has been discovered empirically about such concepts within the cognitive sciences. Only then can we ask what new insights cognitive science and cognitive linguistics have to contribute to an understanding of such basic philosophical ideas.

Our enterprise is the opposite of the common philosophical enterprise of applying a purely philosophical methodology to a given subject matter. For example, there are subfields of philosophy called "the philosophy of _____," in which a traditional form of philosophical analysis is applied to a subject matter. Thus, in the philosophy of language, a philosopher might try to define the meaning of an important philosophical concept such as causation. An analytic philosopher would typically not approach the task empirically in the way, say, a cognitive linguist would. The analytic philosopher would ordinarily use the philosophical techniques he or she has been trained to use: definitions by necessary and sufficient conditions and, perhaps, the tools of formal logic.

By contrast, the cognitive linguist approaches causation by attempting to find all the causal expressions in English and in other languages throughout the world and to state generalizations governing both their meanings and their linguistic forms. Our approach to conceptual analysis uses the tools of cognitive science and cognitive linguistics to study empirically concepts such as time, causation, the self, and the mind. The result in each case is an endeavor of the form "the cognitive science of _____." Here in Part II of the book, we will be doing the cognitive science of philosophical ideas, that is, studying basic philosophical ideas as a subject matter for cognitive science. Thus, there is a cognitive science of time, a cognitive science of causation, a cognitive science of morality, and so on.

In Part III, we go on to apply these empirical methodologies to philosophy itself as a subject matter, considering, for example, Aristotle's theory of causation as a subject matter to be analyzed from a cognitive perspective. This we call "the cognitive science of philosophical theories." It employs methods from cognitive science to study the structure and content of particular philosophical theories. There we will be engaging in the cognitive science of such general topics as metaphysics, epistemology, and moral theory as carried on by great philosophical thinkers. Only after such empirical work has been completed can an empirically responsible philosophy emerge.

The ideas we will be discussing here in Part II are among those that would have to be included in anybody's list of ideas basic to philosophical discourse: time, events, causation, the self, the mind, and morality. We are taking the ideas as they occur in the cognitive unconscious of present-day speakers. This is a very different enterprise than the study of the consciously constructed ideas of great philosophers, like Plato's idea of the good or Kant's idea of autonomy or of the categorical imperative. We will discuss ideas of that sort in Part III of the book as part of our discussion of the overall conceptual structure of philosophical theories.

Each of the abstract ideas we will be discussing—events, causation, time, the self, the mind, and morality—turns out to be largely metaphorical. Although each idea has an underspecified nonmetaphorical conceptual skeleton, each is fleshed out by conceptual metaphor, not in one way, but in many ways by different metaphors. Each of these basic philosophical ideas, we will argue, is not purely literal, but fundamentally and inescapably metaphorical. Moreover, none of them is monolithic, with a single overall consistent structure; rather, each is a metaphorical patchwork, sometimes conceptualized by one metaphor, at other times by another. The metaphors are typically *not* arbitrary, culturally specific, novel historical accidents, or the innovations of great poets or philosophers. Rather, they tend to be normal, conventional, relatively fixed and stable, nonarbitrary, and widespread throughout the cultures and languages of the world. In addition, they are not purely abstract but, rather, are based on bodily experience.

In each case, certain fundamental philosophical questions will arise: If basic philosophical concepts are not completely literal but are largely metaphorical, what becomes of metaphysics and ontology? What does it mean for *knowledge* itself to be constituted of metaphorical concepts? What does it mean to state truths using these concepts? If there is a significant metaphorical component to the concept of causation, does this mean that there are no causes in the world? And if not, what does it mean? If the most basic of the concepts we use to understand and think about the world are not monolithic, but rather inconsistently polythetic, does that mean we can have no consistent understanding of anything? And if our concept of morality is constituted by multiple, inconsistent metaphors, does that mean that there can be no moral constants? If our concept of the self is constituted by multiple inconsistent metaphors, does that mean that there is no single thing that we are? Does it mean that the postmodern idea of the decentered subject is correct? All these questions and many more will arise as we proceed to do the cognitive science of basic philosophical ideas.

What Is Philosophical Inquiry?

To many philosophers, this enterprise may seem irrelevant to philosophical inquiry. When philosophers throughout history have inquired about time, events, causes, the mind, and so on, they have asked what these things are in themselves, metaphysically. Take time as an example. Traditional metaphysical speculation has wanted to know what time is, in itself. The question of how people conceptualize time is taken to be irrelevant to what time *really* is. The way people conceptualize things is supposed to be part of the subject matter of psychology, not philosophy, which sees itself as exploring the nature of things-in-themselves.

Thus, philosophers have asked whether time in itself is bounded or unbounded; whether it is continuous or divisible; whether it flows; whether the passage of time is the same for everyone and everything everywhere; whether time is directional and if so whether its direction is a consequence of change, causation, or possibility; whether there can be time without change; whether it loops back on itself; and so on. How we happen to conceptualize time has been seen as irrelevant to such questions. It is assumed that philosophical inquiry can proceed without knowing or caring about the details of how human beings happen to conceptualize what is being studied.

Research in cognitive science, especially in cognitive semantics, leads us to disagree. Philosophy is carried out by human beings, who have conceptual systems, who think using them, and who use language that expresses concepts in those conceptual systems. When an all-too-human philosopher asks a question like "What is time?" the word *time* has a meaning for that philosopher; that person already has a conceptualization of time in his or her conceptual system. What the question means depends on that conceptualization. The meaning of any philosophical question depends on what conceptual system is being used to comprehend the question. *That* is an empirical issue, an issue to be taken up by cognitive science in general and cognitive semantics in particular.

The same is the case for any proposed answer. An answer to a question like "What is time?" is given relative to a philosophical conceptual system in which that answer is a meaningful answer. Such a philosophical conceptual system is part of the conceptual systems of the philosophers doing the inquiry. The conceptual systems of philosophers are no more consciously accessible than those of anyone else. To understand what counts as a meaningful answer, one must study the conceptual systems of the philosophers engaged in that inquiry. That too is an empirical question for cognitive science and cognitive semantics.

In short, the whole undertaking of philosophical inquiry requires a prior understanding of the conceptual system in which the undertaking is set. That is an empirical job for cognitive science and cognitive semantics. It is the job we begin in this part of the book. Unless this job is done, we will not know whether the answers philosophers give to their questions are a function of the conceptualization built into the questions themselves. For example, if one concludes that time is boundless, is the answer based on a prior conceptualization of time that entails that it is boundless?

One ought to ask whether any of this matters. It could, in principle, be the case that philosophers have had a perfectly adequate understanding of their own conceptual systems and that further cognitive science research would not change anything at all. If we thought that were true, we would not bother to write this book.

We believe that a detailed study of the cognitive science of philosophical ideas drastically changes our understanding of philosophy as an enterprise and should change how philosophy is done as well as the results of philosophical inquiry.

10

Time

This is an inquiry into our concept of time. That is, it is an inquiry into the cognitive mechanisms we use as part of the cognitive unconscious to conceptualize time, reason about it, and talk about it. It begins not with consciously constructed conceptions of time but instead with the unconscious, automatically used, conventional conceptions of time that are part of our everyday conceptual systems.

We have a rich and complex notion of time built into our conceptual systems. To understand what we are talking about when we use a word like *time*, we must first analyze the way we conceptualize time and reason about it. Time is not conceptualized on its own terms, but rather is conceptualized in significant part metaphorically and metonymically.

This finding will raise difficult questions about not only the philosophy of time but about philosophical inquiry in general. What can it mean, if anything, to ask about the metaphysics of a concept that is in significant measure constituted metaphorically and metonymically? As we shall see shortly, we have no fully fleshed-out concept of time-in-itself. All of our understandings of time are relative to other concepts such as motion, space, and events.

Events and Time

Consider how we measure the time something takes, for example, a concert. What we do is compare events—the beginning and end of the event of the concert compared with states of some instrument constructed to "measure time."

137

Each such instrument depends upon regularly iterated events, whose iterations are taken as defining the "same" interval of time. Thus, sundials depend on the regular, repeated movement of the sun. Clocks depend upon either the regular iterated motion of pendulums, or the motion of wheels driven by the release of springs, or the regular iterated release of subatomic particles. In all these cases, the "same" intervals of time are defined by the successive iteration of physical events of the same kind. To say that a concert takes a certain "amount of time" is to say that the event of the concert is compared with some iteration of such events as the motion of a pendulum or the spinning of the wheels of a clock.

This is even true of the brain. The brain has been said to have its own "clock." What could that mean? Forty times a second an electrical pulse is sent across the brain. Some neuroscientists currently believe that this pulse regulates the neural firings in the brain and thus many of the body's rhythms. Whether or not this particular theory turns out to be correct, it gives some idea of what an internal "clock" might be. The motion of this pulse is a regular, iterative event that has been hypothesized as being the basis for the correlations among many of our bodily rhythms, rhythms that give us our intuitive sense of timing and time. The sense of time in us is created by such internal regular, iterative events as neural firings.

We cannot observe time itself—if time even exists as a thing-in-itself. We can only observe events and compare them. In the world, there are iterative events against which other events are compared. We define time by metonymy: successive iterations of a type of event stand for intervals of "time." Consequently, the basic literal properties of our concept of time are consequences of properties of events:

Time is directional and irreversible because events are directional and irreversible; events cannot "unhappen."
Time is continuous because we experience events as continuous.
Time is segmentable because periodic events have beginnings and ends.
Time can be measured because iterations of events can be counted.

What we call the domain of time appears to be a conceptual domain that we use for asking certain questions about events through their comparison to other events: where they are "located" relative to other events, how can they be measured relative to other events, and so on. What is literal and inherent about the conceptual domain of time is that it is characterized by the comparison of events.

This does not mean that we do not have an experience of time. Quite the reverse. What it means is that our real experience of time is always relative to our real experience of events. It also means that our experience of time is dependent on our embodied conceptualization of time in terms of events. This is a major point: Experience does not always come prior to conceptualization, because conceptualization is itself embodied. Further, it means that our experience of time is grounded in other experiences, the experiences of events.

We choose certain canonical events as temporal "yardsticks": the movement of the hands of an analog clock or the sequential flashing of numbers on a digital clock. These in turn are defined relative to other events—the movement of the sun, a pendulum, or wheels, or the release of subatomic particles.

Literal time is a matter of event comparison, but that is only the beginning of our concept of time.

The Metaphorization of Time

When we ask how time is conceptualized, we do not get very far before we encounter conceptual metaphor. As we shall see, it is virtually impossible for us to conceptualize time without metaphor. We use a number of metaphors in conceptualizing time, and each one comes with its own conceptual metaphysics. The conceptual metaphysics introduced by our conceptual metaphors for time raises important philosophical questions. For example, what, if anything, is time "in itself"? How are we to speak of the truth of expressions about time that make use of a conceptual metaphor and its metaphysics? What do metaphysical questions become in such a situation?

Spatial Time

Time is as basic a concept as we have. Yet time, in English and in other languages is, for the most part, not conceptualized and talked about on its own terms. Very little of our understanding of time is purely temporal. Most of our understanding of time is a metaphorical version of our understanding of motion in space.

It should be said at the outset that motion in our conceptual systems is not understood in the same way as in physics. In physics, time is a more primitive concept than motion and motion is defined as the change of location over time.

But cognitively the situation is reversed. Motion appears to be primary and time is metaphorically conceptualized in terms of motion. There is an area in the visual system of our brains dedicated to the detection of motion. There is no such area for the detection of global time. That means that motion is directly perceived and is available for use as a source domain by our metaphor systems.

The metaphor system for time in English has a structure. The most basic metaphor for time has an observer at the present who is facing toward the future, with the past behind the observer. We will refer to this as the Time Orientation metaphor:

The Time Orientation Metaphor

The Location Of The Observer	→	The Present
The Space In Front Of The Observer	→	The Future
The Space Behind The Observer	→	The Past

Linguistic expressions of this metaphorical mapping include:

That's all *behind* us now. Let's put that *in back of* us. We're looking *ahead* to the future. He has a great future *in front of* him.

This is a common way of orienting time in the world's languages. Even Henry David Thoreau in *Walden* described such a case: "I lived like the Puri Indians, of whom it is said that, 'For yesterday, today, and tomorrow, they have only one word, and they express the variety of meaning by pointing backward for yesterday, forward for tomorrow, and overhead for the passing day.'"

Furthermore, as James D. McCawley has pointed out to us, in Hindi /kal/ means both "yesterday" and "tomorrow" and /parsoM/ means both "the day before yesterday" and "the day after tomorrow." The Hindi record jacket for the Beatles' *Yesterday, Today, and Tomorrow* album says *Kal Aaj Kal*. Here the sequence tells you which "kal" is "yesterday" and which is "tomorrow." The point here is that time orientation is cognitively separate from other aspects of time. The Hindi words indicate one day from the present day and two days from the present day, but are neutral as to which direction. Hindi speakers, of course, have no problem knowing which is which, since Hindi has a rich future tense system and the verb indicates future versus past.

Where the Future Is Behind

Some languages, however, put the past in front of the observer and the future behind. Such a language is Aymara, a Chilean language of the Andes (A1, Núñez et al. 1997). The metaphor The Past Is In Front is grounded by the experience of being able to see the results of what you have just done in front of you. Thus "past time" in Aymara is *mayra pacha,* where *mayra* is "eye," "sight," "front," and *pacha* is "time." Future time is *q'ipa pacha,* where *q'ipa* is "back," "behind." For example, *maymara,* literally "eye year" or "front year" means "last year." Similarly, *uka ancha mayra pachan pasiwa,* which literally translates as "that several eye (front) time happened" means "That happened a long time ago." On the other hand, *q'ipüru,* which literally translates as "back-day (or behind-day)," means "some day in the future." Similarly, *q'ipa marana,* which literally translates as "back year-in (or behind year-in)" means "in the coming year (or the next year)."

The Time Orientation metaphor has a spatial source domain, but it says nothing about motion. The observer might be either stationary or moving. As it happens there are two additional metaphors for time that typically are combined with the Time Orientation metaphor. Both involve motion, but in one the observer is stationary and time is moving, while in the other the observer is moving and time is stationary.

The Moving Time Metaphor

The Moving Time metaphor applies to a very specific spatial schema:

> There is a lone, stationary observer facing in a fixed direction. There is an indefinitely long sequence of objects moving past the observer from front to back. The moving objects are conceptualized as having fronts in their direction of motion.

This schema provides the basis for a metaphorical mapping in which elements and structures of this spatial schema are mapped onto the target domain of time.

THE MOVING TIME METAPHOR

Objects	→	Times
The Motion Of Objects Past The Observer	→	The "Passage" Of Time

Put together with the Time Orientation metaphor this gives us the composite mapping:

The Location Of The Observer	→	The Present
The Space In Front Of The Observer	→	The Future
The Space Behind The Observer	→	The Past
Objects	→	Times
The Motion Of Objects Past The Observer	→	The "Passage" Of Time

What the Moving Time mapping does is use information in the spatial schema to give us an understanding of time as moving. For example:

There is only one Observer.	→	There is only one Present.
The Objects all move in the same direction.	→	Times move in the same direction.
Moving Objects face in the direction of motion.	→	Times face in their direction of motion.

The mapping, in general, maps knowledge and inferences about the source-domain schema onto knowledge and inferences about time. For example, suppose we pick two objects from the spatial domain as described by the spatial schema. Since they are moving in sequence, one of them is ahead of the other. Suppose we call them Object 1 and Object 2. The mapping then maps spatial truths onto temporal truths. Here are some examples:

Object 2 is behind Object 1.	→	Time 2 is in the future relative to Time 1.
Object 2 is ahead of Object 1.	→	Time 2 is in the past relative to Time 1.
If Object 1 is ahead of Object 2 and Object 2 is ahead of Object 3, then Object 1 is ahead of Object 3.	→	If Time 1 is in the past relative to Time 2 and Time 2 is in the past relative to Time 3, then Time 1 is in the past relative to Time 3.
If Object 1 is behind Object 2 and Object 2 is behind Object 3, then Object 1 is behind Object 3.	→	If Time 1 is in the future relative to Time 2 and Time 2 is in the future relative to Time 3, then Time 1 is in the future relative to Time 3.

In this manner, the inference structure of the source-domain schema of motion in space is mapped onto the inference structure of the target domain of the passage of time. Here are some linguistic examples of the Moving Time metaphor:

> The time will *come* when there are no more typewriters. The time has long since *gone* when you could mail a letter for three cents. The time for action *has arrived.* The deadline is *approaching.* The time to start thinking about irreversible environmental decay is *here.* Thanksgiving is *coming up* on us. The summer just *zoomed by.* Time is *flying by.* The time for end-of-summer sales *has passed.*

In this metaphor, times, as is typical of moving objects, are conceptualized as facing in their direction of motion. Hence, future times are facing toward the observer at the present.

> I can see the *face* of things to come. I can't *face* the future. Let's meet the future *head-on.*

As a result, times have metaphorical fronts and backs and can be conceived of as preceding and following one another. Times farther in the future follow a given time, and times farther in the past precede a given time.

> In the weeks *following* next Tuesday, there will be very little to do. During the week *preceding* last Tuesday, things were impossibly hectic around here.

In this metaphor, the present time is the time that is at the same location as the stationary observer. That is why we speak of the *here and now.* The observer's location serves as a reference point for the words *preceding* and *following.* Thus, the preceding day is in the past and the following day is in the future.

> On the *preceding* day, I took a long walk. In the *following* weeks, there will be no vacations.

Evidence for the Mapping

The analysis just given has an implicit evidential structure that should be made explicit. The mapping is a statement of a generalization over both inferential structure and language. Thus the same mapping that maps the transitivity of *ahead of* into the transitivity of *in the past relative to* also maps the spatial meanings of the linguistic expressions *behind, precede, follow, come, arrive, approach, pass, zoom by, fly by,* and so on onto their temporal meanings. The same mapping generalizes to poetic metaphors, like the use of *runs* in the following lines from scene 3 of *Macbeth*.

> *Come what come may*
> *Time and the hour runs through the roughest day.*

Thus, the same mapping states generalizations over (1) inference patterns, (2) the senses of polysemous lexical items, and (3) poetic expressions. Those inference patterns, those lexical items, and those poetic expressions constitute evidence for the existence of the mapping.

Priming experiments have provided confirming evidence for the existence of this metaphor (A2, Boroditsky 1997). And Taub has also confirmed its existence through studies in American Sign Language (A3, Taub 1997). As for gesture, you can informally test it yourself. It should be natural for a public speaker to use the following gestures when uttering the following lines.

That's what's in our future! (points ahead). That was in our past! (points behind).

And it should be unnatural for such a speaker to say the same lines with the gestures reversed:

That's what's in our future! (points behind). That was in our past! (points ahead).

This could be tested either by having participants in an experiment judge the naturalness of gestures or by gesture-priming experiments. In one such experiment, an image of someone pointing either ahead or behind is flashed briefly on a screen and subjects then have to press a button as fast as possible to say whether "That's in our future" or "That was in our past" is a sentence of English. The subjects should be faster when pointing ahead is paired with the future sentence and when pointing behind is paired with the past sentence.

The Time-Substance Variation

In a minor variation of the Moving Time metaphor, time is conceptualized not in terms of a multiplicity of objects moving in sequence but instead as a flowing substance. Thus, we speak of *the flow of time* and often conceptualize the linear flow of time in terms of a common linear moving substance—a river.

Since a substance can be measured—there can be a lot or a little of it—we can speak of a lot of time or a little time, a large amount or a small amount of time going by. This variation on the Moving Time mapping looks like this:

Substance	→	Time
Amount Of Substance	→	Duration Of Time
The Size Of The Amount	→	The Extent Of The Duration
Motion Of Substance Past The Observer	→	The "Passage" Of Time

This mapping maps knowledge about amounts of substances onto knowledge about durations of times:

A small amount of a substance added to a large amount of a substance yields a large amount.	→	A "small" duration of time added to a "large" duration yields a "large" duration.
If Amount A of a substance is bigger than Amount B of that substance and Amount B is bigger than Amount C, then Amount A is bigger than Amount C.	→	If Duration A of time is greater than Duration B of time and Duration B is greater than Duration C, then Duration A is greater than Duration C.

This sort of variation, between a multiplicity and a mass, is natural in conceptual systems. In general, we find a systematic relationship between multiplicities and masses, which is called the *multiplicity-to-mass image-schema transformation* (A4, Lakoff 1987, 428–429, 440–444). Such a relationship is based on the commonest of everyday experiences: a group of similar individuals standing near each other looks like a mass when viewed from a distance.

The Moving Observer, or Time's Landscape

The second major metaphor for time is the Moving Observer. Here the observer, instead of being fixed in one location, is moving. Each location in the observer's path is a time. The observer's location is the present. Since times here are conceptualized as locations on a landscape, we could just as aptly have called this the Time's Landscape metaphor.

THE MOVING OBSERVER METAPHOR

Locations On Observer's Path Of Motion	→	Times
The Motion Of The Observer	→	The "Passage" Of Time
The Distance Moved By The Observer	→	The Amount Of Time "Passed"

As before, this combines frequently with the Time Orientation metaphor, in which the future is ahead and the past is behind. Combining these, we get the mapping:

The Location Of The Observer	→	The Present
The Space In Front Of The Observer	→	The Future
The Space Behind The Observer	→	The Past
Locations On Observer's Path Of Motion	→	Times
The Motion Of The Observer	→	The "Passage" Of Time
The Distance Moved By The Observer	→	The Amount Of Time "Passed"

Since time is a path on the ground the observer moves over, it has extension and can be measured. Hence an amount of time can be long or short. An extent of time can also be bounded; therefore, one can perform an action within an allotted time. This metaphorical mapping gives rise to expressions like:

> There's going to be trouble *down the road*. Will you be staying a *long* time or a *short* time? What will be the *length* of his visit? His visit to Russia *extended over* many years. Let's spread the conference *over* two weeks. The conference *runs from* the first *to* the tenth of the month. She arrived *on* time. We're *coming up on* Christmas. We're *getting close to* Christmas. He'll have his degree *within* two years. I'll be there *in* a minute. He left *at* 10 o'clock. We *passed* the deadline. We're *halfway through* September. We've *reached* June already.

In these examples, the following locational expressions have temporal correlates: *long, short, extend, spread, over, on, runs, from, to, come, close to, within, in, at, pass, through, reach,* and *down the road*. Each of these words shows the same systematic polysemy between spatial and temporal senses. And the generalization about what is systematic here is given by the mapping.

Moreover, this mapping of spatial concepts onto temporal concepts will map knowledge and inferences about motion through space onto knowledge and inferences about the passage of time. Here are some examples:

The Observer is at Location 1, and Location 2 is ahead of the Observer.	→	Time 2 is in the future relative to Time 1.
The Observer is at Location 1, and Location 2 is behind the Observer.	→	Time 2 is in the past relative to Time 1.
A short distance added to a long distance is a long distance.	→	A short time added to a long time is a long time.
If Distance A is longer than Distance B and Distance B is longer than Distance C, then Distance A is longer than Distance C.	→	If Duration A is "longer" than Duration B and Duration B is "longer" than Duration C, then Duration A is "longer" than Duration C.

It is by virtue of such mappings of knowledge and inference patterns that the meanings of spatial words and phrases get mapped onto corresponding meanings in the domain of time. Take, for example, *close to*. Part of the meaning of *close to* in its spatial sense is the following:

- If we are close to Location A, then we have only a short distance to travel to reach Location A.
- If Location A is close to Location B and Location B is close to Location C, then it is not very far from Location A to Location C.

If we apply the Moving Observer mapping stated above to these inference patterns about spatial closeness, we get the corresponding inference patterns about temporal closeness, which characterize what temporal closeness means:

- If we are close to Time A, then we have only a short time to wait to reach Time A.
- If Time A is close to Time B and Time B is close to Time C, then it is not very far from Time A to Time C.

A Comparison of the Two Metaphors

The details of the two general metaphors for time are rather different; indeed, they are inconsistent with one another. Take, for example, the *come* of "Christmas is coming" (Moving Time) and the *come* of "We're coming up on Christmas" (Moving Observer). Both instances of *come* are temporal, but one takes a moving time as its subject and the other takes a moving observer as its subject. The same is true of *pass* in "That time has passed" (Moving Time) and *pass* in "He passed the time pleasantly" (Moving Observer).

The difference between these two metaphors for time can be seen in the sentence "Let's move the meeting ahead a week." This sentence is ambiguous, since the expression *move ahead* fits both metaphors. In the Moving Time metaphor, the times are moving. If the meeting had been scheduled for some future time and that time is facing toward the present, then moving the meeting ahead means to move it ahead of the time at which it is scheduled, namely, closer to the present. By contrast, in the Moving Observer metaphor, the observer is moving and facing toward the future. If the meeting has been scheduled for a future time, then moving the meeting ahead means moving the meeting ahead of where the observer will be at that time, namely, farther into the future. Thus, the same expression can have two entirely opposite meanings depending on which conceptual metaphor (Moving Time or Moving Observer) is being used in a particular context. The difference between these two metaphors for time explains the ambiguity of "Let's move the meeting ahead."

The differences in the details of the mappings show that one cannot just say blithely that spatial expressions can be used to conceptualize and speak of time without specifying details, as though there were only one correspondence between time and space. When we are explicit about stating the mappings, we discover that there are two different—and inconsistent—mappings.

Duality

As we have seen, the Moving Time and Moving Observer metaphors can both be paired with the Time Orientation metaphor, in which the observer is at the present, the future is ahead, and the past behind. In addition, the Moving Time and Moving Observer metaphors can be seen as sharing something general in common, namely, that the "passage" of time is conceptualized in terms of relative motion between the observer and times conceptualized in terms of space.

The two metaphors are, strictly speaking, inconsistent with each other: In one, times are objects that move past a stationary observer; in the other, times are locations in a landscape that an observer moves over. But these are actually minimally differing variants of one another. In short, they are *figure-ground reversals* of one another. In the Moving Time metaphor, the observer is the ground and the times are figures that move relative to it. In the Moving Observer metaphor, the observer is the figure and time is the ground—the times are locations that are fixed and the observer moves with respect to them.

As we shall see, it is common for metaphors to come in pairs that are figure-ground reversals of each other. We will refer to such metaphor pairs as *duals* and to the phenomenon as *duality*. Object-location duality occurs for a simple reason: Many metaphorical mappings take motion in space as a source domain. With motion in space, there is the possibility of reversing figure and ground. That possibility, as we shall see, is realized quite often.

There are other forms of duality. We have already seen one, multiplicity-mass duality, which is manifested in the two variations on moving time: moving individual times and the flow of a time substance.

Novel Cases

We have been discussing fixed conceptual mappings and fixed conventional expressions of these mappings. The same fixed conventional mappings can, however, be extended to produce novel metaphorical expressions. Such novel metaphorical expressions can be understood instantaneously using immediately activated conventional mappings, like those given above. Consider the following examples:

> The precious seconds *oozed* through my fingers. The deadline *sneaked* by me. The deadline was *marching* toward me like a brass band. The days *cascaded* by.

Words like *ooze, sneak, march,* and *cascade* are not conventional ways of expressing the passage of time. But they all indicate motion; they all suggest that time is moving by or toward an observer. Because the subject in each sentence is a temporal expression and the verb is a verb of physical motion, the Moving Time mapping is activated, and we understand these expressions in part using that mapping. That mapping is certainly necessary in understanding these sentences, but more is involved. For example, if you describe something as oozing through your fingers, it suggests that it is moving slowly and that you want to slow it down or try to stop it. Describing something as sneaking by you suggests that you don't notice it and that it is not your fault for not noticing. Describing something as marching toward you like a brass band suggests that you can't help but notice it. Something that cascades by you moves quickly, dazzlingly, and perhaps somewhat violently. In each case, time is conceptualized as moving by or toward you, but there is something more to be made sense of in context. In these cases we have novel linguistic expressions to be understood, but the mechanism of understanding is mostly just the activation of an already existing stable correspondence between concepts across conceptual domains.

Time in Other Languages

The Moving Time and Moving Observer metaphors are not limited to English. Although detailed studies remain to be done, a preliminary survey suggests that these metaphors are common in the world's languages. One could choose many examples, but we have chosen our examples from Hopi. Hopi is a celebrated case because of Benjamin Lee Whorf's suggestion (D, Whorf 1956) that Hopi has no concept of time or metaphors. Whorf's claim was not based on any serious fieldwork on Hopi time. Ekkehart Malotki has done that fieldwork and has discovered that Whorf was grossly mistaken. Malotki's classic, *Hopi Time* (D, Malotki 1983), provides more than four hundred pages of Hopi time expressions, more than two hundred pages of which are time metaphors. Here are some examples that give the flavor of Malotki's findings.

Hopi has a verb meaning *arrive,* which has a normal spatial sense, but can also be used for time. In the following case, time is conceptualized as something that moves and can arrive. It appears to be an instance of the Moving Time metaphor.

> *pu' hapi a-w pitsi-w-iw-ta*
> now EMPH 'REF'-to arrive-STAT-IMPERF-(temp. adv.)
> "Now the [appropriate time] for it has arrived."
> HT 22/1.2.1.1/ex. 1

Hopi also has a verb meaning *approach*. Here a time is seen as a location that people can approach, as in the Moving Observer metaphor.

nuutungk talong-va-ni-qa-t a-qw hayingw-na-ya
last daylight-REALZ-FUT-REL-ACC it-to(EX) approach-CAUS-PL
"They approached the last day."
HT 193/1.10.1/ex. 2

Here is a case in which winter is conceptualized as a location that people can move toward:

tomolangwu-y a-qw itam hoyo-yo-ta
winter-ACC it-to(EX) we move-RDP-IMPERF
"We're moving toward winter."
HT 196/1.10.3/ex. 4

For voluminous examples, we refer the reader to Malotki's massive study.

The Embodiment of Time Metaphors and Space-Time Metonymies

Why do we have such metaphors for time? Why should the same ones occur in very different languages around the world? The answer is that these metaphors arise from our most common everyday embodied experience of functioning in the world. Every day we take part in "motion-situations"—that is, we move relative to others and others move relative to us. We automatically correlate that motion (whether by us or by others) with those events that provide us with our sense of time, what we will call "time-defining events": our bodily rhythms, the movements of clocks, and so on. In short, we correlate time-defining events with motion, either by us or by others. For example, we correlate distance traveled with duration. Thus, in a motion-situation, motion is correlated with time-defining events.

In such motion-situations, we are typically looking ahead, either in the direction of our motion or at things or people moving toward us. In motion-situations, those things and people that we will be coming close to and encountering in the immediate future are ahead of us. That is, there is a correlation between future encounters and what is ahead of us.

Motion-situations thus contain the literal correlations that are the experiential bases for the Time Orientation, Moving Time, and Moving Observer metaphors.

Time Orientation

What we will encounter in the *future* is *ahead* of us.
What we are encountering at *present* is *where we are* (present to us).
What we encountered in the *past* is *behind* us.

Moving Time

What we will encounter in the *future* is *moving toward* us.
What we are encountering *now* is *moving by* (passing) us.μ
What we encountered in the *past* has *moved past* us.

Moving Observer

What we will encounter in the *future* is *what we are moving towards*.
What we are encountering *now* is *what we are moving by*.
What we encountered in the *past* is *what we moved past*.

These literal correlations in everyday motion-situations bring together the source and target domains of these metaphors and the elements that are mapped in these metaphors.

In literal motion-situations where these correlations hold, metonymy is possible. The reason is simple: a motion-situation defines a single complex conceptual schema in which the two domains of time (that is, time-defining events) and motion are present together as part of single whole. Where two things are correlated in such a schema, one can stand metonymically for the other. For example, time duration can stand metonymically for distance, as in "San Francisco is half an hour from Berkeley." Here, *half an hour*, the time it takes to travel the distance, stands for the distance. The metonymy can go the other way as well; distance can stand metonymically for time, as in "I slept for fifty miles while she drove." Here *fifty miles* is the distance corresponding to the amount of time slept.

Time metaphors are grounded in literal motion-situations, in which the time and motion domains come together in experience. Consider a literal motion-situation in which you are walking down an alley and you see an intersection up ahead of you. You might say to your companion, "I don't know what's *up ahead of us.*" *Up ahead of us* refers to a spatial location, which is correlated with the time at which you will reach that spatial location. This kind of correlation provides the basis for the Moving Observer metaphor for time, which occurs when you are referring to a nonmotion-situation. In such cases, the literal correspondences of the motion-situations are replaced by mappings from the domain of motion to the domain of time (that is, time-defining events). Thus, when we say "I don't know what's *up ahead of us* in the next century," we are using the Moving Observer metaphor for time.

What all this shows is that the Time Orientation, Moving Time, and Moving Observer metaphors, which occur widely around the world, are not arbitrary, but are motivated by the most basic of everyday experiences. Such commonality of motion-situations and the correlational structure within motion-situations explain why those metaphors should exist in the form they do and why they should be so commonplace.

Why Events Occur in Time and at Times

The Moving Observer metaphor arises spontaneously as part of the cognitive unconscious in conceptual systems around the world, because the motion-situations that give rise to that metaphor occur every day in virtually everyone's experience. In that metaphor, a fixed duration of time is a bounded region on a path along which an observer moves. In short, a duration of time is, in this metaphor, conceptualized as a container. Extended events of less than that duration are therefore conceptualized, via this metaphor, as occurring *within* that *span* of time, as in the sentence "He ran a mile in five minutes," in which *in* locates the event of running a mile within a metaphorical temporal container (that is, a bounded region).

Similarly, events viewed as being instantaneous or as single unextended entities are conceptualized via that part of the Moving Observer metaphor that conceptualizes time as being located *at* time locations, as in a sentence like "The execution occurred at 10:06 P.M."

Philosophy and Common Sense

This spatial metaphor for time seems to be an automatic part of our cognitive unconscious that structures not only the way we conceptualize the relationship between events and time but the very way we experience time. That is why we necessarily think of events as occurring *at* times or *in* time. In short, this metaphor constitutes part of our commonsense understanding of our experience.

If this is so, there is an extraordinarily important consequence for philosophy. Consider the following argument:

> The analysis given in which our very concept of time is defined by regular iterative events (such as the 40-hertz signal in our brains, according to some theories) cannot be true. The reason is that events, according to our commonsense understanding, must occur *in* time, and so time must exist prior to events.

But such a commonsense understanding does not fit the realities of the body and brain. We do not perceive time independently of events. Moreover, time in the brain can only be generated by regular iterated brain events (like the 40-hertz signal). And as we pointed out, we cannot measure time-in-itself, whatever that could mean. We can only define *time* to be that which is measured by regular iterated events. Therefore, we cannot take the commonsense understanding of time at face value from a cognitive perspective. However, as we have seen, if we start from the view that time is conceptualized through the comparison of events, we can arrive at an adequate analysis of the commonsense understanding of time perfectly well.

There is a moral here. If you accept the cognitive semantic enterprise, certain forms of commonsense philosophizing are ruled out. Analyses may well emerge that directly contradict the presuppositions of commonsense philosophical questions.

The Event-for-Time Metonymy

There is a metonymy that can co-occur with the time metaphors we have discussed—the Event-for-Time metonymy. When we say "The Kronos Quartet Concert is approaching," the event of the concert is standing for the time of the concert and the time is conceptualized as approaching. Similarly, when we say "Harry had a heart attack during the rock concert," the rock concert is standing metonymically for the "length" of time during which the rock concert occurred.

But now a question of detail arises. We observed above that the literal, inherent structure of the time domain is characterized in terms of "time-defining events" (regular iterative events like the swinging of a pendulum or the release of subatomic particles or the 40-hertz signal) that define a "clock." Such times (that is, time-defining events) are then conceptualized in terms of motion in space via the Moving Time and Moving Observer metaphorical mappings. Times are then conceptualized as locations or bounded regions in space or as objects or substances that move. Events are then located with respect to those locations in space or objects that move. Once this pairing of events with metaphorically conceptualized times occurs, the Event-for-Time metonymy can apply, yielding sentences like those discussed above.

The Utility of Time Metaphors

One of the themes we will be repeating is that conceptual metaphor is one of our central intellectual tools. It is the principal instrument of abstract reason, the means by which the inferential structures of concrete domains are employed in abstract domains. This is as true of time as it is of any other abstract domain.

We experience only the present. We have to conceptualize past and future. We have memory and we have images of what we expect. But memories and expectations are not in themselves laid out along a time line.

Think of the benefits we reap from, say, the Moving Observer metaphor, in which times are locations in space and temporal intervals are distances. Analog clocks use this metaphor with the hands of the clock as the moving observer and the location of the hands in space as representing the time. Digital clocks make use of this metaphor indirectly, via the intermediate metaphor that numbers are points on a line. The numbers pick out points on a line that metaphorically represent instants of time.

This metaphor is also central to the cultural tradition of establishing histories and calendars—time lines on which events are spread out. It allows us to visualize change with respect to time as change with respect to space. For example, in calculus, the use of Cartesian coordinates allows us to use the metaphor that times are locations in space to visualize time as the x-axis; given a curve plotting distance with respect to time, the instantaneous change at a given time is the slope of the line that is tangent to the curve—a line in space. It thus allows us to mathematicize physics and many, many other endeavors. Moreover, it allows us to visualize future plans, to visualize purposes to be achieved in time as being spread out in space, with measures of time represented metaphorically as measures of space.

The metaphor of time as moving or flowing by us also has an important use: It allows us to gauge the urgency of "upcoming" events. Events "close" to us loom larger and give a greater sense of urgency. It allows us to visualize the order of events that are coming up—which precede and which follow. And it allows us to apply our sense of speed for things moving in space to future and past events.

The Metaphysics of Time Metaphors

Via the Moving Observer metaphor, we naturally, automatically, and unconsciously think of times as being locations or regions in space. Given such a conceptualization of time, we conceive of events as occurring at those time locations or in those regions of time. But doesn't this show that time is metaphysically, or at least cognitively, more basic than events, since containers and locations must exist prior to anything being in those containers or at those locations?

Not at all. All these examples show is that our metaphorical conceptual systems naturally lead us to conceptualize durations of time as a containers for extended events and instants of time as locations for instantaneous events. It says nothing whatever about either the relative metaphysical or ultimate cognitive priority of time over events or events over time. It is only an entailment of the natural, universal, automatic, and unconscious Moving Observer metaphor that times exist prior to events. But this metaphorical entailment does not give times either cognitive or metaphysical priority over events.

Examples like this are extremely important for philosophy. The reason is that the metaphorical nature of our conceptual system, if unrecognized, can lead philosophers astray. Two things lead to such philosophical errors. First, a philosopher may fail to recognize conceptual metaphor and hence may see metaphorical sentences as literal and take them at face value. Once one takes a metaphor as being literal, the second error is to assume the correspondence theory of truth and therefore to regard the objective world as structured by the metaphor.

An example of second type of fallacious reasoning of this sort goes like this: It can be true that "John ran a mile within 5 minutes" or that "The execution occurred at 10:06 P.M." In other words it can be true that an extended event can occur *within* a duration of time or an instantaneous event *at* an instant of time. If X can occur *in* or *at* Y, then Y must exist independently of X. Therefore, durations and instants of time must exist independently of events. Thus, time must have a metaphysical existence independent of events.

What is missed, of course, is that both the sentences and the situations in which they can be true are conceptualized via the Moving Observer metaphor. In the situation, the very understanding of the time that one is *in* or *at* is itself conceived of metaphorically. Therefore, locational time—time with regions to be *in* or locations to be *at*—doesn't exist independently of the spatial metaphors for time. Truth about metaphorically conceived time depends on the metaphorical conceptualization of both the sentence and the situation.

Augustine dramatized these errors in the eleventh chapter of his *Confessions* in his discussion of what constitutes a long time. Just when, he asks, is a time long? Is it long when it is present, or when it is past or future? A. N. Prior (C2, 1993, 38–39) asks the same question about processes. When is a process long? When it is going on, or when it is past or future? As a literal, metaphysical question about time, there is no answer, since only a short part of a process can occur at any present time. Augustine's answer is interesting. Past, present, and future, he says, do all "exist in some sort in the soul, but otherwise I do not see them." A cognitive scientist who speaks about minds instead of souls might echo Augustine in contemporary terms, saying that the very idea of "lengths of time" is conceptual, and indeed metaphoric. Our very notion of a "long time" or "long process" is a product of our use of spatial metaphor.

Zeno's paradox of the arrow can also be seen as pointing out the mistake of taking a metaphor to be literal (though he didn't understand it as such). Suppose, Zeno argues, that time really is a sequence of points constituting a time line. Consider the flight of an arrow. At any point in time, the arrow is at some fixed location. At a later point, it is at another fixed location. The flight of the arrow would be like the sequence of still frames that make up a movie. Since the arrow is located at a single fixed place at every time, where, asks Zeno, is the motion? Time, Zeno argues, is not divided up into instants. In our terms, the idea that time is a linear sequence of points is metaphorical, a consequence of times seen as locations in the Moving Observer metaphor. The mistake, once again, is to take what is metaphorical as literal.

Incidentally, a cognitive response to Zeno's paradox of the arrow is simple. There is a part of the brain that detects motion. Our motion detectors identify the arrow as moving. That is, our brains give us multiple ways of perceiving and conceptualizing the world. Motion is not a metaphorical concept. The idea that time is a linear sequence of finite points is. Our direct nonmetaphorically structured experience provides a simple response: Of course the arrow is moving. But in addition, we have an unconscious metaphorical conceptualization of instants of time as locations in space. We use this, for example, when we comprehend a picture of a moving object at a time: "This is Sam driving by directly in front of our house at 10:06 P.M." In other words, we have more than one way to conceptualize motion—one literal and one metaphorical. We can conceptualize motion directly, as when we think of Sam driving by and the hands of the clock moving. We can also conceptualize motion using a metaphorical conceptualization of time as a line with point locations on it. In the metaphor, and only in the metaphor, there is temporal location. Relative to the metaphor, we can fix a point location in time. Within the metaphor, *at* that point location, there can be no motion, since motion can only occur *over* regions of time in the metaphor. The appearance of paradox comes from attributing real existence to metaphorical point locations. Zeno's brilliance was to concoct an example that forced a contradiction upon us: literal motion and motion metaphorically conceptualized as a sequence of fixed locations at fixed points in time.

Such observations by Zeno and Augustine are not mere conundrums dreamed up in ancient and medieval philosophy, conundrums that are irrelevant today. They are early insights into the fact that our conceptual systems are not literal. They show that the most common concepts that we use every day and in terms of which we state our truths cannot be taken as literally fitting an objective reality.

The Flow-of-Time Metaphysics

The history of the philosophy of time is rife with philosophical errors of this sort, in which philosophers reach metaphysical conclusions from the fact that we judge everyday metaphorical sentences about time to be true. In each case, the reasoning assumes that the metaphorical sentence is literally true and that the correspondence theory of truth is literally true. This, not surprisingly, leads to philosophical puzzles and paradoxes that are taken seriously and discussed interminably.

Take the metaphorical idea of the flow of time, which arises from the substance version of the Moving Time metaphor. In that metaphor, as we saw, time is a fluid substance, like a river, that flows by us. In this metaphor, the present is the part of the river that is passing us, the future is the part of the river flowing toward us, and the past is the part of the river that has already flowed past where we are.

This metaphor has an important entailment if you take it seriously at face value. In the source domain, where there is a literal river flowing, the part of the river moving toward us exists at present. If it flows by us, it must flow from somewhere where it exists at present. The metaphor preserves this logic and maps it onto a metaphorical logic of time: If time flows by us, it must flow from some "place" in the future where it now exists. In other words, the Flow-of-Time metaphor entails that the future exists at the present.

Now, many of us would take a sentence like "Time is flowing by rapidly" to be true. Suppose you take this metaphor as being literal; that is, you assume that there really is a "flow" of time past us. This entails that the future is flowing toward us from somewhere and that it presently exists at the future "place." In short, it implies that the future, at least some of it, must exist at the present. This, of course, is only an entailment of the Moving Time metaphor. If you do not realize the metaphorical nature of the question, you might be led to ask, as some philosophers have, "If time flows, it has to flow at a rate relative to time. Mustn't there be some higher-order time relative to which time itself flows?" The question arises from taking the metaphor literally. To treat it as a deep metaphysical question would be silly.

The Space-Time Metaphor

Now let us consider a more dramatic case. According to the Moving Observer metaphor, times are locations in space. At any present time, the observer is moving ahead toward locations that are future times. In the source domain of the metaphor, any locations you are moving toward must exist before you get to them. Similarly, future locations must exist, as must the past locations that you have already gone over. In short, it is an entailment of this metaphor that the past and the future exist at the present. Thus, we can say things like "We can see far enough into the future to see that it (sic!) looks bleak."

This metaphor also leads us to deny the idea, described eloquently by Stephen Hawking in *A Brief History of Time* (E, 1988), that time began with the Big Bang. Our ordinary metaphor that time is a spatial-like dimension leads us to ask, "But what happened before the Big Bang?" If all events can occur *in* time, then presumably so could the event of the Big Bang. It is a question that makes sense given our metaphor. The idea that time itself started with the Big Bang makes no sense given our common metaphor. The Big Bang would then not be occurring *in* time, but rather defining the start of time.

Einstein's theory of general relativity uses a version of the metaphor that turns time into a landscape, that is, in which time is conceptualized as a space-like dimension. In relativity theory, certain parts of the commonplace metaphor are left out, for example, the moving observer and absolute simultaneity. There is a space-time continuum, a four-dimensional space, with time being one of the dimensions. What we normally call the force of gravity is, in this metaphorical theory, not a force but curvature of space-time. That is, general relativity contains a metaphor by which we conceptualize gravity as curvature of space. The Time As Space metaphor permits the Gravity As Curvature Of Space metaphor. Jointly, they permit the mathematics of Riemannian geometry to be used to describe force in terms of properties of space.

Philosophers have observed that taking the theory of general relativity as literally true entails that the past, present, and future all exist "at once." That is, the theory seems to suggest determinism and the impossibility of free will or even random probabilistic events, as required by quantum mechanics.

Of course, if one recognizes that general relativity uses our common metaphor for conceptualizing time metaphorically in terms of space, one need not reach such metaphysical conclusions. One can see general relativity as metaphorical. This does not make general relativity either false or fanciful or subjective, since its metaphors can still be apt. That is, they can entail nonmetaphorical predictions that can be verified or falsified. In general, to say that a science is metaphorical is not to belittle it. Because metaphors preserve inferences, and because those inferences can have nonmetaphorical consequences, one can often test whether or not a scientific metaphor is apt. Indeed, metaphor is what allows mathematical models to be linked to phenomena in the world and to be regarded as scientific theories.

We will return to these issues below, but let us summarize briefly some tentative conclusions: Our ordinary metaphor that times are locations in space (in the Moving Observer case) partly structures our experience of time and is crucial to such wondrous things as calendars, clocks, histories, and physical theories. Yet it is still a metaphor. If it is taken as literal, all kinds of difficulties arise: general relativity seems to be saying that the future, present, and past all exist together. Taking it as literal seems to rule out the possibility of the most promising of our cosmological theories, the Big Bang. The brief moral is: Yes, Time As Spatial Location is a metaphorical concept. Yet, it can in some cases be a constitutive part of our best understanding of what is true. But, as a metaphor, it can lead us into silliness if we are not careful.

Can't we just eliminate all metaphor—including the metaphor of Times As Spatial Locations? No, we cannot. We do not have a purely literal, fully fleshed-out concept of time sufficient for the inferences we draw concerning our experiences of time.

Time as a Resource and as Money

One of the most striking characteristics of Western culture is that time is conceptualized in general as a resource and in particular as money. Here are some examples from English of the Time Is A Resource metaphor:

> You have some time *left*. You've *used up* all your time. I've *got plenty* of time to do that. I don't have *enough* time to do that. That *took* three hours. He *wasted* an hour of my time. This shortcut will *save* you time. It isn't *worth* two weeks of my time to do that job. Time *ran out*. He *uses* his time efficiently. I *need* more time. I can't *spare* the time for that. You've given me *a lot of* your time. I hope I haven't *taken too much of* your time. *Thank you for* your time.

The Time Is A Resource metaphor is a mapping that applies to a conceptual schema that characterizes what a resource is. This schema consists of a set of elements and a scenario indicating the relationships among these elements. This is what the resource schema looks like:

THE RESOURCE SCHEMA

The Elements of the Schema:
 A Resource
 The User of the Resource
 A Purpose that requires an amount of the Resource
 The Value of the Resource
 The Value of the Purpose

The Scenario Constituting the Schema:

Background:
 The User wants to achieve a Purpose.
 The Purpose requires an amount of the Resource.
 The User has, or acquires the use of, the Resource.

Action:
   The User uses up an amount of the Resource to achieve the Purpose.

Result:
   The portion of the Resource used is no longer available to the User.
   The Value of the Resource used has been lost to the User.
   The Value of the Purpose achieved has been gained by the User.

This schema characterizes what is typically meant by a *resource* (actually, a nonrenewable resource). Given this schema, other concepts are defined relative to it, concepts like Scarcity, Efficiency, Waste, and Savings. Here are some examples:

CONCEPTS DEFINED RELATIVE TO THE RESOURCE SCHEMA

Actual Expenditure: The amount of Resource used
Ideal Expenditure: The least amount of Resource that could have been used
Scarcity: The lack of enough of the Resource to achieve all of one's Purposes
Efficiency: The ratio of the Ideal Expenditure to the Actual Expenditure
Waste: The difference between Actual Expenditure and the Ideal Expenditure
Savings: The difference between the Actual Expenditure and a larger expenditure that would otherwise have occurred
Cost: The Value of the Actual Expenditure
Worthiness (of the Purpose): The degree to which the Value of the Purpose exceeds the Value of the Resource required

The Time Is A Resource metaphor is a mapping that applies to the elements of the Resource schema in the source domain, mapping the Resource schema into a corresponding Time Is A Resource schema in the target. The mapping looks like this:

THE TIME IS A RESOURCE METAPHOR

The Resource	→	Time
The User Of The Resource	→	The Agent (The User Of Time)
The Purpose That Requires The Resource	→	The Purpose That Requires Time

The Value Of The Resource	→	The Value Of The Time
The Value Of The Purpose	→	The Value Of The Purpose

Correspondingly, the concepts defined relative to the Resource schema are mapped onto the corresponding concepts of the Time Is A Resource schema:

Actual Expenditure: The amount of Time used

Ideal Expenditure: The least amount of Time that could have been used

Scarcity: The lack of enough Time to achieve all of one's Purposes

Efficiency: The ratio of the Ideal Expenditure of Time to the Actual Expenditure of Time

Waste: The difference between Actual Expenditure and the Ideal Expenditure of Time

Savings: The difference between the Actual Expenditure of Time and a larger expenditure of Time that would otherwise have occurred

Cost: The Value of the Actual Expenditure of Time

Worthiness (of the Purpose): The degree to which the Value of the Purpose exceeds the Value of the Time required to achieve it

As a result of this mapping, words defined relative to the Resource schema, *waste, save, worth, spare,* and so on, acquire a meaning in the time domain. This is what makes it meaningful in this culture to speak of *wasting time* and *saving time.* In cultures without the Time Is A Resource metaphor, such expressions would be nonsense.

Time as Money

The Time Is Money metaphor arises by taking money as a special case of a resource, that is, by substituting Money for Resource throughout the Resource schema. Words like *budget, spend, invest, profit,* and *loss* are defined relative to the resulting Resource schema for Money. If we then take the Time Is A Resource mapping and substitute Money for Resource, we get the Time Is Money mapping.

THE TIME IS MONEY METAPHOR

Money	→	Time
The User Of The Money	→	The User Of Time (The Agent)
The Purpose That Requires The Money	→	The Purpose That Requires Time

The Value Of The Money   →   The Value Of The Time

The Value Of The Purpose   →   The Value Of The Purpose

This mapping provides time-domain senses for money words like *budget, spend, invest, profit,* and *loss,* and hence allows us to comprehend sentences like the following:

> I have to *budget* my time. I *spent* too much time on that. I've *invested* a lot of time on this project. You don't use your time *profitably.* That mistake resulted in a considerable *loss* of time.

Thus we can see in detail what it means for a metaphorical mapping like Time Is A Resource to have a special case, namely, Time Is Money.

The Reification of Metaphor in Institutions

One of the mistaken views commonly held about metaphor is that metaphors are necessarily untrue. It should be clear by now that metaphorical statements can be as true as any nonmetaphorical statements. The Time Is A Resource and Time Is Money mappings should make that clear. It is true that in this society we have to budget our time. It can be true that someone can waste an hour of our time. It is true that certain household gadgets like washing machines can save us time. It is true that we don't always use our time profitably. And it is true that we have invested a lot of time in writing this book.

Our culture happens to have a great many institutions that reify the Time Is A Resource and Time Is Money metaphors. One of them is the institution of paying people according to the amount of time they work—by the hour or week or year. Another is the institution of appointment books, by which time is budgeted. There are also the institutions of time clocks and business hours, which are ways of pairing income with time worked. And of course, there are deadlines, which define the limitations of time resources.

Not every culture has such institutions, and not every culture has a Time Is A Resource metaphor. According to anthropologist Elizabeth Brandt (personal communication) the Pueblos do not even have in their languages a means of saying the equivalent of "I didn't have enough time for that." They can say "My path didn't take me there" or "I couldn't find a path to that," but those are not instances of time being conceptualized as a resource.

The view in which time is not considered a resource, in which there is no rush to get things done with maximum efficiency, is sometimes viewed mockingly by those who are not part of Native American culture as "Indian time." Western businessmen seeking to set up factories in Third World countries often see indigenous peoples who do not conceptualize time as a resource as being lazy. Part of Westernization is the importation of institutions that reify the Time Is A Resource and Time Is Money metaphors.

Cultures in which time is not conceptualized and institutionalized as a resource remind us that time in itself is not inherently resourcelike. There are people in the world who live their lives without even the idea of budgeting time or worrying if they are wasting it. The existence of such cultures reveals how our own culture has reified a metaphor in cultural institutions, thereby making it possible for metaphorical expressions to be true.

Stealing Time

There are those who believe that Time Is A Resource is not a metaphor, but a basic truth—time is simply a literal, special case of the Resource schema. The following example, discussed earlier in Lakoff (A4, 1987), should make one think twice about such a claim.

On November 14, 1984, the following appeared in the *San Francisco Chronicle*:

THE GREAT EMPLOYEE TIME ROBBERY

Employees across the nation this year will steal $150 billion worth of time from their jobs in what is termed by an employment specialist as the "deliberate and persistent abuse" of paid working hours.

The study, released by Robert Half International, Inc., reported that the theft of working time is America's No. 1 crime against business, surpassing employee pilferage, insurance fraud, and embezzlement combined.

Robert Half, president of the firm bearing his name, said that time is the most valuable resource to business because it "cannot be replaced, recovered, or replenished."

He defined time theft as leaving work early or arriving late, extended lunch hours, excessive personal phone calls, conducting personal business during company hours, unwarranted sick days, and nonstop chitchat at the proverbial water cooler.

The study showed the average weekly time theft figure per employee amounted to 4 hours and 22 minutes.

One of the things that we know about resources is that they can be stolen. If time were simply a special case of a resource, then it should be a simple truth that time too can be stolen. Indeed, that is Robert Half's assertion. Yet most readers cannot take this suggestion seriously. Stealing time seems like a novel metaphor, not a literal truth or even a conventional metaphor. Yet this metaphor could, in principle, be made law, as Robert Half has proposed. The Robert Halfs of this world could, in principle, set up cultural and legal institutions reifying this extension of the Time Is A Resource metaphor. Given such institutions, it could be true that those of you who are reading this book on the job are indeed stealing time and could be subject to prosecution for it.

The philosophical moral: It is not an objective aspect of the world, independent of human beings and human institutions, that Time Is Money. Social institutions can reify metaphors like the Time Is Money metaphor, and via those social reifications of metaphor, metaphorical truths can be created. It can be true that you have budgeted your time well or invested it properly. And if the Robert Halfs of the world have their way, it could become true by passing a law that you can steal time. Truth is relative to understanding, often metaphorical understanding. Human institutions can impose such metaphorical understandings and create metaphorical truths.

A Thought Experiment: Can We Conceptualize Time Without Metaphor?

Try to think about time without any of the metaphors we have discussed. Try to think about time without motion and space—without a landscape you move over and without objects or substances moving toward you or away from you. Try to think about time without thinking about whether it will *run out* or if you can *budget* it or are *wasting* it.

We have found that we cannot think (much less talk) about time without those metaphors. That leads us to believe that we conceptualize time using those metaphors and that such a metaphorical conceptualization of time is constitutive, at least in significant part, of our concept of time. What, after all, would time be without flow, without time going by, without the future approaching? What would time be if there were no *lengths* of time? Would we still be experiencing time if time could not *creep* or *fly* by us? Would time still be time for us if we could not *waste* it or *budget* it? We think not.

Does Time Exist Independent of Minds?

Consider the classic ontological question: Does time exist independent of minds, and if so, what are its properties?

We reject the question. It is a loaded question. The word *time* names a human concept of the sort we have described—partly characterized via the correlation of events and partly characterized via metaphor. Both the correlation of events and the metaphor together structure our experience, giving us temporal experience. That experience, like our other experiences, is real. Thus time is something "created" via our bodies and brains, yet it structures our real experience and allows us an important understanding of our world, its physics, and its history.

Because of the nature of our conceptualization of time, its role in our experience, its utility, and its limits, any answer will have its silliness. If we take our metaphorical concept of time as literal, we get silly results: The future already exists, the past continues to exist, arrows cannot move in flight, passing a law can make it possible for time to be stolen, time both moves and stands still, there was a time before the Big Bang, and so on.

But if we dismiss metaphor as always misleading and having nothing at all to do with reality, we get equally silly results: There can be no such thing as a long time or a long process, the theory of general relativity and any theory like it cannot be taken seriously, time cannot pass rapidly, we can never budget, squander, or lose time, and so on.

Are we saying that our concept of time is only constituted metaphorically? Not at all. Just as in the case of love, there is a certain amount of non-metaphoric structure to our concept of time. As we have seen, the literal aspects of time such as directionality and irreversibility arise from the fundamental characterization of time as a comparison of events, where time-defining events are regular and iterative.

Nonetheless, *we* create the concept of time and conceptualize events naturally and unconsciously as occurring *in* time or *at* times. And we have no choice in the matter. All of us automatically do this because we have human bodies and brains, just as all of us "see" color categories as being in the world because we have human bodies and brains.

What we are suggesting is that our concept of time is cognitively constructed by two processes, one metonymic (based on correlations with events) and one metaphoric (based on motion and resources). From a cognitive perspective, events and motion are more basic than time. The concept of time gets its inherent structure by virtue of time-defining events. The cognitive mechanism that accomplishes this is metonymy: The directional, irreversible, continuous, segmentable, and measurable character of events is imposed upon time by time-defining events. The metaphoric mechanism allows one to use experiences of motion in space to conceptualize this event-correlation domain, giving it our familiar metaphorical conceptualizations of time.

Yet the biological and cognitive construction of time does not make it subjective or arbitrary or merely cultural. Consider our construction of time via the correlation with events or via spatial metaphors. We all do it in the same way, unconsciously and automatically, and in a way that is grounded in our bodies and brains and constant bodily experience. The spatial metaphors are not arbitrary; they are deeply motivated. They permit the measurement of time, our very notion of history, the science of physics, and much more that is invaluable. In many, many ways, the spatial metaphors are apt metaphors. Yet in some ways they are not and, being metaphors, can get us into silliness if we take them literally.

Can we avoid such metaphors and think and talk about time only literally? No. Our conceptual and linguistic systems do not allow it. Would it be a good thing? Not at all. It would eliminate our richest way of thinking about abstract concepts. Can we "regiment" language, form an artificial logical language that gets rid of all the metaphors and characterizes time literally? There are two responses: First, if we got rid of all the metaphors, we would no longer have the concept of time, with lengths of time and the flow of time and the wasting of time. Second, such logical "languages" are just symbol systems. They contain no ideas. Any axioms one invents that one could accurately use the word *time* for must somehow accord with our ordinary metaphorical concept of time if they are to really be axioms for *time*. The metaphors don't disappear. They are hidden somewhere else and their entailments show up in our understanding of that "regimented formal language" and its axioms.

Where Does This Leave Us?

It leaves us with a much more sophisticated understanding of the nature of our concept of time. Does it give us an objective metaphysics of time? No. Indeed, it suggests instead that the very idea in itself is a bit strange. When the concept itself is defined by metonymy and multiple metaphors, it is odd to ask what the objectively real correlate of that concept is. If you insist on asking that question, you will wind up doing one of the things that philosophers have typically done: choosing some aspect of the concept that you want to focus on and claiming that that one aspect really is time, either time as a flow, or time as a continuous unbounded line, or time as a linear sequence of points, or time as a single spatial-like dimension in a mathematical theory of physics. What you will probably not be able to do is arrive at a single, unified, objective, literal understanding of that subject matter that does full justice to all aspects of the concept.

Is the enterprise worthless then? Not at all. The study of time, even within the limits of the metaphorical concepts we have, is a magnificent and enormously useful enterprise. But it is an enterprise that requires serious empirical study of the brain, mind, and language.

11
Events and Causes

I t would be difficult to find concepts more central to philosophy than events, causes, changes, states, actions, and purposes. We will refer to these as *event-structure concepts*. These concepts have traditionally been philosophically important because they are central to what constitutes general knowledge—knowledge of causes, changes, purposes, and so on. We would be hard-pressed to find a newspaper story that was not concerned with causes, actions, changes, and states. What *actions* will *cause changes* in the *state* of the economy? In the Middle East peace process? In our health-care system? And on and on. It's hard to have a discussion of anything at all without using these concepts.

Because we have purposes and act in the world to achieve those purposes, we are very much concerned with causation and its negative counterpart, prevention. In the sciences and the social sciences, in the making of social policy, and in legal decisions, causation takes central stage. We want to know what causes what and what prevents what. Does pornography cause sexual violence? Will toughening sentences prevent crime? Does smoking marijuana lead to drug addiction? Our understanding of causation is absolutely central to any plans we make for acting in the world.

In the usual interpretation of such questions, causation is assumed to be something in the world, an objective matter in which human conceptualization tells us nothing about whether a cause—a real cause—exists or not. Traditional objectivist approaches to philosophy mirror the commonsense view: Causes are causes, no matter how we conceptualize them. Conversely, conceptualizing something as a cause doesn't make it one. The same goes for events, actions, changes, and states. These event-structure concepts have been viewed in the objectivist tradition as central and foundational, which means:

- Our concepts of causes, actions, states, and changes represent objective features of the world; they are mind-independent constituents of reality—part of the basic ontology of what exists. Hence, the concepts of causation, action, state, and change are literal, not metaphorical.
- There is a single, general, literal logic of causation that adequately characterizes the causal structure of the world and all of our causal inferences.

As in the case of time, we will look at the evidence about the nature of the concepts first. On the basis of that evidence, we will argue that all of these statements are incorrect. It appears, instead, that:

- Event-structure concepts, for example, state, action, and cause, are conceptualized metaphorically in terms of more "specialized" notions (e.g., self-propelled motion and force). Metaphor is, in a significant way, constitutive of all event-structure concepts. Moreover, we reason about events and causes using these metaphors. In addition, these metaphors emerge from everyday bodily experience. Patterns of body-based inference are the source of abstract inference patterns characterizing how we reason using such event-structure concepts.
- Consequently, there is neither a single, literal concept of causation nor a single literal logic of causation that characterizes the full range of our important causal inferences. Causation metaphors are central to our causal reasoning, and there are many of them.

In short, what we shall be claiming about causes and events is very much like what we said about time. The concepts of cause and event and all other event-structure concepts are not just reflections of a mind-independent reality. They are fundamentally human concepts. They arise from human biology. Their meanings have a rather impoverished literal aspect; instead, they are metaphorical in significant, ineliminable ways.

As in the case of time, the view of events and causes that will emerge is neither objectivist nor subjectivist. We will be denying that causes, as we conceptualize them, objectively fit an objectivist's mind-independent world. We will also be denying that there are no causes at all as well as denying that all notions of causation are purely subjective, historically contingent, and radically relative. Thus, we will not be making a subjectivist claim. Instead, the evidence will lead us along a third path: an experientialist approach to causes and events.

Two Puzzles About Causation

As we shall see, there are, in our everyday conceptual systems, a great many distinct conceptualizations of causation, each with a different logic. This gives rise to what we will call the Causal Concept Puzzle: How can all these distinct conceptualizations, with distinct logics, be conceptualizations of the same thing? Correspondingly, there are, in philosophy, many distinct philosophical theories of causation, each with a different logic. This constitutes the Causal Theory Puzzle: How can all these distinct philosophical theories, with their distinct logics, be theories of the same thing?

The Causal Concept Puzzle

Let us consider a few of the concepts of causation actually used in causal models in the social sciences. Here are four examples:

Causal paths: Change depends on the "path" of other changes. A common example according to the economic historian Paul David is the QWERTY keyboard. During the early days of the typewriter, the typewriter keys moved more slowly than reasonably fast typists. The result was that the keys regularly got stuck. The QWERTY keyboard was adopted to slow down typists. Once the QWERTY path was chosen, once it was adopted and mastered by typists all over the country, there was no way to go back, even when physical typewriter keys were eliminated altogether.

Similarly, international relations theorists have argued that there is a "path" to democracy, and once a nation starts down the path, the change is irreversible and the ultimate development of democracy needs only a push at the right time. This has been used to argue that certain not so nice governments should get American military and economic aid because, even though they are far from democracies, they are "on the right path."

The domino effect: This theory was given as one of the justifications for going to war in Vietnam: Once one country falls to communism, then the next will, and the next, until force is applied to keep one from falling. It was reasoned that, if the United States used force to keep Vietnam from "falling" to communism, then the spread of communism in Asia would be stopped.

Thresholds: For a while there is a buildup of force with no effect, but once change starts, it becomes uncontrollable. This has been used as an argument to keep giving U.S. aid to nondemocratic regimes even when the aid has resulted in no noticeable effect in democratization.

The plate tectonic theory of international relations: When force is applied to something large, the effect lags until well after the action of the cause. This is similar to a threshold effect, but is limited to large cases and very long-term applications of force. After the breakup of the Soviet Union, this theory was used as a post hoc justification of the massive U.S. cold war spending over many years.

These are metaphorical causal models that have actually been proposed in the social sciences. Each has its own logic, taken from a physical domain. Each of the causal logics is somewhat different. In domino logic, but not in causal-path logic, the change is stopped by the application of force. In domino logic, a change is to be prevented. In plate tectonic logic, a change is to be effected. In domino logic, just enough force is necessary to keep the domino from falling. In causal-path logic, just a little push now and then is necessary. But, in plate tectonic logic, a huge amount of force is necessary over a long time.

The Causal Concept Puzzle is this: What makes all these kinds of causal theories causal? If causation has just one logic, then these cases with different kinds of logics should not be instances of a single causation concept. Yet they are all widely accepted as causal theories. What brings them together under the rubric of causation?

The puzzle has another aspect: All of these models are stated metaphorically. What conceptual metaphors are used in causal models? Is there a theory of metaphorical causal models—a theory of the range of conceptual metaphors and their logics available to express forms of causation? What are the possible logics of causation?

The Causal Theory Puzzle

Over the course of history, philosophers have formulated a wide variety of theories of causation, each substantively different from the others and therefore each with its own distinct logic. The puzzle is very much like the one above: What makes them all theories of causation? Philosophers may disagree as to what is the *right* theory of causation, but the philosophical community recognizes all of them as theories of the same thing. Why should philosophers have come up with this particular range of theories of causation? Other philosophical theories—say, of mind, language, or morality—are never mistaken for theories of causation. What makes these distinct theories theories of the same thing?

Given the range of philosophical theories, the answer is not obvious. Here are some examples of how broad that range is:

- Certain of the pre-Socratics saw causation as material, as residing in the substances that things are made of. Others saw it as residing in form—in shapes or in patterns of change.
- Aristotle listed both these types of causes—material and formal—and two others as well: final and efficient. The final cause was either a purpose governing an individual's action or an objective purpose residing in the world. The efficient cause was an application of force, resulting in a change.
- In much of modern philosophy, material, formal, and final causes have been downplayed or dismissed. Causation has been seen primarily as efficient causation.
- In criticizing the idea of necessary connection between cause and effect, Hume observed that we only experience a "constant conjunction" (that is, a correlation) between what we call "cause" and what we call "effect." Other empiricists have seen causation as residing in necessary conditions or in uniformities of nature. Collingwood saw a cause primarily as a kind of means, or "lever," for achieving a change in the natural world. He included two variations: a voluntary act by an agent and a set of conditions in nature invariably accompanied by some change.
- Hart and Honoré claimed that causes are either abnormal conditions or deliberate human actions that are necessary conditions for some event.
- Ayer saw a cause as being either a necessary or a sufficient condition.

Among the widespread views are:

Causes are material substance.
Causes are forms.
Causes are purposes.
Causes are applications of force or "power."
Causes are necessary conditions.

Causes are temporally prior to effects.
Causes are laws of nature.
Causes are uniformities of nature.
Causes are correlations, or "constant conjunctions."

Here we can see both of our puzzles clearly: Why should so many different things all be called causation? Why has philosophy produced these particular theories of causation and not others?

Beginning with the Empirical Study of Thought and Language

We begin once more with the observation that any questions we ask and any answers we give about causation and other event-structure concepts can only be framed within a human language and a human conceptual system. Any understanding of both questions and answers about event-structure concepts must therefore begin with an empirical analysis of the conceptual system used to ask those questions and give those answers. Thus we must ask, using the empirical criteria discussed above, just what is our everyday conceptual system for events, actions, causes, and changes.

What follows is the present state of our understanding of the conceptual system for event-structure concepts. As we shall see, there is a system of metaphorical mappings that characterizes how event structure is conceptualized. The evidence we will present is, as before, largely of two sorts, generalizations over polysemy and over inference patterns. Evidence from other sources—novel expressions, historical change, gesture studies, and psycholinguistic experiments—has not yet been gathered, though the methodology exists and should not be hard to apply. We will discuss the possibility for additional evidence below.

The Skeletal Literal Concepts of Events and Causes

Linguists studying the expression and the logic of *aspect* in the languages of the world have discovered what appears to be a common structuring of events across the world's languages (D, Comrie 1976). As we observed in Chapter 3, Narayanan has correspondingly discovered a single general structure governing all neural control systems for bodily movements. Moreover, he found that these two structures are the same, suggesting that our structuring of all events, concrete or abstract, arises from the way we structure the movements of our bodies.

The structure that Narayanan found provides a literal skeleton for our conception of event structure. Recall what that structure is:

Initial State: Whatever is required for the event is satisfied
Start: The starting up process for the event
End of Start: The end of the starting up process and the beginning of the main process
Main Process: The central aspects of the event
Possible Interruptions: Disruptions of the main process
Possible Continuation or Iteration: The perpetuation or repetition of the main process
Resultant State: The state resulting from the main process

This is how we structure both the movements of our bodies and events in the world in general.

Though this is a bare skeletal structure, it does come with a rich inferential structure. For example, if you haven't started, you haven't finished. If you're not prepared to start, you can't start. If you're repeating the process, then it's occurred before. If you're in the resultant state, you've been through the main process.

Primary metaphors flesh out this skeleton, not just in language but, as we shall see, in inferential structures. For example, states are conceptualized as containers, as bounded regions in space. Changes are conceptualized as movements from location to location. And so on, with many, many cases. It is both the inherent inferential structure of the skeleton and the rich inferential structure of the metaphors that jointly provide our enormously rich capacity for conceptualizing and reasoning about events.

Skeletal Literal Causation

The conceptualization of causes fits hand in glove with the conceptualization of events. Causation too has a skeletal structure, a very basic structure common to all causation. However, unlike the skeletal structure of events, the skeletal structure of causation is so minimal and impoverished that hardly any significant inferences can be drawn from it. We will see that this is so through a survey of the full variety of cases we categorize as causes. By the time we are done with this survey, it should be clear that what is in common among all the types of causation is, indeed, quite minimal. The richness of the forms of causal reasoning that we actually use arises from two sources: a causal prototype and a wide variety of metaphors for causation.

Here is what we have found to be the literal skeletal concept of causation: *a cause is a determining factor for a situation*, where by a "situation" we mean a state, change, process, or action. Inferentially, this is extremely weak. All it implies is that *if the cause were absent and we knew nothing more, we could not conclude that the situation existed*. This doesn't mean that it didn't; another cause might have done the job. The only implication is entirely negative: Given a lack of such a cause and a lack of any other knowledge, we lack a justification for concluding anything.

To see how the literal skeletal concept is fleshed out, we need to look at the real sources of causal inference—the central prototype of causation and the metaphors, without which most causal reasoning would not be possible. We will start with the central prototype for causation.

Prototypical Causation

At the heart of causation is its most fundamental case: the manipulation of objects by force, the volitional use of bodily force to change something physically by direct contact in one's immediate environment. It is conscious volitional human agency acting via direct physical force that is at the center of our concept of causation (A1, Lakoff and Johnson 1980, chap. 14). Prototypical causation is the direct application of force resulting in motion or other physical change.

The category of forms of causation is radial: less prototypical literal causes are literal variations, say, variations in degrees of directness or in whether the effect is positive or negative. Other noncentral forms of causation are metaphorical, mostly using Causes Are Forces and some metaphor for event structure, as we will see below.

All members of the category of causation share the minimal literal condition that a cause is a determining factor for a situation, but that concept in itself is so weak that we hardly ever use it and it alone in our causal reasoning.

The Structure of the Category of Kinds of Causation

The category of kinds of causation is a radial category with the following structure:

- There is a literal skeleton, the weak condition of being a determining factor for a situation. This holds for all cases.
- The center of the category is object manipulation: the volitional, direct application of physical force to an object that results in a change in it.
- The literal extensions of the central prototype to (a) forced motion of one object by another (billiard-ball causation), (b) indirect causation, (c) causation via an intermediate agent, (d) enabling causation, and so on.
- The extension of the prototype (where physical force is salient) to cases in which abstract causation is conceptualized metaphorically in terms of physical force via the primary metaphor that Causes Are Forces.
- A special case of direct human agency is giving birth. This is the basis of the metaphor that Causation Is Progeneration (see discussion below).
- Some metaphorical extensions of the central prototype are based on common correlations within the central prototype:

> The cause occurring before the effect in the prototypical case is the basis for the primary metaphor that Causal Priority Is Temporal Priority.
>
> The accompaniment of cause by effect is the basis of the metaphor that Causes Are Correlations.
>
> When there is motion from agent to patient, the cause originates at the location of the agent. This is the basis of the metaphor that Causes Are Sources, expressed by the use of the word *from*.

There are more metaphorical extensions, and we will discuss them below.

What we find here is a highly structured concept of causation. It has a literal, though skeletal, necessary condition and a robust central prototype—object manipulation—which extends in systematic ways, both literal and metaphorical, to a wide variety of distinct, noncentral causal concepts.

We will now turn to a survey of our metaphorical concepts of causation. These are the sources of the real richness of our forms of causal reasoning. Without them, causation would be a bare shell. The next eight sections set out our best understanding to date of the system of metaphors for causes and events.

The Basic Event-Structure Metaphors

Our most fundamental understanding of what events and causes are comes from two fundamental metaphors, which we shall call the Location and Object Event-Structure metaphors. Both make use of the primary metaphors Causes Are Forces and Changes Are Movements. They differ, however, in that one conceptualizes events in terms of locations, the other, in terms of objects.

THE LOCATION EVENT-STRUCTURE METAPHOR

States Are Locations (interiors of bounded regions in space)
Changes Are Movements (into or out of bounded regions)
Causes Are Forces
Causation Is Forced Movement (from one location to another)
Actions Are Self-propelled Movements
Purposes Are Destinations
Means Are Paths (to destinations)
Difficulties Are Impediments To Motion
Freedom Of Action Is The Lack Of Impediments To Motion
External Events Are Large, Moving Objects (that exert force)
Long-term, Purposeful Activities Are Journeys

This is a single, complex mapping with a number of submappings. The source domain is the domain of motion-in-space. The target domain is the domain of events. This mapping provides our most common and extensive understanding of the internal structure of events, and it uses our everyday knowledge of motion in space to do so. We have an enormously rich knowledge about motion in space that comes from our own movements and from the movements of others that we perceive.

Some movements are movements to desired locations (called destinations). Some movements begin in one bounded region of space and end in another. Some movements are forced, others are not. The force of a forced movement may be internal or external. If someone moves to a desired location, that person must follow a path. There are various kinds of impediments that can keep someone from moving to a desired location, for example, blockages or features of the terrain.

What this mapping does is to allow us to conceptualize events and all aspects of them—actions, causes, changes, states, purposes, and so forth—in terms of our extensive experience with, and knowledge about, motion in space. To see exactly what the mapping does and what is profound about it, we will go through it, submapping by submapping.

States Are Locations

The mapping:

$$\text{Locations} \quad \rightarrow \quad \text{States}$$

Examples:

> I'm *in* love. She's *out of* her depression. He's *on the edge of* madness. He's in a *deep* depression. She's *close to* insanity. We're *far from* safety.

By "locations," we mean bounded regions in space. Each bounded region has an interior, an exterior, and a boundary. Being *on the edge* of madness means being at the boundary of a state facing toward the interior. The issue is whether you will *go over the edge.*

Bounded regions can also be of various sizes and dimensions. Consider "a *deep* depression." Here, the state of being depressed must be conceptualized as a bounded region that has a vertical dimension. The vertical dimension is imposed by another, well-known conceptual metaphor, namely, Happy Is Up (Sad Is Down). And the concept of a *deep* depression requires that there be a considerable distance from the depths to the boundary, which entails that if you are in the *depths* of a depression, you have a long way to go to get out of it.

Try to imagine conceptualizing a state without its being a bounded region in space. Can you conceptualize a state without an interior and an exterior? Without a boundary––either sharp or gradual? Without interior locations far from the boundary? We have tried to conceptualize a state without these features of bounded regions in space, but we simply cannot do it. In short, the conceptual metaphor States Are Locations (bounded regions in space) seems to be central to the concept of a state. It is not a mere ornamentation or an expendable conceptual extra.

Let us consider three kinds of evidence that states are conceptualized metaphorically as bounded regions in space. The *polysemy evidence* comes from the use of the expressions *in, out, edge, deep,* and so on. Each of these words has a spatial sense and a sense concerning states. This submapping links those senses in a systematic way for each word. Without such a submapping, one would have to list both senses for each word as if they were unrelated. This would miss the systematic nature of the relationship among the senses. *The metaphorical mapping states a generalization over all these cases of polysemic correspondence,* in which a word with a locative bounded region sense has a corresponding state sense. Hence, each word or phrase that shows the systematic relationship between the spatial meaning of a word and the state meaning is a piece of polysemy evidence for the existence of the mapping.

The second form of evidence for a metaphorical mapping is *inferential evidence*. For example, the first member of each of the pairs below is an inference form true of bounded regions in space. The second member of each of the pairs below is an inference form true of states. The metaphorical mapping from bounded regions in space to states maps each of the inference forms true of bounded regions into the corresponding inference forms true of states.

- If you're *in* a bounded region, you're not *out* of that bounded region.
- If you're *in* a state, you're not *out* of that state.

- If you're *out* of a bounded region, you're not *in* that bounded region.
- If you're *out* of a state you're not *in* that state.

- If you're *deep in* a bounded region, you are *far from* being *out* of that bounded region.
- If you're *deep in* a state, you are *far from* being *out* of that state.

- If you are *on the edge of* a bounded region, then you are *close to* being *in* that bounded region.
- If you are *on the edge of* a state, then you are *close* to being *in* that state.

The metaphorical mapping *states a generalization over all these cases of inferential correspondence.* Each such inferential correspondence is therefore a piece of evidence for the mapping. Note that *the same mapping* generalizes over both polysemic correspondences and inferential correspondences.

The third kind of evidence for this submapping is *poetic evidence,* novel expressions that are understandable by virtue of this submapping. Consider the famous line from Shakespeare's seventy-third sonnet:

Death's second self, that seals up all in rest.

There are of course a number of conceptual metaphors in this line (see A1, Lakoff and Turner 1989, 17ff. for details). But consider merely "seals up all in rest." *Rest* here is a state. The preposition *in* naturally indicates that the state is conceptualized as the interior of a bounded region in space. Now consider the words *seal up*. You can only seal something up in a bounded region of physical space. Because a state is normally metaphorically conceptualized as a bounded region in physical space, Shakespeare can extend the metaphor to Death *sealing up* everyone in the state of rest.

Such cases are anything but rare in the body of English poetry. Each such example constitutes a piece of poetic evidence that the conventional metaphor exists, since it can be extended in novel ways.

At present, we know of no psychological experiments done to test for the reality of this mapping, but it would be easy to create such an experiment. Make up two drawings:

Drawing A: A circle with an arrow crossing the boundary and pointing inward.

Drawing B: A circle with an arrow crossing the boundary and pointing outward.

Now set up a task in which participants are presented one of the two sentences:

Sentence A: John is in a depression.
Sentence B: John is out of his depression.

First prime the participants by presenting them with one of the drawings. Then present one of the sentences and ask them to press a button when they understand the sentence.

Now consider two hypotheses:

Hypothesis 1: States are conceptualized metaphorically as interiors of bounded regions in space.

Hypothesis 2: There is no such conceptual metaphor. States are merely abstract concepts. The use of prepositions *in* and *out* with states is thus not connected with their use to indicate interiors and exteriors of bounded regions of physical space.

The drawings specify interiors versus exteriors of bounded regions of physical space. If hypothesis 2 is correct, there should be no priming difference between the drawings, since there would be no metaphorical connection between states and bounded regions in space. If hypothesis 1 is correct, drawing A should make the task quicker with sentence A than with sentence

B, and drawing B should make the task quicker with sentence B than with sentence A.

The point is that it is an empirical issue as to whether the States Are Locations metaphor is cognitively real. We hope experiments of this sort will be carried out.

Our plan was to go through the mapping of the Location Event-Structure metaphor submapping by submapping, providing a discussion of each along the lines of the States Are Locations discussion above. The submapping would be stated. A number of sentences that constitute polysemic evidence would be given. Then various kinds of evidence would be discussed: polysemic, inferential, poetic, and experimental—perhaps with the addition of any existing historical, gestural, or sign-language evidence.

This is doable, but a bit tedious. From the submapping and the sentences constituting the polysemic evidence and the inferential evidence, readers can pretty much put together such arguments for themselves. For this reason, we will shorten the presentation, either by simply giving the sentences or by presenting other forms of evidence only briefly. Only when we feel it is not obvious will we explicitly discuss the forms of evidence in detail.

Changes

Movements   →   Changes

I *came out of* my depression. He *went* crazy. He *went over the edge*. She *entered* a state of euphoria. He *fell into* a depression. He *went deeper* into his depression. In the sun, the clothes *went from* wet *to* dry in an hour. The clothes are *somewhere between* wet *and* dry.

This submapping builds on the States Are Locations submapping and conceptualizes a change of state as a movement from one bounded region in space to another. The polysemy evidence involves verbs and prepositions of motion like *go, come, enter, fall, from, to, into,* and *between.* Each of these has a sense in the domain of spatial movement and another sense in the domain of change of states. The submapping Changes Are Movements maps the movement senses into the corresponding change-of-state senses.

The same submapping maps movement inferences into change-of-state inferences.

- If something *moves from* Location A *to* Location B, it is first *in* Location A and later *in* Location B.

- If something *changes from* State A *to* State B, it is first *in* State A and later *in* State B.

- If something *moves from* Location A *to* Location B over a period of time, there is a point at which it is *between* Location A and Location B.
- If something *changes from* State A *to* State B over a period of time, there is point at which it is *between* State A and State B.

As before, the same mappings that relate the corresponding senses of the words *come, go, fall, enter, from, to, into,* and *between* also relate the corresponding inference patterns. Each such correspondence accounted for by the mapping constitutes a piece of evidence for the mapping.

There are two kinds of special cases of changes of state, which are both conceptualized in terms of movement. The first is *continuous change of state,* in which there is a graded continuum of states. This is conceptualized as continuous motion through a continuum of locations. The second is a *process,* which is conceptualized as a linear sequence of states. Processes are conceptualized in terms of motions through a linear sequence of locations.

Causation

Since changes of state are understood as movements from one bounded region to another, caused changes of state are understood as forced movements from one state to another. Thus, the Location Event-Structure metaphor has the additional submappings:

Forces	→	Causes
Forced Movement	→	Causation

Here are some examples of verbs of forced movement being used to express abstract causation:

FDR's leadership *brought* the country out of the depression. The home run *threw* the crowd into a frenzy. He *drove* her crazy. Their negotiations *pulled* both sides from the brink of war. That experience *pushed* him over the edge. Her speech *moved* the crowd to rage. The news *propelled* the stock market to record heights. The trial *thrust* O.J.'s attorneys into the limelight.

Bring, throw, drive, pull, push, propel, thrust, and *move* are all verbs of forced movement in their central, physical senses. But in the above cases, they are used to indicate abstract causation: caused change from one state to another.

Interestingly, each of these verbs specifies a different kind of forced movement. For example, *bring* in its spatial use requires a constant application of force. "I *brought* Bill a glass of water" entails that I was exerting force on the glass the whole time it was moving to Bill. Similarly, *bring* as an abstract causal verb implies that the causation extended through the period of change of state. For example, "I *brought* the water to a boil" entails that the causal force (in this case, heat) was applied throughout the period of change to a boiling state. Similarly, "FDR's leadership *brought* the country out of the depression" suggests that the causal effect of the leadership extended continuously until the country was out of the depression.

Compare *bring* with *throw*. As a verb of forced motion, *throw* describes a situation in which a force is applied instantaneously or for a very short time, and the movement occurs after the removal of the force: "He *threw* the rock over the fence." Similarly, in its causal sense, the causal force of *throw* applies quickly or instantaneously and the ensuing change of state occurs afterward, for example, "The stock market crash *threw* the country into turmoil."

Verbs like *bring, throw, drive, pull, push, propel,* and *move* provide evidence for existence of the mappings Causes Are Forces and Causation Is Forced Movement. The polysemy evidence resides in the fact that each verb has a central sense in the domain of forced movement as well as a conventionalized causal sense. In each case the causal sense is systematically related to the forced-movement sense by the Causation Is Forced Movement mapping. Each such word is a piece of evidence for the mappings.

Inferential evidence is provided by the systematic relationship between the logic of forced movement and the logic of causation:

- Forced movement: The application of the force precedes or accompanies the movement.
- Causation: The occurrence of the cause precedes or accompanies the change of state.

- Forced movement: The movement would not have occurred without the application of a force.
- Causation: The change of state would not have occurred without a cause.
- Forced movement: The force impinges on the entity that moves.
- Causation: The cause impinges on the entity that changes state.

In this case, the logic of causation *is* the logic of forced movement, under the mappings Causes Are Forces and Causation Is Forced Movement. As we shall see there are other causal logics defined by other metaphors for causation and events.

Evidence from novel examples is not difficult to come by. Verbs of forced movement like *hurl, tear, fling,* and *drag* do not have conventional abstract causal senses. Yet, they can be used in novel metaphorical expressions to indicate abstract causation. Here are some examples:

> A sudden drop in prices *hurled* the farm belt into chaos. The kidnapping of the young girl *tore* the country from its sense of security. The invasion of Pearl Harbor *flung* the United States into World War II. The United States *dragged* its allies into a commitment to the Gulf War.

The Causes Are Forces and Causation Is Forced Movement mappings will productively and systematically map the meanings of these forced-movement verbs onto appropriate causal senses. This is accomplished by the same mappings that link the forced movement senses of verbs like *bring* and *throw* with their conventional abstract causal senses. The same mappings also link the logic of forced movement with the logic of causation.

At this point, we have more than enough evidence for the existence of the following parts of the Event-Structure metaphor:

States Are Locations (interiors of bounded regions in space)
Changes Are Movements (into or out of bounded regions)
Causes Are Forces
Causation Is Forced Movement (from one location to another)

Philosophical Implications

Before we continue with the rest of the metaphors for event structure, we ought to stop for a moment to contemplate why all this matters. The causal uses of verbs like *bring, throw, hurl, propel, drag, pull, push, drive, tear, thrust,* and *fling* are not a mere linguistic curiosity, a supply of many words for the same thing. The point is that these verbs, in their abstract causal senses, do *not* all name the same concept. Each names a somewhat different concept—a different form of abstract causation. Each has its own logic, somewhat different from the others. And each is the product of a form of forced movement mapped onto the abstract domain of events.

To see this, try another thought experiment. Consider the kind of abstract causation characterized by each of the following sentences:

FDR *brought* the country out of the depression. The home run *threw* the crowd into a frenzy. That experience *pushed* him over the edge. The stock market crash *propelled* the country into a depression. The trial *thrust* O.J.'s attorneys into the limelight. A sudden drop in prices *hurled* the farm belt into chaos. The kidnapping of the young girl *tore* the country from its sense of security. The United States *dragged* its allies into a commitment to the Gulf War.

Now try to conceptualize all of these particular kinds of causation purely abstractly, without conceptualizing them in terms of forced movement. We doubt that you can do it any more than we could. It appears that the Causation Is Forced Movement metaphor is necessary to our conceptualization of forms of abstract causation. Since the concept of causation must include all its possible forms—all its special cases—it follows that the Causation Is Forced Movement metaphor is in a crucial way constitutive of the concept of causation.

Actions

In the Location version of the Event-Structure metaphor, actions are seen as movements that an agent carries out under the agent's own force. In other words,

Actions Are Self-propelled Movements

The evidence for this mapping comes from its entailments: If actions are conceptualized as self-propelled movements, then the following aspects of actions should be conceptualized as the corresponding aspects of movement:

Aids To Action Are Aids To Movement
Manner Of Action Is Manner Of Movement
Careful Action Is Careful Movement
Speed Of Action Is Speed Of Movement
Freedom Of Action Is The Lack Of Impediment To Movement
Suspension Of Action Is The Stopping Of Movement

This is, indeed, the case, as the following examples show. Each example takes its language and inferential structure from movement, but each is applicable to actions in general.

- Aids To Action Are Aids To Movement

  It is *smooth sailing* from here on in. It's all *downhill* from here. Getting the grant *gave* us just *the boost* the project needed.

- Manner Of Action Is Manner Of Movement

  We're *skipping right along*. We *slogged through* it. He is *flailing around*. She is *falling all over herself*. He is *out of step*. She is *in step*.

- Careful Action Is Careful Movement

  I'm *walking on eggshells*. He is *treading on thin ice*. He is *walking a fine line*.

- Speed Of Action Is Speed Of Movement

  He *flew* through his work. It is *going swimmingly*. Keep things *moving at a good clip*. Things have *slowed to a crawl*. She is going *by leaps and bounds*. I am moving *at a snail's pace*.

- Freedom Of Action Is The Lack Of Impediment To Movement

  Workers of the world, *throw off your chains!* I don't want anything to *tie me down*. I feel *imprisoned* in this job. I'm *trapped* in my marriage. *Break out* of your daily routine.

- Suspension Of Action Is The Stopping Of Movement

  They *halted* the project. I'll put a *stop* to his carousing. The work came to a *standstill*.

Difficulties

Since action is conceptualized as self-propelled movement, difficulties in acting are conceptualized as anything that can impede movement.

Difficulties Are Impediments To Movement

In English, there are at least five kinds of difficulties, corresponding to five kinds of impediments to movement: blockages, features of the terrain, burdens, counterforces, and the lack of an energy source. Let us begin with blockages.

Blockages:

Harry got *over* his divorce. She's trying to get *around* the regulations. He went *through* the trial. We ran into *a brick wall.* We've got him *boxed into a corner.*

In each case, an aspect of the source-domain inferential structure is imported into the target domain. Let us start with *over* in "Harry got *over* his divorce." In the spatial domain, if you get *over* a physical impediment like a wall or a hill on your way to a destination, you *encounter* the impediment, it *takes extra energy* to get over it, and *then it is no longer an impediment.* The Event-Structure metaphor maps this inference structure to the domain of events. If Harry got over his divorce, then he *encountered* it, it *took extra energy* to get over it, and once over it, it is *no longer a difficulty.*

Similarly, in the physical domain, if you get *around* an impediment, you *do not encounter* it, but *it takes extra energy to avoid* it. The Event-Structure metaphor maps this inference structure to the domain of event as follows: If you get *around* regulations, then you *do not encounter* them, but *it takes extra energy to avoid* them.

Difficulties conceptualized as features of the terrain work the same way. If you are physically stuck in a bog, then it is *very hard to move forward* at all, and it is *uncertain as to whether you will be able to reach your destination.* So, if you are bogged down on a project, then it is *very hard to make progress* at all, and it *uncertain as to whether you will reach your goal,* that is, succeed with the project.

In each of the following cases, both the language and the logic of physical impediments to motion are mapped onto the language and logic of difficulties in achieving purposes:

Features of the terrain:

She's *between a rock and a hard place.* It's been *uphill* all the way. We've been *bogged down.* We've been *hacking* our way *through a jungle* of regulations.

Burdens:

He's *carrying* quite a load. She's *weighed down* by a lot of assignments. He's been trying to *shoulder* all the responsibility. Get *off my back!*

Counterforces:

Quit *pushing me around.* She's *leading him around by the nose.* He's *holding her back.*

Lack of an energy source:

I'm *out of gas*. We're *running out of steam*.

Freedom of Action

When philosophers talk about the concept of freedom, they are usually talking about freedom of the will or political freedom. Both of those concepts are, in turn, based on the more basic notion of freedom of action. That notion is conceptualized in terms of the Event-Structure metaphor. Action is conceptualized in the Event-Structure metaphor as self-propelled movement and difficulties in acting as things that impede self-propelled movement. Freedom of action is therefore the absence of impediments to self-propelled movement.

Freedom Of Action Is The Absence Of Impediments To Movement

Consequently, achieving freedom is conceptualized as eliminating impediments to movement: *casting off chains, escaping imprisonment, tearing down walls, opening doors, clearing fields, getting out of holes, learning to walk,* and so on.

Purposes and Means

In the Event-Structure metaphor, purposes are conceptualized as destinations to be reached.

- Purposes Are Destinations

The evidence for this submapping is the collection of its entailments and the linguistic means of expressing them. In each case, the language and logic of moving toward, reaching, or not reaching a destination are recruited from the source domain of movement through space and used to speak and reason about achieving purposes or the failure to achieve them.

- Achieving A Purpose Is Reaching A Destination

    We've *reached the end*. We are *seeing the light at the end of the tunnel*. We only *have a short way to go*. We're *where we wanted to go*. The *goal is a long way off*.

- Lack Of Purpose Is Lack Of Direction

    He is *just floating around*. She is *drifting aimlessly*. He *needs some direction*.

- Means Are Paths

  Do it *this way*. She did it the *other way*. Do it *any way* you can. However you want to go *about it* is fine with me.

Such simple mappings can be superimposed on one another to yield complex metaphorical mappings:

Purposes Are Destinations

> Plus

Action Is Self-propelled Motion

> Yields

Purposeful Action Is Self-Propelled Motion To A Destination.

Putting together the understanding of purposes as destinations and actions as self-propelled motions, we get a conception of purposeful action as self-propelled motion to a destination. As before, the evidence is the set of entailments and the linguistic forms that express them.

- Starting A Purposeful Action Is Starting Out On A Path

  We are just *starting out*. We have *taken the first step*. We've *marked out a path*.

- Making Progress Is Forward Movement

  We are *moving ahead*. Let's *forge ahead*. Let's keep *moving forward*. We made lots of *forward movement*.

- Amount Of Progress Is Distance Moved

  We've *come a long way*. We've *covered lots of ground*. We've *made it this far*.

- Undoing Progress Is Backward Movement

  We are *sliding backward*. We are *backsliding*. We need to *backtrack*. It is time to turn around and *retrace our steps*.

- Expected Progress Is A Travel Schedule; A Schedule Is A Traveler Who Reaches Prearranged Destinations At Prearranged Times

  We're *behind schedule* on the project. We got *a head start* on the project. I'm trying to *catch up*. I finally *got a little ahead*.

- Lack Of Progress Is Lack Of Movement

We are *at a standstill.* We *aren't getting anyplace.* We *aren't going anywhere.* We are *going nowhere* with this.

External Events

One of the most interesting submappings in the Location Event-Structure metaphor is the one used to conceptualize external events. Imagine that you are trying to achieve a purpose, which is conceptualized as reaching a destination. External events may either help or hinder you. They are thus conceptualized as large, moving objects or substances that can exert a force on you, either help you move to your destination or impede your movement. In English, there are three special cases: very general objects (called "things"), fluids (especially liquids), and horses!

External Events Are Large Moving Objects

Special Case 1: Things:
How're *things going? Things* seem to be *going with* me, rather than *against* me these days. *Things took a turn* for the worse. *Things* are *going my way.*

In "*Things* seem to be *going with* me" external events, conceptualized as objects exerting a force, are moving in the direction I want to go in, helping reach my destinations, that is, achieve my purposes. In "*Things* seem to be *going against* me," the opposite is true. The force is now seen as impeding my motion toward my destinations and thus impeding my progress in achieving my purposes.

Special Case 2: Fluids:
You gotta *go with the flow.* I'm just trying to *keep my head above water. The tide of events . . . The winds of change . . . The flow of history . . .* I'm trying to *get my bearings.* He's *up a creek without a paddle.* We're all *in the same boat.*

If "We are all *in the same boat,*" then we are all subject to the same external forces that move the boat this way and that. The force of the waves will take us to the same final destination—the same final state. If the boat sinks, none of us will achieve our purposes. The entailment is that it is to our collective advantage to help keep the boat afloat.

In "You gotta *go with the flow*," the external events constitute a fluid like a river that is flowing so strongly that you cannot get to a destination that the river doesn't take you to. Thus, any effort to get to any other destination (achieve any purpose of your own) is pointless. You may as well just follow the flow of the external events and see where you wind up.

Here, as in other cases, there is knowledge about a special case in the source domain (being in the same boat or being subject to a river flowing with an overwhelming force) that is mapped by the general mappings onto the target domain of events.

Special Case 3: Horses:
> Try to keep a *tight rein* on the situation. *Wild horses* couldn't make me go. *Hold your horses*! *"Whoa!"* (said when things start to get out of hand).

Here *wild horses* refers to "even the strongest external forces." *Hold your horses* is used when someone is about to act precipitously, as if some strong force were moving the person. This is one of our favorite examples, because it shows how historical images that are preserved through cultural mechanisms (movies showing runaway horses, often pulling buckboards and stagecoaches) can be preserved in the live conceptual and linguistic system. In this case, as in the others, the issue is the control of external events, conceptualized as large moving entities that can exert force on you. Here those entities are horses, which can be controlled with strength, skill, and attention, but which otherwise get out of control. This special case thus focuses on external events that are subject to control, but require strength, skill, and attention if that control is to be exerted.

Long-Term Activities

Long-Term Activities Are Journeys

Finally, let us consider journeys. A journey takes an extended period of time, covers a lot of ground, and usually involves stopping at a number of destinations along the way before one reaches a final destination, if there is one. Given the rest of the Event-Structure metaphor, journeys correspond to long-term activities, usually with a number of intermediate purposes. The intermediate purposes are intermediate destinations, the ultimate purpose is the ultimate destination, the actions performed are movements, progress is movement toward a destination, the initial state is the initial location, and achieving the purpose is reaching the ultimate destination. Every aspect of the source domain of the Event-Structure metaphor may occur in some kind of journey, and hence journeys are very useful for talking about long-term activities of many kinds.

Journeys may or may not have prescribed or ultimate destinations. Some journeys are just wanderings. Some are semistructured with some intermediate, perhaps vague, destinations, but no ultimate destination. And some are very well planned, with a course and destinations completely specified ahead of time. The very flexibility of the concept of a journey makes it extremely useful in metaphorical thought.

Summary

We have just taken a tour through the Location branch of the Event-Structure metaphor. It is one of the most profound and most used metaphors in our conceptual system, since it lays out the fundamental means of conceptualizing our most basic concepts: states, changes, causes, actions, difficulties, freedom of action, and purposes. What we have seen is that our pervasive experience of motion through space is the basis for a vast metaphor system by which we understand events, causes, and purposive action. These metaphors are essential to our reasoning and define the various logics of causation that we have discussed. However, as we shall see directly, there are other systems and other logics of causation.

Event-Structure Duality

In our discussion of time, we saw that the two spatial metaphors for time are duals—metaphors that overlap in content but differ in figure-ground orientation. Thus in the Moving Time metaphor, times move and the observer is stationary, while in the Moving Observer metaphor, the observer moves and time is stationary. Event structure also shows such a metaphor duality.

The Anatomy of Event Structure

What we have been calling the Location Event-Structure metaphor focuses on locations. States are conceptualized as locations (bounded regions in space). This elementary mapping fixes the possibility for what change and causation can be. The Changes Are Movements metaphor combines with States Are Locations to construe a change in an entity as the movement of that entity from one location to another. The Causes Are Forces metaphor combines with these to provide a conceptualization of causation as the forced movement of an entity from one location to another.

With these metaphorical parameters fixed, the other mappings are pretty well determined. Purposes are naturally conceptualized as desired locations—destinations one wants to reach. Self-propelled motions, in which the agent exerts force on him- or herself to move, correspond naturally to actions. Impediments to motion and paths to desired locations then correspond naturally to difficulties and means. In short, the basic parameter is States Are Locations. Given Causes Are Forces and Changes Are Movements, the rest falls into place.

Duality

The States Are Locations metaphor has a dual, the Attributes Are Possessions metaphor, in which attributes are seen as objects one possesses. Thus, you can *have* a headache, an easy-going manner, or a fine reputation. To see a minimal difference, compare:

> Harry's in trouble. (States Are Locations)
> Harry has trouble. (Attributes Are Possessions)

In the first case, trouble is conceptualized as a location you are in; in the second, it is an object you can have. The difference can be seen as a figure-ground shift. In the first case, Harry is a figure and trouble is the ground with respect to which the figure is located. In the second case, Harry is the ground and the figure, trouble, is located with respect to him. Grounds are, of course, taken as stationary and figures as movable relative to them.

The Attributes Are Possessions metaphor combines with Changes Are Movements and Causes Are Forces to form a dual Event-Structure system. If an attribute is conceptualized as a possessible object, then adding Changes Are Movements makes change metaphorically the acquisition of a possessible object (where the object moves to you) or the loss of a possessible object (where the object moves away from you). Hence, Harry cannot only *have* a headache, he can *get* a headache or *lose* his headache; that is, his headache can *go away*.

Given this view of change as acquisition or loss, the Causes Are Forces metaphor yields a conception of causation as the forced movement of a possessible object to or from some entity. That is, causation can be seen as either *giving* or *taking*. Hence, a noise can *give* you a headache and an aspirin can *take* it *away*.

In short, there is an Object branch of the Event-Structure metaphor with the mapping:

> Attributes Are Possessions
> Changes Are Movements Of Possessions (acquisitions or losses)
> Causation Is Transfer Of Possessions (giving or taking)

This mapping yields examples like the following:

> I *have* a headache. (The headache is a possession.)
> I *got* a headache. (Change is acquisition—motion to.)
> My headache *went away*. (Change is loss—motion from.)
> The noise *gave* me a headache. (Causation is giving—motion to.)
> The aspirin *took away* my headache. (Causation is taking—motion from.)

The parallelism between the Location and Object branches of the Event-Structure metaphor extends even further. Where purposes are conceptualized in the location dual as desired locations (destinations), purposes are conceptualized in the object dual as desired objects (things you want to get). The heart of the duality can be seen very simply in the following contrast:

THE LOCATION EVENT-STRUCTURE METAPHOR

States Are Locations
Changes Are Movements (to or from locations)
Causation Is Forced Movement (to or from locations)
Purposes Are Desired Locations (destinations)

THE OBJECT EVENT-STRUCTURE METAPHOR

Attributes Are Possessions
Changes Are Movements Of Possessions (acquisitions or losses)
Causation Is Transfer Of Possessions (giving or taking)
Purposes Are Desired Objects

Purposes in the Object Branch

Let us now consider some details of the Object Event-Structure metaphor. Given the Purposes Are Desired Objects submapping, achieving a purpose is conceptualized as acquiring a desired object.

Acquiring A Desired Object → Achieving A Purpose

Examples:

> They just *handed* him the job. Fame and fortune were *within my grasp,* but once again they *eluded* me. If you have a chance *at* a promotion, *go for* it! I kept *getting close to* a world championship, but it kept *escaping* me. He almost *got his hands on* the Johnson ranch, but it *slipped through his fingers.* She is *pursuing* an impossible dream. He has interesting *pursuits. Latch onto* a good job. *Seize* the opportunity!

This submapping has an extensive special-case substructure. A special case of a desired object is something to eat. This special case is conventionalized in the Object branch of the event-structure system, yielding the submapping:

- Achieving A Purpose Is Getting Something To Eat

  > He *savored* the victory. All the good jobs have been *gobbled up.* She's *hungry* for success. The opportunity has me *drooling.* This is a *mouth-watering* opportunity.

Traditional methods of getting things to eat are hunting, fishing, and agriculture. Each of these special cases is also conventionalized in the Event-Structure system, giving rise to an extensive metaphorical substructure. Thus hunting, fishing, and agriculture are all used to conceptualize the attempt to achieve a purpose.

- Trying To Achieve A Purpose Is Hunting

  > I'm *hunting* for a job. I *bagged* a promotion. The pennant is *in the bag.* I'm *shooting* for a promotion. I'm *aiming* for a career in the movies. I'm afraid I *missed* my chance. (The typical way to hunt is to use projectiles—bullets, arrows, etc.)

- Trying To Achieve A Purpose Is Fishing

  > He's *fishing* for compliments. I *landed* a promotion. She *netted* a good job. I've got *a line out* on a good used car. It's time to *fish or cut bait.*

- Trying To Achieve A Purpose Is Agriculture

  > It's time I *reaped* some rewards. That job is a *plum.* Those are the *fruits* of his labor. The contract is *ripe for the picking.*

The Object Event-Structure metaphor can thus be represented as follows; the indentations indicate special-case hierarchical structure.

THE OBJECT EVENT-STRUCTURE METAPHOR

- Attributes Are Possessions
- Changes Are Movements Of Possessions (acquisitions or losses)
- Causation Is Transfer Of Possessions (giving or taking)
- Purposes Are Desired Objects
- Achieving A Purpose Is Acquiring A Desired Object
  - Achieving A Purpose Is Getting Something To Eat
    - Trying To Achieve A Purpose Is Hunting
    - Trying To Achieve A Purpose Is Fishing
    - Trying To Achieve A Purpose Is Agriculture

Philosophical Implications of Event-Structure Duality

The fact that we can do figure-ground reversals in perception—as in the case of Necker cubes—indicates that figure-ground organization is a separable dimension of cognition. In other words, other aspects of a scene can be held constant while figure-ground organization is changed. But some choice of figure and ground is necessary. Perception requires a figure-ground choice. We do not perceive scenes that are neutral between figure and ground.

The phenomenon of duality reminds us that the same is true of concepts. Metaphorical duals may be figure-ground reversals of one another, but some choice of figure and ground is necessary for human concepts. Thus, there is no neutral Event-Structure metaphor with a source domain neutral between figure and ground. Instead, what we have are two diametrically opposed mappings.

Figure and ground are aspects of human cognition. They are not features of objective, mind-independent reality. Hence, figure-ground orientation tends to be absent from studies of concepts done in the objectivist tradition, in which meaning is based on allegedly "objective" truth rather than human cognition. For human concepts, figure-ground distinctions are crucial.

Their philosophical importance becomes manifest when one looks at event-structure duality. Consider causation. Causation is metaphorically conceptualized differently in the two Event-Structure metaphors:

- Location: Causation Is The Forced Movement Of An Entity (The Affected Entity) To A New Location (The Effect)

  Figure = Affected Entity
  Ground = Effect
  Example: "The home run sent the crowd (Figure) into a frenzy (Ground)."

- Object: Causation Is The Transfer Of A Possessible Object (The Effect) To Or From An Entity (The Affected Entity).

  Figure = Effect
  Ground = Affected Entity
  Example: "The loud music gave a headache (Figure) to each of the guests (Ground)."

These two conceptualizations of causation have opposite figure-ground orientations. In the Location metaphor, the affected entity is the figure; it moves to the new location (the ground). In the Object metaphor, the effect is the figure; it moves to the affected party (the ground).

But the figure-ground difference in these cases leads to an additional difference. In both cases, the figure is conceptualized as moving and the ground as stationary. And in both cases, there is a causal force that is applied to the figure, moving it with respect to the ground. But in the two cases, the causal force is applied to different things.

- In the Location case, the causal force is applied to the affected party, since it is the figure.
- In the Object case, the causal force is applied to the effect, since it is the figure.

In other words, the figure-ground difference induces an additional difference in conceptualization.

What this means is that *there is no conceptualization of causation that is neutral between these two!* One cannot just abstract away the figure-ground difference and say that what remains is identical. It isn't. In the Location metaphor, the causal force is applied to the affected entity, while in the Object metaphor, the causal force is applied to the effect.

This is important because it is usually assumed in philosophical discussions of causation that there is a single abstract causation concept with a single conceptual structure and a single inference pattern. We saw in our discussion of the Location version that there are different versions of forced-movement causation, with different inference patterns. Here we have another different version of causation. As we shall see in the following sections, there are yet others.

The question that naturally arises is: What makes these different concepts both concepts of causation? What unifies them, if it is not the same conceptual structure and not a common set of inference patterns? In the two cases given, it is the submetaphor of Causes Are Forces, which is common to both. As we shall see, this works for some of the other cases of causation, but not all.

The Word Cause and the Causative Construction

One might think that the word *cause* is simply literal and does not involve these metaphors at all. However, when we look at the grammar of *cause* in the light of metaphor research, we can see that *cause* has two syntactic valence structures that correspond to the metaphors (1) Causation As Forced Movement Of An Affected Entity To An Effect and (2) Causation As Transfer Of An Effect To An Affected Entity.

These metaphorical valence structures can be seen in the sentences:

I caused the vase to fall. I caused pain to a great many people.

In "I caused the vase to fall," we have the syntactic correlate of Causation As Forced Movement. The direct object *the vase* is the affected entity; the causal force is applied to it. *To fall* indicates the effect. In "I caused pain to a great many people," the direct object *pain* is the effect, *to* indicates the direction of transfer, and *a great many people* indicates the affected entity, which has the semantic role of recipient.

What is lexicographically interesting about this is that the word *cause* is metaphorical in some ways and not others. It is not metaphorical in phonological form; the phoneme sequence is not taken from some other conceptual domain. But the grammatical valence structure *is* metaphorical. It has two grammatical valence structures reflecting two different metaphorical conceptualizations.

One might think that the same metaphorical valence structures would apply to both the word *cause* and causative constructions. Interestingly, this is not the case. Whereas the word *cause* has syntactic valence structures reflecting both metaphors, the causative construction in English only reflects the Causation As Forced Movement metaphor.

To see this, consider a causative sentence like:

I warmed the milk to body temperature.

Warm is fundamentally an adjective that can function as either an intransitive verb in an inchoative construction or a causative transitive verb in a causative construction with the semantics:

The Cause (*Verbed*) the Affected Entity *to* the Resulting State.

In this case, *I* is the Cause, *the milk* is the Affected Entity, and *body temperature* is the Resulting State, while *to* indicates the direction of change. What we have here is the grammar of Causation As Forced Movement.

Event-Structure Hierarchies

Consider what most people think about an electric car. How do they understand what an electric car is? They take their ordinary idea of a car with a gas engine, replace the gas engine with an electric engine, get rid of everything associated with the gas engine (gas tank, carburetor, exhaust), replace that with what would be associated with an electric engine (batteries, a recharging plug), and then leave as much as possible the same (wheels, tires, seats, body, steering wheel, windshield, and so on).

Cognitive scientists call this process *inheritance*. We "inherit" all the information we can from our prototypical idea of a car, provided it is consistent with the new information. Inheritance of this sort is an extremely common cognitive mechanism.

We have already seen two types of hierarchical structure in the metaphor system—the special-case hierarchy and the entailment hierarchy. Let's begin with special-case hierarchies.

Special-Case Hierarchies

Consider Difficulties Are Impediments To Movement, which is part of the Location Event-Structure metaphor. There are a number of special kinds: blockages, features of the terrain, burdens, counterforces, and lack of energy. Given the general mapping

Difficulties Are Impediments To Movement

the subcategories of impediments to movement create new, special case submappings:

Difficulties Are Blockages
Difficulties Are Features Of The Terrain
Difficulties Are Burdens
Difficulties Are Counterforces
Difficulties Are Lacks Of Energy

Another example is the general mapping Time Is A Resource, and the special-case mapping Time Is Money, in which money is a special case of a resource. Obviously, not all special cases of resources get mapped; time is not coal or oil or timber. Only the special case of money is conventionalized in a mapping. It is at the level of special cases that much of the conventionalization and cultural variation that we find in metaphor systems enters in.

Entailment Hierarchies

A second major type of hierarchical structure that we saw above is the *entailment hierarchy*. In the case of

Actions Are Self-Propelled Motions

we saw that various entailments of that metaphor were also conventionalized as mappings:

Aids To Action Are Aids To Movement
Manner Of Action Is Manner Of Movement
Careful Action Is Careful Movement
Speed Of Action Is Speed Of Movement
Freedom Of Action Is The Lack Of Impediment To Movement
Suspension Of Action Is Stopping Movement

For example, if Actions Are Self-Propelled Movements, it is entailed that Careful Action Is Careful Movement. This entailment happens to be conventionalized as a submapping.

Variations on the Location Event-Structure Metaphor

So far we have looked only at the two major complex metaphors for conceptualizing event structure—the Location and Object Event-Structure metaphors. But ideas as general and important as events, actions, and causes are invariably conceptualized in more than two ways. We now need to examine some of the dazzling complexity of the metaphorical means used for conceptualizing events and related concepts. We will begin with three metaphors that use some of the same primary metaphors used in Location Event Structure yet add different primary metaphors to provide their own distinctive perspective on events.

The Moving Activity Metaphor

The first is a metaphor in which activities are conceptualized as things that move and the completion of the activities as those things reaching a destination. Examples include "The project has *slowed to a crawl*" and "The book is *moving right along.*" The latter example contains an instance of a common metonymy of Product For Process, in which *the book,* the product of the activity of writing, stands for the activity itself. Without the metonymy of the book for the writing of the book, one would say "The writing is *moving right along.*"

This is a primary metaphor that arises from common activities that require the moving of things. For example, when you wipe off a counter, your arm must move in order to wipe. The motion is part of the wiping, and the more your arm moves, the more wiping occurs. The same goes for sweeping the walk, mowing the lawn, or pushing any object at all. The correlation between activity and things that move in performing the activity results in the metaphor Activities Are Things That Move.

This primary metaphor combines naturally with other primary metaphors, States Are Locations, Causes Are Forces, and Purposes Are Destinations, to produce a unique variation, the Moving Activity Metaphor. Examples that show the interaction of this primary metaphor with others include:

The renovation is *at a crossroads* (States Are Locations). We've *accelerated* the building of the new bridge (Causes Are Forces). The council *brought* the project *into* compliance with state regulations (Causation Is Forced Motion). The new stadium is *bogged down* (Difficulties Are Impediments To Motion). The necessary retrofitting is now *at an end* (Purposes Are Destinations).

The Moving Activity Metaphor

Activities Are Things That Move
Completion Of The Activity Is Reaching A Destination

Other primary metaphors that are part of the mapping:

States Are Locations
Causes Are Forces
Causation Is Forced Movement (Or Prevention Of Movement)
Difficulties Are Impediments To Motion

The Action-Location Metaphor

Another primary metaphor for actions is based on the common experience of being able to perform a given action only when in a particular location, for example, you can turn on the stove only when you are at the stove. Such a correlation between performing an action and being in a particular location is so common that the correlation has become the basis for a primary metaphor, namely, An Action Is Being In A Location.

In the metaphor, the action is performed when the actor is at the action location. When the action is seen as a purpose, the location is seen as a destination. A caused action is a forced movement to the location. And a prevented action is a forced stoppage of motion to the location. Here are some examples:

> I'm *leaning toward* leaving. They *pushed* him *into* running for president. They *prodded* me to run. I was *drawn into* the bank robbery. They stopped me *from* leaving. I've *taken steps toward* canceling my policy. She's *close to* resigning. She *backed away from* resigning. She *came near to* resigning. He's *inching toward* invading another country.

Perhaps the most celebrated, and misanalyzed, example is "I'm *going to* leave." This is often misanalyzed as simply being about the future. But it is actually an instance of the Action-Location metaphor. Notice that you can use it in the past: "I was going to leave." It entails an expected future: If you're moving to an action location, you're not there yet, but you expect to be there in the future.

THE ACTION-LOCATION METAPHOR

An Action Is Being In A Location

This primary metaphor forms a complex metaphor by combining with the following other primary metaphors:

Causes Are Forces
Purposes Are Destinations

Additional submappings are entailed:

"Closeness" To An Action Is Closeness To A Location
Causing An Action Is Forcing Movement To A Location
Preventing An Action Is Stopping The Traveler From Reaching A Location

The Existence as Location Metaphor

Next, there is the metaphor based on the fact that objects that exist exist in a location. Moreover, our existence is correlated with our being located right where we are: here! This correlation is the basis for the primary metaphor:

Existence Is Being Located Here

Existence is thus conceptualized as presence in a bounded region around some deictic center, that is, around where we are. Combining this with Change Is Motion, yields:

Becoming Is Coming Here
Ceasing To Exist Is Going Away

Adding Causation Is Forced Movement results in:

Causing To Exist Is Forced Movement Here

> Causing To Cease To Exist Is Forced Movement Away

Consequently, we speak of things being *in* existence, *coming into* existence, and *going out of* existence; correspondingly we speak of creation as *bringing* something *into* existence. The Bible speaks of God as *bringing forth* the heavens and the earth. Causing something to cease to exist is naturally seen as *removal,* or *getting rid of* something, as in "I removed all the errors in the manuscript" and "I got rid of all the evidence."

A special case of this metaphor recognizes life as a form of existence:

> Being Alive Is Being Located Here
> Being Born Is Coming Here
> Death Is Going Away
> Causing Death Is Forced Removal

When a baby is born, we send out announcements of the child's *arrival.* Wishing to live can be expressed as wishing to *be here for a long time.* We speak of death euphemistically as *departure,* or *leaving us,* or *going away.* And killing someone is spoken of as *getting rid of* him, *taking* him *out,* or *blowing* him *away.*

The Full Conceptual Complexity
of Change and Causation

So far, we have seen extensive evidence of the complexity of our metaphor system for events and causes. One thing that should be clear by now is that causes cannot be separated from events. Causes are mostly conceptualized as forces, and each force must apply to something producing some effect as it is applied. The forms of causation depend on the kind of force it is, what the force is applied to, and the kinds of changes it produces. The result is different logics of causation as a whole. These different logics of causation are not a simple product of a single logic of cause or force, combined with a single logic of change. Because there are so many different kinds of changes and because there are so many different ways in which forces can produce them, we have a great number of metaphors for conceptualizing causation and change.

From the examples given above, we have seen that it is unlikely that one could find a single literal logic of causation and change that accounts for all the cases considered. But when one looks at the true profusion of such metaphorical conceptions of causation and change, it becomes clear that such an endeavor would be utterly impossible. What follows in this section is a brief examination of a large number of metaphors that produce such a profusion of logics of causation and change.

We have two purposes in this section. The first is to demonstrate the full complexity of causation and change. The second is to lead up to certain important philosophical conclusions that depend upon an understanding of that complexity. We will discuss nine cases in all.

Direction

Remaining in a state is conceptualized as moving in the same direction, and changing is seen as turning to a new direction.

CHANGING IS TURNING

Remaining In A State Is Going In The Same Direction
Changing Is Turning

Examples include "He *went on* talking" and "The milk *turned* sour." This metaphor combines with Causation Is Forced Motion to produce forced turning, as in "He *turned* the lead *into* gold."

Shapes

States, in addition to being conceptualized as locations, can be conceptualized as shapes. Thus we can ask, "What *shape* is the car in?" This too combines with Causes Are Forces, entailing that causation is conceptualized as the forced changing of shape, as in "He wants to *reshape* the government," "She's a *reformer,*" and "Meditation *transformed* him into a saint." We can state this mapping as:

STATES ARE SHAPES

States Are Shapes
Causes Are Forces
Causation Is A Forced Change Of Shape

Replacement

When there is dramatic change, change so drastic that the thing changed is conceptualized as a different thing in the same place:

Change Is Replacement.
A Thoroughly Changed Entity Is A Different Entity
    In The Same Place.

Examples include:

I'm a *different* man. After the experiment, the water was *gone* and hydrogen and oxygen were *in its place*. Under hypnosis, the sweet old lady was *replaced* by a scheming criminal.

Changing a Category

When we make an object into an object of another kind, we apply force and literally change the object. We express this by sentences like "Harry *made* the log *into* a canoe."

We often change our categories without changing the things in the categories, but we conceptualize a category change as if we did change the things in the categories. Suppose a law were passed making cigarette smoking a criminal offense. We conceptualize this as changing the things themselves, and we use the same grammar as above: "The new law *makes* cigarette smokers *into* criminals."

The fact is that the law would affect cigarette smokers; it would do something to them. Interestingly, such a law would be conceptualized metaphorically according to the following:

Changing The Category Of An Entity Is Changing The
    Entity (into an entity in a different category)

When a law recategorizes cigarette smokers as criminals, it doesn't literally change the cigarette smokers; it changes their categorization with respect to the legal system. But the point of effect of the law *is* to change smokers, to make them into nonsmokers by putting legal pressure on them. This reality of recategorization—that it can change or put pressure for change on what is recategorized—is reflected in the metaphor.

Causing Is Making

There is a minor causation metaphor in which causing is conceptualized as making. When you make something, you apply a direct force to an object, changing it to a new kind of object with a new significance. For example, "He *made* lead *into* gold." When we conceptualize causation as making, we understand there being a causal force directly applied to a person or a situation to change him or it into something of a different kind. That can be either a kind of thing the person wouldn't otherwise do, or a kind of situation that otherwise wouldn't occur. Examples include "I *made* him steal the money" and "The DNA tests *made* it clear that he committed the murder." The mapping is simply:

Causing Is Making
Effects Are Objects Made

Progeneration
Causation can be also understood in biological terms:

Causation Is Progeneration
Causes Are Progenitors
Effects Are Offspring

Examples again are commonplace:

Necessity is the *mother* of invention. Teller was the *father* of the H-bomb. The *seeds* of World War II were *sown* at Versailles.

Causal Precedence
It is commonly the case that causes precede their effects, that is, the causal precedence coincides with temporal precedence. In this case, causal precedence can be conceptualized metaphorically as precedence in time:

Causal Precedence Is Temporal Precedence

Examples include:

You have to pull on the door *before* it will open. The door opened *after* he pulled on it. *When* you don't have enough iron, you're anemic. *Since* he doesn't have enough iron, he's anemic. If you go outside without dressing warmly, *then* you will get a cold.

Each of these temporal expressions expresses a causal relation. Note that there does not have to be literal temporal precedence in these cases. For example, the sentences above with *when* and *since* express simultaneity of cause and effect. Other cases in which there is causal effect but no literal temporal priority are:

When your heart stops beating, you're dead. *Since* he got the most votes, he won the election. She went blind *after* her optic nerve was severed in the accident.

Causal Paths

When people travel, whether they are taking a medium-length walk or driving somewhere, they tend to continue along the paths and roads they have started out on. As a result, people tend to wind up at places that the paths and roads they are already on lead to. Thus, if you are traveling, the path or road you are already on is a crucial determining factor in where you will wind up. In such already-on-a-path situations, the paths are determining factors for final locations.

The Location Event-Structure metaphor conceptualizes States As Locations and Actions As Self-Propelled Motions. These common mappings make it natural to conceptualize already-on-a-path situations as applying to events in the following way:

THE CAUSAL PATH METAPHOR

Self-Propelled Motion	→	Action
Traveler	→	Actor
Locations	→	States
A Lone Path	→	A Natural Course Of Action
Being On The Path	→	Natural Causation
Leading To	→	Results In
The End Of The Path	→	The Resulting Final State

This metaphor maps an already-on-a-path situation onto an already-on-a-course-of-action situation: An actor already following a natural course of action will tend to continue on that course of action till he reaches a resulting state. The course of action one is on therefore plays a causal role in determining the final state of the actor. Here are some examples of the Causal Paths metaphor:

Smoking marijuana *leads to* drug addiction. Young man, you are on a course of action that will *lead* you to ruin. She is *on the path to* self-destruction. If you keep *going the way you're going*, you're going to be in trouble. You're *heading for* trouble.

Causal Links

The last metaphor in this group concerns causal links. Consider the Object Event-Structure metaphor, in which Attributes Are Possessions, Change Is Motion, Causes Are Forces, and Causation Is Transfer (of the effect to the affected party). Apply the Object Event-Structure metaphor to what we know about a situation in which two objects, A and B, are tied or linked tightly together. The metaphor maps the source-domain knowledge:

- If Object B is tied tightly to Object A, and if X has Object A, then the tie between Object A and Object B will result in X having Object B.

onto the target-domain knowledge:

- If Attribute B is causally closely correlated with Attribute A, and if X has Attribute A, then the close causal correlation between Attribute A and Attribute B will result in X also having Attribute B.

In short, the Object Event-Structure metaphor, applied to the commonplace knowledge about a tie or link, results in the Causal Link metaphor.

The logic of causal links shows clearly in sentences like "AIDS has been closely *linked* to the HIV virus" and "AIDS is *tied* to the HIV virus." This expresses a causal relation: The HIV virus causes AIDS.

Note that if the tie or link is loose or weak, the causal inference may not go through. B will definitely go where A goes only if the link is tight and strong. If B is loosely or weakly linked to A, then B may not come along when X acquires A. If the link is weak, it may break and if it is loose (as with a long, loose rope), B may never arrive at the same destination as A. Thus, a sentence like "Breast cancer has only been loosely linked to pesticides" is not a strong causal statement.

These nine cases should provide a sense of the richness of our metaphorical means for conceptualizing different forms of causation and change, each with its own logic. Such a wide array should dispel any illusions that all forms of causation are instances of a single literal concept with a single logic. But the metaphors for causation to do not stop here. There are still major cases to discuss.

Natural Causation and Essence

Nature as Agent

We have a general metaphor in which natural phenomena are conceptualized as human agents. We see this in sentences such as "The wind *blew* open the door," and "The waves *smashed* the boat to pieces." Since Causes Are Forces, human agents exert force, and Natural Phenomena Are Human Agents, natural causes are conceptualized metaphorically as forces exerted by a human agent. Consequently, verbs expressing actions by a human agent—*bring, send, push, pull, drive, thrust, propel, give, take*—can express causation by natural phenomena:

A comet *slammed into* the surface of Mars. The surface of the earth has taken a *beating* from meteorites. Meteorites have *dug out* huge craters on the moon.

THE NATURE AS HUMAN AGENT METAPHOR

Human Agents	→	Natural Phenomena
Force Exerted By Agents	→	Natural Causes
Effects Of Force Exerted By Agents	→	Natural Events

According to this metaphor, the natural event of the door opening in the wind is conceptualized as follows: The wind is a natural phenomenon, which is a metaphorical agent. The force exerted by that agent is a natural cause. The natural event of the door opening in the wind is metaphorically seen as an effect of that force exerted by the agent (the wind). This metaphorical process is so commonplace it is barely noticed.

This metaphor also applies to anything considered to be an instance of a natural phenomenon. We will consider two cases, natural processes and natural properties. Becoming old and dying is a natural process. As a natural phenomenon, this process can be seen as an agent exerting force, and hence as a cause of particular deaths. Thus, we can say "Old age killed him" or "Old age was the cause of his death." Accordingly, we can personify the phenomenon of death as an agent (e.g., the Grim Reaper) and say, conceptualizing death as departure, "Death took her from us" or "The Grim Reaper took her from us" (see A1, Lakoff and Turner 1989). Via this mechanism, it is commonplace to conceptualize the very existence of natural processes as causes and natural events as effects of those causes.

Arisings

Consider sentences like "A problem has *arisen* in Bosnia" and "There *arose* a commotion." The intransitive verb *arise* is used to express the occurrence of an effect—a natural effect. The causal source of the effect, when unexpressed, is taken to be the relevant situation. The causal source may, of course, be expressed, as in "Primary metaphors *arise* from correlations in everyday experience."

Interestingly, natural causation here is conceptualized as upward motion. "What's *up*?" asks what the current effects are of the situation you are in. "A difficulty *came up* in the planning stage" expresses that the difficulty was a natural effect of the planning process. Of course, attributing something to the situation, not to anyone in particular, can be a way of pragmatically hiding a very nonnatural causal source.

NATURAL CAUSATION IS UPWARD MOTION

Upward Motion	→	Natural Causation
Thing Moving Upward	→	A Natural Effect
Original Location	→	The Situation, Taken As Natural Cause

Causal Sources

In the most basic kind of causation, a physical force is applied to move something or change its appearance. In such cases, typically whatever exerts the force must move *from* an initial source location to a position in which force is exerted. In such a situation, there is a correlation between the application of the causal force and motion *from* an initial location. This correlation is the basis of a metaphor for causation in which causes are conceptualized as sources and the word *from* expresses source.

Causes Are Sources

Examples:

She got rich *from* her investments. He got a sore arm *from* pitching too many innings. Harry died *from* pneumonia. I'm tired *from* working all day.

When the sources alone are mentioned, the causation is taken to be natural, given the source.

Emergings

A sentence like "A difficulty *emerged* during the planning process" expresses something very much like "A difficulty *arose* during the planning process." That is, it expresses an effect that is natural given the situation. Again, the natural causation is expressed by motion, but here the motion is outward rather than upward. And once more, in the absence of any mention of causal source, the causal source is taken to be the situation. When expressed, the causal source is conceptualized as a container and the preposition *out of* is used.

> He shot the mayor *out of* desperation. The chaos in Eastern Europe *emerged from* the end of the cold war.

NATURAL CAUSATION IS MOTION OUT

Motion Out	→	Natural Causation
Thing Moving Out	→	A Natural Effect
Original Container	→	The Situation, Taken As Natural Cause

All three of these are cases of natural situational causation, where situations are taken to be natural causes and the occurrence of the effect is a form of motion away from the situation.

Essences

Now let us turn to natural properties. The Folk Theory of Essences is commonplace in this culture and other cultures around the world. According to that folk theory, everything has an essence that makes it the kind of thing it is. An essence is a collection of natural properties that inheres in whatever it is the essence of. Since natural properties are natural phenomena, natural properties (essences) can be seen as causes of the natural behavior of things. For example, it is a natural property of trees that they are made of wood. Trees have natural behaviors: They bend in the wind and they can burn. That natural property of trees—being made of wood (which is part of a tree's "essence")—is therefore conceptualized metaphorically as a *cause* of the bending and burning behavior of trees. Aristotle called this the *material cause.*

As a result, the Folk Theory of Essences has a part that is causal. We will state it as follows:

> Every thing has an essence that inheres in it and that makes it the kind of thing it is.
> The essence of each thing is the *cause* of that thing's natural behavior.

This is a metaphor, by which essences are understood in terms of causes of natural behavior.

Linguistically, the Folk Theory of Essences is built into the grammar of English. There are grammatical constructions whose meaning involves the notion of an essence. One construction is of the form "X will be X," as in "Boys will be boys," which expresses that boys have an essence, that they will act according to their essence, and that what they do is, therefore, natural behavior. Another construction expressing essence is the use of a verb in the simple present tense with a plural or generic subject, as in "Birds sing," "Wood burns," "Squirrels eat nuts," and "The female praying mantis devours her mate." In this construction, the verb expresses a natural behavior of the subject that is a causal consequence of the subject's essence.

As we shall see in Part III, in addition to being so commonplace a folk theory that it is built into the grammar of our language, the idea that essences cause natural behavior has a long and important history in philosophy.

Reason in the World

Why Reasons Are Causes

Consider such expressions as "I was forced to that conclusion," "The data compelled me to change my theory," "The overwhelming evidence dragged him kicking and screaming to a conclusion he didn't want to face," and "He's going to resist that argument no matter how strong it is." Eve Sweetser (A1, 1990) observes that such sentences are a linguistic reflection of a deep conceptual metaphor, Thinking Is Moving combined with Reason Is A Force. In this metaphor Ideas Are Locations and Being In A Location Is Having An Idea. A logical "conclusion" is thus a place that Reason has, by force, moved us to. That is why we can be "*compelled* by the facts" or "try to *resist the force* of an argument" or "be *overwhelmed* by the *weight* of the evidence."

Reasons can be conceptualized as causes via a metaphorical blend, a natural composition of two metaphors: Reasons Are Forces and Causes Are Forces. This explains why the question "Are reasons causes?" is a perennial philosophical question. In our conceptual system, they are causes, but only very indirectly, via two metaphors and a conceptual blend of those metaphors. Hence, the categorization of reasons as causes is not only nonliteral, but triply nonliteral. This is what gives philosophers pause. The links are there and so the proposition can be defended, but their indirectness and nonliteralness allow a defender of literality to claim the opposite. There is, of course, no simple literal answer to the question. From a cognitive perspective, we can either conceptualize reasons as causes or not, depending on whether we use the blend of the two metaphors. However, because the metaphors are so deep in our conceptual system, the blend is a natural one, so much so that it seems to many to characterize a deep truth.

Epistemic Causation

Deducing the existence of causation in the world from evidence proceeds from knowledge of the effect. Knowing the effect, you reason "backward," given what you know about the world, to knowledge of the cause. At the level of the reasoning done, you went *from* the evidence of the effect *to* knowledge of the cause. Metaphorically, what took you from the evidence of the effect to the knowledge of the cause was the *force* of reason (since Reasons Are Forces). Since Causes Are Forces, reasons, as we discussed above, can be conceptualized as causes.

The metaphorical result is a reversal of the direction of causation at the epistemic level, as Sweetser has observed (A1, 1990). If *A* caused the effect *B,* then knowledge of *B* is the metaphorical cause of the knowledge of *A*. Since both worldly and epistemic causation can be expressed by *because*, the following sentences can both be true, one in the world and one epistemically:

My husband typed my thesis because he loves me. (worldly causation)
My husband loves me, because he typed my thesis. (epistemic causation)

In the second case, the speaker reaches the conclusion that her husband loves her, reasoning from the evidence that her husband typed her thesis. Via the metaphor that Reasons Are Causes, the word *because* in the second sentence can be used to express the metaphorical causal direction of the reasoning.

Teleology: Why We See the World as Rational

In constructing a plan of action to achieve a purpose, we use our reasoning, which tells us that we should do action *A* to achieve result *B*. We assume that there is a correlation between this mental plan and what will occur in the world. We actually experience that correlation thousands of times a day when such plans work in our everyday experience. The correlation is between (1) actions taken on the basis of reason to achieve a purpose and (2) the causal relation between the actions taken and their result.

This regular correlation is the basis for one of our most important primary metaphors, Causation Is Action To Achieve A Purpose, where Causes Are Reasons (why the action will in fact achieve the purpose). This is the metaphor that tells us that the world is rational, that what happens happens for a reason. This primary metaphor is the basis of our everyday notion of teleology, *that there are purposes in the world.*

The metaphor is hardly rare. It is anything but limited to philosophical discussions. It occurs throughout everyday discourse, perhaps most frequently in the discourse of science and in other discourse about nature. Consider examples like:

> Trees in a forest grow toward the sun *in order to get* the light they need. Positive ions *need* another electron. The immune system *fights off* disease.

Look through the literature of science writing and you will find examples like these everywhere. For example, consider a sentence like "Scientists have identified a gene *for* aggression." The word *for* here indicates that the gene has a purpose. Here are some examples taken from a recent issue of *Science* (vol. 275, March 14, 1997).

> Bottom-up influences are responsible for certain illusions in which the brain is *tricked* into perceiving something distinctly different from the image perceived by the retinas. (p. 1583)
>
> *To do its job*, however, this cortex *must cooperate with* connected sensory regions that hold and use the information for briefer periods of time. (p. 1582)
>
> Most of the neurons in that very early processing stage merely *report* what is happening on the retina. (p. 1584)

There is, of course, a difference between conceptualizing and reasoning about the world according to this metaphor and actually believing that the metaphor is a truth.

Aristotle, as we shall see in Chapter 18, took this metaphor as a truth. Causes conceptualized according to this metaphor are what Aristotle called *final causes*, that is, causes constituted by purposes, either the purposes of a person or purposes conceptualized as being in nature.

Incidentally, the word *final* is used because, in the Purposes Are Destinations metaphor, purposes are conceptualized as end points on a course of action.

Causation as Correlation and Probabilistic Causation

When a cause produces an effect, it is common to find the effect physically near the cause. This correspondence in common experience between causation and accompaniment is the basis for a metaphor in which causation is conceptualized in terms of accompaniment. Thus, it is common to find sentences like the following expressing causation:

> An increase in pressure *accompanies* an increase in temperature. Pressure goes up *with* an increase in temperature. Homelessness *came with* Reaganomics.

Each of these is a causal statement. In each case, causation is conceptualized as accompaniment and the language of accompaniment is used: *accompany, go with, come with*. In each case, causation is understood in terms of correlation, as in a correlation of pressure with temperature. We will refer to the metaphor generalizing over such cases as Causation Is Correlation. The mapping can be stated as follows:

CAUSATION IS CORRELATION

Correlation Of *B* With *A*	→	Causation Of *B* By *A*
Independent Variable	→	Cause
Dependent Variable	→	Effect

It is extremely common in everyday life for people to conceptualize causation in terms of correlation. A noteworthy example occurred in the 1996 presidential campaign. Senator Robert Dole, in a campaign speech against President Bill Clinton, pointed out a correlation: Drug use among teenagers went up 104 percent during Clinton's term in office. He was blaming Clinton for the increase in drug use. In doing so, he was using the Causation Is Correlation metaphor and thus making a (metaphorical) causal claim by merely stating a true correlation that literally had no causal content. Millions of Americans understood this statement of correlation as a claim about causation.

The Causation Is Correlation metaphor is at the heart of the concept of probabilistic causation. The concept of probabilistic causation is a complex concept. It is composed of Causation Is Correlation plus another very common metaphor, our principal metaphor for probability, namely, Probability Is Distribution. This metaphor is based on a truth of gambling:

> The probability that X will happen in the future for an individual in a population equals the distribution of occurrences of X in the past for the population as a whole (or a large enough sample).

If you roll a die, the number four (as well as each of the other numbers) comes up one time in six in a large enough sample. Its distribution in that sample is one-sixth, or one in six. It is correspondingly true that the probability that you will get a four on one role of the die is also one in six. Probability equals prior distribution in gambling and similar random processes.

This correlation for gambling is the basis for our most common metaphor for probability, namely:

> The Distribution Of An Occurrence In The Past For A Population →
> The Probability Of Such An Occurrence In The Future For An Arbitrary Individual In That Population

This metaphor is so common that it may be perceived as a truth rather than a metaphor. For example, suppose you were to read the following finding:

> One-half of the women whose mothers had breast cancer also developed breast cancer.

If you read such a finding, and you are a woman whose mother had breast cancer, you may very well conclude that you have a one-in-two chance of developing breast cancer. If you draw this conclusion, it will be via the Probability Is Distribution metaphor.

To take another example, suppose you read a different finding:

> One-fifth of the women in the Bay Area develop breast cancer.

If you are a woman living in the Bay Area, you may conclude via this metaphor that you have a one-in-five probability of developing breast cancer. But suppose you read both findings. Which would apply? Do you have a one-in-two probability or a one-in-five probability? Or neither?

The Probability Is Distribution metaphor is a metaphor, not a truth. Probability is about *you*, distribution isn't. The reason that medical researchers are interested in distributions is that they believe there may be some causal factor in breast cancer that is either genetic or regional and that distributional data may help lead them to discover it. The distributional statistics may be suggestive that such a factor exists. But that, in itself, says nothing literal about *you* or the probability that *you* will get breast cancer. *You* may be so different from the people in the studies that the probability that *you* will get breast cancer may be very much smaller.

But you don't know. That uncertainty has everything to do with our everyday use of the Probability Is Distribution metaphor. What links our lack of knowledge and our use of this metaphor is yet another commonplace metaphor, namely, Uncertain Action Is Gambling. Examples of this metaphor are extremely common:

> I'll *take my chances*. The *stakes* are high. She's got *an ace up her sleeve*. The *odds* are against me. He's holding *all the aces*. It's *a toss-up*. If you *play your cards right*, you can do it. Where is she when *the chips are down*? He's playing it *close to his vest*. Let's *up the ante*. That's *the luck of the draw*.

In this metaphor, any action that you have to take when you don't know all the relevant information is conceptualized as a gamble. This is our primary metaphor for action with uncertain knowledge.

The metaphor that Uncertain Knowledge Is Gambling has a metaphorical entailment. Recall that it is true of gambling that probability does exactly correlate with distribution. Because of this, the Probability Is Distribution metaphor is entailed. It is no accident that it is a normal part of our conceptual systems. Nor is it an accident that we use it so commonly in situations in which we lack knowledge of the relevant factors.

Probabilistic causation is a technical concept composed of the commonplace metaphors Causation Is Correlation and Probability Is Distribution. Here is a statement of the Probabilistic Causation metaphor:

Y is correlated with a Difference D in the Distribution of X in the past in a population as a whole	$\rightarrow$	Y causes a Difference D in the Probability of Occurrence of X in the future for an arbitrary member of that population.

Here we can see how Causation Is Correlation is put together with Probability Is Distribution. Of course, there are many people who take this metaphor as a literal truth, namely, that causation of a difference in probability literally can be measured by correlation of a difference in distribution.

Statisticians, of course, know better. They know that mere correlation of distributions proves nothing unless it is considered relative to causal theories. A classic counterexample is that, among schoolchildren, shoe size correlates highly with reading skills. But there is no causal relation between size of feet and ability to read. There is, instead, an intermediate factor: age.

The sophisticated use of probabilistic causation consists of considering possible causal theories and ruling out alternative intermediate factors on the basis of further studies of correlations of distributions. In short, a case must be made that high correlations of distribution do indicate causation and that case cannot rationally be made just on the Probabilistic Causation metaphor.

In the classic case of whether smoking causes cancer, the mere correlation of smoking with cancer was not enough. What was shown additionally was that possible confounding factors could be eliminated.

The Range of Causal Concepts and Literal Causation

Lest you think that the survey so far has exhausted all the varieties of causal concepts, we should point out that there are still many classic cases we have not touched on. There is *emotional causation*, in which a perception or thought is conceptualized as an external stimulus that forcefully produces an emotion in us ("The beauty of the sunset stunned me"). There is *instrumental causation*, in which causality is attributed to an instrument ("A crowbar will open that door"). There is *enabling causation*, in which the causation involves either the absence or forceful removal of an impediment to action ("The body-guards let the assassination happen" or "The fog made it possible for the thieves to escape"). There is the causality involved in the conceptualization of *biological inheritance*, in which traits are seen as possessible objects transferred from parent to child ("He passed on his good looks to his son").

Further complicating the picture are degrees of causal directness. Compare:

John killed Bill.
John caused Bill to die.
John had Bill killed.
John brought it about that Bill died.

The first typically describes direct causation construed as a single event; the second describes indirect causation where there are two events; the third describes an intermediate cause; and the fourth involves very indirect causation. Yet, however diverse, they are all forms of causation.

We have now finished our survey of metaphors for events and causes. Let us now turn to answering the puzzles that we began with and discussing the philosophical importance of this analysis.

Answering the Causal Concept Puzzle

The answers to the puzzles given above fall out of the cognitive analysis of our normal, everyday complex concept of causation, with all of its internal structure, both literal and metaphorical. It is that complex radial structure that unites the various concepts of causation into a single category. What hold that structure together are (1) the literal necessary condition, namely, being a determining factor for a situation; and (2) the literal central prototype, direct human agency, that forms the basis for all extensions.

Let us take the Causal Concept Puzzle first. Note that direct human agency, the central prototype of the category of causation, has the following properties:

Application of force precedes change.
Application of force is contiguous to change.
Change wouldn't have happened without the application of force.

The Causes Are Forces metaphor maps these into properties of central (but not all) forms of causation:

Causes precede changes.
Causes are contiguous to changes.
The change wouldn't have happened without the cause.

Such central cases are good for describing changes

that do not depend on other changes,
that occur simultaneously with the action of the cause,
that are gradual, that is, that unfold with the occurrence of the cause, and
that are controllable.

Noncentral conceptions of causes as forces are good for describing other kinds of changes, namely, changes

that depend on other changes,
that lag after the action of the cause,
that are sudden, and
that, once started, perpetuate themselves and are no longer controllable.

The cases of causal theories in the social sciences that we cited above are of the latter, noncentral kind.

Causal paths: Change depends on other changes.
The domino effect: Changes perpetuate themselves.
Thresholds: Change lags after the action of the cause, perpetuates itself, and becomes uncontrollable.
The plate tectonic theory of international relations: The effect lags long after the continued action of a large cause.

Each of these causal theories is metaphorical. These are all special cases of Causation Is Forced Movement. Each different kind of forced movement has a different logic, which is mapped onto causation by Causes Are Forces. It is the metaphor of Causation Is Forced Movement that unifies all of these very different kinds of phenomena as kinds of causation.

Are Causal Theories Necessarily False When They Are Metaphorical?

A question immediately arises: Does the very fact that these theories are metaphorical and that they have different logics impugn them as being *really* causal theories? Traditional philosophical views of causation might lead to this conclusion. Suppose one assumes, as many philosophers have, what we will call the Theory of the One True Causation. According to this theory, there is only one true concept of causation. It is literal. It has only one logic. Moreover, true causation, as conceptualized, is an objective feature of the world.

If you accept this theory of what causation must be, then certain things follow. First, the only-one-logic condition says that these cannot all be forms of causation, since each has a different logic. Second, take the conditions that causation must be literal and that, *as conceptualized*, causation must be an objective feature of the world. This rules out all of these as true causal theories, because each is metaphorical. The metaphors cannot be an objective feature of the world. Countries are not dominoes, continental plates, or travelers on a path. If causation must be literal and have only one logic, then most causal theories in the social sciences will not count as causal theories.

We disagree with those conclusions that would follow from such traditional philosophical assumptions as these. Our very concept of causation is multivalent: It consists of the entire radial structure, with human agency at the center and many extensions. What we mean by causation is all of those cases with all of their logics. What we take to be the central case is human agency. One might decide that one likes one type of causation better than another, but as far as the cognitive unconscious of ordinary people is concerned, they all count as causation.

Metaphor is central to our concept of causation. Most of our notions of causation use metaphor in an ineliminable way, with the logics of the concepts being imported from other domains via metaphor. There are nonmetaphorical forms of causation, such as human agency, but for the most part, causal concepts make use of conceptual metaphor.

But this does not mean that the use of metaphorical causal concepts is illegitimate. Consider for a moment the literal, though skeletal, necessary condition on causation: the notion of a determining factor. There do appear to be real determining factors in the world, determining factors of many different sorts. Causal theories in the social sciences seek determining factors of various kinds. Each type of determining factor sought may have a different logic. The conceptual metaphors for causation used in the social sciences are chosen for their logics. Each metaphorical causal theory makes a claim about what types of determining factors there are in a subject matter and what the logic of each type of determining factor is.

Such claims may happen to be false (as in the case of the domino theory) or maybe even nonsensical (as we surmise the plate tectonic theory to be), but that does not mean in principle that every time a metaphorical form of causation is used, the theory is false just because it uses conceptual metaphor. As we shall see in our discussion below of general relativity, metaphorical theories can have testable consequences.

The Causal Theory Puzzle

This brings us to the second puzzle, which has an answer similar to the first. The various theories of causation proposed over the centuries by philosophers reflect the radial structure of our ordinary concept of causation. In other words, each philosophical claim about the nature of causation is sanctioned by a corresponding type of causation in our radial causation concept. Many of those, of course, are metaphorical.

The Natural Phenomena Are Human Agents metaphor leads us to a notion of natural causes as forces exerted through human agency. The idea that essences exist as natural phenomena leads to the conception of essences as causes. Essences have traditionally been conceptualized in philosophy as material or formal, and so the ancient Greek view of material and formal causes is a consequence of a common metaphor for causation that has been with us since before the ancient Greeks, namely, that Natural Phenomena Are Human Agents. The Greek gods were personifications of this metaphor.

This metaphor also sanctions other philosophical views of causation. Uniformities of nature are kinds of natural phenomena, and so uniformities of nature have also been set forth as causes. When natural phenomena are characterized by laws of nature that state certain constraints, those constraints on nature are taken as causes.

The fact that human agency requires directly applied force (or "power") corresponds to the common view that causation requires the application of power to effect a change. This is a common philosophical theory.

The metaphor that Causal Priority Is Temporal Priority sanctions the common philosophical view that causes are temporally prior to effects.

The literal, skeletal necessary condition that causes are determining factors sanctions the common philosophical claim that causes are necessary conditions.

The metaphor Causation Is Correlation sanctions the common empiricist view, from Hume (as commonly interpreted) to Nancy Cartwright, that causation is correlation.

There are two cases that we cannot discuss now but must delay until later: final causes and internal causation. To discuss what final causes are, we will first have to go through the metaphor system for the mind in the next chapter. Internal causation must wait till the chapter after that, in which the system of metaphors for the internal structure of the self is described.

From this discussion, we can see how the range of philosophical theories of causation arises from our ordinary literal and metaphorical concepts of causation. Each particular theory of causation picks one or more of our ordinary types of causation and insists that real causation only consists of that type or types. In cases in which philosophers have held the theory of the one true causation, they have been forced to choose, from among the commonplace causation concepts, the one they think is the only right one. They then take it as literal and attribute to it unique objective existence in the world, as required by the theory of the one true causation.

The Overall Philosophical Consequences

We began with a cognitive semantic analysis of the concepts of events and causation. If one accepts that analysis, a great deal follows. Given that causation is a multivalent radial concept with inherently metaphorical senses, the theory of the one true causation becomes not merely false, but silly. Once we know that it is multivalent, not monolithic, and that it is largely metaphorical, it turns out not to be the kind of thing that could have a single logic or could be an objective feature of the world. Since the concept of causation has ineliminably metaphorical subcases, those forms of causation, *as conceptualized metaphorically*, cannot literally be objective features of the world. There can be no one true causation.

That does not mean that causation does not exist, that there are no determining factors in the world. If one gives up the correspondence theory of truth and adopts the experientialist account of truth as based on embodied understanding, then there is a perfectly sensible view of causation to be given. We do not claim to know whether the world, in itself, contains "determining factors." But the world as we normally conceptualize it certainly does. Those determining factors consist in all the very different kinds of situations we call causal.

When we see or hypothesize a determining factor of some kind, we conceptualize it using one of our forms of causation, either literal or metaphorical. If metaphorical, we choose a metaphor with which to conceptualize the situation, preferably a metaphor whose logic is appropriate to the kind of determining factor noticed. Using that metaphor we can make claims about that determining factor. The claims can be "true" relative to our understanding, which itself may be literal or metaphorical.

This does not eliminate all problems of truth with respect to metaphor. It moves many of them to another place, but a more appropriate place. It leads us to ask, "When is a metaphorical conceptualization of a situation apt?" Is it an apt use of metaphor to apply the metaphor of Causal Paths to democracy in the arena of foreign policy? Only relative to a decision concerning the aptness of the metaphor can we draw conclusions on the basis of the Causal Paths metaphor.

The Aptness of Metaphor in Science

In the *Principles of Mathematics*, Bertrand Russell claims that "force is a mathematical fiction, not a physical entity. . . . In virtue of the philosophy of the calculus, acceleration is a mere mathematical limit, and does not express a definite state of an accelerated particle" (C2, 1903, 482).

As a classical scientific realist, Russell is technically right given his assumptions. Newton's laws, formulated mathematically, are equations—constraints. Acceleration is just a mathematical limit, not a physical entity; and so force, which is equal to mass times acceleration, is also not a physical entity. From the perspective of the laws of physics taken as literal truth, Russell argues correctly: Force does not exist. And if force is a fiction, causation cannot be otherwise.

Russell is arguing on the basis of the correspondence theory of truth, taking the mathematical formulation of the laws of physics as literally true. He argues that, on this basis, force cannot exist. The argument is impeccable, though his premises are not. We see a bit better what goes into cases like this by looking at Einstein's general theory of relativity.

In general relativity, as we mentioned in the previous chapter, time is conceptualized metaphorically as a spatial dimension. Gravitational force can then be conceptualized metaphorically in purely spatial terms as curvature in space-time. We have all seen the illustration of this idea with the common drawing of a rubber sheet with a heavy ball in the center stretching the sheet and curving its surface. When a marble is rolled "in a straight line" across the sheet, it follows the geodesic, which corresponds to a straight line on a curved surface. Accordingly, the path the marble takes over the curved surface looks curved to us, as if the ball at the center of the sheet had attracted it by force.

Similarly, Einstein argues, a body with a large mass imposes a curvature on four-dimensional space-time. What we call the force of gravity is "really" a curvature of space-time, and the "pull" of gravity on an object is no pull at all—it is just the object moving along the geodesic through a curved region of space-time.

What are we to make of this? If I knock over a book and it falls on the floor, do I say that there was no force that pulled it to the earth (as Newton had claimed), but rather that it was following a curvature in space-time?

From our perspective, Einstein created a useful metaphorical theory. By using the metaphor of Time As A Spatial Dimension, Einstein could then use the mathematics of Riemannian geometry. Given the Spatialization Of Time metaphor, Riemannian geometry allowed him further to conceptualize force metaphorically in terms of the curvature of space-time and hence create a beautiful mathematical unification, one that allowed the calculations of empirical consequences.

What exactly was proved when Einstein's theory was "confirmed"? Einstein's theory claimed that a large body like the sun should create a significant space-time curvature in its immediate vicinity. If a light ray passed near the sun, it should follow a curved path. This was seen as providing for a test of the gravitational-pull theory versus the space-time–curvature theory. It was assumed that light had no mass; hence there should be no "pull" and the light should travel in a straight line by the sun. But if space-time was curved near the sun, such a light ray should travel along a curved path, mass or no mass.

During an eclipse of the sun, the position was observed of a star that could not normally be seen next to the sun when it was shining. If space-time was curved, the light from the star should move in a curved path by the sun, and the star should appear shifted over a few degrees. The measurement was made during a 1919 eclipse, and Einstein's calculation of where the star should appear was verified. Einstein's theory was taken as confirmed—and interpreted literally: There is no force of gravity. What we've been calling that force is space-time curvature.

Einstein's theory need not have been interpreted literally. One could have said: Einstein has created a beautiful metaphorical system for doing calculations of the motion of light in a gravitational field. The metaphor of space as a temporal dimension allows him to use well-understood mathematics to do his calculations. That is a magnificent metaphorical accomplishment. But that doesn't mean we have to understand that theory as characterizing the objectively true nature of the universe.

What makes one rebel at the literal interpretation of general relativity is the implication that when someone drops a ball from the top of a tall building, the ball is not subject to any force of gravity, but rather is just moving along a geodesic in a curved region of space-time.

The literalist scientific response is that to deny that is like denying that the earth turns and claiming instead that the sun literally does rise and set and go around the earth.

Of course, any physicist will tell you that the last thing you want to do when calculating the trajectory of a ball dropped from a tall building is to use general relativity to make your calculation. For speeds much less than the speed of light, Newton's theory will work just as well for getting the numbers right. But that it not the point of philosophy! The question classical philosophy asks is one of ontology: Does gravitational force really exist as a force or is it really space-time curvature? That is why someone interested in ontology will want to ask about dropping a ball from a tall building.

It is not just the existence of the force of gravity that is at issue here. Suppose one makes the theoretical move of superstring theory, which does for all forces what Einstein's theory did for gravity. Suppose we look at superstring theory from the perspective of the system of Event-Structure metaphors.

> In the Object Event-Structure metaphor, attributes are conceptualized as things, as metaphorical possessed objects.
> In the Location Event-Structure metaphor, attributes are seen as "states," as metaphorical locations that a thing can be at or in.

In the traditional theory of elementary particles, particles are conceptualized in terms of the Object Event-Structure metaphor; particles were seen as possessing attributes like mass, charge, and spin. But the Object and Location Event-Structure metaphors are duals, differing by figure-ground reversal. Superstring theory makes the move of choosing the other dual, the Location Event-Structure metaphor. Elementary particles, via this metaphor, do not *possess* attributes; instead, they are locational, not things separate from the space they are "in," but aspects of space itself. Particles exist in a ten-dimensional space, the usual four of physical space and time and six others that are very small. They are not points in space, but very small "loops"—closed lines with a minuscule, but finite circumference. Hence, the name "string." A single loop, extending over a large region of time can be visualized as being like a long hose, with a small looplike cross-section and a much larger length. Each particle consists of not one loop, but many, each extending in one of six very small dimensions. The multiple loops are not still, but "vibrate"; the different harmonics of the strings correspond to different elementary particles. Notions like charge and spin become aspects of space, especially of the six tiny dimensions through which the strings loop.

Elementary particles can thus be conceptualized in purely geometric terms. They are aspects of space—multidimensional loops in the six tiny curling dimensions in terms of which charge, spin, and so on are represented. In the time dimension, they extend across all the points in time at which they exist. In the three spatial dimensions, they extend along all the locations they pass through over their "lifetimes." Thus they are "long" in these dimensions.

In superstring theory, all forces—gravitational, electromagnetic, and strong and weak nuclear forces—are conceptualized as curvatures in ten-dimensional space. What this does is allow the same mathematics, Riemannian geometry, to be used to calculate all of what we ordinarily call "forces." But of course, if one takes this theory literally, no forces at all exist as forces. What we used to conceptualize as forces are now all curvatures in ten-dimensional space. If we take superstring theory literally, no forces exist at all. And we live in a radically multidimensional universe, one with ten dimensions!

Do we "really" live in a world with ten or more dimensions, many of them very small, with no forces but lots of curvatures in multidimensional space? Or is superstring theory an ingenious and productive technical metaphor that allows all calculations of force to be unified using the same mathematics—Riemannian geometry?

These are not mutually exclusive alternatives. From the perspective of the everyday human conceptual system, superstring theory is metaphorical, as is general relativity, as is Newtonian mechanics. To take any of these theories literally is to say that force, and therefore causation, is nonexistent. But to take these scientific theories metaphorically is to allow for the "existence" of causes from our everyday perspective. Embodied realism allows both perspectives to count as "true" for the same person. Let us explain how both of these perspectives are possible at once.

The Experiential Stance and Embodied Standpoints

The study of human categorization has revealed that our conceptual system is organized around basic-level concepts, concepts that are defined relative to our ability to function optimally in our environment, given our bodies. Concepts of direct human agency—*pushing, pulling, hitting, throwing, lifting, giving, taking,* and so on—are among the basic-level anchors of our conceptual system in general and our system of causal concepts in particular.

We have no more fundamental way of comprehending the world than through our embodied, basic-level concepts and the basic-level experiences that they generalize over. Such basic concepts are fundamental not only to our literal conceptualization of the world but to our metaphorical conceptualization as well.

Our basic-level understanding, which makes use of basic-level concepts, is required for any account of truth at all. Suppose I lift a glass. My most fundamental understanding of such an action will involve a basic-level conceptualization in terms of the concept of lifting, which will in turn involve the general motor programs used in typical cases of lifting and a conceptualization of the spatial-relations concept *up*.

My lifting a glass can be understood from many perspectives. From the perspective of the subatomic level, there is no lifting and no glass. From the perspective of superstring theory, no force entity exists, only curvatures in multidimensional space. But from the human, experiential stance, the optimal way for me to conceptualize the situation, given my normal purposes, is in terms of the basic-level concepts *lift* and *glass*. Lifting an object directly involves the direct application of "force." From this perspective, given the understanding I naturally project onto such a situation, "force" exists. From the standpoint of the human conceptual system in the cognitive unconscious, there is a concept of causation with human agency as the central prototype. From the ordinary human standpoint, force exists and causation exists, and lifting a glass is an instance of both the exertion of force and of causation.

Our conceptual systems also contain metaphorical concepts, as we have seen. Philosophical and scientific theories often make use of those metaphorical concepts. Moreover, our fundamental metaphorical concepts are not arbitrary, subjective, or even for the most part culturally determined. They are largely embodied, having a basis in our embodied experience. Even the most abstruse scientific theories, like general relativity and superstring theory, make use of such fundamental embodied metaphors as the Time As Spatial Dimension metaphor and the Location Event-Structure metaphor.

One important thing that cognitive science has revealed clearly is that we have multiple conceptual means for understanding and thinking about situations. What we take as "true" depends on how we conceptualize the situation at hand. From the perspective of our ordinary visual experience, the sun does rise; it does move up from behind the horizon. From the perspective of our scientific knowledge, it does not.

Similarly, when we lift an object, we experience ourselves exerting a force to overcome a force pulling the object down. From the standpoint of our basic-level experience, the force of gravity does exist, no matter what the general theory of relativity says. But if we are physicists concerned with calculating how light will move in the presence of a large mass, then it is advantageous to take the perspective of general relativity, in which there is no gravitational force.

It is not that one is objectively true while the other is not. *Both* are human perspectives. One, the nonscientific one, is literal relative to human, body-based conceptual systems. The other, the scientific one, is metaphorical relative to human, body-based conceptual systems. From the metaphorical scientific perspective of general relativity and superstring theory, gravitational force does not exist as an entity—instead it is space-time curvature. From the literal, nonscientific perspective, forces exist.

Now, if we take one scientific theory or another as being literally true, and if we insist that there is only one truth and it is the best scientific truth we have, then, as Russell observed, force does not exist, and so neither does causation. If, however, we can allow scientific theories to be recognized for the metaphorical conceptual structures that they are for human beings, then we can allow multiple ways of conceptualizing the world, including both the scientific and nonscientific. Allowing for the multiple perspectives indicated by cognitive analyses allows us to maintain both scientific perspectives, in which causation doesn't exist, and our everyday perspective, in which it does.

Causation and Realism: Does Causation Exist?

When someone asks, "Does causation exist?" that person usually wants to know whether there is a single unified phenomenon (which is called "causation") objectively existing in the mind-independent world and operating according to a single logic. Furthermore, he or she assumes that there is a straightforward simple yes-or-no answer. As we have seen, the situation is more complex than that.

But the presuppositions lying behind this apparently simple question are massively false. First, *causation* is a word in a human language and it designates a human category, a radial category of extraordinary complexity. In that complex radial category, there is no set of necessary and sufficient conditions that covers all the cases of causation. Therefore, causation as we conceptualize it is not a unified phenomenon. It does not simply designate an objectively existing category of phenomena, defined by necessary and sufficient conditions and operating with a single logic in the mind-independent world. Because the presuppositions lying behind the question are so far off base, the question has no simple straightforward answer.

This eliminates a simpleminded realism that assumes that our language is simply a reflection of the mind-independent world, and hence that such questions are simple and straightforward. But eliminating simpleminded realism does not eliminate all forms of realism, and it does not require either idealism or total relativism.

What remains is an embodied realism that recognizes that human language and thought are structured by, and bound to, embodied experience. In the case of physics, there is certainly a mind-independent world. But in order to conceptualize and describe it, we must use embodied human concepts and human language. Certain of those embodied human concepts, the basic-level ones, accord very well with middle-level physical experience and therefore have an epistemic priority for us. It is here that we feel comfortable saying that causation exists for ordinary cases of the direct application of physical force in our everyday lives. The central prototypical case in our basic-level experience gives us no problem in answering the question. He punched me in the arm. He caused me pain. Yes, causation exists.

The question is, however, problematic just about everywhere else, because we are moving away from the central prototypical case of causation to other, very different senses with different logics and different criteria for determining what is true. The question is not so simple for causal paths, causal links, and so on. These cases require an *embodied* correspondence theory of truth, where embodied conceptualizations of the situation, metaphorical and nonmetaphorical, are taken into account. In such cases, causation exists or doesn't depending both on the world and on our conceptualization of it.

Beyond middle-level physical experience—in the microuniverse of elementary particles and the macrouniverse of black holes—our basic-level concepts utterly fail us. To conceptualize such experience requires the magnificent tool of conceptual metaphor. But once we move to the domain of conceptual metaphor in theorizing about the micro and macro levels, any ordinary everyday literal notion of causation fails us. When our theories are metaphorical and contain no concept of causation, we answer the question of whether causation exists depending on how literally we take our theories.

In short, the question "Does causation exist?" is not a simpleminded yes-or-no question. It drastically oversimplifies something that we have seen is massively complex.

12

The Mind

It is virtually impossible to think or talk about the mind in any serious way without conceptualizing it metaphorically. Whenever we conceptualize aspects of mind in terms of grasping ideas, reaching conclusions, being unclear, or swallowing a claim, we are using metaphor to make sense of what we do with our minds.

Our system of metaphors for mind is so extensive that we could not possibly describe it all here. We will include only enough of the system to draw some important philosophical conclusions. Our results will parallel what we found in the case of time, causation, and events: First, the metaphor system conceptualizing thought itself does not give us a single, overall, consistent understanding of mental life. Instead, it provides us with conceptual metaphors that are inconsistent with each other. Second, the metaphor system for the mind provides the raw material for commonplace philosophical theories.

The Mind as Body System

Eve Sweetser (A1, 1990) has shown that there is an extensive subsystem of metaphors for mind in which the mind is conceptualized as a body. The main outline of this subsystem can be seen in the following general mapping:

The Mind Is A Body

Thinking Is Physical Functioning
Ideas Are Entities With An Independent Existence

> Thinking Of An Idea Is Functioning Physically With
>   Respect To An Independently Existing Entity

There are four extensive special cases of this metaphor, with thinking conceptualized as four different kinds of physical functioning: moving, perceiving, manipulating objects, and eating. Let us consider these one at a time.

Thinking Is Moving

One of our major ways of getting information is by moving around in the world. This is the basis for our metaphor Thinking Is Moving, which consists of the following mapping:

> The Mind Is A Body
> Thinking Is Moving
> Ideas Are Locations
> Reason Is A Force
> Rational Thought Is Motion That Is Direct, Deliberate,
>   Step-By-Step, And In Accord With The Force Of Reason
> Being Unable To Think Is Being Unable To Move
> A Line Of Thought Is A Path
> Thinking About X Is Moving In The Area Around X
> Communicating Is Guiding
> Understanding Is Following
> Rethinking Is Going Over The Path Again

That thought can be conceptualized as motion is clear from sentences like "My mind was *racing*," "My mind *wandered* for a moment," and "Harry kept *going off on flights of fancy*." That ideas are locations is clear from sentences like "How did you *reach that conclusion*?" "We have *arrived at the crucial point* in the argument," and "*Where* are you in the discussion?" Note that locations exist independently of the people located at them. Being unable to think is, of course, being unable to move: "I'm *stuck*! I can't *go any farther along this line of reasoning*."

Consider now what it means to be *forced to a conclusion*. Suppose you have reached a certain point in your reasoning. To be forced to a conclusion is to arrive at another idea-location by thinking rationally, whether you want to or not. Reason is thus seen as a strong and typically overwhelming force moving the thinker from one idea-location to another. Reason is very much like a force of nature; any conclusion that you are forced to by reason is a natural conclusion. To refuse to reach a natural conclusion is to resist the force of reason, to be unreasonable or irrational. The force of reason is thus conceptualized as leading one along a certain line of thought. That is why we speak of *being led to* a conclusion. When you are led to a conclusion, you are thinking in accordance with reason, that is, along the lines that reason forces you to think along to the place where reason takes you. When you are led to a conclusion, you are therefore not personally responsible for reaching that conclusion. What is responsible is not you, but the natural force of reason itself.

To think rationally is, first, to think along the lines required by the force of reason. Second, it is to *go step-by-step*, not *skipping any steps,* so that you are sure to be on the path required by reason. Third, it is to move toward a conclusion as directly as possible, without thinking *in circles* or *going off on a tangent* or *wandering away* from the point.

The word *topic*, incidentally, is etymologically derived from the Greek *topos*, meaning "a place." To think about a certain topic is metaphorically to move in the vicinity of a certain place. Thus, we can speak of *returning* to the topic, *straying away from* the topic, and *approaching* a topic.

Communicating is conceptualized as guiding, and understanding what is being communicated is following someone who is guiding you through a terrain. Thus, when we can no longer understand someone, we can say things like: *"Slow down."* "You're *going too fast* for me." "I can't *follow* you." "I can't *keep up with* you." "*Where are you going* with this?" "Can you *go over* that again?"

The metaphor Thinking Is Moving is the basis of our notion of final causes, in which purposes are conceptualized as causes of the actions performed to achieve those purposes. The metaphors Causes Are Forces and Actions Are Locations combine with the metaphor Reason Is A Force to yield the complex metaphor of Rational Causation. Suppose I know that result *B* can be brought about only by performing action *A*. Suppose I want *B* to occur. Then, *the force of reason moves me to* perform *A* and hence achieve result *B*. My desire for *B* is a rational cause of my performing *A*. These are some cases of what Aristotle called *final causes*. The other cases will be discussed in Chapter 18, where we analyze Aristotle's basic metaphors.

To sum up, we have been taking you through the first special case of the metaphor composed of The Mind Is A Body and Thinking Is Physical Functioning. In this case, the type of functioning is moving. Let us now *move on to* the next special case, in which, as we shall *see*, the appropriate type of bodily functioning is perception.

Thinking Is Perceiving

We get most of our knowledge through vision. This most common of everyday experiences leads us to conceptualize knowing as seeing. Similarly, other concepts related to knowing are conceptualized in terms of corresponding concepts related to seeing. In general, we take an important part of our logic of knowledge from our logic of vision. Here is the mapping that projects our logic of vision onto our logic of knowledge.

> The Mind Is A Body
> Thinking Is Perceiving
> Ideas Are Things Perceived
> Knowing Is Seeing
> Communicating Is Showing
> Attempting To Gain Knowledge Is Searching
> Becoming Aware Is Noticing
> An Aid To Knowing Is A Light Source
> Being Able To Know Is Being Able To See
> Being Ignorant Is Being Unable To See
> Impediments To Knowledge Are Impediments To Vision
> Deception Is Purposefully Impeding Vision
> Knowing From A "Perspective" Is Seeing From A Point Of View
> Explaining In Detail Is Drawing A Picture
> Directing Attention Is Pointing
> Paying Attention Is Looking At
> Being Receptive Is Hearing
> Taking Seriously Is Listening
> Sensing Is Smelling
> Emotional Reaction Is Feeling
> Personal Preference Is Taste

This is an extraordinarily common metaphor. When we say "I *see* what you're saying," we are expressing successful communication. A *cover-up* is an attempt to *hide* something, to keep people from knowing about it. To deceive people is to *pull the wool over their eyes, put up a smokescreen,* or *cloud the issue. Clear* writing is writing that allows readers to know what is being communicated; *unclear* or *murky* writing makes it harder for readers to know what is being said.

An attempt to gain knowledge of something is conceptualized as *looking* or *searching* for it, and gaining knowledge is conceptualized as *discovering* or *finding.* Someone who is ignorant is *in the dark,* while someone who is incapable of knowing is *blind.* To enable people to know something is to *shed light* on the matter. Something that enables you to know something is *enlightening;* it is something *that enables you to see.* New facts that have *come to light* are facts that have become known (to those who are looking).

When we speak of someone who *has blinders on,* who can only *see what's in front of his nose,* we mean someone whose focus of attention narrows the range of what he can think about and makes it impossible for him to see certain things. When we speak of *pointing something out* so that you can *see* it, we mean we are directing your attention to something so that you can have knowledge of it. If someone says to you "Do I have to *draw you a picture?*" that person is asking if he or she has to explain something in detail. If I understand, then I *get the picture.*

The notion of a perspective, angle, viewpoint, or standpoint derives from this metaphor. When you are looking at a scene, you have to be looking at it from some location. From a given location, you can only see certain things. If you are far away, small details may be invisible. Some things may be hidden from your view. The implication is that you can know a scene better by taking many viewpoints. Metaphorically, someone who has only one perspective on the world may be ignorant of things that are hidden from that perspective. Closeness matters as well. To know something, you need to be close enough to see the details, but not so close that you can't make out the overall shape of things. You don't want to be someone who *can't see the forest for the trees.*

Since vision plays such a dominant role in our ability to gain knowledge, the Thinking Is Perceiving metaphor is, for the most part, concerned with vision. However, the other senses also play a lesser role. Consider what it means metaphorically to listen to someone, as in "I always *listen to* what my father tells me." That means that, besides attending to his words, you take seriously what he is communicating. *Being deaf to what your father tells you* means not being receptive to the content of what he is saying—you don't *hear* what he is trying to communicate.

When you say that "Something doesn't *smell* quite right here," you are suggesting that you mentally sense that something is out of order. When you *feel strongly* that you are right, you are combining both a mental sense with a strong emotional reaction. The sense of taste is used to convey personal preference. A *sweet thought* is one that you like, while a *bitter* thought is one that you have a negative attitude toward.

Thinking Is Object Manipulation

Another way in which we get information is by examining objects and manipulating them. This forms the basis of another major metaphor for thinking.

> The Mind Is A Body
> Thinking Is Object Manipulation
> Ideas Are Manipulable Objects
> Communicating Is Sending
> Understanding Is Grasping
> Inability To Understand Is Inability To Grasp
> Memory Is A Storehouse
> Remembering Is Retrieval (Or Recall)
> The Structure Of An Idea Is The Structure Of An Object
> Analyzing Ideas Is Taking Apart Objects

In this metaphor, ideas are objects that you can *play with, toss around,* or *turn over in your mind.* To understand an idea is to *grasp* it, to *get* it, to *have it firmly in mind.* Communication is *exchanging ideas.* Thus, you can *give* someone ideas and *get ideas across* to people. Teaching is *putting ideas into* the minds of students, *cramming* their heads full of *ideas.* To fail to understand is to fail to grasp, as when an idea *goes over your head* or *right past you.* Problems with understanding may arise when an idea is *slippery,* when someone *throws too many things at you at once,* or when someone *throws you a curve.* When a subject matter is too difficult for you to understand, it is seen as being *beyond your grasp.*

Just as objects have a physical structure, so ideas have a conceptual structure. You can *put ideas together* to form complex ideas. Complex ideas can be *crafted, fashioned, shaped,* and *reshaped.* There can be many *sides* to an issue. Analyzing ideas is taking them apart so that you can see their component ideas.

This metaphor combines with Knowing Is Seeing, so that we can *turn an idea over* in our heads to *see all sides* of it. We can *hold the idea up* to scrutiny or *put the idea under a microscope*.

The three mappings we have just discussed occur in languages throughout the world. The next four we will discuss are much less widespread.

Acquiring Ideas Is Eating

So far, we have seen three ways in which the mind is conceptualized in bodily terms. In these, thinking is seen as bodily functioning—as moving, perceiving, and manipulating objects. The central concerns of those metaphors were gaining knowledge, reasoning, comprehending, and communicating. We now turn to a very different metaphor in which the mind is again conceptualized in terms of the body, but here the concern is a well-functioning mind, which is conceptualized as a healthy body. Just as a body needs the right kind of food—healthful, nutritious, and appetizing—so the mind needs the right kind of ideas, ideas that are true, helpful, and interesting. Here is the mapping:

A Well-Functioning Mind Is A Healthy Body
Ideas Are Food
Acquiring Ideas Is Eating
   Interest In Ideas Is Appetite For Food
   Good Ideas Are Healthful Foods
   Helpful Ideas Are Nutritious Foods
   Bad Ideas Are Harmful Foods
   Disturbing Ideas Are Disgusting Foods
   Interesting, Pleasurable Ideas Are Appetizing Foods
   Uninteresting Ideas Are Flavorless Foods
Testing The Nature Of Ideas Is Smelling Or Tasting
Considering Is Chewing
Accepting Is Swallowing
Fully Comprehending Is Digesting
Ideas That Are Incomprehensible Are Indigestible
Preparing Ideas To Be Understood Is Food Preparation
Communicating Is Feeding
Substantial Ideas Are Meat

An interest in ideas is conceptualized as an appetite for food, as in having *a thirst for knowledge,* an *appetite for learning,* and *an insatiable curiosity.* But you don't want to *swallow,* that is, accept, the wrong kinds of ideas. The right kinds of ideas are necessary, and we have to test for them. The wrong kinds of ideas are bad in some way: false, disturbing, uninteresting, or incomprehensible. They are conceptualized as food that is unhealthful, disgusting, bland, or indigestible. Since you often don't fully process the consequences of ideas that you accept, testing for their acceptability ahead of time is crucial, and this is largely what this metaphor is about.

Metaphorically you test by smell and taste. If an idea *smells fishy* or *stinks,* it is judged to be unhealthy, that is, false. *Raw facts* are not suitable because they are not digestible; that is, they have not been prepared so that they can be fully comprehended. The same for *half-baked ideas. Warmed-over theories* are leftovers, old ideas not as *palatable* as they once were, presented as new. *Rotten* ideas are unhealthful and therefore unhelpful; a rotten idea is not likely to work. *Fresh ideas* are more likely to be appetizing, that is, interesting. *Bland* ideas are just uninteresting. *Disgusting* or *unsavory* ideas—ideas that *make you want to puke* or *make you sick*—are disturbing and not acceptable for a well-functioning mind. A common special case of a substance that is both disgusting and unhealthy to eat is shit. Thus, the word *shit* can be used to indicate untruthful ideas. Hence, we have expressions like "That's a lot of *shit,*" "That's *bullshit,*" "Don't *bullshit* me," and "You're not *shittin'* me are you?"

Digestion in this metaphor is the full mental processing required for full understanding. Some ideas need further preparation to be digestible; those are ideas that have to be *put on the back burner* or *stewed over.* An idea that cannot be immediately comprehended is one you have to *chew on* for a while. Chewing also gives you an opportunity to check its taste.

Then there are ideas that an eater won't want to accept, ideas that have to be *sugar-coated* or *forced down* his or her *throat.* Thus, the metaphor presents criteria for the acceptability of an idea—it has to smell good, be appetizing, be cooked properly, and be chewed thoroughly. Only then should it be accepted— *swallowed.* When you take in unacceptable thoughts, ideas you shouldn't have swallowed, they can leave *a bad taste in your mouth.* When you get a taste of an unsavory thought, you say *"Yuck!"*

Sometimes ideas are incomprehensible because there is just too much content in them: "There's *too much here* for me *to digest.*" If you want to make sure that people's ability to digest information isn't overloaded, you can *spoon-feed them.* If you want them to accept ideas they might find disturbing, you might *sugar-coat them.* And if you want to deceive them, to get them to accept an idea they would reject if they could chew on it and taste it, then you need to get them to *swallow it whole.* Someone who is gullible is someone who swallows ideas whole.

It should now be clear why we speak of an idea you can *really bite into,* a *meaty idea,* as opposed to just *chewing the fat.* An idea that is substantial is conceptualized as meaty. It is an idea that seen as nutritious, that is, metaphorically helpful. Fat is not nutritious, though it can be tasty. A pleasant conversation in which you don't get any helpful ideas, one where the point of it is not to get ideas but just pass the time pleasantly, is *chewing the fat, fat* because the ideas are not substantial or helpful like meaty ones and *chewing* because you don't ingest, or swallow, any ideas. In addition, there is an image metaphor in *chewing the fat,* in which moving the mouth while talking is imaged as a form of chewing.

And what about *regurgitating* ideas on the final exam? Those are ideas that are not really swallowed or digested—that is, not accepted or comprehended, just *chewed on,* that is, considered. *Food for thought* constitutes appropriate ideas for mental eating, healthful, nutritious, and appetizing—that is, good, helpful, and interesting.

Actually this metaphor is even more complex, and we will discuss the added complexity only in passing. If one is conceptualizing a well-functioning mind as a healthy body, one cannot ignore the issue of mental exercise. Education is not just a matter of feeding students true and helpful ideas. It is also a matter of giving them rigorous mental training to develop *powerful* minds. That is why problems in textbooks are called *exercises.*

The Homunculus and Fregean Intensions

The four metaphors we have examined so far are ordinary, everyday metaphors for the mind. In each one, the mind is conceptualized in bodily terms, as if the mind were a separate person with its own bodily functions: moving, perceiving, manipulating objects, and eating. In short, the philosophical idea of the mind as a homunculus arises from our everyday metaphor system.

These conventional metaphors also contain within them another important philosophical idea. In each of these metaphors, ideas are metaphorical entities that exist independently of the thinker: locations, objects, and food. In addition, each metaphor for ideas is a correspondence between ideas and things in the world: locations, objects, and food.

In these respects, these metaphors all have two of the crucial properties of Fregean senses, or intensions. Ideas in these metaphors have a public, objective existence independent of any thinker. And by virtue of each metaphor, there is a correspondence between ideas and things in the world.

We are not saying that ideas as characterized by these metaphors are Fregean senses. They certainly aren't. For the present, we simply note that they share two crucial properties with Fregean intensions, and they might explain why many philosophers regard Frege's view as "intuitive." Intuitive theories tend to use ideas we already have.

Metaphors for the Mind and the Linguistic Turn in Philosophy

We turn next to three everyday metaphors for mind that have played an important role in defining the approaches to mind and language characteristic of much Anglo-American analytic philosophy. First, there is a metaphor in which thought is conceptualized as language.

THE THOUGHT AS LANGUAGE METAPHOR

Thinking Is Linguistic Activity (Speaking Or Writing)
Simple Ideas Are Words
Complex Ideas Are Sentences
Fully Communicating A Sequence Of Thought Is Spelling
Memorization Is Writing

When we say "Let me make a *mental note* of that," we are using a metaphor in which thoughts are linguistic forms *written* in the mind. Similarly, sentences like "*She's an open book* to me," "I can *read her mind,*" "I *misread his intentions,*" and "She has a whole *catalogue of great ideas* for gardening" use the same metaphor, that thoughts are written linguistic expressions. An important entailment of this metaphor is that, if you can *read* someone's mind, then all their thoughts must be in readable linguistic form.

We also have other metaphorical expressions in which thoughts are conceptualized as *spoken* language, as in:

I can barely *hear myself think*. He's an *articulate thinker*. She doesn't *listen to her conscience. Her conscience told* her not to do it. I don't like *the sound of his ideas.* That *sounds like a good idea.*

We see the Thought As Language metaphor in many cases:

It's *Greek* to me. Liberals and conservatives don't *speak the same language*. She can't *translate* her ideas into well-defined plans. His thoughts are *eloquent*. What is the *vocabulary* of basic philosophical ideas? The argument is *abbreviated*. He's *reading between the lines*. He's *computer literate*. *I wouldn't read too much into* what he says.

In the written language version of this metaphor, the notion of spelling is important. When you are spelling you are carefully communicating the structure of the written word in a deliberate step-by-step fashion. This maps onto carefully communicating the structure of the thought in a deliberate step-by-step fashion, as in sentences like "The theory is *spelled out* in chapter 4" and "Do I have to *spell it out* for you?" the latter said of any thought of some (even minimal) complexity. Just as a single letter is one detail in the structure of the word, so in a sentence like "Follow the *letter* of the law" each *letter* is metaphorical for a detail of the conceptual structure of the law.

Punctuation also enters into this metaphor. A punctuation mark stands both metonymically and metaphorically for the meaning of that punctuation mark. For example, a question mark indicates something unknown, as in "He's *a big question mark* to me." Because a period indicates the end of a sentence, the word *period* in this metaphor indicates the end of what is to be communicated, as in "Be home by midnight—*period!*" Grammatical morphemes are included in this metaphor as well: "I want this done—no *ifs, ands,* or *buts!*"

What this metaphor does is conceptualize thought in terms of symbols, as if a thought were a sequence of written letters. It makes the internal, private character of thought into a public, external thing. It has an important entailment, namely, that thought has a structure that can be represented accurately in terms of linear sequences of written letters. Bear in mind these four aspects of this common metaphor:

1. Thought has the properties of language.
2. Thought is external and public.
3. The structure of thought is accurately representable as a linear sequence of written symbols.

4. Every thought corresponds to a linguistic expression; and hence, every thought is expressible in language.

Now let us turn to another common metaphor of philosophical importance.

THE THOUGHT AS MATHEMATICAL CALCULATION METAPHOR

Reasoning Is Adding
Ideas Considered In Reasoning Are Figures Counted
  In Adding
Inferences Are Sums
An Explanation Is An Accounting

In "I *put* two and two *together*," thinking is conceptualized as adding. Performing a simpleminded, obvious inference is conceptualized as doing just about the most simpleminded, obvious case of addition. We can also see inference being conceptualized as a sum in "What does it all *add up to?*" which expresses puzzlement over what one should infer from a collection of ideas. Similarly, one can express frustration at being unable to make inferential sense of all the information at one's disposal by saying, "It doesn't *add up*," which expresses frustration at not getting a sensible sum from the addition of a given list of figures. Again, this conceptual metaphor shows up in sentences like "What's the *bottom line?*" in which one is asking what one should infer from all the information at hand.

The notion of counting is, of course, central to this metaphor. Counting assumes that one has decided what to count, that is, what to include in the addition. Via this metaphor, the word *count* indicates what information should be included in one's reasoning, as in sentences like "I don't know if we should *count* that" and "That doesn't *count*." Similarly, when one *counts on* someone, one includes a dependence on a person in one's reasoning.

Adding long lists of figures is important because of the tradition of accounting. Accounting is a form of explanation in which you say why you have the funds you have on hand in terms of adding up credits and debits. In this metaphor, in which inference is addition, inferential explanation is a form of an accounting, as in "Give me an *account* of why that happened."

The idea that arithmetic is the ideal form of reasoning goes back at least as far as the ancient Greeks. This metaphor conceptualizes reason itself in terms of mathematical calculation. The metaphor has important entailments:

- Just as numbers can be accurately represented by sequences of written symbols, so thoughts can adequately represented by sequences of written symbols.
- Just as mathematical calculation is mechanical, so thought is also.
- Just as there are systematic universal principles of mathematical calculation that work step-by-step, so there are systematic universal principles of reason that work step-by-step.

Finally, there is a common metaphor that conceptualizes the mind as a machinelike mechanical system.

THE MIND AS MACHINE METAPHOR

The Mind Is A Machine
Ideas Are Products Of The Machine
Thinking Is The Automated Step-By-Step Assembly Of Thoughts.
Normal Thought Is The Normal Operation Of The Machine
Inability To Think Is A Failure Of The Machine To Function

When we speak of someone as really *turning out ideas* at a great rate we are conceptualizing the mind as a machine and ideas as products. In this metaphor, sentences like "Boy, the *wheels are turning* now" indicate that the thinker is producing a lot of thoughts. "I'm a little *rusty*" is an explanation for why the thinker is slow at turning out thoughts. And "He had a *mental breakdown*" indicates that the thinker is no longer capable of productive thought.

The entailments of this metaphor are that thoughts are produced by the mind in a regular, describable, mechanical, step-by-step fashion and that each thought has a structure imposed by the operation of the mind.

No Consistent Conception of Mind

The metaphors we have just given are absolutely central to our conception of what ideas are and what rational thought is. Would ideas be ideas if we could not grasp them, look at them carefully, or take them apart? What would reason be if we could not reach conclusions, or go step by step, or come directly to the point? Would ideas be ideas if you couldn't let them simmer for a while, spoon-feed them, sugar-coat them, or digest them? Would thinking be the same if you could not make mental notes, translate your vague ideas into plans, sum up an argument, or crank out ideas? We think not. These metaphors define our conceptualization of what ideas are and what thinking is.

But the metaphors are not all consistent. It is not consistent to conceptualize ideas as both locations you can be at and objects you can manipulate or transfer. Nor is it consistent to conceptualize thinking as both motion and vision. Nor it is consistent to conceptualize ideas as objects you manufacture and food you consume. Moreover, the entailments that ideas are produced by thinking and exist independently of thinking are inconsistent. Such inconsistency across different metaphors is normal in human conceptual systems.

We have no single, consistent, univocal set of nonmetaphoric concepts for mental operations and ideas. Independent of these metaphors, we have no conception of how the mind works. Even the notion *works* derives from the Mind As Machine metaphor. Even to get some *grasp* of what ideas in themselves might be, we have to conceptualize ideas as graspable objects. To approach the study of ideas from any intuitive point of view is to use metaphors for ideas that we already have. What a theory of mind or a theory of ideas must do is pick a consistent subset of the entailments of these metaphors. In so doing, any consistent theory will necessarily leave behind other entailments, inconsistent with these, that are also "intuitive." Each such theory is metaphorically intuitive and consistent, but not comprehensive. It appears that there is no comprehensive and consistent theory that is also metaphorically intuitive—that is, made up of entailments of the above metaphors.

Metaphors for Mind and
Anglo-American Analytic Philosophy

Anglo-American analytic philosophy is based on technical versions of the metaphors for mind and thought that we have just analyzed. To see how these metaphors fit together, we need to summarize the entailments of the various everyday metaphors for mind that we have been discussing.

As we have seen, the entailments of the metaphors mentioned above include the following:

THE MIND AS BODY

1. Thoughts have a public, objective existence independent of any thinker.
2. Thoughts correspond to things in the world.

THOUGHT AS MOTION

3. Rational thought is direct, deliberate, and step-by-step.

THOUGHT AS OBJECT MANIPULATION

4. Thinking is object manipulation.
5. Thoughts are objective. Hence, they are the same for everyone; that is, they are universal.
6. Communicating is sending.
7. The structure of a thought is the structure of an object.
8. Analyzing thoughts is taking apart objects.

THOUGHT AS LANGUAGE

9. Thought has the properties of language.
10. Thought is external and public.
11. The structure of thought is accurately representable as a linear sequence of written symbols of the sort that constitute a written language.
12. Every thought is expressible in language.

THOUGHT AS MATHEMATICAL CALCULATION

13. Just as numbers can be accurately represented by sequences of written symbols, so thoughts can be adequately represented by sequences of written symbols.
14. Just as mathematical calculation is mechanical (i.e., algorithmic), so thought is also.
15. Just as there are systematic universal principles of mathematical calculation that work step-by-step, so there are systematic universal principles of reason that work step-by-step.
16. Just as numbers and mathematics are universal, so thoughts and reason are universal.

THE MIND AS MACHINE

17. Each complex thought has a structure imposed by mechanically putting together simple thoughts in a regular, describable, step-by-step fashion.

As one might expect, the Thought As Language metaphor plays a central role in the practice of analytic philosophy. The sentence is a unique complex composition made up of words in a particular order. In the metaphor, simple ideas are words and complex ideas are sentences. The metaphor therefore entails that any complex idea is made up of a unique combination of simple ideas. Therefore, there is one and only one correct analysis of a complex concept into its ultimate conceptual parts. In analytic philosophy, such an analysis is a definition of the concept: It provides necessary and sufficient conditions for the constitutive parts of the concept to constitute the whole concept. Thus, for every concept X, there is a correct theory of the one true X—an objectively correct account of the unique internal structure of the concept X. The theory of the one true causation, which we discussed above, is a special case. Much of classical analytic philosophy (not counting Wittgenstein's later work) is concerned with definitions of this sort that are to constitute theories of the one true X, for some concept X.

Readers familiar with contemporary Anglo-American philosophy of language and mind will find many of the above entailments familiar. Michael Dummett, in *The Origins of Analytic Philosophy* (C2, 1993), points out that Frege distinguished between "ideas" and "senses." Frege saw ideas as psychological, subjective, and private—essentially incommunicable and hence not a part of the public, shared meanings that are communicated through language. He took senses, which go together to make up thoughts, not to have anything to do with human psychology, to be free of the subjective. Senses and thoughts, being nonpsychological, public, objective, and communicable, were capable of being the meanings of linguistic expressions. This distinction is what Dummett refers to as "the extrusion of thoughts from the mind." It lies behind virtually all of Anglo-American philosophy of language. Thoughts, freed from the mind, are objective; they are characterizable in terms of direct correspondences to things in the world.

This Fregean view of thoughts corresponds to entailments 1 and 2 above. It is this Fregean notion of a thought as objective and corresponding to things in the world that gives rise to the correspondence theory of truth as it occurs in Anglo-American philosophy of language.

These Fregean notions plus the entailments of the metaphors of Thought As Language and Thought As Object Manipulation define the "linguistic turn" made by analytic philosophy. There, the Thought As Language metaphor is taken literally. Thought is seen as having the properties of language: external, public, representable in written symbols, communicable. As such, thoughts are objects; they are objective, universal, able to be sent to others, and, most important of all, can be "analyzed" into their parts. These notions correspond to entailments 5 through 11 above.

The notion that reason is characterizable as mathematical logic can also be seen as making use of the above entailments. Mathematical logic begins by assuming that one can adequately represent thoughts by sequences of written symbols of the sort one finds in written language. The Thought As Language metaphor, and especially entailment 11, are the correlates in everyday metaphor of the expert metaphor of a "logical language." The notion that reasoning can be seen as a form of mathematics has its correlate in the everyday metaphor of Thought As Mathematical Calculation, especially in its entailments 13 through 16, which conceptualize reason as a universal form of mechanical calculation using sequences of written symbols.

Here we have, in our everyday metaphorical conception of thought, the basis of the notion that thought can be represented by a logical language, with reason as mathematical calculation and the meaning of the logical language given by correspondence with things in the world. This Language of Thought metaphor, with certain variations, constitutes the major worldview of Anglo-American philosophy. That is, we have, in the collective entailments of our everyday metaphors for mind, the basis of analytic philosophy from Russell and Carnap through Quine, Davidson, Montague, and Fodor.

The same assumptions formed the basis, in linguistics, of the generative semantics of Lakoff and McCawley, in which logical forms were taken to be the underlying structures in transformational derivations within a type of transformational grammar. From generative semantics, there developed Jerry Fodor's Language of Thought theory of mind.

Those same assumptions lie behind the idea of artificial intelligence (AI). Classical AI assumed that thoughts can all be adequately expressed in a logical language (a computer "language" like LISP) and that reason is a matter of mechanical calculation and proceeds in a step-by-step fashion. Given these assumptions, it followed that computers could "think rationally." This view was the basis for the metaphor for mind assumed in first-generation cognitive science, The Mind Is A Computer, in which a "computer" is understood in the following way:

A computer is a machine that reasons via mathematical computations using a language whose expressions are objects that are manipulated; it communicates by sending and remembers by storing.

The Mind As Computer metaphor combines various entailments of other metaphors: The Mind Is A Machine, Thought Is Mathematical Calculation, Thought Is Language, Thought Is Movement, and Thought Is Object Manipulation. A "computer" in this metaphor is an abstract device, software not hardware; the mind is the software, the brain is hardware. In Chapter 19, we will discuss the Society of Mind metaphor for faculty psychology that was popular in the Enlightenment. Marvin Minsky has added this metaphor to the Mind As Computer metaphor in his book *Society of Mind* (E, 1986), in which he argues that the computer program of the mind is broken down into subprograms with specialized functions. Daniel Dennett's computational theory of mind (C2, 1991) makes use of this metaphor.

Even an anti-AI philosophy of mind such as John Searle's uses many of the above metaphorical entailments. Searle's philosophy of mind, so far as we can tell, makes use of the following of the above metaphorical entailments: 1 through 3 and 5 through 12. Searle, in his Chinese Room Argument, rails against the version of the Mind As Computer metaphor that does not contain entailment 2 above, the direct link between words and the world that Frege, and following him Searle, assumes is necessary for ideas to be meaningful. But aside from this, Searle accepts many of the same tenets that go to make up the Mind As Computer metaphor.

We do not in any way want to give the impression that these theories are all the same. They differ from one another in many ways. Some of those ways can be expressed in terms of which of the above metaphorical entailments a given theory uses; some cannot. Here are some questions to which Anglo-American philosophical theories give different answers:

> Are the entailments of the Thought As Mathematical Calculation metaphor to be accepted?
> Can rational discourse take place in "ordinary language"? Or is ordinary language to be rejected as illogical and an ideal mathematicized logical language substituted in order to carry on rational discourse?
> Is meaning to be characterized in terms of reference and truth, via entailment 2 and the correspondence theory of truth?

Russell, the early Wittgenstein, Carnap, and Quine all insisted that ordinary language was too ambiguous and vague for rational philosophical and scientific discourse. They accepted all of the Mathematical Calculation entailments, 13 through 16, and assumed that only a logical language of thought formulated within mathematical logic (a "regimented" language, in Quine's terminology) would be adequate for rational discourse. They all assumed that meaning was to be characterized in terms of reference, that is, in terms of the correspondence theory of truth, accepting entailment 2.

In short, they all answered yes to the first question, no to the second, and yes to the third. Here they differed with the later Wittgenstein, Austin, and Strawson, who accepted ordinary language and rejected the need for a mathematical logical language.

A third position was taken by Lakoff and McCawley in their generative semantics, by Montague, and by Fodor. They all answered yes to the first and third questions and accepted the adequacy of ordinary language. They assumed linguistics would provide the link between mathematical logic with the correspondence theory of truth, on the one hand, and ordinary language, on the other. They differed in their assumptions about linguistics, with Lakoff and McCawley assuming that semantics and grammar were inseparable, with grammatical categories arising from semantic categories, and that logical forms could be underlying structures in a transformational grammar; Montague assumed a categorial grammar, in which surface linguistic forms could be assigned truth conditions directly using a higher-order theory of functions; and Fodor assumed a Chomskyan grammar that had no semantic base. There were many other differences as well.

Two more questions about which philosophers differ are:

Are thoughts universal, according to 5 and 16, or are meanings relative, accepting 2, but allowing reference to vary?

Assuming 2, is reference fixed expression by expression or holistically?

With respect to these two questions, Quine argued that reference varies, meaning is relative, and that meaning is fixed holistically. Davidson and Putnam have agreed. Frege, Russell, and Carnap argued that thoughts were universal. More recently, Fodor has argued this as well, although he has made minor concessions to holism.

Does thought have an existence independent of the mind? Or is thought produced by the mind, as the Mind As Machine metaphor entails?

Those in the artificial intelligence tradition have accepted the Mind As Computer metaphor, and with it the Mind As Machine. There are, however, different versions of the Mind As Computer metaphor, one accepting entailment 2, the idea of meaning as reference, and the other rejecting 2. Within first-generation cognitive science, there were those like Dedre Gentner, who saw the mind as a computer and its languagelike representations as "internal representations of an external reality" that were to be given meaning via direct connections to the world. The combination of the Mind As Computer metaphor was seen by many in first-generation cognitive science as perfectly compatible with a Fregean account of meaning in terms of reference. On the other hand, there were those, like Roger Schank and Marvin Minsky, who saw "semantics" as what the machine did, as the inferences and operations it carried out. Searle, who accepts 1 through 3 and 5 through 12, has argued vigorously against such a view, claiming that, without a semantics linking symbols directly to the world, the mind, conceptualized as a computer, could not understand anything. Advocates of AI have replied that meaning and understanding of language come about not through reference (via 2) but through the inferential functions of the machine and the way it operates on the world.

This brief survey is not meant in any way to be complete. It is meant to show only three things: First, our technical philosophical theories of mind and language are built up out of various combinations of the widely shared metaphors for mind. The specific tenets of a theory of mind are typically individual entailments of the metaphors that constitute our ordinary conceptualizations of mind. Second, choices among those metaphorical entailments can sometimes lead to philosophical differences. And third, the fact that such philosophical assumptions are entailments of everyday metaphors makes the philosophical theories that use them seem "intuitive" to many people. An intuitive theory is one that uses ideas already there in the cognitive unconscious.

Ironically, all such philosophical theories in the Anglo-American tradition reject the very idea of conceptual metaphor. Indeed, our very metaphors for mind have entailments that suggest that such conceptual metaphors should not exist. There are two reasons. First, in the Mind As Body system, all the source-domain entities that are mapped onto ideas—locations, objects, and food—exist independently of the thinker. This yields the entailment that ideas have an objective existence independent of thinkers. But conceptual metaphors cannot have an objective existence independent of thinkers. Conceptual metaphors can only be products of embodied minds interacting with environments, not things existing objectively in the world. Therefore such theories require, counter to all that we are discussing, that conceptual metaphors cannot exist.

Second, the Thought As Language metaphor sees ideas as linguistic expressions, that is, unitary entities. But the very notion of conceptual metaphor requires that ideas not be unitary entities; instead, cross-domain conceptual mappings enter into what an idea is. Each metaphorical idea is therefore binary, not unitary—it has both a source and a target that is at least partly structured by that source. Similarly, the Thought As Mathematical Calculation metaphor conceptualizes ideas as numbers, which are also unitary entities. And the Mind As Machine metaphor sees ideas as products of a machine, and such products are unitary entities as well. Thus all these metaphors for ideas are at odds with the very notion of conceptual metaphor.

None of this is surprising. The entities we encounter in daily life typically have existence independent of us and are unitary entities that do not rely on anything else to characterize their nature. Since our everyday conceptual metaphors for mind must take everyday experience as the basis for each metaphor for mind, it is not surprising that ideas should be conceptualized in terms of unitary entities that exist independently of us. It is a consequence of the theory of conceptual metaphor that we should have everyday metaphors for ideas in which ideas cannot be metaphorical!

Given that Anglo-American analytic philosophy has been constructed out of those everyday metaphors for ideas, analytic philosophy could not have sanctioned the existence of conceptual metaphors, and no future version ever will. It would mean giving up all of analytic philosophy's central ideas: the objectivity of meaning, the classical correspondence theory of truth, the notion of an ideal logical language, the adequacy of logical form, definition in terms of necessary and sufficient conditions, and so on.

The Oddness of Anglo-American Philosophy

Consider a cognitive scientist concerned with the empirical study of the mind, especially the cognitive unconscious, and ultimately committed to understanding the mind in terms of the brain and its neural structure. To such a scientist of the mind, Anglo-American approaches to the philosophy of mind and language of the sort discussed above seem odd indeed. The brain uses neurons, not languagelike symbols. Neural computation works by real-time spreading activation, which is neither akin to prooflike deductions in a mathematical logic, nor like disembodied algorithms in classical artificial intelligence, nor like derivations in a transformational grammar.

Cognitive scientists looking for a naturally based account of understanding must turn to the brain and body for empirical reasons. They cannot start a priori with a logician's set-theoretical models. Nor will they start a priori with a theory of meaning in which meaning has nothing to do with mind, brain, body, or experience, but is given in terms of reference and truth. Meaning in a neurally based cognitive theory can only arise through the body and brain and human experience as encoded in the brain.

To a cognitive scientist in the empirical tradition, the approach of Anglo-American philosophy to mind and language seems quite bizarre. This is especially true of what Michael Dummett calls the central idea behind both Anglo-American philosophy and phenomenology in the tradition of Brentano and Husserl, namely, "the extrusion of thoughts from the mind." To a cognitive scientist, what could be stranger than to take thought as being external to the mind, with human biology and psychology irrelevant to the nature of thought? Why would so many philosophers find such a notion intuitive? Why don't most philosophy students and philosophers simply find such ideas of Anglo-American philosophy ridiculous? Why don't they just laugh when they are told that the meanings of sentences in a language have nothing to do with the human mind or any aspect of human psychology? This requires explanation.

We have just given that explanation. The central ideas of Anglo-American philosophy are versions of ideas we already have. They arise from our metaphorical conceptual system for the mind. This is what makes them "intuitive." Indeed, a philosophy that does not depend on the empirical study of its subject matter must always depend on such intuitive notions supplied by our unconscious conceptual system, and especially by our system of conceptual metaphor.

We do not in any way want to suggest that any philosophical or scientific view that uses conceptual metaphor or metaphorical entailments is wrong just because it uses metaphor. On the contrary, conceptual metaphor is necessary for any abstract intellectual undertaking, such as theorizing. The issue of the validity of any theory, philosophical or scientific, is ultimately empirical. That is why we have given so much attention to evidence.

It should be clear now why an experientialist philosophy—one that takes second-generation cognitive science seriously—is utterly different from a "naturalized" version of analytic philosophy. A naturalized analytic philosophy takes analytic philosophy for granted and just adds empirical results consistent with it. But the results of second-generation cognitive science are fundamentally at odds with analytic philosophy in any form, "naturalized" or not.

Metaphorical Thought in the Contemporary Philosophy of Mind

As we noted above, first-generation cognitive science was set within Anglo-American philosophy of mind, in which the main trend through the early 1980s was "functionalism." According to N. Block (C2, 1980):

> Whatever mystery our mental life may initially seem to have is dissolved by functional analysis of mental processes to the point where they are seen to be composed of computations as mechanical as the primitive operations of a digital computer—processes so stupid that appealing to them in psychological explanations involves no hint of question-begging. The key notions of functionalism in this sense are representation and computation. Psychological states are seen as systematically representing the world via a language of thought, and psychological processes are seen as computations involving these representations.

In short, the functionalist program in Anglo-American philosophy of mind consists of two metaphors:

THE MIND AS COMPUTER METAPHOR

Physical Computer	→	The Person (Especially, The Brain)
Computer Program	→	The Mind
Formal Symbols	→	Concepts
Computer Language	→	Conceptual System
Formal Symbol Sequences	→	Thoughts
Formal Symbol Manipulation	→	Thinking
Algorithmic Processing	→	Step-by-Step Thought
Database	→	Memory
Database Contents	→	Knowledge
Ability To Compute Successfully	→	Ability To Understand

THE REPRESENTATION METAPHOR

Relations Between Formal Symbols And Things In The World	→	Meanings Of Concepts

The Mind As Computer metaphor incorporates a version of the Thought As Language metaphor, in which concepts are symbols in some "language of thought." The Representation metaphor is a version of Frege's strange idea that the publicly available meanings in natural language are "objective" and independent of human psychology or biology, that is, independent of human minds and brains. The Representation metaphor conceptualizes meaning in a way appropriate to the Fregean idea—in terms of relations between symbols in a language of thought and things in the world. Searle has referred to this as the "fit" between words and the world. In functionalist philosophy of mind, such relations between symbols and things are seen metaphorically as meanings of concepts; by virtue of such relations, the abstract formal symbols are conceptualized metaphorically as "representing reality." As in Anglo-American philosophy in general, the Representation metaphor assumes that meaning has nothing whatever to do with minds, brains, bodies, or bodily experience, but is defined only as a relation between abstract formal symbols and things in an external mind-independent world.

The Metaphors of Symbol Manipulation

The Mind As Computer metaphor has as its source domain our understanding of what a computer is. But that understanding is itself metaphorical in two important ways, since the concepts of a formal language and of symbol manipulation are themselves conceptualized via metaphor.

THE FORMAL LANGUAGE METAPHOR

Written Signs Of A Natural Language	→	Abstract Formal Symbols
A Natural Language	→	A Formal "Language"
Sentences	→	Well-Formed Symbol Sequences
Syntax	→	Principles For Combining Formal Symbols

THE SYMBOL MANIPULATION METAPHOR

Physical Objects	→	Formal Symbols
Placing Objects In Proximity	→	The Concatenation Relation
Object Manipulation	→	Changing Symbol Sequences Into Other Symbol Sequences

Computer science is based on the mathematical theory of formal languages, an abstract form of mathematics that is conceptualized and taught in terms of the Formal Language and Symbol Manipulation metaphors. In that form of mathematics, abstract formal symbols are not actually physically instantiated signs of some natural language. They are purely abstract entities with no internal structure, they are distinct from each other, and they are meaningless in themselves. "Concatenation" is not actually placing objects next to one another in physical space, but is rather the placing of symbols in some order—a formal mathematical relation satisfying the axioms of semigroups. Concatenated sequences of symbols (symbols placed in a specified order) are not literally complex physical objects that are physically manipulated in real space; they are, instead, abstract mathematical entities that form combinations that can be changed into other combinations. Moreover, this "changing" does not occur in real time.

The Formal Language and Symbol Manipulation metaphors are useful for making sense of this abstract mathematical subject matter. Computer science students are introduced to the subject matter using these metaphors. Just about every computer scientist conceptualizes that subject matter in these terms at least sometimes, though professionals of course know that they are just metaphors.

It is important to see exactly why they are metaphors and not literal statements. A natural language has phonetics, phonology, and morphology. Formal "languages" don't. Intonation in natural language is on a separate plane from phonemic segments; that is, an intonation contour typically extends over many segments and can have meaning separate from the meaning of the segments. Formal "languages" have nothing like intonation. Formal "languages" are not meaningful. They are defined in terms of pure form, and expressions in a formal language are to be manipulated without regard to the meaning of the symbols. By contrast, natural languages are meaningful, and their meaning arises naturally from everyday human experience. Moreover, meaning is built into the grammar and lexical structure of natural languages. Whereas the symbols of formal languages must be assigned unique referents, with all ambiguity eliminated, expressions in natural languages are normally polysemous; that is, they have multiple meanings that are related by cognitive principles (see A4, Lakoff 1987). In short, natural languages are not subcases of formal "languages."

The notion of a "language" in the Formal Language metaphor is modeled on the idea of a hypothetical unknown language written in a linear script that is to be deciphered, as, for example, Linear B was. Any such real language would, of course, be deciphered in terms of phonetics, phonology, morphology, and polysemous semantics. The Formal Language metaphor therefore does not map most of the essential aspects of a natural language. Thus, it cannot be a literal statement. A formal "language" is therefore only metaphorically a language.

Strong Artificial Intelligence

There are three attitudes that one can take toward the conceptualization of the mind as a computer as stated in the Mind As Computer metaphor. First, one can, as we are doing, note that it is a metaphor and study it in detail. Second, one can recognize its metaphorical nature and take it very seriously as a scientific model for the mind (see A2, Gentner and Grudin 1985, for a defense of this view). Many practitioners of what has been called the weak version of artificial intelligence take this position.

A third position has been called "strong AI." When the Mind As Computer metaphor is believed as a deep scientific truth, the true believers interpret the ontology and the inferential patterns that the metaphor imposes on the mind as defining the essence of mind itself. For them, concepts *are* formal symbols, thought *is* computation (the manipulation of those symbols), and the mind *is* a computer program.

For true believers, the essence of mind is computation. In the classical theory of categorization (A4, Lakoff 1987), an essence defines a category containing everything with that essence and no other things. The category defined by computation as an essence is the category of computers, real and abstract. Special cases include both physical computers and "human computers," that is, people who think.

Principles of classical categorization are at work here: Since computer programs and minds share the same essential properties, they must form a higher, more abstract category—intelligent systems. Human minds and computer programs become just special cases of intelligent systems. Once the Mind As Computer metaphor is taken as defining the very essence of mind, it is not consciously seen as a metaphor at all, but rather as "the Truth."

Advocates of strong AI differ from philosophical functionalists in an important way: They typically do not require the Representation metaphor. That is, in strong AI, the idea of "representation" as a relation between abstract symbols and things in the world is not deemed necessary or is consciously excluded. Instead, meaningful understanding arises through the computations themselves. If the computer can get the computations right—if it can correctly manipulate the symbols given to it as input and generate the right output—then the computer has understood. Likewise, strong AI claims that human understanding is just getting the computations—the manipulation of formal symbols—right.

Searle's Metaphorical Chinese Room Argument

For nearly two decades, John Searle's Chinese Room Argument has been one of the most celebrated examples of philosophical argument within Anglo-American philosophy of mind. It is a philosophical argument against strong AI; indeed, it is the best-known and most-cited philosophical argument against strong AI.

As an upholder of the Anglo-American philosophical tradition, Searle has, of course, denied that there is any such thing as metaphorical thought as we have discussed it here. What we will show (following the suggestion of György László) is that this celebrated argument in the Anglo-American tradition is fundamentally and irreducibly metaphoric. This is a preview of Part III of the book, in which we will argue that philosophy in general is irreducibly metaphoric.

Our goal in this section is not to argue against the conclusion Searle reaches, namely, that computers cannot understand anything. We happen to believe that, but for an entirely different reason, namely, that meaning must be embodied. Our goal is merely to show in detail how Searle makes crucial use of metaphorical thought in his Chinese Room Argument and how he could not reach his conclusion without metaphorical inference. Here is the argument as it appeared in *Scientific American* (C2, Searle 1990, 26–27):

SEARLE'S CHINESE ROOM ARGUMENT

Consider a language you don't understand. In my case, I do not understand Chinese. To me, Chinese writing looks like so many meaningless squiggles. Now suppose I am placed in a room containing baskets full of Chinese symbols. Suppose also that I am given a rule book in English for matching Chinese symbols with other Chinese symbols. The rules identify the symbols entirely by their shapes and do not require that I understand any of them. The rules might say such things as "take a squiggle-squiggle sign from basket number one and put it next to a squoggle-squoggle sign from basket number two."

Imagine that people outside the room who understand Chinese hand in small bunches of symbols and that in response I manipulate the symbols according to the rule book and hand back more small bunches of symbols. Now the rule book is the "computer program." The people who wrote it are the "programmers," and I am the "computer." The baskets full of symbols are the "data base," the small bunches that are handed in to me are "questions" and the bunches I then hand out are "answers."

Now suppose that the rule book is written in such a way that my "answers" to the "questions" are indistinguishable from those of a native Chinese speaker. For example, the people outside might hand me some symbols that, unknown to me, mean, "What is your favorite color?" and I might after going through the rules give back symbols that, also unknown to me, mean, "My favorite color is blue, but I also like green a lot." I satisfy the Turing test for understanding Chinese. All the same, I am totally ignorant of Chinese. And there is no way I could come to understand Chinese in the system as described, since there is no way that I can learn the meanings of any of the symbols. Like a computer, I manipulate symbols, but I attach no meaning to the symbols.

The point of the thought experiment is this: if I do not understand Chinese solely on the basis of running a computer program for understanding Chinese, then neither does any other digital computer solely on that basis. Digital computers merely manipulate formal symbols according to rules in the program.

What goes for Chinese goes for other forms of cognition as well. Just manipulating the symbols is not by itself enough to guarantee cognition, perception, understanding, thinking, and so forth. And since computers, qua computers, are symbol-manipulating devices, merely running the computer program is not enough to guarantee cognition.

This simple argument is decisive against the claims of strong AI.

The argument is based on the following idiosyncratic metaphor for understanding the operation of a computer as the person in the Chinese Room:

THE CHINESE ROOM METAPHOR

Person In Chinese Room	→	Physical Computer
Rule Book	→	Computer Program
Chinese Symbols	→	Formal Symbols
Chinese Symbol Manipulation	→	Formal Symbol Manipulation
Chinese Symbol Sequences	→	Computer Language
Baskets Of Chinese Symbols	→	Database In Computer Language
Symbols Handed In	→	Symbolic Inputs
Symbols Handed Out	→	Symbolic Outputs

What makes this metaphor believable is that it is, in turn, based on other commonplace metaphors. We have already seen two of them: the Formal Language metaphor and the Symbol Manipulation metaphor. By the Formal Language metaphor, meaningless abstract mathematical entities are metaphorically conceptualized as the written symbols of a natural language that you can't understand. Searle chooses Chinese, which most English-speaking philosophers can't understand, as the natural language to fit the metaphor. By the Symbol Manipulation metaphor, the formation of abstract combinations of mathematical entities is conceptualized metaphorically as the physical manipulation of physical objects. To fit that metaphor, Searle has the person in the Chinese Room actually manipulating physical objects. These are important aesthetic choices for Searle's grand metaphor. Since they fit the metaphors used in teaching about computer science, they seem natural.

The Chinese Room metaphor also makes use of a very common metaphor in English, the Machine As Person metaphor, which shows up in sentences like:

> This vacuum cleaner can *pick up* the tiniest grains of sand. My new Lexus can *choose* the best route to the ski resort and *take me there* in the shortest possible time. My microwave can *bake a cake* in six minutes. My car *refuses* to start. This vending machine is a bit *recalcitrant;* I'd better give it a kick.

In this metaphor, the functioning of a machine is conceptualized as the performance of that function by a person, and the failure to function is seen as a refusal to perform.

Searle uses a special case of the Machine As Person metaphor in the Chinese Room Argument, when he conceptualizes the computer as a person—himself—physically moving symbols around. This instance of the Machine As Person metaphor, taken together with his special cases of the Formal Language and Symbol Manipulation metaphors, makes much of the Chinese Room metaphor seem natural, because it uses conceptual metaphors that we already have in our conceptual systems. When these metaphors are taken together, it seems natural to conceptualize a computer as a person physically manipulating Chinese characters that he doesn't understand according to rules.

At this point, Searle makes crucial use of the fact that conceptual metaphors map patterns of inference from the source domain to the target domain. With the Chinese Room metaphor set up in this way, there is an entailed metaphorical mapping: Searle's lack of understanding of the meaning of the Chinese characters maps onto the computer's lack of understanding of the meaning of the formal symbols. This is a metaphorical inference par excellence!

Note, incidentally, what is not mapped. Searle—the person in the Chinese Room—does understand a great deal. He understands English. He understands the rule book. He understands that he is in a room, that he is manipulating objects, and that the objects are symbols. And he understands that he does not understand the symbols.

None of this understanding is mapped by Searle's made-up metaphor. He includes the mappings "I am the 'computer'" and "The rule book is the 'computer program,'" but he specifically excludes the mapping "My understanding of the rule book is the computer's understanding of the program." Hence, Searle does not conclude that the computer understands the program, just as he understands the rule book. Nor does he conclude that the computer understands that it is manipulating formal symbols, just as he understands that he is manipulating Chinese symbols. Nor does he conclude that the computer understands that it doesn't understand the symbols that it is manipulating, just as he understands that he doesn't understand the Chinese symbols. Searle has carefully constructed the Chinese Room metaphor so that none of this understanding is mapped onto the computer. All that is attributed to the computer is a lack of understanding.

It is important to understand that Searle's argument cannot be given an interpretation as a literal argument. If it were, the mapping we have stated would not be metaphorical, but a statement of literal subcategorizations. We have seen that such an interpretation does not work, because the Formal Language and Symbol Manipulation metaphors, as we have seen, are not literal statements.

More important, consider Searle's crucial mapping of his degree of understanding of the Chinese characters onto the computer's degree of understanding of the formal symbols. For this to be literal and not metaphorical, Searle's degree of understanding of the Chinese characters must be a literal subcase of the computer's degree of understanding of its formal symbols. Strong AI claims that the computer can understand if it does the computation correctly. If Searle in the Chinese Room is literally a kind of computer, then Searle should be able to understand the Chinese characters that he is successfully manipulating. If he cannot understand, then the computer cannot understand.

The argument form goes as follows:

1. Searle is a kind of computer.

2. If computers can understand via manipulating symbols, then Searle can understand via manipulating meaningless symbols.
3. Since Searle cannot understand via manipulating meaningless symbols, computers cannot understand via manipulating meaningless symbols.

The flaw in this as a literal argument lies hidden in the first statement. As Györgi László (personal communication) has pointed out, Searle's mind in the Chinese Room is not literally any subpart of a computer. There is nothing in a general-purpose digital computer that Searle's mind in the Chinese Room could be a special case of! Searle's understanding the rule book is not a literal subcase of the computer's understanding its program. The computer does not understand the program; it just runs according to it. What the first statement should say is that Searle's overall mechanical functioning, leaving his mind and what he understands out of it, is an instance of a computer's overall functioning. But when the first statement is made explicit so as to exclude Searle's mind and his understanding as being a special case of any aspect of a computer, then the argument fails. The fact that Searle's mind fails to understand is not a literal special case of any part or aspect of the computer failing to understand. For this reason, Searle's argument cannot be a literal form of *modus tollens*.

As we saw, there are other reasons as well that this is not a literal argument, namely, Searle's use of the Formal Language, Symbol Manipulation, and Machine As Person metaphors. The reason that the Chinese Room Argument is compelling for so many people is not its incorrect status a literal *modus tollens* argument, but rather its status as a metaphorical argument. Given that we implicitly use the Formal Language, Symbol Manipulation, and Machine As Person metaphors, the Chinese Room Argument works as a powerful metaphorical argument. It seems compelling because it uses metaphors we already have.

What is interesting to us about Searle's Chinese Room Argument is that so many Anglo-American philosophers of mind, including Searle himself, took it as literal. But then, they could hardly have done otherwise, since Anglo-American philosophy, because of its own deep-seated metaphors, recognizes neither the cognitive unconscious nor conceptual metaphor.

The Metaphorical Mind

As we saw in Chapter 3, the mind is embodied, not in any trivial sense (e.g., the "wetware" of the brain runs the software of the mind), but in the deep sense that our conceptual systems and our capacity for thought are shaped by the nature of our brains, our bodies, and our bodily interactions. There is no mind separate from and independent of the body, nor are there thoughts that have an existence independent of bodies and brains.

But our metaphors for mind conflict with what cognitive science has discovered. We conceptualize the mind metaphorically in terms of a container image schema defining a space that is inside the body and separate from it. Via metaphor, the mind is given an inside and an outside. Ideas and concepts are internal, existing somewhere in the inner space of our minds, while what they refer to are things in the external, physical world. This metaphor is so deeply ingrained that it is hard to think about mind in any other way.

Is there a purely literal conception of mind? There is an impoverished, skeletal, literal conception: The mind is what thinks, perceives, believes, reasons, imagines, and wills. But as soon as we try to go beyond this skeletal understanding of mind, as soon as we try to spell out what constitutes thinking, perceiving, and so on, metaphor enters. The metaphors we have cited above, and others that are too numerous to mention here, are necessary for any detailed reasoning about mental acts.

Our understanding of what mental acts are is fashioned metaphorically in terms of physical acts like moving, seeing, manipulating objects, and eating, as well as other kinds of activities like adding, speaking or writing, and making objects. We cannot comprehend or reason about the mind without such metaphors. We simply have no rich, purely literal understanding of mind in itself that allows us to do all our important reasoning about mental life. Yet such metaphors hide what is perhaps the most central property of mind, its embodied character.

What we call "mind" is really embodied. There is no true separation of mind and body. These are not two independent entities that somehow come together and couple. The word *mental* picks out those bodily capacities and performances that constitute our awareness and determine our creative and constructive responses to the situations we encounter. Mind isn't some mysterious abstract entity that we bring to bear on our experience. Rather, mind is part of the very structure and fabric of our interactions with our world.

13
The Self

The study of the mind, as we have just seen, takes up such questions as what thinking is, what thoughts are, and what thinkers are. The study of the self, on the other hand, concerns the structure of our inner lives, who we really are, and how these questions arise every day in important ways. What we call our "inner lives" concerns at least five kinds of experience that are consequences of living in a social world with the kinds of brains and bodies that we have.

First, there are the ways in which we try to control our bodies and in which they "get out of control." Second, there are cases in which our conscious values conflict with the values implicit in our behavior. Third, there are disparities between what we know or believe about ourselves and what other people know or believe about us. Fourth, there are experiences of taking an external viewpoint, as when we imitate others or try to see the world as they do. And last, there are the forms of inner dialog and inner monitoring we engage in.

What we find striking about such experiences of self is how commonplace they are—so commonplace that they are good candidates for universal experiences. Yet as we shall see, we do not have any single, monolithic, consistent way of conceptualizing our inner life that covers all these kinds of cases. Instead, we have a system of different metaphorical conceptions of our internal structure. There are certain inconsistencies within the system. And there are a small number of source domains that the system draws upon: space, possession, force, and social relationships. What is perhaps most surprising is that the same system of metaphors can occur in a very different culture, as we shall illustrate below with Japanese examples.

Subject and Self in the Cognitive Unconscious

The general structure of our metaphoric system for our inner lives was first uncovered by Andrew Lakoff and Miles Becker (A1, Lakoff and Becker 1992). Their analysis showed that the system is based on a fundamental distinction between what they called the Subject and one or more Selves. The Subject is the locus of consciousness, subjective experience, reason, will, and our "essence," everything that makes us who we uniquely are. There is at least one Self and possibly more. The Selves consist of everything else about us—our bodies, our social roles, our histories, and so on.

What follows is a considerable modification and extension of the system that they uncovered. It differs principally in two ways: It greatly extends the range of cases covered, and it shows how each metaphor arises from a fundamental kind of experience.

What is philosophically important about this study is that there is no single, unified notion of our inner lives. There is not one Subject-Self distinction, but many. They are all metaphorical and cannot be reduced to any consistent literal conception of Subject and Self. Indeed, there is no consistency across the distinctions. Yet, the multifarious notions of Subject and Self are far from arbitrary. On the contrary, they express apparently universal experiences of an "inner life," and the metaphors for conceptualizing our inner lives are grounded in other apparently universal experiences. These metaphors appear to be unavoidable, to arise naturally from common experience. Moreover, each such metaphor conceptualizes the Subject as being personlike, with an existence independent of the Self. The Self, in this range of cases, can be either a person, an object, or a location.

The ultimate philosophical significance of the study is that the very way that we normally conceptualize our inner lives is inconsistent with what we know scientifically about the nature of mind. In our system for conceptualizing our inner lives, there is always a Subject that is the locus of reason and that metaphorically has an existence independent of the body. As we have seen, this contradicts the fundamental findings of cognitive science. And yet, the conception of such a Subject arises around the world uniformly on the basis of apparently universal and unchangeable experiences. If this is true, it means that we all grow up with a view of our inner lives that is mostly unconscious, used every day of our lives in our self-understanding, and yet both internally inconsistent and incompatible with what we have learned from the scientific study of the mind.

The Structure of the Subject-Self Metaphor System

Our metaphoric conceptions of inner life have a hierarchical structure. At the highest level, there is the general Subject-Self metaphor, which conceptualizes a person as bifurcated. The exact nature of this bifurcation is specified more precisely one level down, where there are five specific instances of the metaphor. These five special cases of the basic Subject-Self metaphor are grounded in four types of everyday experience: (1) manipulating objects, (2) being located in space, (3) entering into social relations, and (4) empathic projection—conceptually projecting yourself onto someone else, as when a child imitates a parent. The fifth special case comes from the Folk Theory of Essences: Each person is seen as having an Essence that is part of the Subject. The person may have more than one Self, but only one of those Selves is compatible with that Essence. This is called the "real" or "true" Self.

Finally, each of these five special cases of the general Subject-Self metaphor has further special cases. It is at this third level of specificity that the real richness of our metaphoric conceptions of Subject and Self emerges. Let us now look at the system in detail.

The General Subject-Self Metaphor

It is not a trivial fact that every metaphor we have for our inner life is a special case of a single general metaphor schema. This schema reveals not only something deep about our conceptual systems but also something deep about our inner experience, mainly that we experience ourselves as split.

In the general Subject-Self metaphor, a person is divided into a Subject and one or more Selves. The Subject is in the target domain of that metaphor. The Subject is that aspect of a person that is the experiencing consciousness and the locus of reason, will, and judgment, which, by its nature, *exists only in the present*. This is what the Subject is in most of the cases; however, there is a subsystem that is different in an important way. In this subsystem, the Subject is also the locus of a person's Essence—that enduring thing that makes us who we are. Metaphorically, the Subject is always conceptualized as a person.

The Self is that part of a person that is not picked out by the Subject. This includes the body, social roles, past states, and actions in the world. There can be more than one Self. And each Self is conceptualized metaphorically as either a person, an object, or a location.

The source domain of the basic Subject-Self metaphor schema, which is neutral among all the special cases, is thus very general, containing only a person (the Subject), one or more general entities (one or more Selves), and a generalized relationship. Here is a statement of the general mapping:

THE BASIC SUBJECT-SELF METAPHOR SCHEMA

People and Entities		_The Whole Person_
A Person	→	The Subject
A Person Or Thing	→	A Self
A Relationship	→	The Subject-Self Relationship

We now turn to the special cases of this metaphor. Each of them adds something. For example, in the case we are about to discuss, the "person or thing" in the source domain is narrowed down to a physical object and the source-domain relationship is specified as a relationship of control.

The Physical-Object Self

Holding onto and manipulating physical objects is one of the things we learn earliest and do the most. It should not be surprising that object control is the basis of one of the five most fundamental metaphors for our inner life. To control objects, we must learn to control our bodies. We learn both forms of control together. Self-control and object control are inseparable experiences from earliest childhood. It is no surprise that we should have as a metaphor—a primary metaphor—Self Control Is Object Control.

SELF CONTROL IS OBJECT CONTROL

A Person	→	The Subject
A Physical Object	→	The Self
Control	→	Control Of Self By Subject
Noncontrol	→	Noncontrol Of Self By Subject

The Internal Causation Metaphor

One of the two most common ways to exert control over an object is to move it by exerting force on it. Given that Self Control Is Object Control, the special case of object control as the forced movement of the object gives us the complex metaphor Self Control Is The Forced Movement Of The Self By The Subject.

SELF CONTROL IS THE FORCED MOVEMENT OF AN OBJECT

A Person	→	The Subject
A Physical Object	→	The Self
Forced Movement	→	Control Of Self By Subject
Lack of Forced Movement	→	Noncontrol Of Self By Subject

This in turn has two special cases. In the first, the body is taken as the relevant aspect of the Self. Control of the body is hence seen as the forced movement of a physical object. A good example is "I lifted my arm," which is ambiguous. Take the sense in which I grab my left arm with my right arm, let my left arm go limp, and lift it with my right arm. My left arm then functions literally as an object that I lift with my right arm as I would any other object.

The other sense of "I lifted my arm" is the metaphorical sense. Literally it means that I exercised body control, causing my right arm to rise. But it is conceptualized and expressed metaphorically in terms of the forced movement of an object.

BODY CONTROL IS THE FORCED MOVEMENT OF AN OBJECT

A Person	→	The Subject
A Physical Object	→	The Body (Instance Of Self)
Forced Movement	→	Control Of Body By Subject
Lack Of Forced Movement	→	Noncontrol Of Body By Subject

Examples include:

I *lifted* my arm. I can *wiggle* my ears. The yogi *bent* his body into a pretzel. I *dropped* my voice. I *dragged* myself out of bed. I *held* myself *back* from hitting him. I *plopped* myself down on the couch. After being knocked down, the champ *picked* himself *up* from the canvas.

A second case of Self Control Is The Forced Movement Of An Object arises when this metaphor is combined with the common metaphors Action Is Movement and Causes Are Forces. The result is that the Subject causing an action by the Self is conceptualized as moving an object by force. A good example is "I've got to get myself *moving* on this project," which is understood as my experiencing consciousness having to cause my Self to start acting on the project.

CAUSING THE SELF TO ACT IS
THE FORCED MOVEMENT OF AN OBJECT

A Person	→	The Subject
A Physical Object	→	The Self
Forced Movement	→	Subject's Causing Self To Act
Lack of Forced Movement	→	Failing To Cause The Self To Act

Examples:

> You're *pushing* yourself too hard. It would take a bulldozer to *get him going* on this job. He's just *sitting on* the work order; I can't *budge* him.

What we have just seen is a good example of how metaphorical complexity arises out of primary metaphors. To show the structure of the metaphor system for Subject and Self, we followed a kind of guided tour showing the relationships among the metaphors. Here are the steps we just went through:

Step 1: Starting with Self Control Is Object Control, we added forced movement as a special case of object control to get Self Control Is The Forced Movement Of An Object.

Step 2: Then taking the body as a special case of the Self, we obtained Body Control Is The Forced Movement Of An Object.

Step 3: Alternatively, adding Causes Are Forces and Action Is Movement to Self Control Is The Forced Movement Of An Object, we obtained Causing The Self To Act Is The Forced Movement Of An Object.

Starting again with Self Control Is Object Control, we can see further structure in the system. Another major way of exercising control over an object is to hold on to it, to keep it in your possession. This special case of object control gives rise to another important metaphor for inner life, namely, that Self Control Is Object Possession.

SELF CONTROL IS OBJECT POSSESSION

A Person	→	The Subject
A Physical Object	→	The Self
Possession	→	Control Of Self
Loss Of Possession	→	Loss Of Control Of Self

What does it mean to *lose yourself* in some activity? It means to cease to be in conscious control and to be unable to be aware of each thing you are doing. For example, suppose you are dancing. You might try to consciously control all your movements. But if it is a fast, complex dance, you may not be able to maintain conscious control of each movement. The dance may require you to *let yourself go*, to *lose yourself* in the dance, to allow yourself to just dance and experience the dancing without being consciously in charge of each movement. The effect can be exhilarating and joyful, a very positive experience, especially when losing oneself entails a freedom from the pressures of everyday concerns.

The metaphor Self Control Is Object Possession characterizes the notion of *losing yourself*. But not all losses of control are positive, exhilarating experiences. A loss of control may be scary and negative. Losing control in such cases is often conceptualized as something negative taking possession of the Self—seizing it, gripping it, carrying it away.

For example, you may lose control because of negative emotions, as when you are *seized by anxiety* or *in the grip of fear*. You may do more than you intended to do, with possible negative consequences, as when you get *carried away*. And perhaps the most scary experience of lack of control is when one feels that one's actions are being controlled by someone else, a hostile being, as when one feels *possessed*. The Self Control Is Object Possession metaphor is thus a way of conceptualizing a wide range of very real experiences, both positive and negative.

This metaphor can also be extended to include the "possession" of one's body by another subject, typically, the devil or an alien or a spirit. Here is the extended version of the metaphor:

TAKING CONTROL OF ANOTHER'S SELF IS TAKING ANOTHER'S POSSESSION

A Person	→	The Subject
A Physical Object	→	The Self
Possession	→	Control Of Self
Loss Of Possession	→	Loss Of Control Of Self
Some Other Person	→	Some Other Subject
Taking Possession	→	Taking Control Of Self

There are a number of versions of this "possession" metaphor. In India, to be possessed by a benevolent spirit or god is seen as positive. And in cultures around the world, trance states are seen as forms of possession of the Self by another Subject, perhaps by a powerful or wise spirit, and techniques for inducing them are cultivated. In American culture, possession—loss of control to another Subject—is mostly seen as evil and scary. Since the movie *Invasion of the Body Snatchers*, possession by aliens has become one of the most common themes of American horror movies. But one doesn't have to go to the movies to find instances of this metaphor in American life. Alcoholism is typically conceptualized in terms of possession, say, by Demon Rum: "That was the rum talking, not me." And people who believe they have been possessed are taken as having a symptom of a mental disorder, as seen by this yes-no question from the Minnesota Multiphasic Personality Inventory: "At one or more times in my life I felt that someone was making me do things by hypnotizing me."

The Locational Self

People typically feel in control in their normal surroundings and less in control in strange places. This is the experiential basis for another primary metaphor central to our inner life: The control of Subject over Self is conceptualized as being in a normal location.

SELF CONTROL IS BEING IN ONE'S NORMAL LOCATION

A Person	→	The Subject
Normal Location	→	The Self
Being In A Normal Location	→	Being In Control Of Self
Not Being In A Normal Location	→	Not Being In Control Of Self

There are two special cases of this metaphor, corresponding to the two commonest forms of normal locations. The first has to do with surroundings, some contained or bounded space one normally occupies: one's home, place of business, the earth, and so on. In this case, the Self is conceptualized as a container, whatever defines familiar surroundings. The Subject's being out of control is conceptualized as its being out of the container, namely, away from home, place of business, or the earth, or out of the part of the Self where the Subject is normally understood as residing, namely, the body, the head, the mind, or the skull. For example, in "I was beside myself," the *I* refers to my Subject—my experiencing consciousness. If the Subject is *beside* the Self, then it is also outside the Self, that is, outside the body, which is not where it normally resides. That's why "I was beside myself" means that I was out of normal control. Similarly, being *out to lunch* is being away from one's normal place of business, which is one of one's normal surroundings. Via this metaphor, "He's out to lunch" means that his Subject—his locus of consciousness, reason, and judgment—is not functioning in a way that allows the normal exercise of control over the Self.

THE SELF AS CONTAINER

A Person	→	The Subject
A Container	→	The Self
Located In Container	→	Control Of Subject By Self
Not Located In Container	→	Out Of Control Of The Self

Examples:

> I was *beside* myself. He's *spaced out.* He's *out to lunch.* The lights are on but *no one's home.* Dude, you're *tripping.* Earth to Joshua: *Come in,* Joshua. I'm *out of it* today. Are you *out of* your mind/head/skull?

The second kind of normal location for us is on the ground, where we are in control of the effects of the force of gravity. In this metaphor, *being down to earth* is exercising normal self-control, while being *off in the clouds* indicates the Subject's lack of control over such things as reasoning, judgment, and attention.

SELF CONTROL AS BEING ON THE GROUND

A Person	→	The Subject
Being On The Ground	→	Being In Control Of The Self
Not Being On The Ground	→	Not Being In Control Of Self
Distance From The Ground	→	Degree Of Lack Of Self Control
Being High	→	Euphoria

Examples:

> He's *got his feet on the ground.* He's *down to earth.* The *ground fell out from under* me. We'll *kick the props out from under* him. I kept *floating off* in lecture. He's got *his head in the clouds.* She reached *new heights* of ecstasy. I'm *high as a kite.* I'm *on cloud nine.* Her smile sent me *soaring.*

Incidentally, as A. Lakoff and Becker (A1, 1992) have pointed out, there are two dual metaphors for self-control—being in possession of the Self and being located where the Self is. In both cases, control is indicated in the same way, namely, by the Subject and Self being in the same place. Lack of control is indicated by Subject and Self being in different places. The two metaphors have opposite figure-ground orientations. In the location metaphor, the Self is a ground (a normal location) and the Subject is a figure located there or not. In the possession metaphor, the figure-ground orientation is reversed: The Subject is the ground (where the possession is located or not) and the Self is a figure, a possession that may or may be where the Subject is.

The Scattered Self

It is hard to function normally when there are a lot of demands on your attention—as when you have divergent needs, responsibilities, or interests—or when you cannot focus your attention on any one task, for example, when you are emotionally upset. In such a situation it is difficult to exert normal conscious self-control. In the Subject-Self metaphor system, the ability to focus attention is an ability of the Subject. Control of attention is part of the Subject's normal self-control.

In both of the dual metaphors just discussed, normal self-control is conceptualized as the Subject and Self being at the same place. When the Self is scattered, Subject and Self cannot be in the same place and control is impossible.

ATTENTIONAL SELF CONTROL IS HAVING THE SELF TOGETHER

A Person	→	The Subject
A Unified Container	→	The Normal Self
The Container Fragmented	→	The Scattered Self
Located In One Place	→	Normal Attentional Control
Not Located In One Place	→	Lack Of Normal Attentional Control

Examples:

Pull yourself *together*. She hasn't *got it together* yet.

In such straightforward examples, the Subject is grammatical subject and the Self is grammatical object. However, there is another grammatical pattern used with this metaphor, in which the Self is grammatical subject and the Subject is unexpressed. Examples:

He's *real together*. She's *all over the place*. He's pretty *scattered*.

Getting Outside Yourself

The metaphor that self-control is being located inside the Self has an important entailment. If you are inside an enclosure, you can't see the outside of the enclosure. Given the metaphor that Knowing Is Seeing, vision from the inside is knowledge from the inside—subjective knowledge. If you want to know how your enclosure appears from the outside, you have to go outside and look. Vision from the outside is knowledge from the outside—objective knowledge.

THE OBJECTIVE STANDPOINT METAPHOR

A Person	$\rightarrow$	The Subject
A Container	$\rightarrow$	The Self
Seeing From Inside	$\rightarrow$	Subjective Knowledge
Seeing From Outside	$\rightarrow$	Objective Knowledge

Examples:

> You need to *step outside* yourself. You should *take a good look at* yourself. I've been *observing* myself and I don't like what I *see*. You should *watch what you do*.

The primary metaphors for Subject and Self that we have discussed so far have been based on two basic correlations in everyday experience: (1) the correlation between self-control and the control of physical objects and (2) the correlation between a sense of control and being in one's normal surroundings. We now turn to a third primary metaphor, based on a third kind of correlation: the correlation between how those around us evaluate both our actions and those of others and how we evaluate our own actions.

The Social Self

From birth, we enter into interpersonal and social relationships with other people, initially with parents and other household members. And from birth, what we do is evaluated by our parents and by others: "Don't punch your sister," "Eat your food," "Don't pour your juice on the cat," "Wave bye-bye when Daddy leaves," and on and on. We all learn to evaluate our own actions in terms of how others evaluate what we do and what others do. We also learn that there are implicit values in family roles, that things that are fine for parents to do are not fine for us to do and that our parents never do lots of the things we do.

Throughout childhood, we develop values toward our past actions, our family roles, and our future plans—all aspects of what we have called the Self. What those values are depends upon how they are correlated with the values that our parents and others place on what we and others do. In short, we learn evaluative relationships between Subject and Self on the basis of evaluative interpersonal and social relationships among those around us. There is thus an ongoing everyday correlation from birth between our experience of evaluative social relationships between ourselves and others and the evaluative relationships our Subjects develop toward our Selves.

THE SOCIAL SELF METAPHOR

One Person	→	The Subject
Another Person	→	The Self
Evaluative Social Relationship	→	Evaluative Subject-Self Relationship

The general character of this metaphor, the fact that the source domain is about evaluative social relationships in general, permits a truly remarkable metaphoric richness. It permits us to map our vast knowledge about specific social relationships onto our inner lives.

Consider just a few of the specific social relationships this metaphor applies to: master-servant, parent-child, friends, lovers, adversaries, interlocutors, advisers, caretakers. Let us begin with inner conflict, where the Subject-Self relationship is conceptualized as adversarial. In these cases, some aspect of the Self (e.g., the emotions) will appear in place of the Self as a whole (e.g., "He's fighting the urge to have a second dessert"). Here are some examples.

Subject and Self as Adversaries:

> He's *at war with* himself over whom to marry. He's *struggling with* himself over whether to go into the church. She's *conflicted*. She's *at odds with* herself over whether to leave or stay. He's *giving* himself *a hard time*. Why do you *torture* yourself? Stop *being* so *mean* to yourself. You're just *making* yourself *suffer*. He's *struggling* with his emotions. She's her own worst *enemy*.

In a parent-child relationship, the parent has a range of responsibilities toward the child, which is helpless without the parent. This includes nurturing, caring, comforting, consoling, protecting, educating, disciplining, rewarding, and punishing. When nurturing goes beyond the normal, healthy range, it can become babying, pampering, coddling, and spoiling.

Subject as Parent and the Self as Child include:

> I still haven't *weaned* myself from sweets. She likes to *pamper* herself. I think you *coddle* yourself a bit too much; you need to *give* yourself some more *discipline*. You've earned the right to *baby* yourself. We all need to *nurture* ourselves. I've done my *chores*; I think I'll *reward* myself with an ice cream cone. Everyone needs to *mother* himself now and then. I'm going to *treat* myself to some ice cream.

Since the Subject is supposed to be in control of the Self, the opposite metaphor is ruled out; we do not conceptualize the Subject as Child and Self as Parent. Similarly, we metaphorize the Subject as Caretaker, but never as the Object of Care.

Here are some additional examples of typical kinds of social relationships that flesh out the Subject-Self relationship in this metaphor.

Subject and Self as Friends:

> I think I'll just *hang out with* myself tonight. I *like* myself and *like being with* myself. I need to be a better *friend to* myself.

Subject and Self as Interlocutors:

> I *debate* things with myself all the time. I *talk* things *over* with myself before I do anything important. I was *debating with* myself whether to leave. I *convinced* myself to stay home.

Subject as Caretaker of Self:

> You need to be *kind* to yourself. I *promised* myself a *vacation*. I have a *responsibility* to myself to give myself time to exercise. She *takes good care* of herself. He *nursed* himself back to health.

Subject as Master, Self as Servant:

> I have to *get myself to* do the laundry. I *told myself to* prepare for the trip well ahead of time. I *bawled myself out* for being impolite. I'm *disappointed in* myself.

The Subject is obligated to meet the standards of the Self:

> Don't *betray* yourself. *Be true to* yourself. I *let* myself *down*. I *disappointed* myself.

This last case is particularly interesting. One's social role, one's position in the community, is part of the Self. That social role comes with certain obligations that we have a responsibility to carry out. But since one's judgment and will are part of the Subject and one's social role is part of the Self, there is a split between the part of you that has the social obligations and the part of you that has the judgment and will that determine how to act. In short, the Subject has an obligation to the Self and the Self has no choice but to trust the Subject to carry out those obligations. The Subject can decide to *be true to* the Self (and honor those obligations) or to *betray* the Self (fail to honor those obligations) and, hence, to *let* the Self *down,* to *disappoint* the Self.

Notice the difference between "I disappointed myself" (the Subject fails to meet the standards of the Self) and "I was disappointed in myself" (the Self fails to carry out an obligation to the Subject). These sentences fit different kinds of social relations between Subject and Self, as indicated above.

This list of types of social relations is anything but fixed. It appears that any form of evaluative social relation will work.

The Multiple Selves Metaphor

There is an important difference between a conflict of values and indecisiveness over values. The Multiple Selves metaphor conceptualizes multiple values as multiple Selves, with each Self instantiating the social role associated with that value. Indecisiveness over values is metaphorized as the Subject's indecisiveness about which Self to associate with.

THE MULTIPLE SELVES METAPHOR

A Person	→	The Subject
Other People	→	Selves
Their Social Roles	→	The Values Attached To The Roles
Being In The Same Place As	→	Having The Same Values As
Being in Different Places	→	Having Different Values

Some examples of values as social roles of Selves are:

I keep *going back and forth* between my scientific self and my religious self. I keep *returning to* my spiritual self. I keep *going back and forth* between the scientist and the priest in me.

Projecting onto Someone Else

From earliest childhood we are able to imitate—to smile when some one smiles at us, to lift an arm when someone lifts an arm, to wave when someone waves.

Imitating makes use of an ability to project, to conceptualize oneself as inhabiting the body of another. Empathy is the extension of this ability to the realm of emotions—not just to move as someone else moves, but to feel as someone else feels.

The ability to project is the basis of another central metaphor in the Subject-Self system. In this metaphor, one Subject is projected onto another in a hypothetical situation. For example, when I say "If I were you, . . . " I am metaphorically conceptualizing my Subject, my subjective consciousness, as inhabiting your Self in a hypothetical situation. We can state the mapping as follows:

THE SUBJECT PROJECTION METAPHOR

Real Situation   →   Hypothetical Situation
Subject 1   →   Subject 2

There are at least two possible special cases of such a projection. In one, what we call Advisory Projection, I am projecting my values onto you so that I experience your life with my values. In the other type, Empathic Projection, I am experiencing your life, but with your values projected onto my subjective experience.

ADVISORY PROJECTION

Values of 1   →   Values of 2

Examples:

If I were you, I'd punch him in the nose. I dreamed that I was Brigitte Bardot and that I kissed me. You're too charitable toward me; if I were you, I'd hate me. You're a cruel person with no conscience; if I were you, I'd hate myself.

EMPATHIC PROJECTION

Values of 2   →   Values of 1

Examples:

If I were you, I'd feel just awful too. I can see why you think I'm a jerk. I feel your pain. I dreamed that I was Brigitte Bardot and found me just as unattractive as she does. Given what you think I've done to you and how you feel about it, if I were you, I'd hate me too. I think you're just fine, but, given your attitudes about your actions, if I were you, I'd hate myself too.

The Essential Self

As we saw in our discussion of causation metaphors, we have a Folk Theory of Essences, according to which every object has an essence that makes it the kind of thing it is and that is the causal source of its natural behavior. There is also version of the Folk Theory of Essences that applies to human beings: In addition to the universal essence of rationality you share with all humans, you, as an individual, have an Essence that makes you unique, that makes you *you*. It is your Essence that makes you behave like you, not like somebody else.

We have in our conceptual systems a very general metaphor in which our Essence is part of our Subject—our subjective consciousness, our locus of thought, judgment, and will. Thus, who we essentially are is associated with how we think, what judgments we make, and how we choose to act. According to the folk theory, it is our Essence that, ideally, should determine our natural behavior.

However, our concept of who we essentially are is often incompatible with what we actually do. This incompatibility between our Essence and what we really do is the subject matter of the Essential Self metaphor. In the metaphor, there are two Selves. One Self (the "real," or "true," Self) is compatible with one's Essence and is always conceptualized as a person. The second Self (not the "real," or "true," Self) is incompatible with one's Essence and is conceptualized as either a person or a container that the first Self hides inside of.

There are three special cases of the Essential Self metaphor. The first is *the Inner Self*. It is common for people to be polite in public, to refrain from expressing their true feelings lest they hurt or offend someone. This is our Outer Self. It is also common for people to act very differently in private than they do in public. This is our Inner Self. Metaphorically, our Inner Self hides inside our Outer Self. The Inner Self is the "real" Self, the one compatible with who we really are, with our Essence. It hides either because it is fragile and ashamed, because it is awful and ashamed, or both.

In the second case, *the External Real Self (the "Real Me"),* the Self that the public normally sees is a quite nice Self, the Real Self, the Self that reflects your Essence, who you really are. But inside can lurk an awful other Self, one who is not who you essentially are, but who can come out if your guard is down.

Suppose you are depressed or grumpy or drunk and you say or do something unkind to friend. You may apologize, explaining your behavior by saying "I wasn't myself yesterday," "I'm sorry, but you know that wasn't the real me," or "My mean side came out."

The third case, *the True Self,* goes like this. Suppose that all your life you've been living a lie, acting in a way that does not fit your true nature, your Essence. Suppose furthermore that you don't know how to behave any other way, but that you'd like to change, to find a way to live as you were meant to, compatible with who you really are, with your Essence.

In the Essential Self metaphor, this situation is one in which your Subject has been inhabiting a Self incompatible with your true nature, your Essence. To change, you will have to find another way to be, another Self compatible with your Essence. This is called "finding your true Self."

Here is the Essential Self metaphor and the three special cases, with examples of each.

THE ESSENTIAL SELF METAPHOR

Person 1	→	The Subject, With The Essence
Person 2	→	Self 1, The Real Self (Fits The Essence)
Person 3	→	Self 2, Not The Real Self, (Doesn't Fit The Essence)

Constraint: The Values Of The Subject Are The Values Of Self 1

The Inner Self: Self 1, the Real Self, is hidden inside Self 2, the Outer Self, because the Real Self is fragile and shy, the Real Self is awful and doesn't want anyone to know he is there, or both. Examples:

Her sophistication is a *façade*. You've never seen what he's really like *on the inside*. He is afraid to *reveal his inner* self. She's sweet *on the outside* and mean *on the inside*. The iron hand *in the velvet glove*. His petty self *came out*. He won't *reveal himself* to strangers. She rarely *shows her real self*. Whenever anyone challenges him, he *retreats into* himself. He *retreats into his shell* to protect himself.

The External Real Self (Real Me): Self 2, who is awful in some way, is hidden inside Self 1, the Real Self, who is quite nice. But when the Real Self lets its guard down, the Awful Self comes out. Examples:

I'm not myself today. That *wasn't the real me* yesterday. That *wasn't my real self* talking.

The True Self: All his life, the Subject has been inhabiting Self 2, which is incompatible with the Subject's Essence. Self 1, which is compatible with the Subject's Essence, is somewhere unknown, and the Subject is trying to find his "true" Self, the one compatible with his Essence, with who he really is. Examples:

> He *found himself* in writing. I'm trying to *get in touch with* myself. She went to India to *look for* her true self, but all she came back with was a pair of sandals. He's still *searching for* his true self.

How Universal Is This System? Some Japanese Examples

When this analysis first began to take shape, we believed that it was a peculiarity of either English or the Western mind. But Yukio Hirose, professor of linguistics at Tsukuba, Japan, pointed out to us that Japanese contains examples that both look like and are understood in the same way as the English examples.

Looking through Hirose's Japanese examples is sobering. Anthropologists and social psychologists have written extensively on how different from the Western concept the Japanese conception of the Self is. But what is radically different is the Japanese conception of the relationship between Self and other, not necessarily the Japanese conception of inner life. From Hirose's examples, provided through personal communication, it appears that our metaphorical conception of inner life is remarkably like the Japanese one. Given the radical differences between American and Japanese cultures, this raises the question of just how universal are experiences of inner life and the metaphors used to reason about them. Though we have no access to the inner lives of those in radically different cultures, we do have access to their metaphor systems and the way they reason using those metaphor systems.

We are including Hirose's examples here in order to be provocative and tantalizing. Very little research has been done on the metaphoric systems of inner life in other languages. That research needs to be done before we can even think of drawing serious empirically based conclusions about whether there are universal experiences of inner life.

THE SUBJECT PROJECTION METAPHOR

Boku-ga kimi dat-ta-ra, boku-wa boku-ga iya-ni-naru.
I(MALE)-NOM you COP-PAST-if I-TOP I-NOM hate-to-become
Lit.: "If I were you, I (would) come to hate me."
"If I were you, I'd hate me." (YOU's Subject hates I's Self.)

Boku-ga kimi dat-ta-ra, boku-wa zibun-ga iya-ni-naru.
I(MALE)-NOM you COP-PAST-if I-TOP self-NOM hate-to-become
Lit.: "If I were you, I (would) come to hate self."
"If I were you, I'd hate myself." (YOU's Subject hates YOU's Self.)

THE OBJECTIVE STANDPOINT METAPHOR

Zibun-no kara-kara de-te, zibun-o yoku mitume-ru koto-ga taisetu da.
self-GEN shell-from get out-CONJ self-ACC well stare-PRES COMP-
NOM important COP
Lit.: "To get out of self's shell and stare at self well is important."
"It is important to get out of yourself and look at yourself well."

THE SCATTERED SELF METAPHOR

Kare-wa ki-ga titte-i-ru.
he-TOP spirit-NOM disperse-STAT-PRES
Lit.: "He has his spirits dispersed."
"He is distracted."

Kare-wa kimoti-o syuutyuu-sase-ta.
he-TOP feeling-ACC concentrate-CAUS-PAST
Lit.: "He made his feelings concentrate."
"He concentrated himself."

Kare-wa ki-o hiki-sime-ta.
he-TOP spirit-ACC pull-tighten-PAST
Lit.: "He pulled-and-tightened his spirits."
"He pulled himself together."

THE SELF CONTROL IS OBJECT POSSESSION METAPHOR

Kare-wa akuma-ni tori-tuk-are-ta.
he-TOP an evil spirit-by take-cling to-PASS-PAST
Lit.: "He was taken-and-clung to by an evil spirit."
"He was possessed by an evil spirit."
Kare-wa dokusyo-ni ware-o *wasure-ta.*
he-TOP reading-LOC self-ACC lose[forget]-PAST
Lit.: "He lost self in reading."
"He lost himself in reading."

Kare-wa ikari-no amari ware-o *wasure-ta.*

he-TOP anger-GEN too much self-ACC lose [forget]-PAST
Lit.: "He lost [forgot] self because of too much anger."
"He was beside himself with anger [had no control over himself]."

Note: The pronoun *ware* can be used in Japanese only in expressions that make use of either the Loss of Self metaphor or the Absent Subject metaphor. Note that these two metaphors are the duals in the system.

SELF CONTROL IS BEING IN ONE'S NORMAL LOCATION

Kare-wa yooyaku ware-ni kaet-ta.
he-TOP finally self-LOC return-PAST
Lit.: "He finally returned to self."
"He finally came to his senses."

Ware-ni mo naku kodomo-o sikatte-simat-ta.
self-LOC even not child-ACC scold-PERF-PAST
Lit.: "Not being even in self, (I) have scolded the child."
"I have scolded the child in spite of myself [unconsciously]."

THE MULTIPLE SELVES METAPHOR

Kono mondai-ni tuite-wa watasi-wa kagakusya-tosite-no zibun-no hooni katamuite-i-ru.
this problem-LOC about-TOP I-TOP scientist-as-GEN self-GEN toward lean-STAT-PRES
Lit.: "About this problem, I lean toward (my) self as a scientist."
"I am inclined to think about this problem as a scientist."

THE SELF AS SERVANT METAPHOR

Kare-wa hito-ni sinsetuni-suru yooni zibun-ni iikikase-ta.
he-TOP people-DAT kind-do COMP self-DAT tell-PAST
"He told himself to be kind to people."

THE INNER SELF METAPHOR

Kare-wa mettani hontoono zibun-o dasa-na-i.
he-TOP rarely real self-ACC get out-NEG-PRES.
Lit.: "He rarely puts out (his) real self."
"He rarely shows his real self."

Kare-wa hitomaede-wa itumo kamen-o kabutte-i-ru.
he-TOP in public-TOP always mask-ACC put on-STAT-PRES
"He always wears a mask in public."

THE TRUE SELF METAPHOR

Kare-wa mono-o kaku koto-ni [zibun/hontoono zibun]-o miidasi-ta.
he-TOP thing-ACC write COMP-LOC [self/true self]-ACC find-PAST
"He found [himself/his true self] in writing."

THE SOCIAL SELF METAPHOR

Zibun-o azamuite-wa ikena-i.
self-ACC deceive-TOP bad-PRES
Lit.: "To deceive self is bad."
"You must not deceive yourself."

THE EXTERNAL REAL SELF METAPHOR

Boku-wa kyoo-wa zibun-ga zibun de-na-i yoona kigasu-ru.
I(MALE)-TOP today-TOP self-NOM self COP-NEG-PRES as if feel-PRES
Lit.: "I feel as if self is not self today."
"I feel as if I am not my normal self today."

Conclusions and Questions Raised

As we have just seen, we have an extraordinarily rich range of metaphorical concepts for our inner life, and yet they arise from just five basic metaphors, one based on the Folk Theory of Essences and four growing out of basic correlations in our everyday experience since early childhood:

1. The correlation between body control and the control of physical objects.
2. The correlation between being in one's normal surroundings and experiencing a sense of control.
3. The correlation between how those around us evaluate our actions and the actions of others and how we evaluate our own actions.
4. The correlation between our own experience and the way we imagine ourselves projected onto others.

From these simple sources, we obtain an enormously rich conceptual system for our inner lives. Moreover, the fact that the system arises from such basic experiences provides a possible explanation for the occurrence of the same metaphors in a language and culture so different from ours as Japanese. It also raises the question of just how widespread around the world this metaphor system is.

Given such an analysis, what can we conclude? For example, can a conceptual metaphor analysis of how we understand our inner lives tell us anything about what our inner lives are really like?

These metaphors do seem to ring true. They appear to be about real inner experiences, and we use them to make statements that to us are true of our inner lives, statements like, "I'm struggling with myself over whom to marry," "I lost myself in dancing," or "I wasn't myself yesterday." The fact that we can make true statements about our inner lives using these metaphors suggests that these metaphors conform in significant ways to the structure of our inner lives as we experience them phenomenologically. These metaphors capture the logic of much of inner experience and characterize how we reason about it.

We are, of course, acutely aware that these modes of conceptualizing our phenomenological experience of the Self do not entail that the structures imposed by these metaphors are ontologically real. They do not entail that we really are divided up into a Subject, an Essence, and one or more Selves.

One of the most important things that we learn is that there is in this system no one consistent structuring of our inner lives, since the metaphors can contradict one another. Consider two subcases of the Social Self metaphor. In the Master-Servant case, the standards of behavior are set by the Subject, the Master. In the case in which the Subject has an obligation to the Self, it is the Self that sets the standards of behavior for the Subject. We can see this in the minimally different cases, "I was disappointed in myself" (Subject sets the standards) versus "I disappointed myself" (Self sets the standards). In short, we have no single, monolithic concept of the Subject or of the structure of our inner lives, but rather many mutually inconsistent ones.

This study does raise an interesting question. Consider the fact that many of these metaphors seem apt, that they seem to capture something of the qualitative feel of inner life. When we conceptualize a difficult decision metaphorically in terms of inner struggle, many of us experience aspects of such a struggle. When we conceptualize acting sensibly about our bodies metaphorically as caring for ourselves, it is common to experience the affect of caring and being cared for. When we do something we shouldn't have done and *bawl ourselves out,* many of us experience a sense of shame. And when we *betray ourselves,* we can experience a sense of guilt. Such phenomena raise a chicken-egg question: Does the metaphor fit a preexisting qualitative experience, or does the qualitative experience come from conceptualizing what we have done via that metaphor.

The answer is not obvious. It is possible that the activation of the metaphor, that is, of the neural connections between the source and target domains, also activates the source-domain concept (e.g., betrayal), which in turn activates the affect associated with that source-domain concept (e.g., guilt). We do not know whether this is so, but it is one of the intriguing questions raised by the knowledge that we conceptualize our inner lives via metaphor.

14
Morality

Morality is about human well-being. All our moral ideals, such as justice, fairness, compassion, virtue, tolerance, freedom, and rights, stem from our fundamental human concern with what is best for us and how we ought to live.

Cognitive science, and especially cognitive semantics, gives us the means for detailed and comprehensive analysis of what our moral concepts are and how their logic works. One of the major findings of this empirical research is that our cognitive unconscious is populated with an extensive system of metaphoric mappings for conceptualizing, reasoning about, and communicating our moral ideas. Virtually all of our abstract moral concepts are structured metaphorically.

The Experiential Grounding of the Moral Metaphor System

Another striking finding is that the range of metaphors that define our moral concepts is fairly restricted (probably not more than two dozen basic metaphors) and that there are substantial constraints on the range of possible metaphors for morality. These metaphors are grounded in the nature of our bodies and social interactions, and they are thus anything but arbitrary and unconstrained. They all appear to be grounded in our various experiences of well-being, especially physical well-being. In other words, we have found that the source domains of our metaphors for morality are typically based on what people over history and across cultures have seen as contributing to their well-being. For example, it is better to be *healthy,* rather than sick. It is better if the food you eat, the water you drink, and the air you breathe are *pure,* rather than contaminated. It is better to be *strong,* rather than weak. It is better to be *in control,* rather than to be out of control or dominated by others. People seek *freedom,* rather than slavery. It is preferable to have sufficient *wealth* to live comfortably rather than being impoverished. People would rather be *socially connected, protected, cared about,* and *nurtured* than be isolated, vulnerable, ignored, or neglected. It is better to be able to function in the *light,* rather than to be subjected to the fear of the dark. And it is better to be *upright* and *balanced,* than to be off balance or unable to stand.

Around the world and over history, people for the most part have valued these kinds of experiences over their opposites insofar as they believe them to contribute to their well-being. These views appear to constitute a widespread folk theory of what physical well-being is. It is, of course, only an idealized folk theory, since one can easily think of situations in which one of these general norms turns out to be contrary to our actual welfare. For example, a wealthy child may not get the necessary attention from his or her parents; having excessive freedom may actually turn out to be harmful; social ties that are too close can become oppressive; and nurturance overdone may sour into smothering constraint.

Morality is fundamentally seen as the enhancing of well-being, especially of others. For this reason, these basic folk theories of what constitutes fundamental well-being form the grounding for systems of moral metaphors around the world. For example, since most people find it better to have enough wealth to live comfortably than to be impoverished, we are not surprised to find that well-being is conceptualized as wealth. An increase in well-being is a *gain;* a decrease, a *loss.* Since it is better to be healthy than to be sick, it is not surprising to find *im*morality conceptualized as a disease. Immoral behavior is often seen as a contagion that can spread out of control. Since nurturance is an absolutely essential condition for human development, it is not surprising to find an ethics of empathy and care. And because strength enables us to achieve our goals and overcome obstacles, we see moral strength—strength of will—as what makes it possible to confront and overcome evil.

It might initially seem quite unlikely, as it originally did to us, that such a simple list of physical goods, given all of the exceptions to it that one can imagine, could be the basis for virtually all of our metaphors by which we understand our abstract moral concepts. But that is exactly what our cognitive analysis reveals. When we began to analyze the metaphoric structure of these ethical concepts, again and again the source domains were based on this simple list of elementary aspects of human well-being—health, wealth, strength, balance, protection, nurturance, and so on.

To get the basic idea of how our moral understanding is thoroughly metaphoric, we need to look at some of the details of several of the most important metaphors for morality that define the Western moral tradition. (For a fuller treatment of these metaphors, see C1, Johnson 1993 and A1, Lakoff 1996a.) Once we have had a look at the internal logic of each of these metaphors, we can then ask the crucial question of what, if anything, binds these metaphors together into a more or less coherent moral system in our culture.

The Moral Metaphor System

Well-Being Is Wealth and Moral Accounting

We all conceptualize well-being as wealth. We understand an increase in well-being as a *gain* and a decrease of well-being as a *loss* or a *cost*. We speak of *profiting* from an experience, of having a *rich* life, of *investing* in happiness, and of *wasting* our lives. Happiness is conceived as a valuable commodity or substance that we can have more or less of, that we can earn, deserve, or lose.

As we shall see, Well-Being Is Wealth is not our only metaphorical conception of well-being, but it is a component of one of the most important moral concepts we have. It is the basis for a massive metaphor system by which we understand our moral interactions, obligations, and responsibilities. That system, which we call the Moral Accounting metaphor (A1, Taub 1990), combines Well-Being Is Wealth with other metaphors and with various accounting schemas, as follows. Recall that in the Object version of the Event-Structure metaphor, causation is seen as giving an effect to an affected party (as in "The noise *gave* me a headache"). When two people interact causally with each other, they are commonly conceptualized as engaging in a transaction, each transferring an effect to the other. An effect that helps is conceptualized as a gain; one that harms, as a loss. Thus moral action is conceptualized in terms of financial transaction.

The basic idea behind moral accounting is simple: Increasing others' well-being is metaphorically increasing their wealth. Decreasing others' well-being is metaphorically decreasing their wealth. In other words, doing something good for someone is metaphorically giving that person something of value, for example, money. Doing something bad to someone is metaphorically taking something of value away from that person. Increasing others' well-being gives you a moral credit; doing them harm creates a moral debt to them; that is, you *owe* them an increase in their well-being–as–wealth.

Justice is when the moral books are balanced. Just as literal bookkeeping is vital to economic functioning, so moral bookkeeping is vital to social functioning. Just as it is important that the financial books be balanced, so it is important that the moral books be balanced.

It is important to bear in mind that the source domain of the metaphor, the domain of financial transaction, itself has a morality: It is moral to pay your debts and immoral not to. When moral action is understood metaphorically in terms of financial transaction, financial morality is carried over to morality in general: There is a moral imperative to pay not only one's financial debts but also one's moral debts.

The Moral Accounting Schemes

The general metaphor of Moral Accounting is realized in a small number of basic moral schemes: reciprocation, retribution, revenge, restitution, altruism, and so on. Each of these moral schemes is defined using the metaphor of Moral Accounting, but the schemes differ as to how they use this metaphor; that is, they differ as to their inherent logic. Here are the basic schemes.

Reciprocation

If you do something good for me, then I *owe* you something, I am *in your debt*. If I do something equally good for you, then I have *repaid* you and we are even. The books are balanced. This explains why financial words like *owe*, *debt*, and *repay* are used to speak of morality and why the logic of gain and loss, debt and repayment, is used to *think* about morality.

Even in the simple case of reciprocation, two distinct principles of moral action arise from the Moral Accounting metaphor:

1. Moral action is giving something of positive value; immoral action is giving something of negative value.
2. There is a moral imperative to pay one's moral debts; the failure to pay one's moral debts is immoral.

Thus, when you do something good for me, you engage in the first form of moral action. When I do something equally good for you, I engage in *both* forms of moral action. I do something good for you *and* I pay my debts. Here the two principles act in concert.

Retribution and Revenge

Moral transactions get complicated in the case of negative action. The complications arise because moral accounting is governed by a moral version of the arithmetic of keeping accounts, in which gaining a credit is equivalent to losing a debit and gaining a debit is equivalent to losing a credit.

Suppose I do something to harm you. Then, by Well-Being Is Wealth, I have given you something of negative value. You owe me something of equal (negative) value. By the metaphor of Moral Arithmetic, giving something negative is equivalent to taking something positive. By harming you, I have given something of negative value (harm) to you and correspondingly taken something of positive value (well-being) from you. That is why, when one person harms another, the issue arises of whether that person will "get away with it."

By harming you, I have placed you in a potential moral dilemma with respect to the first and second principles of moral accounting given above. Here are the horns of dilemma:

> *The first horn:* If you do something equally harmful to me, you will have done something with two moral interpretations. By the first principle, you have acted immorally since you did something harmful to me ("Two wrongs don't make a right"). By the second principle, you have acted morally, since you have paid your moral debts ("paid me back in kind").
>
> *The second horn:* Had you done nothing to punish me for harming you, you would have acted morally by the first principle: You would have avoided doing harm. But you would have acted immorally by the second principle: In "letting me get away with it" you would not have done your moral duty, which is to make "make me pay" for what I have done.

No matter what you do, you violate one of the two principles. You have to make a choice. You have to give priority to one of the principles. Such a choice gives two different versions of moral accounting: The Morality of Absolute Goodness puts the first principle first. The Morality of Retribution puts the second principle first. As might be expected, different people and different subcultures have different solutions to this dilemma, some preferring retribution, others preferring absolute goodness. For example, in debates over the death penalty, opponents rank absolute goodness over retribution, insisting that evil should never be returned. Those who approve of the death penalty typically give priority to retribution: a life for a life.

The difference between retribution and revenge is one of legitimate authority. When the balancing of moral books is carried out by a legitimate authority, it is retribution. When it is carried out vigilante-style without legitimate authority, it is revenge. If a judge sentences someone to death for the murder of your brother, it is retributive justice, since the judge has legitimate authority. But if you take it upon yourself to balance the moral books by killing the murderer of your brother, you are taking revenge.

The retribution system also plays a central role in defining our concept of honor. Honorable persons are those who can be counted on to pay their moral debts. In other words, an honorable person does what's right and fair, never letting moral debts mount up. Honor is a form of social capital that people get because they are the kind who pay their moral debts. Dishonor is a form of social debt that one accrues by not paying moral debts. Respect is what you get for preserving your honor.

This retributivist system is also the basis for a "morality of honor." Societies that place a great emphasis on honor develop a code according to which a person whose honor is challenged has a duty to defend it. To insult someone is to inflict a metaphorical harm on that person. The "injured" party then has a moral duty to rebalance the moral books by inflicting an equal harm on the person who issued the challenge.

Restitution

If I do something harmful to you, then I have given you something of negative value (harm) and, by moral arithmetic, taken something of positive value (well-being). I then owe you something of equal positive value. I can therefore make restitution—make up for what I have done—by paying you back with something of equal positive value. Of course, in many cases, full restitution is impossible, but partial restitution may be possible.

An interesting advantage of restitution is that it does not place you in a moral dilemma with respect to the first and second principles. You do not have to do any harm, nor is there any moral debt for you to pay, since full restitution, where possible, cancels all debts.

Altruism

If I do something good for you, then by moral accounting I have given you something of positive value. You are then in my debt. In altruism, I cancel the debt, since I don't want anything in return. I nonetheless build up moral credit.

Turning the Other Cheek

If I harm you, I have (by Well-Being Is Wealth) given you something of negative value and (by Moral Arithmetic) taken something of positive value. Therefore, I owe you something of positive value. Suppose you then refuse both Retribution and Revenge. You allow me to harm you further or, perhaps, even do something good for me. By Moral Accounting, either harming you further or accepting something good from you would incur an even greater debt: By turning the other cheek, you make me even more morally indebted to you. If I have a conscience, I should feel even more guilty. Turning the other cheek involves the rejection of retribution and revenge and the acceptance of basic goodness—and when it works, it works via this mechanism of Moral Accounting.

Karma: Moral Accounting with the Universe

The Buddhist theory of *karma* has a contemporary American counterpart: What goes around comes around. The basic idea is that some balance of good and bad things will happen to you, with the bad balancing the good. You can affect the balance by your actions: You will get what you deserve. The more good things you do for people, the more good things will happen to you. The more bad things you do to people, the more bad things will happen to you.

In another version of moral balance with the universe, the good and bad things that happen to you are balanced out. Thus, we occasionally find people saying things like, "Things have been rotten for a long time. They're bound to get better." or "Too many good things have been happening to me. I'm starting to get scared."

Fairness

According to the Moral Accounting metaphor, justice is the settling of accounts, which results in the balancing of the moral books. Justice is understood as fairness, in which people get what they *deserve,* that is, when they get their *just deserts.*

However, there are several different conceptions of what the basis should be for tallying up the moral books in a fair way. From the time we are toddlers we learn what is and isn't fair. It's fair when the cookies are divided equally, when everybody gets a chance to play, when following the rules gives everyone an equal chance at winning, and when everybody does his or her job and gets paid equally for the work.

In general, fairness concerns the equitable distribution of things of value (either positive or negative values) according to some accepted standard. The objects that get distributed may be either physical objects and goods (such as cookies, balloons, or money) or else metaphorical objects (such as job opportunities, the option to participate in some activity, tasks to be done, punishments, or commendations).

Thus, fairness is assessed relative to any of a number of models:

Equality of distribution (one person, one "object")
Equality of opportunity
Procedural distribution (playing by the rules determines what you get)
Rights-based fairness (you get what you have a right to)
Need-based fairness (the more you need, the more you have a right to)
Scalar distribution (the more you work, the more you get)
Contractual distribution (you get what you agree to)
Equal distribution of responsibility (we share the burden equally)
Scalar distribution of responsibility (the greater your abilities, the greater your responsibilities)
Equal distribution of power

In one of the most basic conceptions of morality we have, moral action is fair distribution and immoral action is unfair distribution. As the above list indicates, there are a number of different models for working out the details of fairness. Most people operate with several or all of these models at once, even though they are not all mutually consistent. Many of our moral disagreements arise from conflicts between two or more of these conceptions of fairness. For example, there are a myriad of cases in which people generally agree on the necessity of some procedural rules of distribution, but find at times that following those "fair" procedures results in a distribution of goods or opportunities that conflicts with their sense of rights-based fairness or equality of distribution fairness. In such cases there is typically no overarching neutral conception of fairness that can resolve the conflict of values.

Rights as Moral I.O.U.'s

There are two basic conceptions of rights, one defined relative to the Moral Accounting system and the other defined relative to our view of Moral Bounds (see below). According to the Moral Accounting metaphor, rights are letters of credit that entitle you to possess certain moral goods, namely, certain conditions and aspects of well-being.

In the source domain of the Moral Accounting metaphor (i.e., financial accounting), rights are conceived as rights to one's property. If the bank is keeping your money, you have a right to get it back upon request. If someone has borrowed money from you, you have a right to be paid back. Combining this notion of financial property rights with the Well-Being Is Wealth metaphor yields a notion of a broader right—a right to one's well-being, special cases of which are life, liberty, and the pursuit of happiness. Having a specific right is equivalent to holding an I.O.U. redeemable for various specific forms of human well-being, such as the freedom to vote, equal access to public offices, and equal opportunities for employment.

A right is thus a form of metaphorical social capital that allows you to claim certain debts from others. A duty is conceived as a standing debt: You have to pay whatever your moral debts are. The concepts *right* and *duty* are therefore second-order metaphorical concepts. They are abstract debts and credits we have regarding the specific moral debts and credits we accrue.

Well-Being as Wealth and Moral Self-Interest

Morality typically concerns the promotion of the well-being of others and the avoidance or prevention of harm to others. Thus, it might seem that the pursuit of one's self-interest would hardly be seen as a form of moral action. Indeed, the expression "moral self-interest" might seem to be a contradiction in terms. Yet there is a pair of metaphors that turns the pursuit of self-interest into moral action.

The first is an economic metaphor: Adam Smith's metaphor of the Invisible Hand. Smith proposed that, in a free market, if we all pursue our own profit, then an Invisible Hand will operate to guarantee that the wealth of all will be maximized. The second is Well-Being Is Wealth. When combined with Well-Being Is Wealth, Smith's economic metaphor becomes a metaphor for morality: Morality Is The Pursuit Of Self-Interest.

For those who believe in the Morality of Self-Interest, it can never be a moral criticism that one is trying to maximize one's self-interest, as long as one is not interfering with anyone else's self-interest.

The Morality of Self-Interest fits very well with the Enlightenment view of human nature, in which people are seen as rational animals and rationality is seen as means-end rationality—rationality that maximizes self-interest.

Moral Strength

One can have a sense of what is moral and immoral and still not have the ability to *do* what is moral. An essential condition for moral action is strength of will. Without sufficient moral strength, one will not be able to act on one's moral knowledge or to realize one's moral values. Consequently, it is hard to imagine any moral system that does not give a central role to moral strength.

The metaphor of Moral Strength is complex. It consists of both the strength to maintain an upright and balanced moral posture and also the strength to overcome evil forces. The uprightness aspect of this metaphor is experientially grounded in the fact that, other things being equal, it is better to be upright and balanced. When one is healthy and in control of things, one is typically upright and balanced. Thus, *moral* uprightness is understood metaphorically in terms of physical uprightness: Being Moral Is Being Upright; Being Immoral Is Being Low. Examples include:

> He's an *upstanding* citizen. She's on the *up and up*. She's as *upright* as they come. That was a *low* thing to do. He's *under*handed. I would never *stoop* to such a thing.

Doing evil is therefore moving from a position of morality (uprightness) to a position of immorality (being low). Hence, Doing Evil Is Falling. The most famous example, of course, is the Fall from grace.

Since physical uprightness requires balance, there is an entailed metaphor: Being Good Is Being Balanced. Someone who cannot control himself enough to remain balanced is likely to fall, that is, to commit immoral acts at any moment. Thus, an *unbalanced* person cannot be trusted to do what is good.

The second aspect of the Moral Strength metaphor concerns control over oneself and over evil. Evil is reified as a force, either internal or external, that can make you fall and lose control, that is, make you commit immoral acts. Thus, metaphorically,

Evil Is A Force (Either Internal Or External)

External evil is understood metaphorically either as another person who struggles with you for control or else as an external force (of nature) that acts on you. Internal evil is the force of your bodily desire, which is conceived metaphorically as either a person, an animal, or a force of nature (as in "*floods of emotion*" or "*fires of passion*"). Thus, to remain upright, one must be strong enough to *stand up to* evil. Hence, morality is conceptualized as strength, as having the *moral fiber* or *backbone* to resist evil. Therefore,

Morality Is Strength

But people are not simply born strong. Moral strength must be built. Just as building physical strength requires self-discipline and self-denial ("No pain, no gain"), so moral strength is also built through self-discipline and self-denial. One consequence of this metaphor is that punishment can be good for you when it is in the service of moral discipline. Hence, the admonition "Spare the rod and spoil the child."

By the logic of the metaphor, moral weakness is in itself a form of immorality. The reasoning goes like this: A morally weak person is likely to fall, to give in to evil, to perform immoral acts, and thus to become part of the forces of evil. Moral weakness is thus nascent immorality—immorality waiting to happen.

There are two forms of moral strength, depending on whether the evil to be faced is external or internal. *Courage* is the strength to stand up to *external* evils and to overcome fear and hardship. *Willpower* is the strength of will necessary to resist the temptations of the flesh. The opposite of self-control is *self-indulgence*—a concept that only makes sense if one accepts the metaphor of moral strength. Self-indulgence is seen in this metaphor as a vice, while frugality and self-denial are virtues. The seven deadly sins is a catalogue of internal evils to be overcome: greed, lust, gluttony, sloth, pride, envy, and anger. It is the metaphor of moral strength that makes them "sins." The corresponding virtues are charity, sexual restraint, temperance, industry, modesty, satisfaction with one's lot, and calmness. It is the metaphor of moral strength that makes these "virtues."

To summarize, the metaphor of Moral Strength consists of the following mapping:

THE MORAL STRENGTH METAPHOR

Being Upright	→	Being Good
Being Low	→	Being Bad
Falling	→	Doing Evil
A Destabilizing Force	→	Evil (Internal Or External)
Strength (To Resist)	→	Moral Virtue

The metaphor of Moral Strength has an important set of entailments:

- To remain good in the face of evil (to "stand up to" evil), one must be morally strong.
- One becomes morally strong through self-discipline and self-denial.

- Someone who is morally weak cannot stand up to evil and so will eventually commit evil.
- Therefore, moral weakness is a form of immorality.
- Lack of self-control (the lack of self-discipline) and self-indulgence (the refusal to engage in self-denial) are therefore forms of immorality.

Moral Authority

Authority in the moral sphere is modeled on dominance in the physical sphere. The moral authority of the parent over the child is metaphorically modeled on the physical dominance of the parent over the young child. The father has the authority to issue commands that must be obeyed by his children. Parental authority is moral authority in the family, which is the capacity to define the moral principles governing the family (hence the term *paternalism*). This is not a might-makes-right literal model. It is a metaphorical model in which the logic of moral authority makes use of the logic of physical dominance. To understand the issues surrounding moral authority, one must look at the two common versions of the notion of parental authority.

Version 1: Legitimate Authority

Young children need to be told what will hurt them and when they are hurting others. They need to learn what's safe and what's harmful, what's right and what's wrong, or they are likely to get hurt or hurt others. Parents have the responsibility of protecting and nurturing their children, teaching them how to protect and care for themselves and how to act morally toward others. Being a good parent requires wisdom in all these matters. That wisdom—or the lack of it—is manifest every day. Parents also have the responsibility of acting morally themselves, setting an example for their children.

It is responsibility, wisdom, and moral action by parents that justifies parental authority and creates the moral imperative for children to obey their parents. Children should obey their parents because their parents have the responsibility of nurturing, protecting, and educating them, because their parents care about them, because their parents have the knowledge and wisdom to carry out their responsibilities of nurturance, protection, and education, and because their parents themselves set an example through moral action. Parents earn the respect and obedience of their children by nurturing, protecting, and educating effectively and by acting morally. Such earned respect is what makes their authority legitimate.

Children have a *right* to adequate nurturance, protection, and education, and parents have a moral *duty* to provide it. When parents perform their moral *duty*, they earn the *right* to be respected and obeyed. It is the performance of the parents in nurturing, protecting, and educating that imposes on their children the moral *duty* to obey them. If parents fail to nurture, protect, and educate, then they have not earned the respect and obedience of their children. Abusive, neglectful, or immoral parents earn no such respect and have no legitimate parental authority.

Version 2: Absolute Authority
Parental authority is absolute. Children have a moral obligation to obey their parents and show them respect, simply because they are their parents, no matter what they are like or what they do.

These are, of course, two extremes. Variations of all sorts exist, but the extreme cases highlight the general issue of the legitimation of authority in the more restricted arena of parental authority, which is the metaphorical source of our general concept of moral authority. Versions of general moral authority will vary with versions of parental authority. The metaphor that characterizes moral authority in terms of parental authority is as follows:

MORAL AUTHORITY IS PARENTAL AUTHORITY

An Authority Figure Is A Parent
A Moral Agent Is A Child
Morality Is Obedience

Knowledge mapping:

- Your parents have your best interests at heart and know what is best for you; therefore you should obey them and accept their teachings.
- A moral authority has your best interests at heart and knows what is best for you; therefore you should obey and accept teachings of the moral authority.

There are many kinds of moral authorities—the gods, prophets, and saints of various religions; people (e.g., spiritual leaders, dedicated public servants, people with a special wisdom); texts (e.g., the Bible, the Qur'an, the *Tao Te Ching*); institutions with a moral purpose (e.g., churches, environmental groups). What counts as a moral authority to a given person will depend on that person's moral and spiritual beliefs as well as his or her understanding of parental authority.

Moral Order

Closely related to the notion of moral authority is the idea of an ideal moral order that justifies the moral authority of certain individuals. This metaphor is based on the Folk Theory of the Natural Order, according to which the natural order is the order of dominance that occurs in the world. Key examples of this hierarchy of dominance are:

God is naturally more powerful than people.
People are naturally more powerful than animals, plants, and natural objects.
Adults are naturally more powerful than children.
Men are naturally more powerful than women.

In nature, according to this folk theory, the strong and better-endowed tend to dominate the weak. In the metaphor of the Moral Order, this natural order of domination is mapped onto a moral order:

The Moral Order Is The Natural Order

This metaphor transforms the folk hierarchy of "natural" power relations into a hierarchy of moral superiority and authority:

God has moral authority over people.
People have moral authority over nature (animals, plants, objects).
Adults have moral authority over children.
Men have moral authority over women.

The Moral Order metaphor does not merely legitimize power relations and establish lines of moral authority. It also generates a hierarchy of moral responsibility, in which those in authority at a given level have responsibilities toward those over whom they have that authority. Thus, according to this metaphor, people have responsibilities toward nature and adults have responsibilities toward children.

The consequences of the metaphor of Moral Order are sweeping, momentous, and, we believe, morally repugnant. The metaphor legitimizes a certain class of existing power relations as being natural and therefore moral. In this way it makes certain social movements, such as feminism, appear to be unnatural and therefore counter to the moral order. It legitimizes the view that nature is a resource for humans and that humans should be good stewards of that resource. Accordingly, it undermines alternative views of nature, for example, the view that nature has inherent value and ought not to be subject to human will.

The Moral Order hierarchy is commonly extended in this culture to include other relations of moral superiority: Western culture over non-Western culture; America over other countries; citizens over immigrants; Christians over non-Christians; straights over gays; the rich over the poor. Incidentally, the Moral Order metaphor gives us a better understanding of what fascism is: Fascism legitimizes such a moral order and seeks to enforce it through the power of the state.

Moral Bounds

According to the Event-Structure metaphor action is conceptualized as a form of self-propelled motion and purposes, as destinations we are trying to reach. Moral action is seen as bounded movement, movement in permissible areas and along permissible paths. Immoral action is seen as motion outside of the permissible range, as straying from a prescribed path or transgressing prescribed boundaries. To characterize morally permissible actions is to lay out paths and areas where one can move freely. Immoral actions are those that in some way violate these boundaries, either by interfering with other people's morally permissible actions or else by entering bounded areas that are morally *off limits.*

According to this metaphor, "deviant" behavior is immoral because it moves in unsanctioned areas and toward unsanctioned destinations. Since action is conceived as self-propelled motion along a path, someone who deviates from the normal paths is saying that it is all right to go "another way." Moreover, some deviant paths will lead to entirely new destinations, that is, to entirely new ends. This goes a long way toward explaining the extreme hostility evoked in some people by any behavior that they regard as deviant. According to the Moral Bounds metaphor, someone who moves off of sanctioned paths or out of sanctioned territory is doing more than merely acting immorally. She is rejecting the purposes, the goals, the very mode of life of the society she is in. In so doing, she is calling into question the purposes that govern most people's everyday lives. Certain people regard such "deviation" from social norms as, therefore, threatening to the entire moral order, since it suggests that their ends, purposes, and boundaries are not absolute and are not the only morally permissible ones.

Constraints on Freedom

Since freedom of action is understood metaphorically as freedom of motion, moral bounds can be, and often are, seen as restraints on freedom. In general, we seek maximum freedom to pursue our different ends. In the Western moral tradition, morality has often been conceived as the maximizing of individual freedom. Freedom of this sort cannot be absolute, however, since some of our free actions might interfere with a like freedom for other people. Consequently, the question of legitimate constraints on freedom lies at the heart of many ethical and political debates. For instance, people who want to impose their moral views on others are seen as restricting the freedom of others, and the question thus arises as to whether there is any morally justified basis for setting such bounds.

Rights as Right-of-Ways

As we saw earlier, one of our two basic conceptions of rights is as an I.O.U., a form of metaphorical moral credit redeemable for various aspects of well-being. The other major conception is defined relative to the Moral Bounds metaphor. When actions are understood metaphorically as motions along paths, then anything that blocks that motion is a constraint on one's freedom. Accordingly, a right becomes a *right-of-way*, an area through which one can move freely without interference from other people or institutions. Moving freely is not just physical motion but, by the Event-Structure metaphor, action of any sort. Since moral bounds leave open and close off areas of free movement, they define rights to free action and freedom from interference.

Those rights impose a corresponding duty on others not to limit that freedom of action. For example, proponents of property rights, such as real estate developers, see environmental regulations as restrictions on the free disposition of their property and therefore want to eliminate governmental regulations as a restriction on their rights. On the other hand, people who see human beings as having a right to a clean, healthy, and biologically diverse environment see unregulated development as "encroaching" on *their* rights. Moral and legal boundaries can thus be seen from two perspectives: What is one person's constraint on free movement is another person's protection against encroachment. This is the metaphorical logic by which moral and legal bounds define conflicts of rights.

Moral Essence

According to the Folk Theory of Essences, objects have natures, defined by sets of properties, that determine their behavior. So, too, with people: Each person has a moral essence that determines his or her moral behavior. That moral essence is called someone's "character."

Imagine judging someone to be inherently stubborn or reliable. To do so is to assign that person an inherent trait, an essential property that determines how he or she will act in certain situations. If the trait is a moral trait, then we have a special case of the metaphor of Essence—the metaphor of Moral Essence. In social psychology, there is an expert version of this metaphor called the "trait theory of personality." We are discussing the folk version here.

According to the metaphor of Moral Essence, people are born with, or develop in early life, essential moral properties and habits that stay with them for life. Such properties are called *virtues* if they are moral properties and habits and *vices* if they are immoral ones. The collection of virtues and vices attributed to a person is called that person's "character." When people say "She has a *heart of gold*," "He doesn't have a *mean bone in his body*," or "He's *rotten to the core*," they are making use of the metaphor of Moral Essence. That is, they are saying that the person in question has certain essential moral qualities that determine certain kinds of moral or immoral behavior.

The metaphor of Moral Essence has three important entailments:

- If you know how a person has acted, you know what that person's character is.
- If you know what a person's character is, you know how that person will act.
- A person's basic character is formed by the time they reach adulthood (or perhaps somewhat earlier).

These entailments form the basis for certain currently debated matters of social policy.

Take, for example, the "Three strikes and you're out" law now gaining popularity in the United States. The premise is that repeated past violations of the law indicate a character defect, an inherent propensity to illegal behavior that will lead to future crimes. Since a felon's basic character is formed by the time he reaches adulthood, felons are *rotten to the core* and cannot change or be rehabilitated. They therefore will keep performing crimes of the same kind if they are allowed to go free. To protect the public from their future crimes, they must be locked up for life.

Or take the proposal to take illegitimate children away from impoverished teenage mothers and put them in orphanages or foster homes. The assumption is that these mothers are immoral, that it is too late to change them since their character is already formed. If the children stay with these mothers, they will also develop an immoral character. But if the children are removed from the mothers before their character is formed, the children's character can be shaped in a better way.

The same premise has been used to justify social programs such as Project Head Start: If you get children early enough, before their basic character is formed, then you can instill in them virtues like responsibility, self-discipline, and caring that will last a lifetime.

The ubiquity and power of the metaphor of Moral Essence has been most manifest in the O. J. Simpson trial. Simpson was a hero, and heroes are conceptualized as being inherently good people. The question that people kept repeating was "How can a good person do bad things?" The very idea that a hero, someone defined as inherently good, could commit two brutal murders simply does not fit the metaphor of Moral Essence. The fact that the Simpson trial has created so much cognitive dissonance is testimony to the power of the metaphor.

Moral Purity

A substance is pure when it has no admixture of any other substance within it. A common impurity is dirt. Thus, substances that are pure are typically clean, and substances that are dirty are usually considered impure. This correlation between purity and cleanliness gives rise to the metaphor Purity Is Cleanliness. Thus, when morality is conceptualized as purity and purity as cleanliness, we get the derived metaphor Morality Is Cleanliness.

There is nothing inherent in the notion of purity that aligns it with goodness. There can be *pure evil* just as well as *pure goodness*. However, in the moral realm purity takes on a positive value—remaining *pure* is a good and desirable thing, while being *impure* (e.g., having *impure thoughts*) is seen as being bad.

Purity is thus contrasted with being soiled, tainted, blemished, and stained. For the most part, in the metaphor of Moral Purity, it is the body that is the source of impurity. In more extreme versions of the metaphor the body is seen as disgusting and even evil.

A well-known philosophical version of this folk theory appropriates a metaphorical faculty psychology and regards the *will* as the source of moral action. The will must remain pure in its moral deliberation and choice. Being *pure* here means being rational, following only the commands of reason, and not letting itself be *tainted* by anything of the body, such as desires, emotions, or passions. The *will* or the *heart* are pure when they act under the guidance of reason and not under the influence of the body, which is seen as an alien force that struggles with reason for the control of the will.

Moral purity is thus contrasted with impurity (i.e., immorality) and with anything that is disgusting. This gives rise to expressions such as:

She's *pure* as the driven snow. He's a *dirty* old man. O Lord, create a *pure* heart within me. Let me be without *spot* of sin. That was a *disgusting* thing to do! If elected, I will *clean up* this town!

There are far-reaching entailments of this metaphor. Just as physical impurities can ruin a substance, so moral impurities can ruin a person or a society. Just as substances can be purged of impurities, so people and societies must be purged of corrupting elements, individuals, or practices. Within an individual, Moral Purity is often paired with Moral Essence. If a person's moral essence is pure, then that person is expected to act morally. If someone's essence is corrupt, that is, if it has been made impure by some evil influence, then he or she will act immorally. In this context, the question of moral rehabilitation amounts to the question of whether it is possible to *clean up* one's act and restore purity of will. The doctrine of original sin is the view that the human moral essence is inherently tainted and impure, and that people will therefore act immorally when left to their own devices.

Morality as Health

Health, for most people, plays an important role in their living a full and happy life. It is not surprising, therefore, that there exists a basic metaphor of Well-Being Is Health, by which we understand moral well-being in general by means of one particular aspect of it, health.

One crucial consequence of this metaphor is that immorality, as moral disease, is a plague that, if left unchecked, can spread throughout society, infecting everyone. This requires strong measures of moral hygiene, such as quarantine and strict observance of measures to ensure moral purity. Since diseases can spread through contact, it follows that immoral people must be kept away from moral people, lest they become immoral, too. This logic underlies guilt-by-association arguments, and it often plays a role in the logic behind urban flight, segregated neighborhoods, and strong sentencing guidelines even for nonviolent offenders.

Many people in this culture tend to regard impurities as causes of illness. This establishes a conceptual link between Moral Purity and Morality As Health. This connection between moral purity and moral health is evident in the *Book of Common Prayer* of the Church of England, in which people confess, "We are by nature sinful and *unclean*, and there is no *health* in us." The *health* here is moral health.

Moral Empathy

Empathy is the capacity to take up the perspective of another person, that is, to see things as that person sees them and to feel what that person feels. It is conceptualized metaphorically as the capacity to project your consciousness into other people, so that you can experience *what* they experience, the *way* they experience it. This is metaphorical, because we cannot literally inhabit another person's consciousness.

The logic of moral empathy is this: If you feel what another person feels, and if you want to feel a sense of well-being, then you will want that person to experience a sense of well-being. Therefore, you will act to promote that person's well-being. The morality of empathy is not merely that of the Golden Rule ("Do unto others as you would have them do unto you"), because others may not share your values. Moral empathy requires, instead, that you make their values your values. This constitutes a much stronger principle, namely, "Do unto others as they would have you do unto them."

There are thus two basic conceptions of moral empathy. *Absolute empathy* is simply feeling as someone else feels, with no strings attached. But very few people would ever espouse this as moral doctrine, since we recognize that other people sometimes have values that are inappropriate or even immoral. Most of the time, we project onto other people not just our capacity to feel as they feel, but also our own value system. This is *egocentric empathy*, which is a way of trying to reach out to other people while preserving your own values.

Moral Nurturance

In order to survive and develop into a normal adults, children need to be nurtured. They need to be fed, protected from harm, sheltered, loved, kept clean, educated, and cared for. Besides being essential for their very existence, such nurturance teaches them how to care for other people. Learning how to care for others requires empathy, concern for the other, responsibility, caring for oneself, and so on. Empathy is necessary in order to understand what children need. Concern for their well-being moves you to act on their behalf. Children are seen as having a right to nurturance, and parents have a responsibility to provide it. A parent who does not adequately nurture a child is thus metaphorically robbing that child of something it has a right to. For a parent to fail to nurture a child is immoral.

The Morality As Nurturance metaphor maps this practical necessity for nurturance onto a moral obligation to nurture others. In conceiving of morality as nurturance, the notion of family-based morality is projected onto society in general via the following mapping:

THE MORALITY AS NURTURANCE METAPHOR

Family Nurturance		*Moral Nurturance*
Family	$\rightarrow$	Community
Nurturing Parents	$\rightarrow$	Moral Agents
Children	$\rightarrow$	People Needing Help
Nurturing Acts	$\rightarrow$	Moral Actions

Morality as Nurturance has a different logic and different entailments from a morality based on absolute principles and corresponding duties. The core of nurturance is empathy and compassion for the other. It focuses not on one's own rights but on the fundamental responsibility to care for other people.

Just as there are different conceptions of empathy, so there are different views of what moral nurturance requires. On the model of *absolute empathy,* moral nurturance requires that you act so as to make it possible for others to realize their goals according to their own value system. On the model of *egocentric empathy,* by contrast, you must understand how others see things and how they feel about them, but your careful concern is guided by your own value system. You strive to help them grow and to appropriate your basic values.

Another crucial dimension of moral nurturance is the responsibility you have to nurture yourself. You cannot care appropriately for others if you haven't cared for yourself. This is both a psychological fact and a moral obligation. The psychological fact is that without proper self-respect, self-esteem, and modest concern for your own well-being, you simply cannot know how to nurture other people. The moral obligation stems from the Moral Nurturance metaphor itself. You have a responsibility to nurture human beings, of which *you* are one. According to the logic of the metaphor, it is just as immoral not to care for yourself properly as it is not to care for other people.

There is nothing intrinsically selfish about moral self-nurturance. A selfish person puts his self-interest ahead of the needs and well-being of those he has a duty to nurture. But someone who simply attends to his most basic needs, who makes self-nurturance a prerequisite to the care of others, is not selfish.

Finally, there are two basic versions of Moral Nurturance, one concerning individuals and the other concerning social relations. In this latter version, what one nurtures are the *social ties* that bind people together into communities. As Gilligan (E, 1982) has observed, an "ethics of care" places special emphasis on cooperation and compromise in the service of maintaining the social and communal bonds that unite us. There are cases in which the individual and social versions of Moral Nurturance can conflict. For example, the obligation of social nurturance might require you to work to preserve social ties to people in your community who do not themselves believe in nurturance. Or you might encounter situations in which you must sacrifice some of your commitments to individuals in order to hold the community together in times of social upheaval.

What Binds Our Metaphors for Morality Together?

Our survey of metaphors for morality is by no means exhaustive. A more complete list would also, for instance, include Morality Is Light/Immorality Is Darkness, Moral Beauty, Moral Balance, and Moral Wholeness, but it contains the most important and representative examples. These metaphors define a large part of the Western moral tradition, but they are not unique to occidental culture. They are also widespread around the world, since their source domains come primarily from basic human experiences of well-being. The cross-cultural research has not been done yet to determine whether any of them are truly universal, but some of them, such as Moral Strength and Moral Accounting are good candidates.

What we have seen so far is clear: Our abstract moral concepts are metaphorical, and we reason via those metaphors. We believe that the evidence for the metaphorical character of moral understanding is quite substantial. We have surveyed only a small part of it.

We now turn from this relatively well-established claim to one that is far less obvious and more highly speculative. This concerns the issue of what, if anything, binds these several metaphors together into a coherent moral view. As you read through the list you may sense that, in your own experience, they somehow must fit together. But how? What connects them, gives certain metaphors priority over others, and makes them form a coherent system that a person can actually act on?

What we are about to propose does not have the massive body of convergent evidence to support it that is available for conceptual metaphor. But the thesis we propose does explain how our metaphors might get organized into the systems we have described, and it also shows how they can be criticized.

In studying the metaphors that underlie Western political liberalism and conservatism, Lakoff (A1, 1996a) proposed that these two political orientations are ultimately based on different models of the family. Mainstream conservatism, he claimed, is grounded on what he called a "strict father" model, whereas mainstream liberalism is based on a "nurturant parent" model. Since each family model includes it own morality, political liberalism and conservatism express different views of morality. Each family model organizes the culturally shared metaphors for morality in different ways, giving priority to certain metaphors and downplaying others. Moreover, each particular metaphor for morality (e.g., Moral Strength or Moral Nurturance) gets a unique interpretation depending on which family model it is identified with. In the Strict Father model, as we shall see below, moral strength is given top priority as the key to acting morally, whereas in the Nurturant Parent model moral strength is also important, but it does not override empathy and responsibilities for nurturance.

Lakoff's political analysis raises the interesting possibility that morality, too, might also be based on models of the family. This thesis makes good sense for two main reasons. First, it is within the family that a child's moral sensibility and understanding are first formed. For the infant and young child, morality just *is* family morality. Second, the vast majority of a child's moral education stems from the family situation. Obviously, there are massive societal influences on the child's development and values. But all of these get filtered through the child's family morality. For example, if the child does not experience respect for other people, as well as self-respect, within the family, it is extremely difficult to incorporate these values from society at large.

Our hypothesis about moral understanding, then, is that it is models of the family that order our metaphors for morality into relatively coherent ethical perspectives by which we live our lives. To see how this works, we need to investigate the two fundamental models of the family to see how each one assigns different priorities to certain metaphors and thereby creates different moral orientations. We need to describe both the Strict Father and the Nurturant Parent models of the family, along with the moral systems that each one entails. Then we can ask whether these two models of family morality can become the bases for our understanding of morality in general.

The Strict Father Family Morality

We live in a world full of dangers, pitfalls, and conflict. To survive in such a world we need to be strong and we need to have our values firmly in place. The Strict Father family model emerges in response to this perception of life as hard and dangerous. It is a model of the family geared toward developing strong, morally upright children who are capable of facing the world's threats and evils. Here is the basic Strict Father family model.

> The family is a traditional nuclear one, with the father having primary responsibility for supporting and protecting the family. The father has authority to determine the policy that will govern the family. Because of his moral authority, his commands are to be obeyed. He teaches his children right from wrong by setting strict rules for their behavior and by setting a moral example in his own life. He enforces these moral rules by reward and punishment. The father also gains his children's cooperation by showing love and by appreciating them when they obey the rules. But children must not be coddled, lest they become spoiled. A spoiled child lacks the appropriate moral values and lacks the moral strength and discipline necessary for living independently and meeting life's challenges.
>
> The mother has day-to-day responsibility for the care of the household, raising the children, and upholding the father's authority. Children must respect and obey their parents, because of the parents' moral authority.

Through their obedience they learn the discipline and self-reliance that is necessary to meet life's challenges. This self-discipline develops in them strong moral character. Love and nurturance are a vital part of family life, but they should never outweigh parental authority, which is itself an expression of love and nurturance—tough love. As children mature, the virtues of respect for moral authority, self-reliance, and self-discipline allow them to incorporate their father's moral values. In this way they incorporate their father's moral authority—they become self-governing and self-legislating. In certain versions, the children are then off on their own and it is inappropriate for the father to meddle in their lives.

This model is an idealization meant to capture the basic structure and values that define Strict Father families. It will have variants, such as when the "strict father" is replaced by a "strict mother" who instantiates the moral authority, moral strength, and self-discipline necessary for governing the family.

The Strict Father family model carries its own distinctive cluster of moral values. It defines a Strict Father family morality. As you would expect, it gives top priority to the metaphors of Moral Authority, Moral Strength, and Moral Order. The Strict Father family is seen as manifesting an appropriate moral order in which the father is naturally fitted to run the family and the parents have control over their children. The strict father's moral authority comes from his natural dominance and strength of character. His moral strength and self-discipline make him the fitting embodiment of morality, a model for his children.

Moral Empathy and Moral Nurturance have a place in this family morality, but they are always subservient to the primary goal of developing moral strength and recognizing legitimate moral authority. In other words, moral nurturance is always for the sake of the cultivation of moral strength. Moral empathy has its place, but it can never be permitted to conflict with the need to discipline children for their own good.

Reward and punishment are moral in this scheme, not just for their own sake, but rather because they help the child succeed in a world of struggle and competition. To survive and compete, children must learn discipline and must develop strong character. Children are disciplined (punished) in order to become self-disciplined. Self-discipline and character are developed through obedience. Obedience to authority thus does not disappear when the child grows into adulthood. Being an adult means that you have become sufficiently self-disciplined so that you can be obedient to your own moral authority—that is, being able to carry out the plans you make and the commitments you undertake.

Nurturant Parent Family Morality

Consider now a contrasting moral system built around a second idealized model of the family—a Nurturant Parent family.

The primal experience behind this model is that of being cared for and cared about, having one's desires for loving interactions met, living as happily as possible, and deriving meaning from mutual interaction and care.

Children develop best in and through their positive relationships to others, through their contribution to their community, and through the ways in which they realize their potential and find joy in life. Children become responsible, self-disciplined, and self-reliant through being cared for and respected and through caring for others. Support and protection are part of nurturance, and they require strength and courage on the part of parents. Ideally, as children mature, they learn obedience out of their love and respect for their parents, not out of the fear of punishment.

Open, two-way, mutually respectful communication is crucial. If parents' authority is to be legitimate, they must tell children why their decisions serve the cause of protection and nurturance. They must allow their children to ask questions about why their parents do what they do, and all family members should participate in important decisions. Responsible parents, of course, have to make the ultimate decisions, and that must be clear.

Protection is a form of caring, and protection from external dangers takes up a significant part of the nurturant parent's attention. The world is filled with evils that can harm a child, and it is the nurturant parent's duty to ward them off.

The principal goal of nurturance is for children to be fulfilled and happy in their lives and to become nurturant themselves. This involves learning self-nurturance as a necessary condition for caring for others. A fulfilling life is assumed to be, in significant part, a nurturant life—one committed to family and community responsibility. What children need to learn most is empathy for others, the capacity for nurturance, cooperation, and the maintenance of social ties, which cannot be done without the strength, respect, self-discipline, and self-reliance that comes through being cared for and caring.

Though this model is very different from the Strict Father model, they both have one very important thing in common. They both assume that the system of child-rearing will be reproduced in the child. In the Strict Father model, discipline is incorporated by the child so that, by adulthood, it has become *self-discipline* and the ability to discipline others. In the nurturant parent model, nurturance is incorporated into the child to eventually become self-nurturance (the ability to take care of oneself) and the ability to nurture others.

Nurturant Parent morality thus has a very different set of priorities in its metaphors for morality than those built into Strict Father morality. The dominant metaphor is Morality Is Nurturance. Nurturance is seen as the basis for all moral interactions within the family. Moral Empathy is also given special emphasis as a necessary condition for appropriate caring for other family members. Thus we ask, "How would *you* like it if your sister did to you what you did to her?" "How do you think *he* feels when you treat him that way?"

Moral Authority is subservient to, and is legitimized by, the parents' nurturant character and behavior. The metaphor of Moral Order plays little or no role in this model. Moral Strength is important, but it is understood relative to the obligation of the nurturant parent to be morally strong and to exercise that strength in protecting and caring for the children. It is part of the responsibility of nurturance to develop moral strength in the child. Nurturant Parent morality is thus a specific version, adapted to the family setting, of what Gilligan (E, 1982) calls an "ethics of care."

Nurturant Parent morality is not, in itself, overly permissive. Just as letting children do whatever they want is not good for them, so helping other people to do whatever they please is likewise not proper nurturance. There are limits to what other people should be allowed to do, and genuine nurturance involves setting boundaries and expecting others to act responsibly. Of course, there are what Lakoff has called "pathological" versions of Nurturant Parent morality that are excessively and imprudently permissive.

Just as there exist Strict Father versions of Judaism and Christianity, likewise there are Nurturant Parent versions of both of these religious traditions. Most notably, in the kabbalistic tradition in Judaism the Shekhinah is understood as a nurturant female manifestation of God. In Catholicism the Virgin Mary is often seen as providing a female model of divine nurturance.

Is All Morality Based on Models of the Family?

The Strict Father and Nurturant Parent models of the family, each with its own distinctive morality, are idealizations. In reality, the family models that people actually experience will seldom measure up to these idealizations. More often than not, one's family situation is either some particular version of either model or else it is a blending of various elements from both models. The range of existing variations and blends of these models is extremely wide (A1, Lakoff 1996a). Pure instances of either form of family morality are rare.

However, we believe that these two models capture something very important about human morality, namely, that it is ultimately based on some conception of the family and of family morality. To think of morality in general as some form of family morality requires another metaphor, in which we understand all of humanity as part of one huge family, which has traditionally been called the "Family of Man" (i.e., the family of all humans). This metaphor entails a moral obligation, binding on all people, to treat each other as we ought to treat our family members.

THE FAMILY OF MAN METAPHOR

Family	→	Humankind
Each Child	→	Each Human Being
Other Children	→	Every Other Human Being
Family Moral Relations	→	Universal Moral Relations
Family Moral Authority	→	Universal Moral Authority
Family Morality	→	Universal Morality
Family Nurturance	→	Universal Moral Nurturance

Since the metaphor projects family moral structure onto a universal moral structure, the moral obligations toward family members are transformed into universal moral obligations toward all human beings. Just as each child in the family is subject to the same moral authority and moral laws, so each person in the world is subject to the same moral authority and moral laws. Just as each family member is responsible for nurturing every other family member, so every person is obliged to nurture every other.

The Family of Man metaphor is so general that it does not specify exactly how we ought to behave. It only generates specific moral duties when the "family morality" side of the mapping is filled in by a specific model of the family, either the Strict Father or the Nurturant Parent.

What the Family of Man metaphor does is to provide the crucial step for moving out of the family to a universal morality. The question then becomes whether our universal moral scheme will be understood metaphorically as a Strict Father morality or a Nurturant Parent morality. Before we examine the basic outline of each of these two models, we first need to understand the range of candidates available to fill the slot of the metaphorical Father in either model.

Who Is the Parent?

In the Family of Man, who is the parent? The answer to this question determines the nature of one's ultimate view of moral authority. Typical candidates for the role of universal parent are God, Universal Reason, Universal Moral Feeling, and Society as a whole.

God as Father (or Mother)

For most religious believers, God the Father is the ultimate moral authority, the absolutely all-powerful and perfect Being who established the moral order, is the source of all moral law, and who punishes immorality and rewards moral behavior. The crucial differences among religious ethical views, therefore, depend primarily on different views of the family and different conceptions of the Father, as either strict, nurturant, or some combination of both. The idea of God As Mother is almost never used to present the strict parent model. God As Mother is typically regarded primarily as a nurturant parent.

Strict Father religious morality has defined large segments of the Western moral tradition. On the Strict Father model, God the Almighty created all that is according to his divine plan and moral order. He issues moral commandments in the form of moral laws binding on all rational creatures. Our duty is to learn God's laws and to develop the moral strength to obey them in a world filled with evil, both internal and external. In the Final Judgment, God will punish the wicked and reward the morally good and obedient.

God as Nurturant Parent

By contrast, the prototypical case of God As Nurturant Parent emphasizes the metaphor of God As Love. This is usually not the "tough love" of the Strict Father God, but rather the nurturant, compassionate, suffering love of various New Testament interpretations. Here there is no talk of reward and punishment, but only of unconditional, all-encompassing love that flows to us undeserved. In the version of this that is central to Christianity, God is the nurturant parent to all humanity. Christ is the bearer of God's nurturance, and God's grace is that undeserved, freely given nurturance.

This is not the morality of obedience to moral laws given by divine authority. Instead, it is a morality of caring for others out of compassion and empathy. People love others, on this view, ultimately because they are first loved and nurtured by God. Moral action is understood as nurturant action, that is, as helping others through feeling empathy, showing compassion, and acting out of love. The idea of God as all-loving, all-suffering nurturer is often aligned with the conception of God as Mother.

Universal Reason as Strict Father

Historically, the Enlightenment crisis of faith led to the emergence of the view that morality is not based on the commands of an all-powerful God, but rather on another type of father, Universal Reason (as the ultimate moral authority). God's Reason is replaced by Universal Reason. We will examine the details of this metaphor more carefully in Chapter 20, on Kant's theory of morality. Briefly, the key to this view is the idea that the father's moral authority can be internalized as Universal Moral Reason. This move requires, as we shall see in detail in the Kant chapter, a metaphorically defined faculty psychology, according to which Reason is understood as a person who has moral authority within a "Society of Mind" consisting of the various mental capacities (e.g., reason, will, feeling, sensation). Universal Moral Reason, by virtue of its moral authority, issues commands that, for us, are moral laws. The faculty of Will receives those commands and yet has the freedom to act either according to or against them.

Universal Moral Feeling

One variation on the morality of faculty psychology simply shifts the mantle of moral authority from Reason to Feeling. Some Enlightenment moral theories held that it is not impotent Reason that runs the show, but rather Feeling, or Passion, that moves us to action. When Feeling calls the shots, the other faculties' jobs are redefined relative to the power of Feeling to produce action. There are typically two forms of feeling: (1) desire, conceived as a bodily force that drives us to act to satisfy our needs and wants; and (2) moral sympathy, which is a sentiment of benevolence toward other people. Moral sympathy is conceived as a feeling that is based on empathy and that moves us to seek the well-being of others.

Society as Family

The fourth major candidate for moral parenthood is society in general. Society is understood as a family, in which the metaphorical Strict Father sets social norms. Social norms are conceived as family norms. Thus, we say "Society *frowns* on public indecency," "Society absolutely *condemns* child abuse," "Society won't *tolerate* obscene behavior," "It is *not allowed* to treat people that way," and so on. The Strict Father in these cases may not be God or Reason, but rather society's values as they are objectified and conventionalized over history. When this happens, we even come to speak of the "General Will" of the people. In some cases the Strict Father may be embodied in particular people who have authority in a society, such as elected officials or clergy.

These represent the most common metaphorical instantiations of the Father (or Mother) in universal morality. There are others, but these cover the vast majority of moral traditions and theories. Once the "father" (or "mother") role is specified in the Family of Man metaphor, the question then arises concerning whether the father is understood within a Strict Father or a Nurturant Parent framework.

Moral Theories as Family Moralities

It is clear that many moralities are family-based. That is, they are grounded on and motivated by a particular family model that organizes some set of our metaphors for morality into a more or less coherent ethical perspective. This raises the question of how many of our moral theories work this way. We know of some cases in which a moral theory is clearly family-based, but we need to examine the less obvious cases. What about rationalism, sentiment theories, virtue ethics, egoism, existentialist ethics, utilitarianism, and various forms of relativism? What about this-worldly versus otherworldly ethical views? What about theological versus humanistic perspectives? Can all of these possibly be covered by two prototypical family models?

We are certainly not contending that our analysis actually covers all forms of moral understanding and experience, or even that it precisely describes every possible view within the Western moral tradition. However, the "family model" appears to be an extremely comprehensive and insightful explanatory hypothesis with considerable psychological motivation and analytic power.

What follows is a very brief account of how some of some classical moral theories can be understood as versions of either Strict Father or Nurturant Parent Universal Morality.

Christian Ethics

In monotheistic religions, as we suggested above, the moral authority is God the Father Almighty, creator and sustainer of all that is and source of all that is good. On the Strict Father interpretation, God is the stern and unforgiving lawgiver who rewards the righteous and punishes wrongdoers. The key to living morally is to hear God's commandments and to align one's will with God's will. This requires great moral strength, because one has to overcome the assaults of the Devil and the temptations of the flesh.

When God is conceived as Nurturant Parent (sometimes as Mother), he is the all-loving, all-merciful protector and nurturer of his people. God is Love, and, in the Christian tradition, Jesus is the bearer of that nurturant and sacrificing love for all humankind. Although there is a place for moral law ("Think not that I come to abolish the law and the prophets; I come not to abolish, but to fulfill," Matthew 5:17), moral commandment and law are not the central focus. Instead, morality is about developing "purity of heart" so that, through empathy, we will reach out to others in acts of love.

Rationalist Ethics

Again, as we saw above and will see in detail later, moral rationalism conceives of the Father as Universal Reason, possessed by all people and telling each person what is morally required of him or her. Rationalism tends to underwrite a Strict Father morality. Reason is a stern lawgiver and judge. It requires absolute obedience to its commands, and it holds us responsible for our willful failures to obey. We are rewarded, not by eternal life or external well-being, but rather by our own inner sense of self-esteem and self-respect, which comes from knowing that we have done our duty. Our punishments for moral wrongdoing are, likewise, internal—guilt, shame, and lack of self-respect. Moral strength takes top priority, since it is the essential condition for our being able to do what Reason morally commands.

While it is not impossible to have a Nurturant Parent rationalist morality (perhaps certain versions of utilitarianism are of this sort), Reason is not typically understood as a nurturer. Reason *commands, lays down the law, gives orders, judges, reprimands*, and so on. We almost never conceive of it as nurturing, feeling, caring, and so forth.

Utilitarianism

Utilitarianism may not seem as though it could have anything to do with family morality. It is often seen as a rational principle, set within Enlightenment economic theory, that focuses on the maximizing of happiness according to a moral calculus. But, of course, the classical utilitarians, Jeremy Bentham and John Stuart Mill, did indeed regard this as the ultimate nurturant morality. It is a morality geared toward the realization of human well-being. It requires each person to act so as to realize the maximum happiness possible in a given situation. Sometimes this might require personal sacrifice of one's own well-being in order to promote the well-being of others as a whole. Hume and Mill are both explicit in seeing morality as motivated ultimately by a broadly shared moral sentiment they called benevolence, fellow-feeling, and sympathy.

Thus, humankind is a large family. Just as individual family members ought to care for and nurture other family members, so should we all do the same for humankind. Just as self-sacrifice may be called for within the family, likewise the individual is not the bottom line in society. The "principle of utility" might sound only like an absolute command of a Strict Father (Reason), but it is also realized by us via our basic empathy and feelings for the happiness of others.

Virtue Ethics

Virtue is about character. An ethics of virtue is based on developing a strong, wise, and even-tempered moral character that will lead you to choose what is best and to act morally. As we will see in Chapter 20, Kant had a Strict Father morality, and so he tends to understand virtue very narrowly as moral strength to do one's duty, as commanded by reason. But the father of virtue ethics, Aristotle, has a far more expansive Greek conception of virtue.

For Aristotle, virtue is about developing habits and states of character that will lead one to naturally choose what is good and right. Morality is about *growth*, about the person's developing his or her capacities and exercising them to the fullest extent in order to realize what is best in them. Aristotle's ethics is thus about nurturance, the nurturance necessary to help a person become a well-balanced, temperate, fully actualized human being. This moral nurturance and education begins with the family, without which morality is doomed to failure. It must be carried on, however, by the continual nurturance of the larger community. That is why the *Politics* is but an extension of the *Nichomachean Ethics*. What this sustained care produces, when it succeeds, is human excellence (Greek *aretē*). Aristotle's ethics is therefore teleological, focused on growth, and absolutely dependent on the nurturing activity of the entire community. Our moral relations toward our fellow citizens ought to be nurturant and ought to aim for the realization of human well-being and excellence.

Permissive Family Moralities

All of the previous moralities recognize clear sets of values, strict standards of behavior, and the necessity of moral strength and discipline. Whether they are Strict Father or Nurturant Parent moralities, they place stringent constraints on our actions. In contrast, there appear to be two other major ethical perspectives that are neither Strict Father nor Nurturant Parent. Instead, they are modeled on what is known as the "permissive family." In a permissive family there are no strict rules and children are not held responsible for their actions. The permissive family is what Lakoff calls a "pathological" form of the nurturant parent family, since it mistakenly thinks that letting the children do whatever they please is an appropriate form of nurturance. The following two types of ethical view seem to be versions of this pathological model.

Ethical Egoism

Ethical egoism is the view that an act is right if it maximizes my own well-being. Crude forms specify well-being as nothing more than pleasure, whereas sophisticated forms recognize a broad range of human activities that give rise to individual flourishing. They also recognize that social interaction is an important part of this pattern of human living.

There are at least two major interpretations of egoism. The first treats it as a nurturance morality that has shriveled up into itself by reducing nurturance to nothing more than self-nurturance. On this reading, egoism is a perversion of the nurturance model, one in which the spoiled and selfish child mistakenly thinks that other people don't really matter, unless they serve his or her self-interest.

The second reading regards egoism as a form of moral self-interest. That is, if each person pursues his or her individual self-interest, then, by some Invisible Hand, the moral interest of all will be served. This second version appropriates some of the values of the Strict Father model, especially moral strength, self-discipline, and self-control.

Existentialist Ethics

At first, existentialism might appear to provide the ultimate challenge to the idea of family-based morality. Above all else, it presents itself as a form of moral relativism that rejects the very notion of moral essence, absolute values, and rational commands. It denies that there is any preexisting human essence or any ultimate human end that could define moral action. All we have is our freedom and the necessity of making choices (since not to choose is to choose). *Freedom* and *authenticity* are its catchwords. You are being inauthentic when you let someone else's morality determine your actions. It is not clear, however, that authenticity can make any sense in this framework. There is no "self" to be true to. How can you be "inauthentic'" when there is no "authentic" you?

This leaves you with freedom. The Father (God, Reason, General Will, Essence) is dead, so you are free (within situational limits) to choose what you will be. Say you choose to act nurturantly. Fine, but this cannot, on this view, be part of any larger framework for justifying your action. The doctor in Camus' *The Plague* chooses to stay in the City of Death out of concern for his fellow creatures who are dying of the plague. *We*, the readers, might regard this as noble, since *we* have a value system that regards nurturance as important. From the perspective of the existentialist, however, there is no basis, one way or the other, for applauding this action. One just *chooses* to approve of such care and concern.

Thus, existentialism appears to be a form of Permissive Family morality. We, the children not under the authority of any parent, make our choices without help or guidance from our metaphorical parents (God, Reason, Feeling, Society). That does not mean that we act with no ethics at all. Rather we act with an ethics of our own choice, one that is not imposed on us. Existentialism might be seen as an instance of the rebellious child rejecting the parent altogether and finding his or her own way in the world.

Moral Relativism

However different their models of the family might be, all of the previous moralities and ethical theories except existentialism share the grounding assumption that there exist universal moral standards. Moral relativism rejects this foundational assumption. It claims that there exist no universal essences upon which universal, absolute moral values rest. All moral standards are seen to be relative to the specific communities in which they arise.

Within the framework of the family morality models that we have been examining, moral relativism presents a challenge to the assumptions underlying the Family of Man metaphor. Moral relativism stems from the denial of this metaphor. It says that there is no universal family of all humankind. Instead, there are only scores of different moral "families," each one having its own family values (i.e., each having its own distinct morality). In other words, each family (i.e., each moral community) gives its own family morality, and there is no universal standpoint above all of these particular families for judging their particular values and ideals.

Is All Morality Metaphoric?

The answer to this question, as we have seen, is no. There is nothing inherently metaphoric about such claims of basic experiential morality as "Health is good," "It is better to be cared for than uncared for," "Everyone ought to be protected from physical harm," and "It is good to be loved."

However, as soon as we develop such claims into a full-fledged human morality, we find that virtually all of our abstract moral concepts—justice, rights, empathy, nurturance, strength, uprightness, and so forth—are defined by metaphors. That is why there is no ethical system that is not metaphorical. We understand our experience via these conceptual metaphors, we reason according to their metaphorical logic, and we make judgments on the basis of the metaphors. This is what we mean when we say that morality is metaphoric.

Because our metaphorical moral concepts are grounded in aspects of basic experiential morality, they tend to be stable across cultures and over large stretches of time. The question of whether they are universal is an empirical one, and the research has not yet been carried out to make this determination. The evidence available so far does suggest that they are very good candidates for universal moral concepts.

However, it is extremely important to qualify this claim about universality with the point that the way each metaphor is developed in a particular setting may vary widely from culture to culture. For example, generally speaking, balance may be universally regarded as a good thing, and moral balance, too. But *what* gets balanced and precisely what it means to achieve balance may well vary across cultures. Moral Balance is a good thing in America and Europe, but in some cultures, such as the Japanese, it takes on an importance far beyond anything found in the West.

Our moral concepts, then, are not absolute, but they are also not arbitrary and unconstrained. To think of these polar opposites as the only two alternatives is to miss the most important dimensions of our moral understanding.

The fact that our moral concepts are grounded and situated gives them relative stability, while their imaginative character makes it possible for us to apply them sensibly to novel situations. Winter (A1, forthcoming) has shown in great detail for legal concepts how this situatedness and flexibility makes legal reasoning possible in the face of ever-changing conditions.

Does Cognitive Science Contribute to Moral Understanding?

We have been suggesting that knowing about the metaphoric nature of our moral concepts makes a huge difference in our moral understanding. At the very least, the cognitive sciences provide us with analytical tools that give us a far deeper understanding of morality than has previously been available. But what, precisely, does it mean to have this deeper moral understanding? How would it affect the way we think and live?

Most philosophers think that such knowledge from the cognitive sciences actually has little or no bearing on moral reasoning or on how we ought to live. Sceptics dismiss results from the cognitive sciences by claiming that empirical research on the mind is irrelevant to ethics in two ways. First, they insist that knowledge of how people actually reason is irrelevant to how they ought to reason. There are no normative claims, they say, derivable from empirical knowledge of our moral understanding. Second, they claim that analysis of conceptual metaphor can be, at best, nothing more than a useful tool for clarifying our moral concepts; they assume that those moral concepts exist independent of the metaphors.

Let us consider each of these objections in turn. The first charge, the assertion that empirical knowledge of our moral cognition can have no normative implications, is based on a false dichotomy between facts and values. Owen Flanagan (C1, 1991) has demonstrated the relevance of moral psychology for moral theory by showing that no morality can be adequate if it is inconsistent with what we know about moral development, emotions, gender differences, and self-identity. Johnson (C1, 1993) argues that facts about human conceptualization and reasoning place normative constraints on what we can morally demand of ourselves and others. For example, any view of morality that involves absolute moral principles defined by literal concepts cannot be cognitively realistic for human beings, whose moral categories often involve radial structure, conceptual metaphor, and metonymy. Damasio's (B1, 1994) work with brain-damaged patients who have lost the ability to perform certain kinds of practical reasoning because their emotional experience is impaired suggests that moral deliberation cannot be the product of an allegedly pure reason. Moral deliberation always requires emotional monitoring and an interplay of affect and reason.

The point here is that empirical knowledge about human psychology and cognition does place constraints on what a cognitively realistic morality will look like. A good example of this is Lakoff's examination (A1, 1996a, chap. 21) of developmental studies that indicate major problems with the family-based foundations of the Strict Father conception of morality. Evidence from three areas of psychological research—attachment theory, socialization theory, and family violence studies—shows that the Strict Father model does not, in fact, produce the kind of child that it is supposed to foster. It is supposed to develop children who have a conscience and who are morally strong, capable of resisting temptations, independent, able to make their own autonomous decisions, and respectful of others. But such research, especially socialization research, shows the Strict Father family tends to produce children who are dependent on the authority of others, cannot chart their own moral course very well, have less of a conscience, are less respectful of others, and have no greater ability to resist temptations.

Though these three current research paradigms produce such convergent evidence at present, future research may change the picture somewhat. However, the crucial point is that empirical research of this sort *is* relevant to our normative assessment of various moral views. The question of whether Strict Father morality is typically successful in developing the kind of moral agents it prizes is an empirical issue subject to testing and confirmation.

The second deflationary argument against the relevance of cognitive science for morality is that, once we have determined the fundamental metaphors by which we understand our most basic moral concepts, we simply proceed with traditional moral theorizing. The way we understand those moral concepts is held to be entirely separate from the concepts themselves. The basic moral concepts are alleged to be literal, even though we happen to understand them via metaphor. The crux of this view is that the metaphors are not themselves the ethical concepts and play no role in the logic of ethical discourse. The traditional moral theorist would thus say such things as, "Of course there is a metaphor of Moral Strength, and it exists because moral strength is essential to morality. Of course, there is a metaphor of Moral Uprightness, because being morally upright is what morality is all about. Of course, there is a metaphor of

Moral Boundaries, since morality defines constraints on morally permissible human actions. The metaphors are just our way of grasping the absolute moral values and imperatives that exist in themselves."

It is extremely important to see that this entire deflationary strategy is based on a profound misunderstanding of what it means to say that morality is metaphoric. It presupposes a deeply mistaken view about where moral concepts and rules come from.

The traditional view of moral concepts and reasoning says the following: Human reasoning is compartmentalized, depending on what aspects of experience it is directed to. There are scientific judgments, technical judgments, prudential judgments, aesthetic judgments, and ethical judgments. For each type of judgment, there is a corresponding distinct type of literal concept. Therefore, there exists a unique set of concepts that pertain only to ethical issues. These ethical concepts are literal and must be understood only "in themselves" or by virtue of their relations to other purely ethical concepts. Moral rules and principles are made up from purely ethical concepts like these, concepts such as *good, right, duty, justice,* and *freedom.* We use our reason to apply these ethical concepts and rules to concrete, actual situations in order to decide how we ought to act in a given case.

Why the Traditional View Can't Work

The traditional view of moral concepts and reasoning is predicated on denying that our moral concepts are metaphoric. Therefore, the empirical question of whether our moral concepts and reasoning are metaphoric is all-important. If they *are* metaphoric, then they *are not* univocal, they *are not* understood in their own terms, and there *is not* some autonomous, monolithic "ethical" domain with its own unique set of ethical concepts. Let us consider each of these crucial points more carefully in order to see why we need to reject the traditional view.

No Pure Moral Concepts

There is no set of pure moral concepts that could be understood "in themselves" or "on their own terms." Instead, we understand morality via mappings of structures from other aspects and domains of our experience: wealth, balance, order, boundaries, light/dark, beauty, strength, and so on. If our moral concepts are metaphorical, then their structure and logic come primarily from the source domains that ground the metaphors. We are thus understanding morality by means of structures drawn from a broad range of dimensions of human experience, including domains that are never considered by the traditional view to be "ethical" domains. In other words, the *constraints* on our moral reasoning are mostly imported from other conceptual domains and aspects of experience.

We are not claiming that there are *no* nonmetaphorical ethical concepts. Some of our moral concepts appear to have a minimal nonmetaphorical "core." However, this core is typically so thin, so underspecified, that it can play little or no role in our reasoning without being fleshed out by various metaphors. Thus, any comprehensive analysis of a moral concept will reveal one or more metaphorical structurings that serve as the basis for our reasoning. For example, consider our concept of *rights*. Its minimal nonmetaphoric core appears as early as infancy and toddlerhood.

1. Very early on, infants and toddlers acquire the idea that something (such as a toy or pacifier) belongs to them—they *possess* it, and it is *theirs* to do with as they wish. Taking away a possession they see as fundamentally theirs leads them to protest loudly.
2. Infants and toddlers react vigorously against undue constraints on the movements of their bodies. Inhibiting normal bodily movement is protested against.
3. From infancy, we react against the infliction of pain.

Basic possessions, normal bodily movement, and freedom from the infliction of pain seem to be, literally, where our notion of rights begin.

Correspondingly, abstract rights in adulthood are based on metaphorical versions of these earliest of rights. Abstract rights are conceptualized as (1) property rights, (2) freedom of action (via Action Is Self-Propelled Movement), and (3) freedom from harm (both literal and metaphorical harm). Locke's rights to "life, liberty, and the pursuit of property" are versions of these abstract rights. Thomas Jefferson's substitution of "happiness" for "property" is based on the common metaphor Achieving A Purpose Is Acquiring A Desired Object. Without these various metaphors, our concept of rights is meager indeed.

Next, consider our notions of *ought* and *should*. Both minimally contain a force-dynamic model in which a moral agent is morally or rationally *forced* or *under pressure* to act in a certain way. But even here, the *moral force* and the *force of reason* are already metaphoric, insofar as we can only understand them via conceptual metaphors based on physical force.

Beyond this minimal sense of moral force, *ought* and *should* get their meaning from various additional metaphors such as Morality Is Empathy, Well-Being Is Wealth, and Moral Strength. In some cases, for instance, the metaphor of Moral Strength will dictate one course of action, while Moral Empathy might dictate a very different course of action. The priority of different metaphors for morality yields different *oughts* and *shoulds*. *What we do not, and cannot, have is some metaphor-free way of conceptualizing abstract moral concepts or entire moral positions.*

No Pure Moral Reason

This view of moral concepts as metaphoric profoundly calls into question the idea of a "pure" moral reason. The whole point of having a pure moral reason was supposedly to generate pure concepts and rules that could define an absolute, universal morality. However, if we have no pure moral concepts that are defined only on their own terms, then the idea of a pure moral reason becomes superfluous. Given that most of our moral concepts are structured by metaphor, then the inference patterns of our moral reasoning come, for the most part, from the source domains of the metaphors. So, even if there were to be such a thing as a "pure practical reason," which we deny, it would not be doing the primary work of our moral thinking.

No Monolithic Morality

We do not have a monolithic, homogeneous, consistent set of moral concepts. For example, we have different, inconsistent, metaphorical structurings of our notion of well-being, and these are employed in moral reasoning. Which one we use, such as Well-Being Is Wealth versus Well-Being Is Health, will depend on the hierarchical structuring imposed by family-based moral systems as well as our purposes, interests, and the particular context we find ourselves in. There is no single, internally consistent concept of well-being that incorporates both of these metaphorical structurings.

Moreover, if you look at the entire system of metaphorically defined concepts that make up our moral understanding, they also do not all fit neatly together in a consistent way. Moral Strength, as we have seen, is not always compatible with Moral Empathy, and that is why one or the other is typically given priority within a particular moral system. In Strict Father morality Moral Strength takes precedence over Moral Empathy, whereas in Nurturant Parent morality the priority is reversed.

Morality Is Grounded

Consequently, our very idea of what morality is comes from those systems of metaphors that are grounded in and constrained by our experience of physical well-being and functioning. This means that our moral concepts are not arbitrary and unconstrained. It also means that we cannot just make up moral concepts *de novo*. On the contrary, they are inextricably tied to our embodied experience of well-being: health, strength, wealth, purity, control, nurturance, empathy, and so forth. The metaphors we have for morality are motivated by these experiences of well-being, and the ethical reasoning we do is constrained by the logic of these experiential source domains for the metaphors.

It is for this reason that at least the most extreme postmodern views of ethics are mistaken. It is sometimes claimed that morality is nothing more than a fabric of arbitrarily chosen narratives that we impose on our experience and that all our values are arbitrary constructs. The grounding of our metaphors for morality shows why such extreme forms of social constructivism are wrong. We have seen some of the ways that the source domains import constraints into our moral concepts, and we have seen how these source domains are tied up with our basic bodily well-being. Even though such constraints allow for a very wide variety of moralities, they establish the general form and substance of human morality. That is, they give general constraints on what a morality will look like.

No Deontological Ethics

The metaphoric character of morality has another far-reaching implication—it calls into question the very notion of a "deontological" basis of ethics. At least since Kant, it has been traditional to distinguish between moral theories that are *deontological* and those that are *teleological*, or *consequentialist*. Consequentialist views are those in which right action is defined by the good ends or consequences that it produces. An action is morally right if it results in a state of affairs that produces more good than any alternative action would. John Rawls characterizes such theories as claiming that "the good" (i.e., consequences) comes prior to "the right" (i.e., correct moral principles). What is "right" is thus defined as that which has the best consequences.

By contrast, so-called deontological theories claim that our moral principles are *sui generis,* having a source independent of our calculations of consequences, ends, and good states of affairs that we may want to realize. Typically, our moral concepts and rules are believed to come straight out of a universal reason and are thought to be binding universally on all people, regardless of their ends and purposes. This is what Rawls means when he says that "the right" (moral principles) precedes "the good" (consequences). The right can be defined entirely independent of the good.

Such an alleged split between principles and ends looks highly problematic in light of the metaphorical character of much of our moral reasoning. We have seen that, in most of our reasoning about morality, the inference patterns we use come from source domains by which we metaphorically understand well-being. Our very modes of stating abstract moral principles and engaging in abstract moral reasoning arise from modes of well-being, that is, "consequences." When we use such metaphorically derived inference patterns to reason about morality, the principles we get and use are inextricably tied up with ends, goals, and purposes. In such cases, therefore, the deontological picture of ethical deliberation just doesn't fit.

The deontologist will no doubt respond by insisting that we can keep morality (as a source of moral principles) entirely separate from other domains (such as well-being) whenever we are reasoning about morals. This view entails that learning morality is just learning preexisting patterns of moral reasoning and learning how to apply them to concrete cases.

However, it is important to see that this is an empirical issue about the nature of human reasoning, and it cannot be decided a priori. We have cited in this book some of the kinds of empirical evidence for the cognitive reality of conceptual metaphor. We have argued that conceptual metaphor is real in our moral understanding and that it is the basis for much of our ethical reasoning. Determining whether this is true for all cases would be an unending task, but the cases of moral reasoning that we have examined so far call into question the deontological idea of reason giving rise to ethical principles without any reference to some conception of the good (i.e., of consequences). It just doesn't seem plausible to think of moral principles springing full-blown and unmotivated from pure reason, as though they were not defined relative to human purposes, goods, and ends. Moreover, the cognitive evidence is against it.

No Compartmentalized Morality

Since there are few, if any, "purely ethical" concepts that are defined solely "in themselves," we ought to be highly suspicious of the idea of a purely "ethical" domain. We know from our analysis of the role of metaphor in moral reasoning that it depends on inferential structure imported from domains that are not typically thought of as "ethical." The Moral Accounting metaphor is a good example, as are all of the metaphors with source domains tied to our bodily experience, such as being upright, balanced, in control, healthy, and pure.

We then apply such patterns of reasoning to other domains that are not typically thought of as ethical. For example, Lakoff (A1, 1996a) has analyzed the role of Strict Father and Nurturant Parent moralities in the formation of political conservatism and political liberalism. If he is right, then morality is not a domain separate from politics. Such metaphors for morality also strongly influence our thinking about education and our social concepts, which means that our conceptual system for morality must enter into educational and social theorizing and policy making.

Since most of our moral understanding comes, via metaphor, from a broad range of other domains of experience, and since we apply those metaphors to a number of different experiential domains, we should be wary of trying to compartmentalize ethics. The cross-domain mappings of the metaphors suggest the intricate web of connections that impose our moral ideas on other aspects of our lives, including considerations that are technical, scientific, political, aesthetic, religious, and social.

Being aware of the vast reach of our moral systems and the complex intertwining of these various strands of our experience need not turn us into moral fanatics who treat *every* trivial decision as having moral import. Instead, it recognizes a scale of moral importance ranging from issues that have little or no moral weight, such as what grade of lead you prefer in your pencil or whether you prefer jam on your toast, up to questions of the utmost ethical weight, such as whether or not to be a pacifist. Some of our decisions thus have little or no impact on the well-being either of ourselves, other people, or the other-than-human world we inhabit, and so they are not thought of primarily as ethical choices. However, we need to be aware of just how much of what we think and do actually does have such moral effects.

To sum up these previous six points, once we take seriously the metaphoric character of our moral understanding, we are forced to abandon the traditional view of moral concepts and reasoning. We can never again proceed with business-as-usual, either in our moral reasoning or in our moral theorizing. There is no pure moral reason and there are no pure moral concepts that are understood solely "in themselves" or in relation only to other pure ethical concepts. Our moral understanding is metaphorical, drawing structure and inference patterns from a wide range of experiential domains that involve values, goods, ends, and purposes. Our system of moral concepts is neither monolithic, nor entirely consistent, nor fixed and finished, and certainly not autonomous.

What cognitive science brings to moral understanding are two absolutely essential things: first, a deeper understanding of what moral reasoning is and where it comes from; second, the ability to look at the fine details, to know which particular moral metaphors you and others are using and the role each metaphor plays in the moral conclusions reached.

As important as it is to be able to notice the role metaphorical morality plays in the overt moral decisions you and others make, it is equally important to recognize when our moral system enters in a hidden way into vital areas of our culture: politics and religion (A1, Lakoff 1996a) and even educational theory and the understanding of such scientific matters as evolutionary biology (see Chapter 25). Moral judgments are implicit in virtually every aspect of our culture, and it is vital to become consciously aware of them.

The End of Innocence

This is the end of cognitive innocence in moral understanding. The outlines of metaphorical moral thought are now fairly clear. Unconscious metaphorical moral thinking can now be brought to moral reflection. We can now notice when we are using Moral Accounting and what forms of it we are using where. We can see when we (or others) use the Moral Strength metaphor to make moral judgments, and we can ask if it is appropriate. We can notice when we use (or do not use) the Moral Empathy metaphor, and we can ask what form our empathy takes. At a higher level, we can explore what model of the family organizes our overall moral outlook or if we sometimes use one model and sometimes another.

This type of moral knowledge makes us responsible not only for our own moral judgments and their consequences but for noticing implicit forms of moral judgment throughout our culture.

Part III

The Cognitive
Science of Philosophy

15

The Cognitive
Science of Philosophy

Philosophical Theories and Folk Theories

Philosophical theories are attempts to make sense of our experience—to figure out why things are the way they are, to learn who we are, and to decide how we ought to live. A philosophical theory tries to answer such "big" questions by seeking a comprehensive, internally consistent, rational account of the world and our place in it. There are many different views concerning what an adequate philosophy would look like, but all of them involve extensive conceptual analysis and rational argument. Since the nature of human concepts and reason is studied scientifically by the cognitive sciences, we should expect them to have a direct bearing on our understanding of the nature of philosophy itself. They do—and in a major way.

We have already seen a few of the more important implications of embodied cognitive science for philosophy, such as a new understanding of basic concepts like time, causation, events, the self, and the mind. These new conceptions are the result of using the tools and methods of embodied cognitive science to analyze fundamental philosophical concepts.

But there is another, equally enlightening, application of embodied cognitive science to philosophy. The history of philosophy itself can be taken as the subject matter to be studied from the perspective of what we are learning about conceptualization and reasoning. Since philosophy makes use of the same conceptual resources possessed by all human beings, it, too, can be studied as a form of human conceptual activity. We can figure out what makes it possible for us to comprehend a given philosophy, how that philosophy makes use of the various kinds of imaginative devices that make up human understanding, and why certain philosophical views will or will not seem intuitively correct for members of a particular culture. We will call this enterprise "the cognitive science of philosophy."

When philosophers construct their theories of being, knowledge, mind, and morality, they employ the very same conceptual resources and the same basic conceptual system shared by ordinary people in their culture. Philosophical theories may refine and transform some of these basic concepts, making the ideas consistent, seeing new connections and drawing out novel implications, but they work with the conceptual materials available to them within their particular historical context. As an example, let us consider for a moment what makes it possible for us to read and understand a well-known passage from John Locke's *An Essay Concerning Human Understanding,* in which he is describing how the mind comes to have ideas:

> Whence has it all the *materials* of reason and knowledge? To this I answer, in one word, from EXPERIENCE. . . . Our observation employed either about external sensible objects, or about the internal operations of our minds perceived and reflected on by ourselves, is that which supplies our understandings with all the *materials* of thinking. (C2, Locke, *Human Understanding,* bk. 2, chap. 1, p. 2)

When we read this, we understand immediately that Locke sees the mind as a container, with some objects entering into the container from outside (i.e., the sensations we get from external physical objects affecting our senses) and others arising "internally" as the mind looks at its own operations. In addition, Locke sees perceptions as materials, that is, basic resources for constructing complex ideas. There are two distinct metaphors here. The first is The Mind As Container for basic perceptions that come in from the outside. The other is The Mind As Builder, in which the mind takes these perceptions and constructs complex ideas out of them.

Locke next surveys the various ways in which ideas can enter the mind from the external world. One of these ways involves only one of the five senses:

> There are some ideas which have admittance only through one sense, which is peculiarly adapted to receive them. Thus light and colours, as white, red, yellow,

blue . . . come in only by the eyes. . . . And if these organs, or the nerves which are the conduits to convey them from without to their audience in the brain,—the mind's presence room (as I may so call it)—are any of them so disordered as not to perform their functions, they have no postern to be admitted by; no other way to bring themselves into view, and be perceived by the understanding. (C2, Locke, *Human Understanding*, bk. 2, chap. 3: 1).

Locke here uses our ordinary metaphor Perceiving Is Receiving, in which the mind passively receives objects of perception through the senses. The nerves are conduits through which sensation-objects pass from outside into the "presence-room"—the anteroom—of the mind. There, the Mind As Person inspects and manipulates (i.e., performs operations on) those perceptions.

These metaphors for mind are neither incidental, nor disposable; they are constitutive. They define the ontology of mind that Locke is using. The ontology has a container, a mental locus with an interior and passages (the senses) from the exterior into the interior. Sense data are inert objects that (secondary qualities aside) have an objective external existence independent of any observer. The mind does not play any role in creating these sense data; it just receives them passively as given. The faculty of understanding is a person who actively puts these objects together to form complex ideas.

All of this can make sense to us, because it uses many of the very same deep conceptual metaphors and folk theories of mind that we use ordinarily every day. Those ordinary metaphors play a vital role in defining Locke's ontology, his metaphysics of mind.

The very possibility that the homunculus-type faculty of understanding (relying on the Reason As Person metaphor) can "view" idea-objects presupposes the Knowing Is Seeing metaphor. The metaphor of an internal viewing space where a personified faculty of mind inspects idea-objects is what Dennett has named the "Cartesian Theater." But this is not an exotic model to be found only in the abstruse meditations of Descartes and other philosophers. On the contrary, it is a metaphor (or cluster of metaphors) deeply embedded in our ordinary conceptions of mind, so much so that it is nearly definitive of how we think about mind.

There is nothing either unusual or remarkable that needs to be called into play from our ordinary conceptual systems and capacities in order to understand such abstract philosophical ideas and theories. They grow out of the soil of our common imaginative understanding, however much they may be creative and transformative of our basic shared folk theories, cognitive models, and metaphors. Human invention and originality, as Mark Turner argues in *Reading Minds* (A1, 1991), is accomplished with the ordinary cognitive resources we all share, using conventional conceptual devices and forms of understanding.

The cognitive science of philosophy can thus apply the conceptual tools, methods, and results of the cognitive sciences to help us understand the nature and to assess the adequacy of philosophical theories. For instance, any philosophical theory of mind, such as those elaborated by Thomas Aquinas, Descartes, Kant, or Patricia Churchland, is complex conceptual structure that presupposes views about the nature of concepts, reason, and what counts as an argument. As such, it can be studied to see what makes it tick. That is, we can analyze its basic concepts and the forms of argument and inference that it employs, and we can investigate the frames, idealized cognitive models, metonymies, and metaphors that define its concepts and forms of reasoning.

What Cognitive Science Offers Philosophy

The cognitive science of philosophy thus promises to give us insight into philosophy in three important ways. It can provide conceptual analysis, critical assessment, and a means of constructive philosophical theorizing.

Conceptual Analysis

Using what we are learning about the nature of concepts, conceptual structure, and reasoning, we can acquire a profound understanding of why we have the kinds of philosophical theories we do, what their presuppositions are, and what implications they have for various parts of our lives. For example, the cognitive science of philosophy gives us the means for understanding why philosophers since the beginning of recorded time have asked such seemingly strange questions as "What is Being?" "What is Truth?" and "What is the Good?" These are unusual questions, not the sort of questions we normally ask in our everyday affairs. It is necessary to look very deeply into the assumptions that make it possible even to frame such questions. When we look into these questions we discover that they can't even be formulated without first presupposing a large number of folk theories, idealized cognitive models, and metaphorically defined concepts. The same holds for the philosophical theories that arise as answers to these questions.

It is our claim that philosophical theories are attempts to refine, extend, clarify, and make consistent certain common metaphors and folk theories shared within a culture. Philosophical theories, therefore, incorporate some collection (perhaps in more precise form) of the folk theories, models, and metaphors that define the culture that they emerge in. If philosophy wasn't tied in to the culture in this way—if it didn't make use of the broadly shared conceptual and imaginative resources of the culture—then it couldn't possibly make sense to ordinary people or have any bearing on their lives.

Philosophical theories, like all theories, do not and cannot spring full-blown from some alleged pure, transcendent reason. Instead, philosophy is built up with the conceptual and inferential resources of a culture, even though it may transform and creatively extend those resources. These cognitive resources are not arbitrary or merely culturally constructed. They depend on the nature of our embodied experience, which includes both the constraints set by our bodily makeup and those imposed by the environments we inhabit.

Critical Assessment

The cognitive science of philosophy is not limited merely to studying what goes into making a philosophical theory what it is. It also gives us a cognitive basis for criticizing and evaluating theories. For example, it is very important to understand that the ordinary folk theory of categories, as defined by the metaphor A Category Is A Container for its members, has been made into a philosophical theory that we have called the Classical Theory of Categorization. It is constituted from a set of defining metaphors and image schemas, including the container schema and various metaphorical mappings, such as A Category Is A Container.

But the analysis does not stop here. We can go further to challenge the adequacy of the classical theory based on empirical evidence concerning category structure, such as prototypes, radial categories, and metonymic and metaphoric principles of category extension (A4, Lakoff 1987). The classical theory can neither account for these cognitive phenomena nor explain the conceptual and inferential structure of a wide range of basic concepts.

This critical perspective recognizes that second-generation cognitive science, like any empirical approach, has its own defining set of philosophical assumptions, as we discussed in Chapter 6. Since there are no philosophically neutral conceptual schemes, theories, or methods for any empirical discipline, second-generation cognitive science has made minimal methodological assumptions that do not predetermine the outcome of the investigation.

For cognitive science to be appropriately self-critical, it must repeatedly critique its own conception of cognitive science, of empirical testing, and of scientific explanation. There is no way out of this problem, but this does not mean that every theory, method, or concept is equally good or that it is all merely a "matter of interpretation." The cognitive sciences must rely on stable *converging* evidence from a number of different sciences, methods, and viewpoints. Only in this way can an empirical approach minimize the problem, so well documented by Thomas Kuhn, of a scientific theory defining what counts as evidence in such a way as to guarantee the truth of the theory in advance.

Cognitive science should be based on an appropriately self-critical methodology, one that makes minimal methodological assumptions that do not determine a priori the details of any particular analysis. Only if this condition is met can a cognitive science of philosophy be appropriately critical of philosophical theories.

Constructive Philosophical Theorizing

Understanding, explaining, and evaluating philosophical theories is only preliminary to the main task of constructing a philosophical orientation that can help us deal with the real and pressing problems that confront us daily in our lives at the personal, communal, and global levels. People want a philosophical understanding that provides realistic guidance for their lives. We are social, moral, political, economic, and religious animals, and our philosophy ought to help us in all of these areas and more.

An empirically responsible philosophy informed by an appropriately self-critical cognitive science can give guidance in two chief ways. First, it can give us a tremendous amount of self-knowledge and an understanding of other people by showing us how we create our sense of reality, why we believe what we believe, and how our conceptions and experiences chart the course of our lives. It can also help us see where we are making false assumptions and seeking answers to bogus questions that ought to be rejected. Dewey and Wittgenstein, prior to the age of cognitive science, helped us achieve this latter kind of insight, helped us stop asking questions predicated on false dichotomies, false presuppositions, and mistaken views about conceptualization and reasoning. Second-generation cognitive science helps us do such tasks even better, because it provides a methodology for doing detailed analysis.

The second kind of constructive understanding is positive guidance for living. The reason that such knowledge is so difficult to achieve is that, as we know from cognitive science, it cannot take the form of absolute rules or principles that tell us how to act in every concrete situation. Our concepts don't typically work this way, and our reasoning doesn't work this way.

The guidance this knowledge gives comes from the intricate details of what we learn about the kinds of creatures we are, how we experience our world, and what the limits are on our cognitive capacities. To act morally, for instance, we must, at the very least, understand our unconscious moral systems and how they function.

The kinds of insight and guidance that can come from doing the cognitive science of philosophy are only possible as a result of highly detailed and rigorous conceptual analysis. It is no small task to apply the methods and tools of embodied cognitive science to even a small part of the history of philosophy. In what follows we offer suggestive and representative analyses of certain episodes in the history of Western philosophy. We could never pretend to satisfy the historian of philosophy who knows every line of every text, every problem of translation, and every controversy over interpretation of key ideas. Our goal, instead, is to provide a sufficiently rich and rigorous analysis to reveal precisely how everyday unconscious metaphors, models, and folk theories contribute to the content of philosophical theories, especially their metaphysical claims.

The kind of analysis we will be doing is not classical text interpretation. It is, instead, typical of the kinds of empirical analysis done in various cognitive sciences. It attempts to account in detail for regularities governing the unconscious inferential structure on which the comprehension of texts is based. It uses the analytic tools discussed above: prototypes, frames, image schemas, metaphors, and so on. And, as is common in cognitive science, it pays special attention to what is not overtly and consciously discussed in the text, but rather to what must be unconsciously taken for granted in order to make sense of the text.

Our analyses are thus in one way like what is called "rational reconstruction" in that they give the details of what has to be assumed in order to make sense of a position. They are, however, unlike classical versions of rational reconstruction in certain ways: First, rational reconstruction classically assumes that the reconstruction is done entirely with the tools of classical logic, whereas we assume that it is done with general cognitive mechanisms such as prototypes, metaphors, and folk theories. Second, our analyses are constrained by empirical studies of the nature of cognition. We do not simply assume a priori that logic is the correct mechanism of reason. Third, as cognitive scientists, we seek generalizations over the phenomena, both synchronically and diachronically. Being required to state generalizations can affect analysis considerably.

In trying to discover generalizations, we look for a minimal set of folk theories, metaphors, and cognitive models that define the specific philosophical positions and that best generalize to all the positions considered. While our arguments may seem pedestrian to a cognitive semanticist, they are likely to be alien to most philosophers, who employ virtually none of these fundamental criteria from embodied cognitive science in their argument forms. As a result, we will be producing analyses that go beyond what classical rational reconstruction can do. In addition, we are implicitly challenging the traditional method of rational reconstruction as not being empirically valid, since it fails to meet these criteria.

Our goal, in this exercise, is not to produce full-blown interpretations of any one philosopher's whole work or even a completely thorough analysis of any one aspect of it. Rather, we have certain limited but philosophically important purposes. They are (1) to demonstrate that vital aspects of each philosopher's metaphysics arise from certain of his central metaphors and folk theories; (2) to show how the logic of his reasoning comes out of the entailments of those metaphors and folk theories; (3) to illustrate how a relatively small set of metaphors and folk theories can make a complex philosophical theory hang together; and (4) to show how the very enterprises of metaphysics, epistemology, and moral theory arise from such metaphors and folk theories. Our point, obviously, is that second-generation cognitive science and, especially, its theory of conceptual metaphor are necessary if philosophy is to understand itself.

The texts we have selected for analysis are taken from two fateful periods in the history of Western philosophy. We will first go back to the beginning of recorded philosophical thinking in ancient Greece. We start with pre-Socratic metaphysics, which first asked the questions about the nature of being, knowledge, and morality that have shaped the course of Western philosophical thought. The cognitive science of philosophy gives us new analytic tools to see just what metaphors and folk theories make it possible to ask a question like "What is the essence of Being?" and to put forth coherent doctrines on the issue. What we will see is that each philosopher's metaphysics arises from his metaphors, that the logic of each position is a metaphorical logic, and that the very concept of metaphysics itself arises from a peculiar combination of metaphors and folk theories.

Once we discern the cognitive structure of pre-Socratic metaphysics, we can see how that structure was elaborated by additional metaphors in Plato's middle-period doctrine of the Forms and in Aristotle's view of the nature of metaphysics, the science of Being qua Being. Our analysis shows that the metaphors that shaped the metaphysical and epistemological doctrines of the Greeks have guided philosophical and scientific thinking ever since.

We will then go on to analyze the Enlightenment conception of mind, a view that directly shaped first-generation cognitive science. This metaphysics of mind, which has become the common sense of much contemporary philosophy of mind, is, like all metaphysics, metaphoric, and its metaphors are very much with us today. Though metaphorical thought is normal and metaphors may or may not be apt, the Enlightenment metaphors for mind, as we shall see, are inconsistent with the results of second-generation cognitive science. Here is a case in which the common sense of contemporary philosophy is at odds with science.

Next, we will turn to one celebrated achievement of Enlightenment moral theory, which rests on the Enlightenment theory of mind. Our analysis will focus on Kant, who provided the archetype of what is allegedly "pure, rational moral theory." Again, we will see that every aspect of Kant's ethics, from his view of agents to his conception of virtue, is defined by the very same metaphors that define much of our modern-day moral worldview. Kant's morality, like all morality, is irreducibly metaphoric. This requires us to abandon Kant's claim that morality issues from a transcendent, universal, and purely literal practical reason. This is not a loss to be mourned; rather, it gives us the possibility of a cognitively realistic view of morality.

Finally, we analyze the fundamental assumptions underlying three massive influential bodies of contemporary philosophical theory. We begin by showing the embodied, metaphorical underpinnings of mainstream Analytic Philosophy, which, ironically, rests on denying the very existence of conceptual metaphor and embodied meaning. We then take a highly critical, analytic look at the philosophical assumptions underlying Noam Chomsky's generative linguistics, contrasting this with a sketch of the cognitive linguistics from which much of the empirical evidence in this book is drawn. We conclude with a critical examination of the theory of rational action that shapes so much recent economics, ethics, political theory, and international relations theory.

What do these case studies of metaphysics, mind, language and morality tell us? First, that all philosophical theories, no matter what they may claim about themselves, are necessarily metaphoric in nature. Second, that the metaphorical thought is ineliminable: It is metaphoric thought that defines the metaphysics and unifies the logic of each philosophical theory. Third, this is simply a consequence of the fact the philosophical theories make use of the same conceptual resources that make up ordinary thought. Because we ordinarily think metaphorically, and because our everyday metaphysics derives from our metaphors, it should be no surprise that philosophical thought works the same way. Metaphor, rather than being an impediment to rationality, is what makes rational philosophical theories possible.

16

The Pre-Socratics:
The Cognitive Science of
Early Greek Metaphysics

What Is at Stake in Metaphysics?

It is natural to ask questions about the nature of things. As Aristotle said at the beginning of the *Metaphysics,* "All men by nature desire to know." We desire to know, for practical reasons, if the mushroom we are about to eat is poisonous. We desire to know, for ethical reasons, if there is some natural difference between men and women. We desire to know, for purely intellectual reasons, whether the universe will come to an end someday.

The very project of seeking knowledge assumes that the world makes systematic sense, that it is not just a random collection of individual phenomena. It is not just determined by the capricious whims of gods who are fickle, mischievous, and cruel, but, rather, it is a "cosmos," a rationally structured whole. In other words, to seek knowledge, we must assume that the world is not absurd. It also assumes that we can gain knowledge of the world.

These two assumptions together define what has come down to us as a commonplace folk theory that we take for granted any time we seek any kind of systematic knowledge:

THE FOLK THEORY OF THE INTELLIGIBILITY OF THE WORLD
The world makes systematic sense, and we can gain knowledge of it.

Thus it is natural for us to ask what things are like and why they behave the way they do. Moreover, we seek general knowledge, knowledge about *kinds* of things, not just particular knowledge that pertains only to a single entity. We want to know whether *this* mushroom is edible, but our knowledge of it depends on our knowledge of the general kind of mushroom it is. We want to know whether men and women are somehow fundamentally different and not just whether this man differs in certain particular ways from this woman. And we assume that such questions have answers, that if we can formulate such a question, there is a fact of the matter that answers it.

In other words, much of the time we assume two particular folk theories about things in general:

THE FOLK THEORY OF GENERAL KINDS
Every particular thing is a kind of thing.

THE FOLK THEORY OF ESSENCES
Every entity has an "essence" or "nature," that is, a collection of properties that makes it the kind of thing it is and that is the causal source of its natural behavior.

The Folk Theory of Essences is metaphorical in two ways. First, the very idea of an essence is based on physical properties that compose the basis of everyday categorization: substance and form. For example, a tree is made of wood and has a form that includes a trunk, branches, leaves, roots, bark, and so on. It also has a pattern of change (another kind of form) in which the tree grows from seed to sapling to mature specimen. These are the physical bases on which we categorize an object as a tree: substance, form, and pattern of change. Where an essence is seen as a collection of physical properties, it is seen as one or more of these things. In the case of abstract essences, these three physical properties become source domains for metaphors of essence: Essence As Substance, Essence As Form, and Essence As Pattern Of Change.

The second way in which the concept of essence is metaphorical concerns its role as a causal source. The intuition is this: If a tree is made of wood, it will burn. Because it has a trunk and stands erect, it can fall over. The idea is that the natural behavior of a tree is a causal consequence of the properties that make it the kind of thing it is: The tree burns *because* it is made of wood. We have the same intuition about abstract essences, like a person's character. Honest people will tell the truth. Their essence as honest is the causal source of their truth telling. In such cases, we are clearly in the domain of the metaphorical, because we are attributing to a person a metaphorical substance called "character," which has causal powers.

An immediate consequence of these two folk theories is the foundational assumption behind all philosophical metaphysics:

The Foundational Assumption of Metaphysics
Kinds exist and are defined by essences.

It is important to see how a natural desire to know leads so easily to metaphysical speculation, for as soon as we believe that kinds exist, what we shall call the metaphysical impulse takes over. We can apply the Folk Theory of Essences to kinds themselves, from which it follows that there are kinds of kinds and that these kinds of kinds themselves are defined by essences. This iteration is a fateful step; it is the first step toward metaphysics in Western philosophy.

This metaphysical impulse lies at the heart not only of Western philosophy but of all Western science, leading physicists to seek a general field theory, or as it has come to be known, "a theory of everything." In biology, there is a similar quest for a theory of life. Such theories seek to find some essence that characterizes the behavior of things in some general domain of study: physical phenomena, life, the mind, language, and so on. Questions like "What is the mind?" or "What is life?" presuppose the meaningfulness of such a quest for general knowledge.

Whether we like it or not, we are all metaphysicians. We do assume that there is a nature of things, and we are led by the metaphysical impulse to seek knowledge at higher and higher levels, defined by ever more general categories of things. Once we have started on this search for higher and higher categories and essences, there are three possible alternatives:

1. The world may not be systematically organized, or we may not be able to know it, above a certain level of generalization, which might even be relatively low in the hierarchy of categories. In other words, there may be a limit to the systematicity of the world or to its intelligibility.

2. The hierarchy of categories may go on indefinitely, with no level at which an all-inclusive category exists. In this case, the world might be systematic, but not completely intelligible. The process of gaining knowledge of the world would be an infinite, and hence uncompletable, task.

3. The iteration up the hierarchy of categories and essences might terminate with an all-inclusive category, whose essence would explain the nature of all things. Only in this case would the world be totally intelligible, at least in principle.

This third alternative is what we call:

THE FOLK THEORY OF THE ALL-INCLUSIVE CATEGORY
There is a category of all things that exist.

From the Folk Theory of Essences, it follows that this all-inclusive category has an essence, and from the Folk Theory of Intelligibility, it follows that we can at least in principle gain knowledge of that essence. This all-inclusive category is called Being, and its essence is called the Essence of Being.

This third alternative, that the world is completely systematic and knowable, is the most hopeful, least skeptical attitude that someone concerned with seeking general knowledge can take. However, such optimism brings with it a substantial ontological presupposition, that there is a category of Being, and that, since it must have an essence, there is an Essence of Being.

As we will see below, there is a profound problem that arises from this ultimate metaphysical impulse, as defined by these four commonplace folk theories. They lead us to ask a set of questions that may not be meaningful. And they give us a view of the world and of knowledge that may be misleading.

To see why this is so, we propose to apply the tools of embodied cognitive science to the emergence of explicit metaphysical thinking in the Western philosophical tradition. The foundational metaphysical projects of Western philosophy were formulated by Aristotle. In early pre-Socratic philosophy, there are hints of this way of thinking about nature that, in Aristotle, finally and explicitly become the quest for an understanding of Being as the ultimate form of knowledge. This sets the stage both for Western science and for theological interpretations of God as Ultimate Being.

Although metaphysics found its explicit formulation in Aristotle, various strands and threads of what came to be the fabric of Aristotelian metaphysics appeared a bit at a time much earlier, in the early Greek philosophers. Looking at the emergence of Greek metaphysics allows us to see each thread in that fabric in clear relief.

Our project is to explain what must be presupposed from a cognitive point of view in order to ask the metaphysical questions that Aristotle bequeathed to Western philosophy. We approach this not as historians of philosophy, but rather as cognitive scientists seeking to make sense of the conceptual system that underlies mature Greek metaphysics.

We are well aware of the pitfalls of anachronistic readings of early Greek philosophy, that is, reading back into the early philosophers doctrines that only came into existence later. Thus, for example, saying that Milesian philosophers were materialist, as we do, will strike the historian as misleading, since the term for matter in general did not even exist at that time. But as any linguist knows, concepts must exist before there can be words for them, and we will argue that such a concept was coming into existence in a way that prefigured its full expression in Aristotelian philosophy. What follows is not traditional historical or textual scholarship, but something new: the cognitive science of early Greek metaphysics.

The Beginnings

Western philosophy emerged in the sixth century B.C. on the Ionian coast of what is today Turkey. The character of much of the Western philosophical tradition was fatefully determined by its origins in the writings of a small group of Greek scientist-philosophers. These early philosophers were struggling to develop "rational" accounts of events in nature as a way of supplementing, or even supplanting, the traditional mythic stories that attributed natural occurrences to the willful, unpredictable, and sometimes even frivolous actions of the gods. These early Greek philosophers thought they had discovered fundamental principles of nature (Greek *phusis*) that could explain how things come into being, why things have the properties they do, and why things behave as they do. In other words, they were driven by their belief in the Folk Theory of Intelligibility. They optimistically sought answers to the ultimate questions "What is Being?" and "How can we know what its Essence is?" taking for granted the Folk Theory of the All-Inclusive Category. But Being for these nature philosophers meant nature—the dynamic material world. For their successors, Plato and Aristotle, Being meant much more.

The revolutionary idea that there might be principles or laws governing the way all things happen, and that human reason could discern these principles, was at once exhilarating and dangerous. It was exciting because it suggested that human beings might no longer be mystified about why things occur as they do, subject always to the capricious will of this or that god. Knowing the nature of all things would give guidance for living. Yet it was also dangerous, because it threatened the established order and called into question the traditional modes of wisdom.

The idea that there might be principles underlying natural events took form gradually. The supreme ontological question, in its full generality, "What is Being?" did not simply emerge fully formed one day in sixth-century Miletus. Rather, a series of attempts to frame a conception of something called "nature" (*phusis*) gradually made it possible to think of nature as a unified whole in which things came into existence, changed, and passed out of existence in an orderly fashion, governed by laws or principles.

As we shall see, it is not until Plato and Aristotle that the question "What is Being?" or "What is the nature of What Is?" is asked in its full generality. Yet even in the earliest recorded philosophical investigations of the Milesian nature philosophers (Thales, Anaximander, and Anaximenes), we see hints of this question and a range of basic answers to it. At this point, what we are calling the question of Being (which they didn't) is limited to physical nature. Their answers are predicated on metaphor and folk theories, and that is where the cognitive science of pre-Socratic metaphysics begins.

The Milesian Nature Philosophers

No later than the sixth century B.C. the scientist-philosophers living near Miletus on the Ionian coast shared, along with their fellow Greeks, a very basic folk theory about the elements of nature:

THE FOLK THEORY OF THE ELEMENTS

Things in nature are made up of some combination of the basic elements: Earth, Air, Fire, and Water. Each element is defined by a unique combination of heat and wetness values, as follows:

Earth = Cold and Dry
Water = Cold and Wet

> *Air = Hot and Wet*
> *Fire = Hot and Dry*

The apparent properties of the objects we experience are the result of various combinations of these elements and their properties.

We call this a "folk theory" because it was a basic explanatory model shared by most Greeks, beginning in the Archaic period. It was a model so intuitively clear that it was still a commonplace throughout most of Europe through the sixteenth century. We intend nothing at all derogatory by describing this knowledge as a "folk theory." On the contrary, it is just such models that make up a culture's shared common sense. There are often good reasons for these models, and in many cases folk theories work sufficiently well to serve everyday purposes.

Some folk theories, like that of the Elements, are explicit and seem to have been widely held as matters of conscious public knowledge. Other folk theories, like those of Essences and General Kinds, are implicit, that is, unconscious and automatic, taken as background assumptions and used in drawing conclusions.

Based on the scanty and fragmentary textual evidence we have of early pre-Socratic thought, it appears that the Milesian philosophers appropriated the Folk Theory of the Elements within a framework defined by a basic metaphor according to which nature—all that exists—is understood as being composed of material "stuff."

The Milesian Metaphor: The Essence of Being Is Matter

Although no term for generalized matter existed at this time, the concept was emerging and lay behind all this metaphysical speculation. The metaphor provides an understanding of the totality of nature (*phusis*) in terms of the properties of one aspect of nature, the material aspect. This metaphor defines the view known as "materialism," which has persisted throughout all Western philosophy to the present day and, as we shall see, contrasts with views that claim that what is real is not matter, but form.

We should not think of the concept of matter as it is understood today. The Milesians conceived of nature as being in an active, dynamic process of change, and so its material dimension was not considered static, lifeless, or mechanistic.

Our claim, that the Milesian nature philosophers adopted the metaphor The Essence Of Being Is Matter, may seem anachronistic for the following reason:

It is not clear that the Milesians even had any clearly defined notion of "Being" that went beyond physical nature. However, the concept of nature, as the totality of natural events, is the precursor to the more general, abstract notion of Being. Therefore, even at this earliest stage, to understand Nature metaphorically as material is to answer the question of the nature of "what is."

The Milesian metaphor carries with it its own peculiar problem for the philosopher. If The Essence Of Being Is Matter, then was the material dimension one of the four elements, all of them, or something else of which the four elements were special cases? Thales realized that the Folk Theory of General Kinds, that every particular thing is an instance of a more general kind of thing, required that the particular kinds of things—earth, air, fire, and water—all be instances of one general kind of thing. His solution was that one of the particular elements was more general than the others: water. "The first principle and basic nature of all things is water."

Various sources suggest possible reasons for this novel hypothesis: All things contain water. All life depends on water. The earth rests on water. Therefore, water is that which most clearly reveals the ultimate nature of what exists.

Thales' Metaphor:
The Essence of Being Is Water

Source Domain: Water
Target Domain: The Essence of Being
Mapping:
    The Essence Of Being Is Water
    The Essences Of Specific Kinds Of Being Are Forms Of Water

This conceptual metaphor is used to comprehend the natures of all the kinds of things that exist. Thales was conceptualizing all kinds of things in terms of water. In addition, he believed that this conceptualization was true. That is what makes this conceptual metaphor into a metaphysical statement. Conceptual metaphors are ways of conceptualizing one kind of thing in terms of another. Such a conceptualization, if taken to be true, imposes an ontological commitment.

Thales' metaphor presupposes the Folk Theory of Essences: Every entity has an "essence" or "nature," that is, a collection of properties that makes it the kind of thing it is and that determines how it will behave. Thales' metaphor thus claims that whatever it is that makes water what it is, must also be the same principle that makes every thing (and every kind of thing) what it is.

Thales' near contemporary, Anaximander, recognized an inconsistency in Thales' reasoning. He argued that any of the four elements will possess two of the four possible qualities (Cold, Hot, Wet, Dry). If everything that is real were only one of these four elements, then it could not be the source of any particular thing that possesses the opposite qualities. This assumes that nothing can be the causal source of its own opposite. For instance, water is cold and wet. If everything that is real is water, then everything that is real must be cold and wet. Yet fire is real, and fire is hot and dry. If the essence of all real things were coldness and wetness, fire could not exist. Since fire does exist, the essence of all that is real cannot be water. Therefore, the essence of all that is real (i.e., of *phusis*) cannot be any determinate kind of thing that has particular values for heat and wetness. It must be an indeterminate material (that is, a material with none of the four determinate qualities).

Anaximander's Metaphor:
The Essence of Being Is Indeterminate Matter

Source Domain: Indeterminate Material (the *Apeiron*)
Target Domain: The Essence of Being
The Mapping:
    The Essence Of Being Is Indeterminate Matter
    The Essences Of Each Kind Of Being Are Determinate Forms Of Matter

Anaximander saw that the essence of all existing things would have to be some indeterminate principle (the *Apeiron*—the Unbounded, the Indeterminate). Since this is the essence of all that is, it must, by the Folk Theory of Essences, cause the behavior of all that is. Somehow, all determinate properties must emerge naturally from indeterminate matter. This left Anaximander with an unsolvable problem: How can specific determinate properties come out of an unbounded, indeterminate matter?

Let us summarize the problem at this point. The folk theories of General Kinds and of Essences lead one to keep generalizing until one finds an ultimate General Essence of everything that exists: the Essence of All Being. By the Folk Theory of General Kinds, nothing specific can be the Essence of all Being, since there must be something still more general than anything specific. That is why Anaximander came to the conclusion that the Essence of all Being is indeterminate matter. However, by the Folk Theory of Essences, essences are causal.

They cause the specific behaviors of all the things they are essences of. But the behavior of matter is such that there are different determinate elements—earth, air, fire, and water—each with its own determinate properties. Since something cannot cause its opposite, indeterminate matter cannot be a causal source of determinate properties. This dilemma is generated by just two folk theories and the logical principle that something cannot be the causal source of its opposite.

Anaximenes dealt with this problem by arguing that the Essence of all Being can be a particular kind of material, provided that there is a principle by which all other particular entities, each with its particular properties, can be seen as forms of the ultimate material. In what appears to be a retrograde move given the argument of Anaximander, Anaximenes chooses Air as the ultimate material.

Anaximenes' Metaphor: The Essence of Being Is Air

Anaximenes' reasons that Air is rarefied matter and that it is the most common form of matter. He is able to take a specific element as the ultimate essence of all that exists by adding the idea that *form* is what distinguishes one kind of thing from another. His theory is that the four elements are different forms of Air.

Water = condensed Air (since water is denser than air)
Earth = condensed Water (since earth is denser than water)
Fire = rarefied Air (since fire is less dense than air)

To solve the problem completely, he must give a causal theory of how the other three elements can come into existence from Air. He notes the existence of condensation and rarefaction (that is, evaporation) and argues for a single fundamental principle of all change of states: All change is either rarefaction or condensation. All things come to be, have the properties they have, and pass out of existence through the condensation and rarefaction of Air.

Why Care About Pre-Socratic Metaphysics?

There is an important reason why we have chosen to discuss these Greek philosophers. They implicitly utilized two common folk theories that most of us still use today when we seek any kind of general knowledge. They reasoned with these two folk theories in a way that philosophers and scientists alike have reasoned ever since. Their reasoning shows in very clear form what the logic of metaphysical questions is. It is the logic of seeking ever more general categories in nature to which one can apply ever more general principles to account for the behavior of things in the world.

In addition, we are witnessing the birth in Western philosophy of kinds of explanatory principles. What can an essence of a kind of thing be? Here we see the three most basic answers: It can be matter (i.e., a kind of substance), or form, or a pattern of change, or all three. These answers make sense because basic physical objects around us are made of substance and manifest forms and patterns of change.

Consider a tree. What makes it the kind of thing it is? First, take substance. A tree is made of wood. If it were not made of wood, if it were made, say, of iron, it would not be a tree. So being made of wood is part of its essence, part of what makes a tree a tree. But not everything made of wood is a tree. Second, to go a step farther, there is form. A tree has a trunk, roots, branches. If it did not have this form, say, if it where spherical, it would not be a tree. So form too is part of its essence, what makes it a tree. Third, consider patterns of change. Trees grow from seeds to saplings to mature trees and then die and fall. If they did not, they wouldn't be trees. So a pattern of change is part of what makes a tree a tree. In short, what we mean by essence in simple, obvious cases like a tree is at least substance, form, and a pattern of change.

What the pre-Socratic nature philosophers did was take these aspects of essence and apply them iteratively to ever more general categories via the two folk theories of General Kinds and of Essences. The same form of reasoning has been applied in philosophy and in science ever since. It is hard to imagine how we could do philosophy or science without them.

The Strange Question of Being

Let us stop and take stock for a moment, not just of the metaphors and folk theories of these early pre-Socratic scientist-philosophers, but also of the strange character of the question they were asking—the question of the nature of Being. To most nonphilosophers the question "What is Being?" must sound extremely odd. We do not, after all, see ourselves as experiencing "Being" and requiring views of what it is. Rather, we experience chairs, trees, people, natural events, language, and human actions. We take chairs and trees and even people as having essences that make them the kinds of things they are. We may even wonder about what makes us human, that is, about the essence of humanity. But we don't go around asking about the essence of Being.

To ask this question one must assume not only the folk theories of Intelligibility, Essences, and General Kinds, but we must especially assume the Folk Theory of the All-inclusive Category. In other words, one must make two optimistic assumptions: first, that there is a reason why everything is the way it is and, second, that we can know that reason. If one makes these assumptions, the Essence of Being is the ultimate reason why things are the way they are. Therefore, the Essence of Being, properly understood, explains why things are the way they are.

Many people do indeed make the optimistic assumptions that there is a reason why everything is the way it is and that we can know it, namely, those with a faith in a monotheistic God as traditionally conceived. A great many people today still have questions about Being and the Essence of Being, but these metaphysical questions take the form of questions about God and the Nature of God. The reason for this is that God has come to be metaphorically conceptualized in much of theology as Being, and the Nature of God as the Essence of Being. Most people who believe in God believe that it is because of the Nature of God that everything is the way it is. The Folk Theory of the All-inclusive Category is very much alive in contemporary religion.

The Task of Metaphysics

We have seen how the four folk theories discussed so far—Intelligibility of the World, Essences, General Kinds, and the All-Inclusive Category—generate the grand tradition of metaphysics in Western philosophy. Whereas particular sciences seek to understand the essence of particular aspects of being, such as matter, life, the mind, and so on, metaphysics saw itself on the greatest quest of all, an understanding the nature of Being itself.

But Being, as we have seen, is definable and makes sense only relative to those four folk theories. The most crucial of these are the folk theories of Essences and of the All-Inclusive Category, without which the abstract notion of Being cannot even be conceptualized. Indeed, many of the antimetaphysical philosophies of the twentieth century are defined by their denial of the validity of these folk theories. It should also be clear that the results of embodied cognitive science that we are focusing on in this book are inconsistent with these folk theories.

But if you do hold all four folk theories, then a certain strategy for doing metaphysics emerges. Since everything in our sense experience is ephemeral and changeable, while Being is eternal and unchangeable, certain pre-Socratics argued that knowledge of Being cannot be based solely on sense experience. Instead, one must search behind the world of our sensible experience (the Realm of Becoming) and probe into the Realm of Being that underlies and makes possible all that we experience.

How does one get at Being itself? One possible strategy, the one employed by the Milesian nature philosophers, was to assume that Being shows itself more clearly in some phenomena than others. The task is then to comprehend Being in terms of those phenomena in which the Essence of Being is most evident. Another possibility, the one taken up by Parmenides and Zeno, is to give priority to the commonplace metaphor that Ideas Are Objects, to extend it naturally to metaphorically conceptualize Thinking As Being and to take that metaphor literally. On the assumption that we know our own thoughts better than external phenomena, it follows that Being reveals itself most clearly in Thinking, not in any physical phenomena. In both of these cases, the general strategy is to find something else that is more directly grasped and better understood, in order to use it to understand Being.

This general strategy is the basis for our hypothesis that pre-Socratic metaphysics (and, ultimately, all metaphysical inquiry) is based on metaphor and folk theories. We have seen the nature of the metaphorical mappings for the conceptions of Being set forth by Thales, Anaximander, and Anaximenes. In each case, the metaphysical view rests on an assumption of the following form: Being shows itself most fully in phenomena of kind X, where X is fairly well understood. X can therefore be used to understand Being in general, via a metaphor of the form: The Essence of Being Is X. Each way of conceptualizing the Essence of Being in terms of something else is a conceptual metaphor through which we understand something general, namely Being, in terms of something specific, some particular form of Being.

What our analysis shows is that the very project of metaphysics—the very idea of understanding Being and finding the Essence of Being—depends on folk theory and is defined by metaphor. This has a profound implication for our understanding of what the philosophical project of metaphysics is. The target domain for ultimate metaphysical metaphors, Being, is a conceptual creation, a product of the Folk Theory of the All-Inclusive Category and the other folk theories that it depends on. Someone who believes all those folk theories will, of course, assume that the Realm of Being is real and that the problems of metaphysics (concerning the nature of Being) are real problems. Those who do not believe one or more of those folk theories will take the Realm of Being as a fictional construct, and they will therefore see the Problem of Metaphysics (the

Problem of Being) as a pseudoproblem. Pre-Socratic metaphysics gave rise to the Western tradition of metaphysics precisely by taking seriously these folk theories and by taking as literal any metaphorical statements of the form, "The Essence of Being Is *X*."

Heraclitus

With all of this as background, we can now turn to the philosopher who is undoubtedly the most enigmatic and puzzling of the early Greek thinkers. Heraclitus employed unconventional, jarring, and paradoxical statements designed to shock his fellow Ephesians into an awareness of Being. He felt that they were so dogmatic and unreflective that only the most paradoxical and challenging views would attract their attention and lead them to insight. In what follows we do not pretend to capture Heraclitus's great rhetorical force; nor can we preserve the rich and provocative enigma that gives life to his sayings. Instead, we attempt to describe the metaphors and folk theories that structure a small, but central, part of Heraclitus' argument.

Heraclitus saw the problem of Being as that of finding the stable and unchanging behind the ever changing flux of human experience. He gives poetic voice to the changeability inherent in everyday life:

Everything flows and nothing abides; everything gives way and nothing stays fixed.
You cannot step twice into the same river, for other waters and yet others go ever flowing on. (C2, Wheelwright 1966, 70–71)

The problem Heraclitus raised was this: Being is assumed to be unchangeable, and knowledge of Being, like any general knowledge, was assumed to be stable. How can Being, which is unchanging, reveal itself through perceptions and experiences that are always in flux?

Heraclitus' solution makes use of the Folk Theory of Essences. Consider the example of the acorn and the oak. There is a general fixed pattern of growth that every acorn follows in growing into an oak. Every particular instance of the change of acorn to oak is an instance of this general pattern. This fixed pattern is part of the nature of oaks; in other words, it is one of the essential properties that defines what it means to be an oak.

Similarly, all specific changes in nature, Heraclitus reasoned, are instances of such fixed general patterns of change. What is crucial here is Heraclitus' original use of an existing folk theory that had previously been applied only to things that were static.

One gets insight into such general patterns of change by studying specific changes. By that means, one can discern not just specific changes, but patterns of a change that constitute essences. Thus the essence of each natural kind in the world is defined in part by the way that kind of thing naturally changes.

The question "What is the Essence of Being?" is answered by asking what all the particular fixed patterns of change have in common. The answer is change itself. Since they are all specific patterns of change, it follows, from the Folk Theory of General Kinds, that what they share is a general pattern of change. That pattern is the Essence of Being. As Heraclitus says, "All things come to pass in accordance with this Logos." (C2, Wheelwright 1966, 69). Heraclitus' view of Being takes the form of a metaphor for understanding Being in terms of patterns governing change in the perceptible world:

HERACLITUS' METAPHOR:
THE ESSENCE OF BEING IS CHANGE

Source Domain: Patterns Of Change
Target Domain: The Categories Of Being
Mapping:
  The Essence Of Being Is Change
  The Essence Of Each Specific Category Is A Specific Pattern Of Change

Since patterns of change are stable and unchanging, they can be unchanging objects of knowledge. As such, they can be grasped by the mind. Thus, even though the world is in flux, we can know the Essence of Being, which is Change itself: "Wisdom is one—to know the intelligence by which all things are steered through all things" (C2, Wheelwright 1966, 79). This conception of Being As Change makes sense to our contemporary ears, once we understand the underlying metaphors and folk theories presupposed by the theory.

The Pythagoreans

For the Milesian nature philosophers, versions of the metaphor The Essence Of Being Is Matter explained why things are the way they are in terms of the material dimension of nature (*phusis*). The chief problem with this metaphysical view is that seeing the Essence of Being in terms of matter alone made it impossible to distinguish among the different kinds of things made of the same matter but having different forms. As Anaximenes discovered, form is needed additionally to differentiate among specific kinds of things.

The disciples of the mathematician Pythagoras believed that they had found a conception of form that would make the world rational, that is, that would explain why all things exist with the characteristics they have. In mathematics, they discovered universal principles of form that they believed could distinguish among different kinds of physical objects. The Pythagoreans thus argued that specific mathematical forms constituted the essences of specific kinds of objects. Mathematical form in general, that is, number, therefore had to be the Essence of Being.

Based on later testimony about the doctrines and practices of the Pythagoreans, we can discern a second argument, from the nature of knowledge, that number must be the Essence of Being. The argument goes as follows: Our knowledge of the Essence of Being must be stable and unchanging. Mathematical knowledge is the only stable, unchanging knowledge. Therefore, the only knowledge of the Essence of Being must be mathematical knowledge. Mathematical knowledge is knowledge about number. Therefore, The Essence Of Being Is Number.

THE PYTHAGOREAN METAPHOR: THE ESSENCE OF BEING IS NUMBER

Source Domain: Number
Target Domain: Being
Mapping:
   The Essences Of Specific Categories Are Particular Numbers
   The Essence Of Being Is Number In General

The later Pythagorean Philolaus sums up the Pythagorean view:

The Knowledge Argument:

Whatever can be grasped by the mind must be characterized by number; for it is impossible to grasp anything by the mind or to recognize it without number. (fragment 4)

The Reality Argument:

Number is the ruling and self-creating bond which maintains the everlasting stability of the things that compose the universe. (fragment 23)

The Pythagoreans saw ample confirmation for their metaphorical hypothesis when they found numerical relations everywhere they looked in nature. Numbers could be used to measure the earth (geometrics). Music seemed to operate according to numerical relations (scales, chords, meter, rhythm, pitch). The heavens were believed to move in numerically describable paths. Health was a matter of proper balance and proportion of elements.

The Pythagorean philosophy had strong metaphorical support within the Greek conceptual system. Then, as now, there was the major metaphor Thinking Is Mathematical Calculation. Within this systematic metaphor, Ideas Are Numbers, Thinking Is Calculating with those idea-numbers, and Logical Conclusions Are Sums. Once this arithmetical metaphor for thought is taken seriously, it can be combined with some of the folk theories discussed above to produce the Pythagorean Metaphor. If ideas are numbers, then all real knowledge is in the form of numbers. Therefore, knowledge of the Essence of Being is in the form of numbers. Hence, the Essence of Being must be number.

Moreover, if the Essence of Being is number, it is an *unseen* essence—a nonphysical essence that transcends physical phenomena. That means that, through studying mathematics and discovering the mathematical forms in physical objects, we can grasp their unseen essences and hence get in touch with the unseen nonphysical causal powers that govern our existence.

This mystical dimension of Pythagorean philosophy is still with us today. It is common to find articles in popular scientific journals about how the arrangement of petals on a flower forms a Fibonacci series or how palm leaves have fractal structure. Such facts capture the popular imagination precisely because our commonplace folk Pythagoreanism is very much with us: To see the fractal structure in a palm leaf is to grasp a mystic truth. Similarly, theoretical physics, especially in its popular presentation, utilizes the same mystical tradition when it suggests that the causal essence of the universe can be found in mathematical formulas.

Pre-Socratic Metaphorical Metaphysics

Let us take stock. We have tried to show, selectively, how the pre-Socratic philosophers developed metaphysical views defined by metaphor and folk theory. Here are the folk theories that all the pre-Socratics shared:

The Folk Theory of the Intelligibility of the World
The world makes systematic sense, and we can gain knowledge of it.

The Folk Theory of General Kinds
Every particular thing is a kind of thing.

THE FOLK THEORY OF ESSENCES

Every entity has an "essence" or "nature," that is, a collection of properties that makes it the kind of thing it is and is the causal source of its natural behavior.

THE FOLK THEORY OF THE ALL-INCLUSIVE CATEGORY

There is a category of all things that exist.

As we shall see, Plato and Aristotle shared these as well. These are the folk theories that jointly characterize the enterprise of Greek metaphysics and make sensible the question, "What is the Essence of Being?" The answer to this question is metaphorical.

We are arguing that any answer to any such question is necessarily metaphorical. Metaphor is a necessary aspect of any philosophical theory, because philosophical theories make use of human conceptual resources and conceptual systems, all of which involve a range of imaginative devices, especially metaphor.

The range of pre-Socratic metaphors for Being is quite instructive, for it covers most of the major metaphysical views we still have today. For the Milesian nature philosophers, The Essence Of Being Is Matter. For Heraclitus, The Essence Of Being Is Change. For the Pythagoreans, The Essence Of Being Is Number.

This same kind of analysis could be extended to cover all of the major pre-Socratics. For Parmenides and his disciple Zeno, The Essence Of Being Is The Logical Form Of Thought. For the Atomists Democritus and Leucippus The Essence Of Being Is Atoms And The Void. For the Sophists, such as Protagoras, The Essence Of Being Is Appearance.

We have also found in the pre-Socratics the three great metaphors for essence that have carried down to the present day: Essence Is Matter, Essence Is Form, and Essence Is Patterns Of Change. These define classical materialist, formalist, and process metaphysics, all of which have their contemporary versions.

It did not take very long for these metaphors for essence, and the folk theories they presuppose, to be firmly in place in the Greek worldview. In the works of Plato and Aristotle that worldview finds its full and explicit expression. It is finally in Aristotle that we find the formulation of these doctrines that has been handed down to the present day.

17
Plato

Throughout his philosophical career Plato is concerned, again and again, with the nature of rational knowledge. His account of such knowledge makes use of the four major folk theories—of Intelligibility of Being, General Kinds, Essences, and the All-Inclusive Category—that we have seen to be operative already in pre-Socratic philosophy.

Plato's concern with what it means to know something is expressed even in his very early dialogue *Euthyphro,* in which Socrates inquires into the nature of human piety, understood as right action. Because the young, cocky Euthyphro believes himself to be pious, he offers a series of successive definitions of piety that are each found by Socrates to be inadequate in some important way. Euthyphro's first definition is that piety is "to do what I am doing now, to prosecute the wrongdoer" (5e). Socrates responds with his classic demand for a true definition that would give the form—or essence—of piety itself, rather than merely giving a specific instance of alleged pious action:

Bear in mind then that I did not bid you tell me one or two of the many pious actions but that form itself that makes all pious actions pious, for you agreed that all impious actions are impious and all pious actions are pious through one form, or don't you remember? (6d)

In demanding that a definition must give the form that makes something the kind of thing it is, Socrates here presupposes the folk theories of General Kinds and of Essences.

(1) The Folk Theory of General Kinds: Socrates reminds Euthyphro of their assumption that "all impious actions are impious, and all pious actions are pious." Every specific pious action must be an instance of a more general kind of thing, piety.

(2) The Folk Theory of Essences: There must be some one thing shared by all instances of pious action: "Pious actions are pious through one form," which is the essence they all share that makes them acts of piety. The category "piety" is defined by a set of characteristics possessed by all pious actions. Socrates continues: "Tell me then what this form itself is, that I may look upon it, and using it as a model, say that any action of yours or another's that is of the kind is pious, and if it is not that it is not" (6e). To know piety is to know what characteristics together must be present for some act to be pious.

In a number of celebrated arguments in the *Republic*, Plato elaborates this view of knowledge in a way that brings out its metaphysical implications. In *Republic VI,* Socrates is discussing with Glaucon the difference between the philosopher—the "lover of wisdom" who seeks rational knowledge—and the person who dabbles in opinions only and who therefore has no genuine knowledge. Socrates begins by stating the same general theory of the Forms we have seen to be operative in *Euthyphro*, and he thus presupposes exactly the same folk theories:

> And in respect of the just and the unjust, the good and the bad, and all the ideas or forms, the same statement holds, that in itself each is one, but that by virtue of their communion with actions and bodies and with one another they present themselves everywhere, each as a multiplicity of aspects. (476a)

In other words, justice is defined by a single essence. There are many specific just actions, but they all share the same essence (here called a "form").

Based on this doctrine of the Forms, Socrates argues that it is the philosopher alone who seeks knowledge, not by attending to the multiplicity of perceptible things, but rather by using reason to discern the ultimate Forms (the essences) that underlie this multiplicity of things. For example, the philosopher does not delight in "the beautiful tones and colors and shapes and in everything that art fashions out of these." Instead, the philosopher's "thought recognizes a beauty in itself, and is able to distinguish that self-beautiful [beauty-in-itself] and the things that participate in it, and neither supposes the participants to be it nor it the participants" (476d, brackets added).

Socrates then goes on to argue that philosophical knowledge, what the philosopher knows, must necessarily be knowledge of Being, or in other words, knowledge of what is:

SOCRATES: Is (what he knows) something that is or is not?

GLAUCON: That is. How could that which is not be known?

SOCRATES: We are sufficiently assured of this, then, even if we should examine it from every point of view, that that which entirely is is entirely knowable, and that which in no way is is in every way unknowable? (477a)

Socrates gets Glaucon to agree that one can only have knowledge of what is. There can be no knowledge of what is not. This then leads directly to the conclusion that reality has a rational structure that can be known, since "that which entirely is, is entirely knowable."

The logic of these two folk theories forces Socrates to the rather odd-sounding view that opinion, which is something midway between knowledge and ignorance, must have as its "object" something that neither is nor is not: "Then since knowledge pertains to that which is and ignorance of necessity to that which is not, for that which lies between [i.e., opinion] we must seek for something between nescience and science, if such a thing there be" (477a, brackets added).

Plato's argument rests on the metaphors Ideas Are Objects and Knowing Is Seeing. He then extends those metaphors to form a new complex metaphor, the Degrees of Being metaphor, in which there is a correspondence between degrees of knowledge and degrees of being: If knowing is seeing an object, and you see whatever is there to be seen, then your degree of knowledge depends on how substantial the object is. Solid knowledge is mental vision of a solid object. Insubstantial knowledge is the mental vision of an insubstantial object. Ignorance is the apparent vision of an object that does not exist. The person who knows is seeing a substantial object (what is). The person who is ignorant sees no object at all (i.e., sees "what is not"). But the person who opines sees only some shadowy object that neither is nor is not: "Then neither that which is nor that which is not is the object of opinion" (478c).

Plato has metaphorically extended the notion of a physical object, which is real, so that there can be objects of intermediate reality. The problem of vague visual perception is attributed not to seeing (which metaphorically is knowing), but to the objects themselves. The set of quasi objects is now metaphorized as a "realm"—the Realm of Becoming. Why "becoming"? Because if something is coming into existence, it is only partway there and therefore does not completely exist. Here Plato uses the common metaphor that Existence Is Location Here, according to which things *come into* existence and *go out of* existence. Plato calls the metaphorical location where existent things are "located" the Realm of Being; the Realm of Becoming is the region of metaphorical space where things in the process of "coming into being" are metaphorically "located." In the metaphor, the Realm of Becoming lies between Being and Not-Being, since, according the metaphorical logic, becoming is motion from Not-Being toward Being. These metaphors jointly entail Plato's Degrees of Being metaphor, according to which there are degrees of reality for objects. As we have seen, degrees of knowledge arise by combining the Degrees of Being metaphor with Knowing Is Seeing: Your degree of knowledge depends on the degree of Being of the object of knowledge. This establishes a correlation between degrees of Being and degrees of knowledge.

Using the Knowing is Seeing metaphor once more, Plato asks what we know best. By the logic of the metaphor, what we can know best is what the mind can see best, namely, ideas. Since, according to the Degrees of Being metaphor, those things that we know best are most real, our *ideas* of physical objects are as real as anything can be. The ideas, which are directly present in the mind, have more reality than the objects themselves, which are not. And as we saw, the objects themselves are more real than images of the objects. These metaphors jointly entail a hierarchy of reality, with ideas being the most real—that is, having the most Being—with physical objects next, and images, shadows, and reflections of objects having the least degree of Being.

The Essences Are Ideas Metaphor

Central to Plato's metaphysics is the metaphorical conceptualization of the essences of things in the world as ideas perceivable by the mind. The Essences Are Ideas metaphor brings the mind and the world together and thereby makes certain knowledge of the world possible: We know the essences of things in the world because we know our ideas directly.

Plato is not appropriating a commonplace everyday metaphor in this case. Instead, he is consciously constructing a metaphor to serve a philosophical purpose, namely, to explain how knowledge is possible.

The entailments of this metaphor, given Plato's other metaphors, are striking: First, if essences are ideas, they cannot be material in nature and therefore must be forms. Hence the celebrated notion of the Platonic form. The essence of a chair is the form of chair. Next, as the Folk Theory of Essences says, essences are real entities. According to this metaphor, the mind can grasp and look at ideas, and if essences of objects in the world are ideas, then the mind can grasp and look at the essences of objects. It is by virtue of this metaphor that the *essence* of a chair is the *idea* of a chair.

The Essences Are Ideals Metaphor

Cognitive linguistics has recognized that there are various kinds of prototypes for conceptual categories. For example, there is a typical-case prototype that characterizes what we take to be typical, for example, the typical politician or the typical used car. Another type of prototype is the ideal—the best example of the category and the standard against which all category members are to be judged. Thus there is a considerable difference between the typical husband and the ideal husband! The essence prototype is a third kind. The essence of a category is taken to be a defining collection of properties, those properties that make something the kind of thing it is. For instance, what makes someone a husband (the essence, or defining properties, of being a husband) is very different from what makes someone an ideal husband or even a typical husband. Essence prototypes and ideal prototypes are very different things.

But not for Plato. Plato brings these distinct ideas together via the metaphor Essences Are Ideals. In reasoning, Plato uses this metaphor together with the metaphor Essences Are Ideas. Many things follow. It follows first that the essence of something is the best example of that thing, the standard against which all (lesser) real things are measured. Next, since Essences Are Ideas, it follows that the idea of a physical object is the ideal version of that object. Hence, the idea of chair not only characterizes the essence of a chair, but also sets the standard for what a chair should be. That is why Plato claims that a physical object—this particular chair—"imitates" the idea of a chair, which is also both ideal and essence.

There is another important consequence of this metaphor for Plato's metaphysics. Consider courageous and cowardly actions. Courage and cowardice are opposite ends on the same scale. Courage is positive: it is good; it is an ideal to be striven for. There is an essence to courage and that essence, via this metaphor, defines the ideal of courageous action. Cowardice, at the negative end of the scale, is not good and hence is not an ideal. Since essences are ideals, only positive qualities can be essences. Therefore, there can be no essence of cowardice. There is no Platonic ideal of cowardice, because ideals have to be good. Cowardice is therefore a deficit—a lack of courageous essence, an absence of an essential form of Being that is appropriate to human beings.

The Essences Are Ideals metaphor provides Plato with a virtue theory of ethics. What the Essences Are Ideals metaphor does is link metaphysics with morality. A virtue is a positive essence (not a lack of one) and therefore is also an ideal. A fully realized human being has all the essential properties that make a human being ideal. For Plato, the pious person, that is, the virtuous person who performs actions that are good, is realizing an ideal of what a human being can be. That is, he or she is realizing one of the essences that make human beings human. The thief or coward, on the other hand, in lacking one of the ideal traits of human beings, is also lacking in one of the essences that make us human. Lack of virtue thus makes one less than fully human.

Plato's Idea of the Good

Let us take stock of Plato's conceptual system: It contains the folk theories of Intelligibility, General Kinds, Essences, and the All-Inclusive Category. These jointly characterize a hierarchy of categories from shadows, reflections, and images of objects at the bottom, up through specific physical objects in the middle, all the way to Being itself at the top. They also characterize a hierarchy of essences corresponding to the categories, with the Essence of Being at the top.

Plato also has the commonplace metaphors that Knowing Is Seeing, Ideas Are Objects, and Existence Is Location Here. He combines these in such a way as to derive from them his metaphor of Degrees of Being. Finally, Plato has two original metaphors that characterize what is innovative in his philosophy: Essences Are Ideas and Essences Are Ideals. These two metaphors and the Degree of Being metaphor are Plato's signature metaphors. They give his metaphysics its distinctive character.

From this conceptual system, the full richness of Plato's metaphysics can be derived. We can now make sense of, and show the logic behind, the most enigmatic of Plato's metaphysical doctrines: the Idea of the Good.

From the folk theories of General Kinds, Essences, and the All-Inclusive Category, we can form an ascending hierarchy with Being above other categories of existing objects and the Essence of Being above the essences of existing objects. But Plato takes this hierarchy one step further than the pre-Socratics did.

He observes that the essences themselves are specific things in the world and hence (by the Folk Theory of General Kinds) form a category—the category of essences. The Essence of Being is only one of the members of this category of all essences. This category of essences, by the Folk Theory of Essences, must itself have an essence—the Essence of Essence. Since Essences Are Ideas, this essence must be an idea. Moreover, it is not merely real, it is the most real possible thing because, by the Degrees of Being metaphor, things get more real (acquire more Being) as you go up the hierarchy, and the Essence of Essence is at the very top of the hierarchy. Moreover, since Essences Are Ideals, this essence must also be an ideal, indeed it is the most ideal—the greatest good. Since Essences Are Ideas, the Essence that is the most ideal, the greatest good, is the Idea of the Good.

By the Folk Theory of Essences, an essence is the causal source of the behavior of all members of its category. The Idea of the Good is therefore the causal source of everything. It is first the causal source of all essences, including the Essence of Being. It is therefore the causal source of all other causal sources. But most important of all, the Idea of the Good is at the absolute pinnacle of the causal chain. One can go no higher. Why? Because, as the Essence of Essence, it is the causal source of all causal sources. It cannot itself have another causal source.

Furthermore, since Essences Are Ideas, the Idea of the Good is the causal source of all other ideas. It is therefore the causal source of all knowledge. To know the Idea of the Good would be to know everything.

> In the region of the known, the last thing to be seen and hardly seen is the idea of good, and that when seen it must needs point us to the conclusion that this is indeed the cause for all things of all that is right and beautiful; giving birth in the visible world to light, and the author of light and itself in the intelligible world being the authentic source of truth and reason, and that anyone who is to act wisely in private or public must have caught sight of this. (517c)

It is only by virtue of these metaphors that this passage makes sense. First, this passage uses the Knowing Is Seeing metaphor, according to which the knowable world is understood as the visible world and the causal source of knowledge is light, which is the causal source of vision. According to this passage the Idea of Good is the causal source of all things and the causal source of all knowledge. The reason is as we said above. Because essences are both ideas and ideals, the essence of all particular essences is the idea of the good. Because essences are causal sources (they make things what they are), the essence of all particular essences is the cause of everything being what it is. Because essences are ideas, the essence of all particular ideas is the causal source of all particular ideas.

The Idea of the Good is unique. It is the causal source of all particular essences, that is, the causal source of all essences of existing physical objects. Plato is careful to distinguish the unique idea of the good from ordinary essences, essences of existing objects. To insist on this difference he even denies that the good is an essence, even though it plays the role of the essence of essences in his metaphysics.

> The objects of knowledge not only receive from the presence of the good their being known, but their very existence and essence is derived to them from it, though the good itself is not essence but still transcends essence in dignity and surpassing power. (509b)

Plato seems here to be reserving the term *essence* for essences of existing objects, not for what we have been calling the essence of all essences (of existing physical objects). He has good reason for insisting on this distinction. Ordinary essences are both causes and things caused. But the essence of essence—the causal source of all that is—is a cause but not a thing caused. For this reason it does not share all the properties of ordinary essences. As the cause of ordinary essences, it is "beyond essence." As the cause of all that is, of all Being, it is beyond Being.

Now the question arises whether the Idea of the Good can itself be known. As we have seen, it can be known that the Idea of the Good exists and is the causal source of all that is—of all Being and of all ordinary essences. Since Essences Are Ideas, the Good, as the causal source of all essences, is the causal source of all ideas, and hence of all knowledge. Therefore, the question becomes: Can the causal source of all knowledge itself be known?

Plato's answer is yes and no. As an essence, the essence of essence is knowable: "As the cause of the knowledge and truth, you can understand it [the Idea of the Good] to be a thing known" (508e, brackets added). Indeed, it is what must be known by "anyone who would act wisely in private or public." Since it is the philosopher who can glimpse the Idea of the Good, it is only the philosopher who can "act wisely in private or public," and so the philosopher should be king.

On the other hand, knowledge of the Good is not like ordinary knowledge, since it is the ultimate causal source of all ordinary knowledge. Here Plato uses the Knowing Is Seeing metaphor again. Just as light, the causal source of vision, cannot be seen like ordinary objects, so the Good, the causal source of knowledge, cannot be known in the way that we ordinarily know things.

These same metaphors provide the logical coherence to Plato's Allegory of the Cave. In that allegory, the Good is conceptualized metaphorically as the sun, the causal source of light, which is metaphorically the causal source of knowledge. The movement from ignorance to knowledge is depicted metaphorically as an ascent from darkness to light. At the beginning, in the stage of ignorance and opinion, the dwellers in the cave know (i.e., see) only faint shadows on the cave wall. Their ascent to knowledge moves in stages from seeing shadows, to seeing firelight, to seeing objects by sunlight, and finally to gazing on the blinding light of the Sun itself.

By the Degrees of Being metaphor, the shadows are less real than physical objects, and seeing them is metaphorically mere opinion, not knowledge. Seeing actual physical objects gets us closer to knowledge, but is still opinion, because physical objects have a lesser degree of reality than do their causal sources—essences. The Idea of the Good, as the causal source of all knowledge and all that is, is the ultimate in reality, and one achieves ultimate knowledge only by seeing it.

Plato's Idea of the Good is not just a quaint archaic notion. It has been articulated in a theological context through a long historical tradition from Plotinus and other Neoplatonists, to Augustine, Anselm, and Thomas Aquinas. In medieval theology, Plato's Idea of the Good became the concept of God—the prime mover, the ultimate causal source of all things, the source and locus of all knowledge, and the Perfect Being, the origin of all that is good. That view of God is still with us today.

18

Aristotle

More than any other philosopher, Aristotle is responsible for our conception of metaphysics. He defined metaphysics as a science, a systematic search for the nature of Being and essence—the word was *episteme* and it meant knowledge based on observation together with an understanding of why things are as they are. He carried out his investigations into the nature of Being in a systematic way that was without precedent. Much of our contemporary conception of metaphysics, theology, and the very nature of science itself stems from Aristotle's mode of thought. It is therefore crucial to understand what gives rise to and unifies Aristotle's way of thinking. In many ways, it remains as much ours as it was his.

Aristotle shares the same folk theories used by Plato and the pre-Socratics, namely, the folk theories of Intelligibility, General Kinds, Essences, and the All-Inclusive Category. As a result, Aristotle too assumes that all specific things are instances of more general kinds of things and that the kinds themselves exist as specific things, which are in turn kinds of still more general things, and so on up the ladder of generality to the all-inclusive category of Being. Consequently, Aristotle sees the world as having a hierarchical category structure, with all things contained in the ultimate category of Being. But, as we will see, he also notes that Being has properties distinct from all other categories. He thus reserves the term *category* for the ten highest subcategories of Being, although he treats Being logically as a category in that he sees it as an object of study having an essence.

Like Plato and the pre-Socratics, Aristotle sees each thing as having an essence that makes it the kind of thing it is and that is the causal source of its natural behavior. He assumes that each category in the hierarchy of Being has an essence, and that there is an essence of Being itself. Moreover, like Plato, Aristotle recognizes that the essences are specific things in the world that form a category of essences, and that that category also has an essence—the essence of essence. That highest essence is necessarily the causal source of all essences, and hence of all things that exist.

Metaphysics, therefore, is the highest philosophical enterprise, since it studies the essence of Being (Being qua Being). In other words, metaphysics studies the highest necessary causal principle by which everything comes necessarily to be the way it is. That principle is what we have called the "essence of essence."

> There is a science which investigates being as being and the attributes which belong to this in virtue of its own nature. Now this is not the same as any of the so-called special sciences; for none of these others treats universally of being as being. . . . Now since we are seeking the first principles and the highest causes, clearly there must be some thing to which these belong in virtue of its own nature. If then those who sought the elements of existing things were seeking these same principles, it is necessary that the elements must be elements of being not by accident but just because it *is* being. Therefore, it is of being as being that we also must grasp the first causes. (*Metaphysics* 1003a21–32)

But Aristotle differs from Plato in a fundamental way. Where Plato had the metaphor Essences As Ideas, Aristotle has the converse metaphor, Ideas Are Essences. These opposite metaphors comprise the fundamental difference between the Platonic and Aristotelian views of philosophy. Where Essences Are Ideas makes Plato an idealist, Ideas Are Essences makes Aristotle a realist. For Plato, the highest reality consists of ideas, which are the essences of things. An essence for Plato is an *eidos*, a form, that is, the "look" of a thing that makes it what it is. Aristotle, by contrast, locates reality ultimately in the world, and he thus sees our thought as dependent upon the nature of the world. For Aristotle, essences exist only in the objects that they are essences of. Thus, for both Plato and Aristotle, there is no separation between the mind and the world. The difference lies in whether the world takes its shape from ideas (as in Plato) or whether the ideas take their shape from the world (as in Aristotle).

The Ideas Are Essences metaphor is Aristotle's way of explaining the intelligibility of the world. If ideas are essences, then the mind can see and grasp the very essences of physical things in the world. There is no fundamental gulf between the mind and the world. What is in the mind depends on what is in the world. Our ideas actually present us with the essences of things as they are in the world. The structure of our rationality is *in the world* :

He whose subject is existing things *qua* existing must be able to state the most certain principles of all things. This is the philosopher, and the most certain principle of all is that regarding which it is impossible to be mistaken, for such a principle must be both the best known (for all men may be mistaken about things which they do not know), and non-hypothetical. For a principle which every one must have who understands anything that is, is not a hypothesis; and that which every one must know who knows anything, he must already have when he comes to a special study. Evidently then such a principle is the most certain of all; which principle this is, let us proceed to say. It is, that the same attribute cannot at the same time belong and not belong to the same subject and in the same respect. (*Metaphysics* 1005b10–20)

Aristotle's principle here has become the primary axiom of classical symbolic logic: "NOT [F(a) and NOT F(a)]"; that is, "It is not the case that an entity *a* both has the property *F* and does not have the property *F*." This is not merely a truth of reason but a truth about the world. This fundamental principle in Aristotelian logic is a principle about the world and becomes a principle of human reason only via the metaphor Ideas Are Essences. For Aristotle, the father of logic, logic is the logic of the world. What we can logically think depends on the way things are in the world. Logic for Aristotle is not an abstract issue: It occurs as part of the world and has a locus in space, in time, and in objects. As a result, there is transcendent reason, a reason of the world. Because Ideas Are Essences, human beings can partake of this transcendent reason. *Logos,* by virtue of the Ideas Are Essences metaphor, is both the logic (the rational structure) of the world as well as human logic (the rational structure of correct thought). The logical law of noncontradition is a logical principle because it is an ontological principle—it is true of the world. And logic (correct reason) is transcendent because it transcends human beings—it is part of the structure of the world.

Aristotle realized that if Ideas Are Essences, then essences can't be substances, since physical substances cannot be grasped by the human mind. He therefore adopted the metaphor Essence Is Form. Because the world, for Aristotle, has a logic, essences must be part of the world. Hence, form must be in the world, instantiated in the substance of the things. And since Ideas Are Essences and Essence Is Form, it follows that the human mind can grasp the forms of things in the world, the forms that make those things the kind of things they are. It is this ability to grasp the forms of things in the world directly that, for Aristotle, guarantees us the possibility of knowledge and rules out skepticism.

A sense faculty is that which has the power to receive into itself the sensible form of things without the matter, in the way in which a piece of wax takes on the im-press of the signet ring without the iron or gold; what produces the impression is a signet of bronze or gold, but not qua bronze or gold: in a similar way the sense is affected by what is colored or flavored or sounding, not insofar as each is what it is, but insofar as it is of such and such a sort and according to its *logos*. (*On the Soul* 424a17–24)

In other words, to perceive something is to incorporate its form into one's mind, to actualize that form in the mind. Aristotle here is using the common metaphors The Mind Is A Container, Understanding Is Grasping, and Ideas Are Physical Objects with a structure of their own. He further uses the common metaphors that perceiving is receiving objects from the world outside the mind, and that these objects leave a "sense impression," the nature of which depends on the structure of the object.

Important things follow from these metaphors. We can get correct ideas from the world. Here's how: Things in the world all have a form, that is, a physical structure. The senses are metaphorically like wax tablets. Things in the world *impinge* on the senses, leaving their impressions as a signet ring does in wax. These impressions are metaphorically physical objects—the very structures of the physical objects perceived. The mind is conceptualized as a person capable of grasping and holding things. When the mind metaphorically grasps the form (the physical structure) of the object perceived, it understands (via the metaphor that Understanding Is Grasping).

What we have here is a metaphorical ontology and metaphorical logic that arises from putting together a number of common metaphors and reasoning in terms of them. The logical structure of Aristotle's reasoning is metaphorical. His ontological commitments come out of the metaphors. One example is his commitment that ideas of things are the structures of the very things in themselves. It follows from his metaphors that there is no gap between knowledge in the mind and things in the world; you can grasp the very structure of the object itself. The sensible forms are actually in the mind! From this metaphor flows the consequence that skepticism—the view that we cannot have knowledge of the way things are in the world—cannot be a problem.

The hallmarks of Aristotelian philosophy flow from the four folk theories (Intelligibility, General Kinds, Essences, and the All-Inclusive Category) plus the metaphors Ideas Are Essences and Essence Is Form. To get some sense of how these metaphors and folk theories shape Aristotle's thought overall, let us consider several classic Aristotelian doctrines: the Categories, Causation, the Classical Theory of Definitions, Logic, the Literalist Theory of Meaning, the Theory of Metaphor, and his Physics.

The Categories

From the above metaphors, as we have seen, it follows that there is a category of Being at the top of a hierarchy of categories. Then and only then does a natural question arise: What are the immediate subcategories of the category of Being? That is, what are the most basic specific forms of Being? Aristotle's answer was his famous enumeration of categories: substance, quantity, quality, relation, place, time, position, condition, action, and passivity.

As an empiricist, he arrived at these empirically via an early form of linguistic investigation into the nature of semantics. Aristotle observed that there was an "equivocation" in sentences like "The musical note and the knife are sharp." He observed that the predicate *sharp* could not be applied to both musical notes and knives in the same sense. Therefore, he reasoned, musical notes and knives must be in different categories. A musical note is a quality, while a knife is a substance. Because of the lack of a gap between the mind's categories and the world's categories, and because an "equivocation" marks a difference in the mind's categories, the systematic study of such equivocations can tell us about differences among the world's categories. Thus, by studying language that reflects the categories of mind, he is studying the world. The idea that we can study the world by studying language has come down to us in Anglo-American analytic philosophy. It was by such a method that Aristotle distinguished his ten basic categories. It was the structure of categories as reflected in language that revealed to Aristotle the fundamental forms of Being.

Causation

For Aristotle, the search for causes is the attempt to answer the ultimate question of Being, of why things come to be the way they are. Aristotle gave us the doctrine of the four types of causes: material, formal, efficient, and final.

The *material cause* is "that out of which a thing comes to be and which persists." Examples are the bronze of the statue or the silver of the bowl, that is, the material from which something is made.

The *formal cause* is "the form or the archetype, i.e., the statement of the essence." He gives as an example the two-to-one ratio that defines an octave.

The *efficient cause* is "the primary source of the change or coming to rest." The sculptor is the efficient cause of the statue.

The *final cause* "is the sense of end or 'that for the sake of which' a thing is done." For example, health is the cause of walking about; one walks about to be healthy.

Each of these types of causes is metaphorical and follows from either commonplace metaphors or those that are central to Aristotle's philosophy. The notion of an efficient cause uses the Location Event-Structure metaphor in which States Are Locations, Changes Are Motions, and Causes Are Forces. For Aristotle, the efficient cause (such as the action of the sculptor) is the force (literal or metaphorical) that brings about motion or change (which is metaphorically understood as motion).

Aristotle accepts the Folk Theory of Essences, in which essences are causal sources of natural behavior. Via the two basic metaphors for essence, Essence Is Material Substance and Essence Is Form, it follows that material substance and form are causal sources.

Aristotle's notion of final causation uses a metaphor discussed in Chapter 11, Causation Is Action To Achieve A Purpose, where the Causes Are Reasons (why the action will in fact achieve the purpose). Recall how this metaphor is grounded in experience. When we construct, through reasoning, a plan of action to achieve a purpose, our reasoning tells us that the actions, if performed, will indeed achieve the desired result. And this works time after time, almost all the time for simple things, every day of our lives. The correlation is between (1) actions taken on the basis of reason to achieve a purpose and (2) the causal relation between the actions taken and their result. This regular correlation is the basis for the primary metaphor Causes Are Reasons, the metaphor that tells us that the world is rational.

Aristotle took this metaphor as a literal truth, and it is the basis of his strong metaphysical doctrine of teleology. A *telos*, for Aristotle, is a purpose that arises naturally as part of the world. Thus, the world contains objectively given purposes (or "ends") that exert a causal pull on natural objects.

As an astute observer of nature, Aristotle noticed over and over that there are natural courses of change in the world. For Aristotle, by the Event-Structure metaphor, change is motion, and every motion has to result from a causal force. Natural changes, which result in natural end states, are "motions" that result in the changed object being in a new final "location" (metaphorically, a final state). Natural changes thus result from natural pulling forces that move objects from their initial state, through intermediate states, to a final state, according to a regular pattern of change. Such a force is a *telos*, an objectively existing purpose in nature.

Most important, such a natural behavior of an object must, for Aristotle, be a consequence of the essence of that object. Since essences are forms, that essence too must be a form, in particular, a pattern of change. That essence, that pattern of change, must reside in the object. Moreover, the *telos*, the final cause that brings about that change, must reside in that object as part of its essence.

Aristotle's classic example is the acorn, which he sees as containing within it a *telos*—a natural end, which is to become an oak—and a regular pattern of change brought about by that *telos*.

This remarkable realist view of teleology is a consequence of the Folk Theory of Essences plus Aristotle's central metaphor Ideas Are Essences. Since human beings have purposes, and purposes are ideas, and ideas are essences, and essences are in the world, it follows that purposes themselves must be in the world! And since all natural behavior is caused by essences, such purposes must be part of the essences of objects.

Definitions

"A definition," says Aristotle, "is a phrase signifying a thing's essence" (*Topics* 102a). In short, it is a collection of necessary and sufficient conditions for an object to be a particular kind of thing, what we would call today a member of a conceptual category. This definition of *definition* is still commonplace in logic and philosophy.

Such a definition expresses what philosophers today would call a concept. That is, a definition expresses an idea, which (via Ideas Are Essences), specifies an essence that characterizes a kind of thing existing objectively in the world. Thus, from Aristotle's central metaphor Ideas Are Essences, plus the Folk Theory of Essences, we get the mainstream contemporary philosophical notion of a concept.

Definitions for Aristotle are anything but trivial matters. They are not mere stipulations of how to use words. Because of the metaphor Ideas Are Essences and the folk theory that essences are causal sources of natural behavior, definitions for Aristotle characterize essential aspects of things in the world that explain why those things behave naturally the way they do. In short, for Aristotle, correct definition is central to the scientific enterprise, because it gives us real causal knowledge of why things behave as they do.

Logic

Aristotle gave us the classical formulation of what we will call "container logic," which arises from the commonplace metaphor that Categories Are Containers for their members.

> That one term should be included in another, as in a whole, is the same as for the other to be predicated of all of the first. (*Prior Analytics* 24b)

Aristotle thus equates predication with "inclusion" in a category. This is the expression of one of the central metaphors of Aristotelian logic, Predication Is Containment. It is by virtue of this metaphor that statements are linked to the general logic of containers.

Containers are image schemas with logical constraints built into their very structure. They are not physical containers, but rather conceptualizations that we impose upon space. Here are some of those logical constraints:

- Given a container and an entity, the entity is either inside or outside and not both at once.
- If Container A is inside Container B, and Entity C is inside Container A, then Entity C is inside Container B.
- If Container A is inside Container B and Entity C is outside Container B, then Entity C is outside Container A.

The metaphors that Categories Are Containers and Predication Is Containment map these spatial truths (from the logic of spatial containment) onto the classic Aristotelian logical principles:

- *The Law of The Excluded Middle*:

  An object cannot both have a property and its negative (in the same respect at the same time).

- *Modus Ponens*

  (version 1): If all B's are C's and all A's are B's, then All A's are C's.
  (version 2): If all B's are C's and some A is a B, then that A is a C.

- *Modus Tollens*

  (version 1): If all B's are C's, and no A's are C's, then no A's are B's.
  (version 2); If all B's are C's and some A is not a C, then that A is not a B.
  Version 1 in each case arises if A in the container logic is itself a container, and version 2 arises if it is not.

These principles are the basic ones that Aristotle uses in what he calls demonstrative syllogisms. The syllogism, for Aristotle, is the most fundamental form of reasoning, and Aristotle is well aware that this uses the logic of containment.

> Whenever three terms are so related to one another that the last is contained in the middle as in a whole, and the middle is either contained in, or excluded from, the first as in or from a whole, the extremes must be related by a perfect syllogism. I call that term "middle" which is itself contained in another and contains another in itself. (*Prior Analytics* 25b)

Such syllogisms, for Aristotle, did not characterize mere tautologies giving no new knowledge. On the contrary, they were seen as the major means of providing new knowledge. The reason is this: The predications that define a category/container are those that define the essence that makes each category member what it is. By the Folk Theory of Essences, an essence is the causal source of the natural behavior of everything with that essence and in that category. A demonstrative syllogism can therefore produce causal knowledge of behavior, given knowledge of essences. This, for Aristotle, is the major mechanism by which we gain scientific knowledge. The syllogism is thus the central engine of scientific explanation. The middle terms in the syllogisms provide causal connections, and a chain of syllogisms characterizes a causal chain.

> By demonstration I mean a syllogism productive of scientific knowledge, a syllogism, that is, the grasp of which is *eo ipso* such knowledge. Assuming then that my thesis as to the nature of scientific knowing is correct, the premises of demonstrated knowledge must be true, primary, immediate, better known than and prior to the conclusion, which is further related to them as effect to cause. (*Posterior Analytics* 71b)

This is why classification is such an important scientific enterprise for Aristotle. Putting things in the right categories allows one to apply syllogistic logic to produce new causal knowledge.

This view of logic as expressing causal relations is not present in modern formal logic. Students of contemporary logic, for this reason, often have a difficult time making sense of Aristotle's causal claims for his syllogistic reasoning. Modern logic does not contain two of Aristotle's central metaphors: Ideas Are Essences and Essences Are Causal Sources (from the Folk Theory of Essences). It is those metaphors that give syllogistic logic a significance for Aristotle that it does not have for us today.

It is vital to remember that for Aristotle logic was primarily in the world and only secondarily in the mind. For Aristotle, logic was not a projection of the mind onto the world, but the opposite: a direct grasping by the mind of the rational causal structure of the world. The demonstrative syllogism was thus his primary means of avoiding skepticism: It provided real causal knowledge.

Finally, we should recall that Aristotle accepted the metaphor Essence Is Form. As a result, his syllogistic logic is a formal logic. It is a logic of spatial containment that is metaphorically applied, via the metaphor that Categories Are Containers, to all categories, regardless of their specific content. It is the form of the syllogism that makes it valid, regardless of its content. This idea, that logic is universal and formal and independent of specific content, has come down to us in contemporary formal logic. It is a consequence of the metaphor Categories Are Containers, which is for us still a commonplace metaphor.

Literal Meaning and Metaphor

Aristotle's theory of knowledge, as we have seen, rests on his metaphors Ideas Are Essences and Essence Is Form. We can have certain knowledge because we can directly grasp the forms, that is, the essences of things as they really are in the world. Via these metaphors, our ideas can not only correspond to things in the world, but they can actually *be* the essences of things in the world.

Since we can express and communicate this knowledge, we must have a conventional language in which linguistic expressions properly designate the appropriate ideas. There must be a conventional proper correspondence between linguistic expressions and ideas. This amounts to what we might call a literalist theory of meaning: Each term properly designates at least one (and perhaps more than one) idea, which in turn is a form characterizing an essence in the world. When terms are used so as to properly designate what they are conventionally supposed to designate, meaning is literal. Terms used in their proper literal senses are necessary for demonstrative reasoning via syllogisms and thus are necessary for communicating scientific knowledge. Scientific knowledge, on Aristotle's view, cannot be communicated if terms are not used in their proper literal senses.

Aristotle, of course, noticed the existence of metaphoric uses of language. But given his central metaphors and the overall conceptual structure of his philosophy, he could not have come up with anything like the contemporary theory of conceptual metaphor that we have been using. Aristotle could not possibly have seen metaphor as a conceptual mapping from one conceptual domain to another, where the inferential structure of one domain is mapped onto the other.

The reason should be clear. Consider his metaphors Ideas Are Essences and Essences Are Forms of things in the world that can be directly grasped by the mind. Ideas therefore are aspects of the physical world. It is not possible for one idea to be conceptualized in terms of another. It is not possible for part of the logic of one idea to come from another idea. The logic of an idea, for Aristotle, is part of the structure of the external world. Because a domain is in the world, not just in the mind, a cross-domain mapping would have to be part of the world. But that is impossible. In the world, things exist as distinct kinds, as part of distinct categories. Each essence has its own inherent logic and not that of another kind of thing. The idea that the essential form of a thing could be that of another kind of thing makes no sense at all in the Aristotelian worldview.

Given the conceptual metaphors that defined Aristotle's worldview, he had to have a very different account of the phenomenon of linguistic metaphor. First, a metaphor had to be linguistic, not conceptual—a mere use of words, not a matter of concepts. The very notion of a metaphorical concept could have made no sense at all to Aristotle, since concepts for him are defined in terms of kinds of things in the mind-independent world. Second, metaphors had to be deviant uses of words, since they were applied to things they do not properly apply to. Any use of a word in its proper sense would be an ordinary literal use of language. Third, if a metaphorical linguistic expression (and that's all metaphor could be!) was to have any meaning at all, it had to be some other literal meaning. That's the only kind of meaning there is for Aristotle.

And fourth, if a metaphorical expression had a meaning at all, there would have to be some consistent basis for determining what the appropriate literal sense was. Aristotle chose similarity as the most general consistent basis for a metaphorical use of language. For him, the most general reason for using the name of one kind of thing to designate another kind of thing is to point out some similarity between the kinds of things.

> Metaphor consists in giving the thing a name that belongs to something else; the transference being either from genus to species, or from species to genus, or from species to species, or on grounds of analogy. (*Poetics* 1457b)
>
> But the greatest thing by far is to be a master of metaphor. It is the one thing that cannot be learnt from others; and it is also a sign of genius, since a good metaphor implies an intuitive perception of the similarity in dissimilars. (*Poetics* 1459a)

There was a good reason why Aristotle valued linguistic metaphor, conceived of in this way. As a scientist concerned with discovering the true essences of things in the world, he saw that an ability to find real similarities was necessary for being a good scientist.

As a result of Aristotle's literalist theory of meaning and his corresponding theory of metaphor, he was led to very bizarre analyses of concepts. Consider what he did with the data supporting the Event-Structure metaphor given in Chapter 11. Consider the submappings States Are Locations and Change Is Motion. In the theory of conceptual metaphor, these are cross-domain conceptual mappings. But Aristotle had to take them as literal statements of similarities. That is, he took states and locations as being similar, two special cases of the same kind of thing: a generalized location that could be either spatial or nonspatial. Correspondingly, he took general change and motion as similar, also two special cases of the same kind of thing. He did the same for causation in general and forced motion. Thus, he conceptualized change in general as a kind of literal motion, and special cases included motion through space, change of state, change of size, change of shape, and so on.

Since Aristotle saw conceptual analysis as part of science, as providing the correct analyses of essences upon which causal explanation could be based, he naturally applied these and related conceptual analyses in theorizing about physics. Consider Aristotle's explanation of why a stone that is thrown into the air falls to the ground and why flames shoot up into the air. Since motion and change, for Aristotle, were just two special cases of the same general thing, Aristotle could appeal to the properties of natural change to explain the properties of natural motion. As species of the same genus, change and motion would have to have the same general properties.

Aristotle then reasoned as follows: People are naturally healthy; occasionally they get sick, but they tend to become healthy again. In general, things have natural states. When they are removed from those natural states they tend to return to them. By Aristotle's theory of metaphor as expressing similarity, states and locations are similar and so are two species of the same genus, that is, two special cases of the same general category, and therefore will have the same general properties. Consequently, what we know about states can be applied with certainty to locations. Since things have states they are naturally in, they must therefore have locations they are naturally in. Since things removed from their natural states tend to return to them, so things removed from their natural locations tend to return to them. A stone's natural location is on the ground. When thrown in the air, it is removed from its natural state to which it seeks to return. In falling to the ground it is simply going to its natural location. Fire is a form of air. Air is naturally located above the earth. Fire moves upward because it is moving to its natural location.

Take another case from Aristotle's physics. Aristotle asks whether there can be a vacuum. The answer is no, and for the following reason: According to his theory of meaning as literal and his theory of metaphor as expressing similarities, states and locations must be similar, that is, they must be two species of the same general kind of thing and must therefore have the same general properties. Therefore one can discover the general properties of locations by looking at the general properties of states. Aristotle asks whether it is possible to have a state with nothing in that state. The answer is no. A state is always a state of a thing. Similarly, a location must always be the location of a thing. Since you cannot have a location with nothing in that location, it follows that a vacuum is impossible.

Aristotle's Metaphoric Philosophy

Aristotle, the founder of logic, followed his own logic exactly in formulating his theory of metaphor and in doing his physics. Aristotle's logic, his literalist theory of meaning, his theory of metaphor as expressing similarity, and his theory of physics were all consequences of his central folk theories and metaphors, especially the metaphors Ideas Are Essences and Essence Is Form.

Aristotle's theory of metaphor could not allow him to see his own conceptual metaphors. His theory could not allow him to look into his own cognitive unconscious and see that he was using conceptual metaphors, that is, mappings across conceptual domains. Blind to his own metaphors, he was forced by his own consistent application of his metaphors to a theory of metaphor that was inadequate to describe either his own metaphors or anyone else's. And that, in turn, led the greatest logician of all time, by his own inexorable logic, to a theory of physics that from a contemporary perspective is strange, to say the least.

What is truly remarkable is that, while Aristotle's theory of physics fell by the wayside after two thousand years, his theories of logic, of literal meaning, and of metaphor have lasted nearly twenty-five hundred years. The continued acceptance within philosophy of some version of literalist theories of meaning and the continued inability to recognize conceptual metaphor are due to the continued existence today of versions of the original Aristotelian worldview.

Literalist theories of meaning and theories of metaphor like Aristotle's are not of mere historical interest. They dominate much of philosophy today. The crucial point here, which has been emerging throughout this analysis, is that with only a literalist theory of meaning and without a theory of conceptual metaphor, philosophy cannot possibly understand its own nature and its own rational structure.

Why Being Is Different

Aristotle, the supreme logician, recognized that Being is different from all other categories of objects in the world. Consider Aristotle's principle NOT [F(a) and NOT F(a)]—"the same attribute cannot at the same time belong and not belong to the same subject and in the same respect" (*Metaphysics* 1005b20). In the modern logical formula, "F" is an attribute predicated of an existing entity named by "a." Via the fundamental metaphor of Aristotelian logic, Predicates Are Categories, F corresponds to a category, the category of objects whose essence is defined by the attribute F.

But although predicates are categories for Aristotle, not all categories are predicates. The exception is the category of Being. Recall that attributes are essences that entities can possess or not. Can Being—existence itself—be such an attribute? What would NOT F(a) mean if "F" meant "EXIST"? Predicates are predicated of existing objects. Hence, "a does not exist" predicates "NOT EXIST" of the existing object "a." It says that "a" has the attribute of nonexistence, which is impossible. Not impossible in some abstract sense, but impossible in the world. In short, Being itself is not a category that a thing can be either in or out of. If it is a thing, it is in the category of Being—period. Recall that logic is the logic of the world, the logic of all things that exist, of all things in the category of Being. For Aristotle, it makes no sense to apply the logic of attributes outside the category of Being, since only existing things do or do not have specific attributes. And so existence itself cannot be a predicate—an attribute that may or may not characterize the essence of some existing object.

This has a crucial consequence for Aristotle. Consider the question "Can Being be known?" For Aristotle, you get to know things by observation, by observing the attributes of things and discovering those attributes that are the essences of a thing. To distinguish among the essences, you must carefully distinguish among the behavior of things, since the essences are taken as causing the behavior that distinguishes one kind of thing from another. Could the essence of Being be discovered in this way? The answer is no. Since all existing things by definition share the essence of Being, they do not differ in this respect. Empirical observation cannot lead to the knowledge of the essence of Being, since there is no differentiation to be observed.

It follows that one cannot literally describe Being as one can describe ordinary categories, namely, by listing literal distinguishing attributes. Literal language simply fails for the description of Being. Moreover, analogies and metaphor (according to Aristotle's theory of metaphor) will fail too because they can only state literal similarities. But there is no literal similarity to Being, that is, there are no shared attributes that are in common between Being and anything else.

This does not mean that the notion of Being is meaningless or that we can know nothing at all about Being.

> But since the unqualified term "Being" has several meanings, of which one was seen to be the accidental, and another the true ("nonbeing" being the false), while besides these there are the figures of predication (e.g., the "what" quality, quantity, place, time, and any similar meanings which "being" may have), and again besides all these there is that which "is" potentially or actually. (*Metaphysics* 1026a, b)

As Aristotle points out, we do know a lot about Being and what we know makes the notion of Being meaningful. For example, predicability is one aspect of Being, since only existing things can have attributes that can be predicated of them: only existing things can populate the categories of quality, quantity, place, time, and so on. Since only existing things can have attributes, only they can have accidental attributes, and so the accidental is another aspect of Being. Since truth can only hold of a predication to an existing entity, truth is another aspect of Being. And finally, since possibility is the possibility of existence, both possibility and actuality are aspects of Being. Thus Being is meaningful because of all its manifestations in all these phenomena.

All of this is a consequence of Aristotle's fundamental metaphor, Ideas Are Essences, and therefore logic is the logic of the world as it exists. The situation is very much like what Stephen Hawking has described for the physicist's concept of time. Time, in physics, is characterized by the physical laws of the universe. For the physicist, functioning as physicist and looking only at the laws of physics, time makes sense only relative to those laws. If there was a Big Bang, which created the physical universe, then the laws of physics did not exist before the Big Bang and time as defined by the laws of physics came into existence with the Big Bang. In terms of physical laws, it makes no sense to speak of something "before the Big Bang." Since "before uses the concept of time, there was no 'before the Big Bang.'"

Hawking, like many other physicists, is taking the laws of nature as defining the essence of the universe, the essence of all that is, the essence of Being. The idea of the Big Bang predicates a beginning of Being. Just as logic for Aristotle makes sense only for things that exist, time for Hawking makes sense only for events since the Big Bang. Just as it makes no sense for Hawking to speak of "before the Big Bang" so it makes no sense for Aristotle to attribute nonexistence to an existing thing.

The issues raised by Aristotle about Being are still with us today, since modern science is commonly interpreted in the Aristotelian mode as characterizing the essence of Being.

The Remarkable Category of Being

"Being" is, on the face of it, a very odd category indeed. As we observed earlier, in order for people to get along in life they need to be able to identify things like chairs, people, light switches, friendships, political institutions, and harmful objects. They also need to have a great deal of basic knowledge about these things, if they are going to survive and flourish. But it seems extremely odd to say that they need to identify and have knowledge of "Being." And yet this is what metaphysics defines as our most noble philosophic task.

We have been suggesting that Being, like every other basic philosophic concept, is a human category, the very articulation of which depends on a cluster of common folk theories and conceptual metaphors. Being, regarded as *the* fundamental ontological category, emerged historically, as we have seen, in pre-Socratic philosophy and was given an elaborate articulation and refinement in Plato and Aristotle. We have argued that Aristotle was able to create the field of metaphysics only by adopting and adapting these shared folk theories and metaphors. The logic of Plato's and Aristotle's doctrines of Being, and indeed their entire philosophic positions, are significantly based on metaphorical concepts and are made possible by folk theoretical assumptions.

Many of these folk theories and conceptual metaphors are so deeply rooted in our Western philosophical tradition that they may seem to us not to be folk theories or metaphors at all. Many people, for instance, take it as a self-evident metaphysical fact that things consist of matter organized by form, or that everything has an essence that makes it the kind of thing it is, or that reality is organized in a hierarchy of categories, with the category of everything that exists at the top.

Many people think it obvious that the world must consist of basic substances that underlie the properties we experience. But there is nothing ontologically absolute about either the form/matter distinction or the idea of substance-attribute metaphysics. Many philosophers, such as Merleau-Ponty, Dewey, Whitehead, and, more recently, Rorty, have shown that the form/matter model is only one possible way of understanding things, and a mostly distorting way at that. Likewise, the idea that *substance* must be the ontologically basic entity is today almost totally discredited by a large number of philosophical traditions.

Nevertheless, the quest for Being goes on, and it is still regarded in many quarters as the ultimate philosophical project. The metaphysical impulse remains strong because the metaphors and folk theories defining it are so deeply embedded in our shared cultural understandings. As long as we believe that the world consists of general kinds of things defined by essences, that essences are the source of all natural behavior, that the world is intelligible, and that there is an all-inclusive category also defined by an essence, we will continue the search for Being.

The search for Being is for many people the search for God. The issues surrounding the quest for Being have always been at the center of Western theology and are still there today. God is widely regarded by theologians and laypeople alike as the ultimate causal source and sustainer of all that is, as the ultimate source of all that is good, as present in every existing thing, as having a plan that gives purpose to the world and meaning to human beings, and as being not merely all-powerful but also all-knowing. As we saw, most of these are the properties of Plato's Idea of the Good, that is, of the essence of essence. This is no accident. Most of the medieval conceptions of, and arguments for, the existence of God stem directly from Greek metaphysics, partly from Plato's

Idea of the Good, but especially from Aristotelian views of causation and change.

The forms of thought that we saw as emerging in the pre-Socratics and finding their most sophisticated expression in Plato and Aristotle are thus anything but quaint and archaic. They exist not only in contemporary philosophy and theology, but they lie at the heart of Western science. The Folk Theory of the Intelligibility of the World is a precondition for any form of rational inquiry. The Folk Theory of General Kinds is required in order to state any generalizations at all. Otherwise, all knowledge would be utterly specific and could never be projected to new cases. The Folk Theory of Essences is commonplace in virtually every science, because science is always looking for the properties of things that make them what they are and explain their behavior. The Folk Theory of the All-Inclusive Category is present in every mode of scientific explanation that seeks ever more comprehensive explanations to cover ever greater ranges of phenomena, for example, theories of everything in physics and theories of life in biology.

Since we cannot do without such folk theories and metaphors, and since they can in some cases be extremely useful, while in other cases extremely misleading, it would be wise to get as clear an understanding as possible of how they operate—when they are present, how they control our reasoning and our very perceptions, how they hang together to form complex conceptual systems, what they entail and what they hide.

19

Descartes and the
Enlightenment Mind

I t is virtually impossible to conceptualize the mind without metaphor. Therefore, it should not be surprising that there is a long history in philosophy in which ontological commitments about the nature of mind are a consequence of common conceptual metaphors. The Knowing As Seeing metaphor, for example, was present in Plato, as it has been for virtually every conception of mind in the history of Western philosophy. The Understanding As Grasping metaphor was present in Aristotle.

Versions of the Platonic and Aristotelian views of knowledge persisted through the Middle Ages. The idealist Platonic views kept the metaphor of Essences As Ideas, while the realist Aristotelian views preserved the metaphor of Ideas As Essences. In both cases, what was preserved was an explanation of the possibility of knowledge via the most intimate of direct linkages between ideas and essences of physical objects—identity. For Plato, the essence of a physical object *is* the idea, and for Aristotle the idea *is* the essence of the object. As we shall see, this kind of identity becomes impossible for Descartes once mind is severed from body and mental substance is seen as utterly different from physical substance.

What emerged from Descartes' philosophy was a new metaphoric view of mind as *representing* in some "inner" realm the objects existing in the "external" world. Since the objects in the mind were nothing like the objects in the world, the problem of knowledge became the problem of how we could know

392 PHILOSOPHY IN THE FLESH

that the internal ideas (representations) in our minds actually corresponded to the "things in themselves." Whereas Plato and Aristotle could claim real knowledge because, according to their metaphors, the mind could grasp the forms of the things themselves, that source of knowledge became unavailable once the mind/world split came to be taken for granted.

René Descartes claimed to have found a philosophical method that would guarantee absolutely certain foundational knowledge on which a philosophy could be based. Descartes' method produced a cluster of novel ideas that have shaped much of philosophy from his time to ours. Many of his conclusions are still with us in contemporary philosophical theories, and they were the basis for much of first-generation cognitive science. Among the most important of Descartes' ideas are the following:

- The mind can know its own ideas with absolute certainty.
- All thought is conscious.
- The structure of the mind is directly accessible to itself.
- No empirical research is necessary to establish certain knowledge of the mind.
- The mind is disembodied. It consists of mental substance, while the body consists of physical substance.
- The essence, and only essence, of human beings is the ability to reason.
- Imagination and emotion, which are bodily and therefore excluded from human reason, are not part of the essence of human nature.
- Certain of our ideas represent external reality, and their origin lies in the perception of external objects.
- Other ideas are innate; they are free of anything bodily and are not representations of anything external.
- Mathematics is about form, not content; because of its formal nature, mathematical knowledge can be certain.
- Thought is formal, just as mathematics is.

What makes these ideas fit together into coherent view of mind and knowledge? What made Descartes' arguments convincing? Why should anyone ever have come up with such a philosophical method? And why should such a method lead to such conclusions?

We will argue that what makes the whole argument fit together, make sense, and seem intuitively appealing to so many generations of philosophers is a peculiar metaphorical logic in which Descartes has pieced together a number of common metaphors about the mind and then followed out their entailments.

We will further argue that, without the metaphors, the logic of the argument is simply not present; nor is the metaphysics of his theory of mind. Descartes' philosophy is necessarily metaphorical. The metaphors are not a mere embellishment, but are constitutive of his theory of mind.

Descartes' Mental Vision

Descartes' method to ensure certain and indubitable knowledge was to "reject all such merely probable knowledge and make it a rule to trust only what is completely known and incapable of being doubted" (C2, Descartes, *Rules* 3). His method, he claimed, worked. "Little by little I was delivered from many errors which might have obscured our natural vision and rendered us less capable of listening to Reason" (C2, Descartes, *Discourse on Method* 87). Other methods did not. "It is very certain that unregulated inquiries and confused reflections of this kind only confound the natural light and blind our mental powers" (*Rules* 9). The only method that worked was to take advantage of the natural light of Reason and to follow precise rules for inquiry.

Descartes conceived of the mind as what Daniel Dennett calls the "Cartesian Theater"—an inner mental stage in which metaphorical objects (our ideas) are illuminated by an inner light (the "Natural Light of Reason") and are observed by a metaphorical spectator (our faculty of understanding). Descartes gives the name *intuition* to such mental vision, which allows him to see the idea-objects clearly and to distinguish them from one another.

Descartes' Metaphorical Logic of Knowing

How is it for Descartes that the mind can know anything at all? His entire account is based on a tightly interwoven cluster of metaphors that define a logic about the possibility of certain knowledge. The most fundamental metaphor Descartes uses is the commonplace Knowing Is Seeing metaphor.

KNOWING IS SEEING

Visual Domain		*Knowledge Domain*
Object Seen	→	Idea
Seeing An Object Clearly	→	Knowing An Idea
Person Who Sees	→	Person Who Knows
Light	→	"Light" Of Reason

Visual Focusing	→	Mental Attention
Visual Acuity	→	Intellectual Acuity
Physical Viewpoint	→	Mental Viewpoint
Visual Obstruction	→	Impediment To Knowing

The Knowing Is Seeing metaphor defines the core of a folk theory about how the mind works that is so widely shared in our intellectual tradition that it virtually defines our public understanding of intellectual operations. That this conceptual metaphor should be so pervasive makes perfectly good sense, given that vision plays such a crucial role in so much of our knowledge of our world. Our language about our mental activity is thus pervaded with expressions based on this underlying vision metaphor. Knowing Is Seeing examples:

> I *see* what you mean. Could you *shed some light* on chaos theory for me? You have a great deal of *insight* into social relations. That's about as *obscure* an idea as I've ever *seen*. We just can't seem to *get clear* about gender roles. Talk about a *murky* argument!

The Knowing Is Seeing metaphor is so firmly rooted in the role of vision in human knowing and is so central to our conception of knowledge that we are seldom aware of the way it works powerfully to structure our sense of what it is to know something. It is the commonality and experiential grounding of this ubiquitous metaphor that makes it an ideal candidate for sophisticated philosophical elaboration in a wide variety of theories of mind and knowledge.

Descartes takes the Knowing As Seeing metaphor as a philosophical truth. This allows him to formulate the fundamental problem of knowledge as a problem concerning how it is possible to obtain clear and unobscured (intellectual) vision. The problem of philosophical method becomes a problem about how to *see clearly* the idea-objects that are present to the mind for *inspection* and also to *discern* the relations existing among these ideas. The mind's ability to *see clearly* is what Descartes calls "intuition."

> By *intuition* I understand, not the fluctuating testimony of the senses, nor the misleading judgment that proceeds from the blundering constructions of imagination, but the conception which an *unclouded* and attentive mind gives us so readily and *distinctly* that we are wholly freed from doubt about that which we understand. Or, what comes to the same thing, *intuition* is the undoubting conception of an *unclouded* mind, and springs from the *light of reason* alone. (*Rules* 7 [emphasis added])

And what is it that gives certainty and removes any possibility of doubt? Descartes' answer is that we cannot doubt "what we can clearly and perspicuously behold and with certainty deduce" (*Rules* 5).

And what is involved in beholding something clearly and perspicuously? It is to have an intuition that is clear and distinct. And what is clear and distinct intuition?

> I term that clear which is present and apparent to an attentive mind, in the same way as we assert that we see objects clearly when being present to the regarding eye, they operate on it with sufficient strength. But the distinct is that which is so precise and different from all other objects that it contains within itself nothing but what is clear. (C2, Descartes, *Principles* 237)

Therefore certain and evident knowledge is guaranteed by the "natural light of reason" shining upon mental objects (ideas), illuminating them in such a way that we cannot help but see every feature of them and how they are distinct from every other idea-object, just as the bodily eye cannot help but see what is before it when there is sufficient light.

Let us pause to examine the metaphorical logic of Descartes' view of intuition. He assumes that the mind is a container for ideas. The interior of the container is Dennett's Cartesian Theater, a locus in which the idea-objects exist and can be put in the spotlight and examined.

Descartes further assumes the metaphors of faculty psychology (discussed at length below), which personify the capacities of the mind as people who perform various mental tasks. Since perceiving is understood metaphorically as receiving sense-impressions from the external world, perception is conceptualized as a person who does the receiving. The faculty of imagination is then conceptualized as a person, Imagination, who forms images from sense-impressions. Reason is the personification of the capacity to think and to know.

Descartes combines the Knowing Is Seeing metaphor with these metaphors to produce the complex metaphorical system that characterizes his notion of intuition. Reason, the person in the Cartesian Theater who is capable of knowing, is conceptualized by the Knowing Is Seeing metaphor as a person who can see. The Ideas Are Objects metaphor, added to the metaphor The Mind Is A Container for ideas, produces the entailment that ideas are objects in the mind that can be seen by Reason. All this metaphorical apparatus is applied to our everyday knowledge of seeing. Here is part of our source-domain knowledge.

> If an object is in the field of vision of an observer who can see, and if the object is sufficiently illuminated by light and no other object hides or obscures the object, then the observer will see the object as it really is, with all its detail, and will be able to distinguish it from other objects.

Descartes' system of metaphors maps this common folk theory of vision into a philosophical theory of knowledge that includes the following metaphorical consequences:

> If an idea is in the field of mental vision of Reason, who can know, and if the idea is sufficiently "illuminated" by the "light" of Reason and no other idea hides or obscures the idea, then Reason will know the idea as it really is, with all its detail, and will be able to distinguish it from other ideas.

Notice that there is no literal way to translate into the mental realm the notion of "illumination" by the "light of reason." Mental illumination and light are part of the ontology introduced by the metaphor. Note especially that Descartes' reasoning does not go through without what we know about illumination and light from the source domain of vision. If there is no light and illumination, then Reason cannot know its own ideas at all, let alone know them with certainty as they really are.

Given these metaphors, their ontology, and the entailments derived from putting them together in just the right way and applying them to the folk theory of vision, we arrive at one of Descartes' most celebrated conclusions:

> *The mind can know its own ideas with absolute certainty.*

As we have seen, Descartes' argument for this conclusion cannot be made without the metaphors just discussed.

Moreover, a further startling conclusion follows from these same metaphors:

> *All thought is conscious.*

This conclusion, which has been invalidated by virtually all of cognitive science, arises from the metaphors as follows: Thought consists of ideas. Since Ideas Are Objects and Knowing Is Seeing, thought can be seen by Reason. At this point two further commonplace folk theories of vision enter:

You are conscious of what you see.
Every object is capable of being seen.

If all ideas are objects and all objects can be seen consciously, and if Knowing Is Seeing, then all ideas can be known consciously. The notion that there could be an idea or thought process that could not be accessible to consciousness would be like an object that was by nature invisible. The metaphors plus the folk theories of vision do not allow this. Therefore, all thought must be conscious.

Another striking consequence follows from Descartes' system of metaphors:

The structure of the mind is directly accessible to itself.

Since all thought consists of ideas, and since Ideas Are Objects and Knowing Is Seeing, then all idea-objects are accessible to vision and hence able to be known by Reason. In other words, the structure and nature of thought processes can be known directly to the mind because they are made up of idea-objects, which can be directly seen (known). Thus Descartes concludes that Reason can reflect directly and successfully on its own nature and is therefore in no need of the aid of empirical research. From this, an additional startling conclusion follows:

No empirical research is necessary to establish certain knowledge of the mind.

These four conclusions remain pillars of present-day philosophy of mind in the Anglo-American tradition. It is still widely assumed that the philosophy of mind can be done without empirical research as an armchair pursuit in which Reason reflects directly on the structure of the mind, all of which is allegedly available to consciousness in such a clear form that, provided close attention is paid to one's ideas, one's conclusions cannot be doubted.

This conception of the philosophy of mind has been inherited from Descartes and has not received justification substantially different from what Descartes provided. Remarkably enough, Descartes' view is not a quaint seventeenth-century oddity of mere historical interest. It is very much with us today, and the only justification for it today is Descartes' argument, which is constituted by his metaphors and which could not be made without them.

Descartes' Logic of Deduction

Descartes realized that mere intellectual ogling of individual ideas is not suffi-cient for knowledge of the world. In addition, we must see the relationships be-tween ideas as they go together in propositions, which are our means of expressing truth claims about reality. Moreover, we must see how one proposi-tion follows from another. This form of knowledge, in which we trace connec-tions among ideas, is called "*deduction*, by which we understand all necessary inference from other facts that are known with certainty" (*Rules* 8).

Deduction is what allows the mind both to "move" from one idea to another and to "see" their connections. This requires a complex metaphor system in which at least three common metaphors are woven together, metaphors that exist independently of Descartes' philosophy.

First, there is Knowing Is Seeing, in which the faculty of reason is conceived metaphorically as a person who is able to see idea-objects. Second, there is Thinking Is Moving, in which the thinking mind is conceptualized as a person in motion. The careful logical thinker *moves step-by-step from* premises *to reach* a conclusion. Third, these two metaphors are joined by another metaphor containing elements of both Seeing and Moving, namely, the Seeing Is Touching metaphor. Here the eyes are conceptualized as limbs that extend outward in a direction. In the metaphor, the touching of objects by the limbs corresponds to the seeing of objects through the eyes. Common examples are: "Her eyes *picked out* every flaw in the carpet," "Our eyes *met* across the room," and "She *undressed* him *with her eyes*." The Seeing Is Touching metaphor maps physical movement of our fingers (as in, "She *ran* her fingers *over* the carpet") onto motion in vision (as in "She *ran* her eyes *over* the car-pet").

If Knowing Is Seeing and Seeing Is Touching, then the faculty of intellectual vision (i.e., reason) can be said to "move" over and among the things it sees (namely, ideas), as in "She *ran over* every detail in the argument one more time." It is in this sense, via these connections of metaphors, that Descartes is able to link mental vision and mental motion.

Descartes makes crucial use of the Thinking Is Moving and Seeing Is Touch-ing metaphors in an attempt to solve a problem about the nature of thought that cannot be handled by Knowing Is Seeing alone. The problem arises in de-duction. If deduction is also to be a form of certain knowledge, then it must provide the same clarity and distinctness that intuition supplies, although it must take place over time, as we connect ideas in our thought. However, since knowing for Descartes is seeing, which is instantaneous, he must somehow model deduction, which occurs over time, as a sort of instantaneous act of seeing something. This is no mean feat.

As Reason moves, mentally, down a train of deductions, we must be certain that our previous deductions—the ones on which our present deduction rests—are themselves infallible and certain. But for this we cannot rely merely on our notoriously feeble and fallible memories. There would be no problem with memory if Descartes could simply make up a metaphor out of thin air like Memory Is Writing. Such a metaphor would allow idea-objects to remain as they were previously recognized. But this metaphor did not exist, so far as we can tell, in the conceptual system of Descartes. That leaves him with a problem.

Descartes' metaphor of Knowing As Seeing, which is instantaneous, forces him to the rather bizarre model of deduction as a single act of vision that encompasses what is really a *series* of cognitive acts (of intellectual seeing) taking place over a period of time:

> For this deduction frequently involves such a long series of transitions from ground to consequent that when we come to the conclusion we have difficulty in recalling the whole of the *route by which we arrived at it*. This is why I say that there must be a *continuous movement* of thought to make good this weakness of the memory. Thus, e.g., if I have first found out by separate mental operations what the relation is between the magnitudes A and B, then what between B and C, between C and D, and finally between D and E, that does not entail my *seeing* what the relation is between A and E, nor can the truths previously learnt give me a precise knowledge of it unless I recall them all. To remedy this I would *run over them* from time to time, keeping the imagination *moving continuously* in such a way that while it is intuitively *perceiving* each fact it *simultaneously passes on to the next*; and this I would do until I had learned to *pass from the first to the last* so quickly, that no stage in the process was left to the care of the memory, but I seemed to have the whole in intuition *before me* at one time. (*Rules* 19 [emphasis added])

What Descartes is trying to do here—what the available metaphors force him to do here—is to compress a course of intellectual motion over time into one all-encompassing instantaneous act of intellectual seeing. The mind has to move (via the metaphors of Knowing Is Seeing, Seeing Is Touching, and Thinking Is Moving) from one idea to another "so quickly" that the mind has "the whole in intuition" before it "at one time." Strictly speaking, of course, this is impossible. The impossibility results from the limits of short-term visual memory. Instantaneous vision with a short-term visual memory cannot accomplish the kind of seeing that Descartes requires, namely, seeing that covers events taking place over a significant period of time.

Descartes' actual philosophical method of weaving together existing metaphors and drawing inferences from the way they are put together cannot succeed here because the stock of existing metaphors is not sufficient. If he could just make up a metaphor like Memory Is Writing, then mental deductions would be like written proofs and the problem would disappear.

In this case, Descartes' definition of the problem is the result of his initial metaphors. The attempted solution he offers comes from his making the best possible use of additional conventional metaphors that are available, and his ultimate inability to solve the problem comes about because of a limitation of the metaphors available to him.

Descartes' Disembodied Mind

Descartes' view of knowledge has left its fateful mark on much contemporary epistemology, philosophy of mind, and philosophy of language. We want to examine it briefly in order to show how its underlying metaphors have led many philosophers to adopt highly problematic and unsatisfactory views about how the mind works. Of special catastrophic significance is the way in which Descartes' vision metaphors, put into the service of his quest for certainty, led to a disembodied conception of the mind.

His argument is well known. In his *Meditations on First Philosophy*, Descartes applies his method to his own beliefs to determine whether anything he believes is certain and indubitable. His famous conclusion is that what he can never doubt is that when he thinks, he exists. After all, if he didn't exist, he could not be thinking or even doubting.

But Descartes reaches conclusions beyond this: first, that being able to think constitutes our essence; second, that the mind is disembodied; and third, therefore, that the essence of human beings, that which makes us human, has nothing to do with our bodies. These three elements of Cartesian philosophy have had a profound effect on the character of much contemporary philosophical thinking. They have affected not only phenomenology, but also a good deal of Anglo-American philosophy of mind. But their influence is not limited merely to philosophy. They have also made their way into other academic disciplines, into our educational system, and into popular culture as well, as in the pervasiveness of the computer metaphor for the mind. These beliefs, in the popular imagination, have led to the dissociation of reason from emotion and thus to the downplaying of emotional and aesthetic life in our culture.

The philosophical respectability of these views still rests on versions of the original Cartesian arguments. These in turn rest on Descartes' peculiar weaving together of commonplace metaphors and his use of folk theories. Let us consider how Descartes arrives at the views that our essence is only to think, that the mind is disembodied, and that therefore what makes us human has nothing to do with our bodies.

Let us begin with his argument that our essence is only to think. First, Descartes reaches the conclusion that he exists because he is thinking; if he didn't exist, he couldn't think. From this, he concludes that his essence is to be a thinking thing.

> I am, I exist, that is certain. But how often? Just when I think; for it might possibly be the case if I ceased entirely to think, that I should likewise cease altogether to exist. . . .
>
> But what then am I? A thing which thinks. (C2, Descartes, *Meditations* 152–153)

The idea that there is a common human nature that determines how people naturally act is based on the long-standing Folk Theory of Essences, discussed earlier in great detail in Chapters 11, 16, 17, and 18. This folk theory is taken for granted by Descartes. It consists of two theses:

THE FOLK THEORY OF ESSENCES

Every kind of thing has an essence that makes it the kind of thing it is.
The way each thing naturally behaves is a consequence of its essence.

Essences therefore have a certain causal capacity, since they result in the natural behavior of things, that is, the behavior of things in the absence of any other cause.

Descartes' reasoning seems to go as follows: He knows he exists as long as he thinks. If he exists, he has an essence (by the Folk Theory of Essences). He assumes that his thinking is spontaneous, that is, not caused by anything else. If it is not caused by anything else, it must come from his nature. That is, it must be a consequence of his essence. Therefore, to be a thinking thing must be at least part of his essence.

The question now arises as to whether there is anything else that is part of his essence. Here is how he argues:

> Just because I know certainly that I exist, and that meanwhile I do not remark that any other thing necessarily pertains to my nature or essence, excepting that I am a thinking thing, I rightly conclude that my essence consists solely in the fact that I am a thinking thing [or a substance whose sole nature or essence is to think]. And although possibly (or rather certainly, as I shall say in a moment) I possess a body with which I am very intimately conjoined, yet because, on the one side, I have a clear and distinct idea of myself inasmuch as I am only a thinking and unextended thing, and as, on the other, I possess a distinct idea of body, inasmuch as it is only an extended and unthinking thing, it is certain that this I [that is to say, my soul by which I am what I am], is entirely and absolutely distinct from my body and can exist without it. (*Meditations* 190)

This argument is based on the Knowing Is Seeing metaphor. The argument rests on his ability to "see," that is know, ideas of two different kinds of things: (1) the idea of thinking; and (2) the idea of extended bodily substance. He has to "see" that there is nothing in his idea of thinking that includes any aspect of embodiment. In other words, the idea of thought is utterly distinct from the idea of embodiment. Applying his method of intuition, he simply notes (no further argument is needed if his method is correct) that this is what he mentally "sees," and therefore that he cannot be mistaken. His monumental conclusion is, therefore, that the essence of thought is utterly distinct from the essence of embodiment.

> It is certain that this I [that is to say, my soul by which I am what I am], is entirely and absolutely distinct from my body and can exist without it. (*Meditations* 190)

To reach this conclusion, he must assume an additional folk theory: the Folk Theory of Substance and Attributes, which goes back at least to the time of Aristotle.

THE FOLK THEORY OF SUBSTANCE AND ATTRIBUTES

A substance is that which exists in itself and does not depend for its existence on any other thing.

Each substance has one and only one primary attribute that defines what its essence is.

Descartes' argument is that when he introspects, that is, when he turns the natural light of reason on his ideas of thinking and bodily activity, here is what he "sees":

There are two kinds of substance, one bodily and the other, mental.
The attribute of bodily substance is extension in space.
The attribute of mental substance is thought.

On the basis of this intuitive observation, he concludes that thought is disembodied; it is the primary attribute of mental substance. Since he has previously concluded that being a thinking thing is at least part of his essence, it follows that at least part of his essence consists of mental substance. But since mental substance is totally distinct from physical substance, mental substance must be all of his essence. He concludes that being a thinking thing is the only essence of human nature and therefore that our bodies have nothing to do with what we essentially are.

Here then are two fateful conclusions that have been carried down to us through three hundred years of philosophy:

The mind is disembodied. It consists of mental substance, while the body consists of physical substance.
The essence, and only essence, of human beings is the ability to reason.

From these two conclusions, an additional one follows immediately:

Imagination is not essential to human nature.

The argument for this is straightforward. Imagination, the capacity to form and put together images, is a capacity of the body, since it is tied to sense perception, which is the source of images. Since any aspect of the body is not part of human nature, imagination is not part of human nature. As he says:

I remark besides that this power of imagination which is in one, inasmuch as it differs from the power of understanding, is in no wise a necessary element in my nature, or in [my essence, that is to say, in] the essence of my mind; for although I did not possess it I should doubtless ever remain the same as I now am, from which it appears that we might conclude that it depends on something which differs from me. (*Meditations* 186)

Similarly, it also follows that:

Emotion is not essential to human nature.

The reasoning is virtually identical. Since emotion is part of our bodily experience, it is utterly distinct from the essence of human nature, which is only thinking substance.

Formalism, Representation, and Innateness

Descartes also held four additional influential notions still widely believed today:

- Mathematics is about form, not content; because of its formal nature, mathematical knowledge can be certain.
- Certain of our ideas represent external reality, and their origin lies in the perception of external objects.
- Other ideas are innate; they are free of anything bodily and are not representations of anything external.
- Thought is formal, just as mathematics is.

The idea that the mind has nothing to do essentially with the body creates a particular problem for a theory of ideas. There are two kinds of ideas. Some of our ideas arise from bodily sensations and are ideas *of* external objects. But other ideas, according to Descartes, are in no way tied to the body and concern only the structure of the mind itself. The first he calls "adventitious ideas" or "representations." The second he calls "innate." Of the latter he gives three types of examples: mathematical ideas, the structure of thought itself, and the idea of God. What is distinctive about all of these is that they do not arise from the senses and therefore must be already present in the infant at birth.

Given the ubiquitous Knowing Is Seeing metaphor, the innate ideas are unproblematic, since there are no bodily factors involved in their creation. Therefore, no question can arise as to whether they accurately represent external reality. Since their content is mental in origin, they are not tainted by the body and can be "seen clearly" by the mind.

The problem arises for ideas whose source lies in sense perception. Since the mind knows only the ideas that are present to it, it cannot step outside the mental realm to inspect the relation between an idea and an external object. Wouldn't there be a problem reasoning with an idea that might be so tainted by the senses?

Descartes saw the solution to this problem in his understanding of mathematics. Descartes was the founder of analytic geometry, in which numbers are conceptualized as points on a line and geometrical figures can be expressed as algebraic equations. Descartes interpreted his own work on analytic geometry as being a model for the study of thought in general. In analytic geometry, the same formal symbolic notation can be used either for arithmetic, geometry, astronomy, or harmony in music. Descartes saw mathematics as being inherently a matter of the formal symbolism, which could be applied to concrete special cases.

> All those matters only were referred to mathematics in which order and measurement are investigated, and that it makes no difference whether it be in numbers, figures, stars, sounds, or any other object that the question of measurement arises. I saw consequently that there must be some general science to explain that element as a whole which gives rise to problems about order and measurement, restricted as these are to no special subject matter. (*Rules* 13)

Mathematics, thus freed from specific subject matters, became for Descartes a general and purely mental subject matter, falling under the domain of innate ideas. What makes this possible is that mathematics is free of specific concrete content and hence free of any bodily taint. Mathematics thus is only a matter of "clear and distinct" ideas, which can be "seen clearly" and about which there can be no mistake. The application of mathematics, say, to astronomy or music, becomes less certain because the body is needed to link the mathematical ideas to things in the world. Though such applications may be problematic, mathematics, which is purely formal, is unproblematic with respect to the certainty of its claims.

Thought Is Mathematical Calculation

One might think that Descartes' views on analytic geometry and mathematics might be irrelevant to his theory of mind. What makes his mathematical views centrally relevant is a common metaphor, the metaphor that Thinking Is Mathematical Calculation. This metaphor seems to have arisen with the Greeks, who saw mathematics as the quintessence of reason. This conceptual metaphor is still with us today. We can see it in everyday expressions such as:

> When you think about it, that just doesn't *add up*. She *put two and two together* and concluded that he had been unfaithful. It all *adds up to* a real problem for us. What's the *bottom line* here? I *reckon* that we'll have the hay in by nightfall. *That figures!* He's messed up again! If you *subtract* that evidence, you get a different conclusion. Could you just *sum up* your argument?

The contemporary version of this metaphor is as follows:

THINKING IS MATHEMATICAL CALCULATION

Mathematical Calculation	→	Thinking
Numbers	→	Ideas
Equations	→	Propositions
Adding	→	Putting Ideas Together
Sum	→	Conclusion

We can see that this metaphor was common in Europe in Descartes' time from the notorious use of it that we find in the writings of his contemporary, Thomas Hobbes:

> When a man *reasons*, he does nothing else but conceive a sum total from *addition* of parcels, or conceive a remainder from *subtraction* of one sum from another; which, if it be done by words, is conceiving of the consequence of the names of all the parts to the name of the whole, or from the names of the whole and one part to the name of the other part. . . . These operations are not incident to numbers only, but to all manner of things that can be added together and taken one out of another. For as arithmeticians teach the same in *lines*, *figures* solid and superficial, *angles, proportions, times,* degrees of *swiftness, force, power,* and the like; the logicians teach the same in *consequences of words,* adding together two *names* to make an *affirmation,* and two *affirmations* to make a *syllogism,* and *many syllogisms* to make a demonstration; and from the *sum* or *conclusion* of a *syllogism* they subtract one *proposition* to find the other. (C2, Hobbes, *Leviathan* 45)

Hobbes sums up this celebrated use of the Thinking Is Mathematical Calculation metaphor with his claim that "Reason, in this sense, is nothing but *reckoning*—that is, adding and subtracting—of the consequences of general names agreed upon for the *marking* and *signifying* of our thought" (*Leviathan* 46).

Hobbes' use of this metaphor is anything but incidental to his philosophical views. For Descartes too, it plays a central role in his philosophy of mind. Here is Descartes writing to Mersenne, November, 20, 1629:

> All the thoughts which can come to the human mind must be arranged in an order like the natural order of the numbers. In a single day one can learn to name every one of the infinite series of numbers, and thus to write infinitely many words in an unknown language. The same could be done for all the other words necessary to express all the other things which fall within the purview of the human mind. . . .
>
> If someone were to explain correctly what are the simple ideas in the human imagination out of which all human thoughts are compounded, and if his explanation were generally received, I would dare to hope for a universal language very easy to learn, speak, and to write.

Here Descartes is using not only the metaphor of Thought As Mathematical Calculation but also the Thought As Language metaphor, in which complex ideas are made up of simple ideas, just as sentences are made up of words. Here Descartes is prefiguring the Language of Thought metaphor of twentieth-century analytic philosophy (see Chapter 21).

Mathematics as a Model for Thought

Descartes applies the Thought As Mathematical Calculation metaphor to his theory of ideas in general. It provides an immediate solution to the problem of how certain reasoning is possible with ideas that are not innate but arise through the senses. As ideas, they can be treated formally, that is, the same way as innate mathematical ideas. Reasoning is akin to mathematical proof.

Ideas that are not innate need not be tainted by the senses at all from the point of view of Reason's ability to inspect them and calculate with them. The only problem is how they relate to things in the external world. Descartes does not answer this in any interesting way. He merely assumes that God is no deceiver and that God gives us accurate ideas to reason with so that we can arrive at certain knowledge as long as we inspect our ideas carefully and calculate with them with precision and mathematical rigor.

But despite this, Descartes created a theory of mental representation—essentially the view inherited by first-generation cognitive science. In this theory, you can separate the problem of how we think with ideas from the problem of what the ideas are supposed to designate. Even if the second problem is only solvable for restricted special cases, the first problem at least can have a general, mathematically precise solution. That was the attraction of a representational theory of mind for first-generation cognitive science.

Our Cartesian Inheritance

We are now in a position to sum up the incredible legacy of Descartes' unique pastiche of common metaphors and folk theories. He has left us with a theory of mind and thought so influential that its main tenets are still widely held and have barely begun to be reevaluated. It has been handed down from generation to generation as if it were a collection of self-evident truths. Much of it is still taught with reverence.

In brief, the Cartesian picture of mind that we have inherited is this:

- What makes human beings human, the only thing that makes them human and that defines their distinctive nature, is their capacity for rational thought.
- Thought is essentially disembodied, and all thought is conscious.
- Thought consists of formal operations on ideas without regard to the relation between those ideas and external reality.
- Ideas thus function like formal symbols in mathematics.
- Some of our ideas are innate and therefore exist in the mind at birth, prior to any experience.
- Other ideas are internal representations of an external reality.
- We can, just by thinking about our own ideas and the operations of our own minds, with care and rigor, come to understand the mind accurately and with absolute certainty.
- Nothing about the body, neither imagination nor emotion nor perception nor any detail of the biological nature of the body, need be known in order to understand the nature of the mind.

We have tried to be as literal as possible in summarizing the Cartesian view of mind. We have done this to raise a question: Do Descartes' metaphors play any essential role in the Cartesian theory of mind or are they incidental to the theory?

Any answer to this question has an adequacy constraint on it: It will have to show how the statements in this theory fit together into a coherent whole. For example, it will have to show what the claim that all thought is conscious has to do with his other claims, such as the claim that certain knowledge of the mind is possible. What does the claim that thought is disembodied have to do with the claim that there are innate ideas or with the claim that rationality constitutes our essence?

The discussion given above answers these questions and many more. As we have seen, for Descartes, it was his metaphors that made these claims into a coherent whole. It is his metaphoric model of the mind, with its metaphoric ontology and metaphoric mode of inference, that unites these claims about ideas, thought, knowledge, imagination, and so on into an organic whole that has been seen to have genuine explanatory value for many generations of philosophers.

Without those metaphors, the claims look like a list of random statements. Yet we all "feel" that they fit together, and once we discover the metaphorical logic that binds them together, we understand why. We know of no account that can bind them together coherently other than the metaphorical account.

In addition, any attempt to cash out this literal-sounding theory in detail will have to make use of our metaphor system for the mind, even if it uses different metaphors than Descartes uses. For example, as soon as you start to spell out what "formal operations" means in detail, you will need to use some conceptual metaphor, either mechanical or mathematical. As soon as you start characterizing "distinctive nature," you will be using the Folk Theory of Essences. When you try to flesh out "internal representation of external reality," you get some version of the Mind As Container metaphor and one of the standard metaphors for "representation."

We most certainly do not mean to suggest that the need for metaphor, either to spell out the Cartesian theory or to explain how it hangs together, in itself in any way invalidates the theory. On the contrary, there can be no such thing as a nonmetaphorical theory of mind.

The body of evidence that supports second-generation cognitive science requires us to reject every tenet of this Cartesian view of mind. This has nothing to do with the fact that it was arrived at via metaphor, that it is held together via metaphor, and that it can only be fleshed out via metaphor.

Faculty Psychology

Although each philosopher approached the problem of knowledge differently, they all shared a general metaphorical model of the mind—what we shall call the metaphorical Folk Theory of Faculty Psychology—that determined the way they defined the problem of knowledge and constrained the kinds of answers they could give to the question of how knowledge was possible. Every Enlightenment epistemological theory is a specific elaboration and refinement of this shared cultural model of the mind.

The Metaphorical Folk Theory of Faculty Psychology

The Cartesian view of mind is very much with us today. But as rich as it is, it is only part of the metaphorical inheritance from the Enlightenment that still defines much of the philosophy of mind and cognitive science. That larger picture, called "faculty psychology," is a model of the mind as divided into discrete "faculties." That model is also metaphoric through and through. It too has an ontology and a logical structure that arise from a network of interrelated metaphors. Those metaphors are so deeply a part of the way we ordinarily think of mind that it seems natural to use them to conceptualize mental operations.

The Society of Mind metaphor is basic to faculty psychology. In the metaphor, the mind is conceptualized as a society whose members perform distinct, nonoverlapping tasks necessary for the successful functioning of that society. The capacities of the mind are thereby conceptualized as autonomous, individual people, each with a different job and each with a distinct, appropriate personality.

The folk model of faculty psychology, built on the Society of Mind metaphor, goes as follows:

THE FOLK THEORY OF FACULTY PSYCHOLOGY

1. The world consists of an external realm of material objects and an internal, mental realm containing as "mental entities" ideas, sensations, feelings, and emotions. The external realm is the "objective" world; the internal realm is the "subjective" world.
2. The internal, mental realm contains a Society of Mind with a least seven members, the "faculties." Each faculty, that is, each capacity of the mind, is conceptualized as a person. The names of these people are Perception, Imagination, Feeling, Will, Understanding, Memory, and Reason.
3. Each faculty-person has a particular personality. Depending on the nature of the personality, that person can be further conceptualized by common metaphors. For example, a methodical, reliable, dispassionate

person is commonly conceptualized as a machine, while a wild, unruly, unpredictable person is commonly conceptualized as a wild animal or a force of nature.

4. Perception is methodical and mostly reliable. He is a kind of receiving clerk, routinely performing the task of passively taking in sense impressions from the body and passing them on to a kind of assembly line on which the other faculties work.

5. Imagination is typically a reliable craftsman, who can at unpredictable moments be playful, mischievous, or out of control. Imagination takes the sense impressions it gets from Perception and constructs from them images that represent things in the external world. Normally it does this in a routine methodical fashion, but sometimes it puts together contents in novel ways to form fantastical images that do not correspond to any existing thing.

6. Feeling is undisciplined, volatile, and sometimes out of control. It can be "aroused" by ideas originating either from outside or inside the mind. When aroused, Feeling can act forcefully to influence Will. Because of its personality, Feeling is often metaphorized further as a wild animal or a force of nature.

7. Understanding is always calm, sober, predictable, under control, and reliable. His job is to function as a judge. He receives images from Imagination and inspects them to see what their internal structure is. If he judges that the structure of an image fits an existing concept, then he assigns that image to the concept. If he judges that the structure of the image does not fit an existing concept, then he forms a new concept for it. Each assignment of a specific image to a general concept is a proposition, referred to as a "judgment."

8. The assembly line so far goes like this: Perception receives sense impressions from the outside and passes them to Imagination, which combines them into images and passes them on to Understanding. Understanding judges how those images are to be assigned to concepts. Understanding thus produces propositions ("judgments") and passes them on to Reason.

9. Reason has good judgment, is cool, controlled, wise, and utterly reliable, and follows procedures explicitly. He acts as lawgiver, judge, and administrator. Reason decides what kinds of things are to be done and sets down the rules for doing them. He judges whether the others are carrying out those rules properly. He also assembles and analyzes the

information made available to him from Understanding and carefully calculates on the basis of this information what needs to be done. He then gives orders to Will.

10. Memory is usually methodical and is expected to be reliable, though he isn't always. Memory functions as a warehouse keeper. He takes items from Perception, Imagination, Understanding, and Reason and stores then for future use. He also keeps records of everybody's actions. He is constantly called upon to produce these objects and records for use by the other faculties and can easily be overburdened.

11. Will is the only person in the society who can move the body to action. Will gets orders as to what to do from Reason and is subject to the pressures and entreaties of Feeling, which may conflict with what Reason commands. Will is free to act as he pleases, provided he is strong enough. Will is strong enough to resist the force of Reason, and he may choose to do so or not. Will may or may not be strong enough to resist Feeling. The stronger Will is, the better he can overcome Feeling. Feeling and Reason commonly struggle for control over Will. If Feeling wins, it is unfortunate, because Reason alone knows what is best for the society as a whole.

This is an elaborate folk theory replete with conceptual metaphors. First, there is the Perceiving Is Receiving metaphor, through which Perception is conceptualized as a passive receiver of information. Ideas Are Objects allows ideas to be taken in by Perception, grasped and worked on by Imagination, and placed under concepts by Understanding. By virtue of Knowing Is Seeing, Understanding inspects the images and concepts to "see" how they best fit. Concepts (or alternatively Categories) are here conceptualized as containers for images.

Thinking Is Object Construction on the assembly line of the mind; Imagination constructs images, and Understanding constructs propositions. Thinking Is Judging, too. Understanding and Reason both act as judges of how images best fit under concepts and how concepts best fit together. In addition, Thinking Is Mathematical Calculation when Reason methodically calculates what the information given him adds up to. Feeling and Reason are both Forces, acting upon Will.

Each of these individual conceptual metaphors has a long history in European culture, and they are still commonplace today. Together they constitute a significant percentage of the metaphors we have for conceptualizing various aspects of mind. Part of the genius of the folk theory is that it combines all these individual metaphors into a morality tale. It incorporates stereotypes about the social roles of various people: the sober judge, the methodical administrator, the receiving clerk, the craftsman, the storage clerk, the hysteric, the independent man of action.

This folk theory, made up of these metaphors and stereotypes, imposes a structure on the mind, producing a metaphorical conception of what the mind is and how it operates. For example, since each person in the Society of Mind is a separate autonomous agent, each faculty of mind is separate and autonomous. Since each person has a one specific task, each faculty of mind has one specific task. Since society is structured hierarchically with an executive giving orders, so too the mind has a hierarchical structure and an executive in control. Just as a society has unruly and uncontrollable individuals, so there are specific isolatable faculties of the mind that can be unruly and uncontrollable. Just as a well-ordered society should not be governed by people out of control, so a properly functioning mind should be governed in a calm, rational, methodical manner.

After several hundred years, a version of this folk theory of the mind is still influential in philosophy of mind, as well as in the various cognitive sciences. The first thing to notice is that it still defines the distinct mental phenomena whose character and functioning is to be studied separately; only then are the interactions between the parts put together. Thus, perception, reason, and emotion are seen as separate phenomena, each requiring its own field of study and capable of being studied independently of the others. They are seen as separable, but interacting, agencies. This is especially obvious in cognitive psychology done from the information-processing perspective. There, the methodical people in the Society of Mind (Perception and Reason) are replaced by machinelike models that each perform a single function.

The assembly line becomes a flow chart with the boxes representing various separate faculties carrying out their functions. The flow of objects from one person to another becomes the flow of information from one box to another. The operations on the objects carried out by a faculty become operations on the information carried out by the mechanism specified by each box in the flow chart. The task of the study of mind is to figure out the mechanical procedures that go on within each box and the structure and directions of the links among the boxes. Generative linguistics uses flow-chart models of the same kind, consciously modeled on Enlightenment faculty psychology.

Of course, the faculties in the Society of Mind that do not act methodically and predictably are not modeled as machines, that is, boxes in the flow chart with functions carried out by algorithms (metaphorical machines). Only the methodical people in the Society of Mind (Perception and Reason) are modeled by metaphorical machines. Emotion, which is not methodical and predictable, is not modeled in this way; nor is the unpredictable aspect of Imagination.

Though we now have overwhelming evidence that the mind does not work like this, the model is still used. We know that reason and emotion go hand in hand, with reason possible only if emotion is present (B1, Damasio 1994). But most people still believe that emotion disrupts reason, and models of reasoning, planning, problem solving, and rational action do not include emotion. We know that the mind has no central executive center, yet there are still models of the mind with such a central processing unit. We know that memory does not merely store and retrieve items and that there is no central place where it is located. Yet there are still models of language in which lexical memory is conceptualized as a bunch of items in a warehouse.

There is a reason why faculty psychology has had such a hold on our imagination and why it continues to be our default mode of thinking about the mind, even in the face of massive counterevidence. The Folk Theory of Faculty Psychology is part of the cognitive unconscious. It is not something that we can willfully control. It fits together into an organic whole many of our most commonplace metaphors for mind and social stereotypes. It is part of our automatic cultural heritage and is embodied in our synapses. We are unlikely to dislodge it anytime soon.

Eventually it may be possible to revise or replace this folk theory, but we will only be able to replace it with still other metaphors. Like time, events, and causation, the mind can only be comprehended metaphorically.

20

Kantian Morality

For over two centuries now, Kantian moral theory has been the paradigm of pure rational morality. Kant believed that he had shown how absolute, universally binding moral principles can be derived from the essence of what he called "pure practical reason." But if, as we have argued, there is no such thing as "pure reason," then Kant must have actually been doing something quite different from unpacking the essence of pure practical reason. What he was doing, we shall argue, was brilliantly working out the entailments of a close-knit cluster of conceptual metaphors that he inherited from Western philosophy and the Judeo-Christian moral tradition.

Morality in this tradition is based on what we have been calling the Strict Father model of the family. As we saw, when this Strict Father model is fleshed out with a number of independently motivated metaphors for morality, we then have Strict Father family morality, which has long been dominant both within the Western moral tradition and in conservative versions of Christianity (A1, Lakoff 1996a, chap. 14).

Kant uses this Strict Father family morality as the key element of a theory of morality in general. In other words, Kant derives all morality as a version of Strict Father family morality. Kant understood this model perfectly, if only implicitly, and he worked with unparalleled insight to develop the implications of the basic metaphors that define this morality. Obviously, Kant thought that he was doing something quite different, namely, analyzing the essence of pure practical reason, and he would have vehemently denied the metaphorical character of morality, at least at the level of the fundamental moral principles he claimed to have identified. Nonetheless, as we will see, his moral theory does not reveal the a priori rational foundations of a universal morality. Rather, it is a working out of the logic of a small set of conceptual metaphors that define mainstream Western morality and that are based on the Strict Father model of the family.

Our goal here is to lay out in detail the conceptual superstructure of Kant's moral theory. We hope to show that Kant's most characteristic ethical doctrines all arise from his unique integration of four folk theories and metaphors common to his age. Our analysis of Kant's conceptual system reveals that his moral theory derives from the following sources:

1. The Folk Theory of Essences
2. Strict Father family morality
3. The Society of Mind metaphor
4. The Family of Man metaphor

It is remarkable that, from such an ordinary collection of commonplace metaphorical ideas, Kant developed his most striking and original moral doctrines, for example:

- Morality must be based on pure reason alone.
- The source of morality is our capacity to give moral laws to ourselves.
- All moral laws are universally binding.
- We have an absolute duty to treat rational creatures as ends-in-themselves and never as means only.
- Morality can consist only of categorical imperatives such as "Act only on that maxim by which you can at the same time will that it become a universal law."

It is Kant's genius that such deep, complex, and subtle doctrines could come from such simple, and at the time intuitive and commonly understood, origins. Let us now turn to the task of showing exactly how these and other key tenets of Kantian moral theory arise from those four common metaphors and folk theories. We will begin by recalling the central elements of the Strict Father view of family morality.

Kant's Strict Father Morality

Kant's ethical theory is a rationalist version of Strict Father morality, which Kant combines with the Family of Man metaphor and the Society of Mind metaphor from faculty psychology (Chapter 19). In this way, as we will see, Universal Reason becomes the Strict Father who issues universal moral commandments that are to be followed by all rational creatures.

Recall that the Strict Father model embodies its own very distinctive morality—a morality defined by such metaphors as moral authority, moral strength, moral obedience, moral boundaries, moral freedom, moral essence, moral purity, moral self-interest, and moral nurturance.

As difficult and complicated as Kant's moral theory is, its conceptual structure is actually Strict Father family morality (1) tied to rationality by faculty psychology, with Reason playing the role of the Strict Father, and (2) universalized to all human beings via the Family of Man metaphor. Understanding Kant's moral theory in this way makes it possible to explain three things: first, what sense it makes to regard Reason as the author of moral precepts; second, how Strict Father family morality can come to be internalized as the basis for a universal rational morality; and, third, what it means to give moral laws to yourself.

Reason as a Strict Father in the Society of Mind Metaphor

Let us think back for a moment to our discussion of the Society of Mind metaphor. We saw that there was a moral component built into the very nature of faculty psychology. Reason, which governs the Society of Mind and is responsible for its well-being, is a moral authority; it knows what is best for the society as a whole and has the duty to issue directives to the members of the society specifying what each needs to do to ensure the well-being of the community. Correspondingly, it is the duty of other community members to obey the dictates of Reason.

Will, who is responsible for what the body does, has a moral obligation to obey the commands of Reason. Passion, who does not typically act morally and who is Reason's antagonist, struggles with Reason over the control of Will. To resist Passion, Will must be strong. This requires that Will be disciplined, and it is the duty of Reason to do everything it can to provide that discipline.

What we see here is the metaphorical imposition of a version of the Strict Father model of family morality onto the Society of Mind. The metaphorical mapping is as follows:

THE REASON AS STRICT FATHER METAPHOR

Strict Father Family		_Society Of Mind_
Family	→	The Mind
Father	→	Reason
The Child	→	Will
External Evil	→	Passion

This mapping applies to the following knowledge about the Strict Father family:

> The father knows what is best and is thus a moral authority; he has to teach the child right from wrong, to make the child as disciplined as possible so that the child will become strong enough to overcome external evils, and to tell the child in specific cases what to do. The child has the moral obligation to obey the father.

The metaphor maps this commonplace knowledge of Strict Father family morality onto the moral aspects of the Society of Mind metaphor, as follows:

> Reason knows what is best and is thus a moral authority; he has to teach Will right from wrong, to make Will as disciplined as possible, to make Will strong enough to overcome passion, and to tell Will in specific cases what to do. Will has the moral obligation to obey Reason.

Given the Reason As Strict Father metaphor, which is built into the Society of Mind metaphor, we can now make sense of two important aspects of Kant's moral theory. (1) Why is Reason the author of moral precepts? Reason is a metaphorical strict father in the Society of Mind and has the moral authority as well as the responsibility to issue moral precepts. (2) What does it mean to give moral precepts to yourself? Recall that each person's mind is conceptualized as a Society of Mind. Reason and Will both reside within each of us. When Reason gives moral precepts to Will, that is equivalent to each of us giving moral precepts to ourselves. This is the capacity Kant calls "autonomy." Each of us is morally autonomous, on Kant's view, insofar as we do not get our moral precepts from others but instead get them from our own capacity to reason. It is in this sense that we are "self-legislating": we give laws to ourselves.

Kant contrasts such moral autonomy with "heteronomy," that is, having someone or something other than your own Reason tell you what you should do and cause you to act. This "other" could be another person, a government, God, or even your own body, that is, your own feelings, passions, habits, desires, and so on. It is only by being rational, by obeying the dictates of Reason, that you become free—autonomous—and independent of any alien influence, including that of your own body.

This is the epitome of rationalism in morality. The body is seen as a foreign influence, which is not really your essential self. For this reason, acting out of empathy (or "fellow-feeling") is a mode of action that, for Kant, "is without any moral worth," since it is based on feeling and does not follow from any directive of Reason alone.

The Role of the Folk Theory of Essences

One of the hallmarks of Kant's theory is the insistence that all moral laws are universal and must issue from a universal moral reason. What makes it natural for Kant to reach such conclusions?

As an Enlightenment figure, Kant accepted the Folk Theory of Essences as an obvious truth. The essence of human beings was, of course, the capacity for Reason. Since an essence is the same for all the members of the category defined by that essence, it follows that all human beings have the *same* capacity for Reason; that is, we all have the same Reason, and so Reason is universal.

Since all human minds were conceptualized in the Enlightenment via the Society of Mind metaphor, Reason, in that metaphor, is therefore Universal Reason. And since Reason, in that metaphor, is the author of moral precepts, those moral precepts must be universal—Universal Moral Laws!

Here we see the Folk Theory of Essence, Strict Father morality, and faculty psychology all working together to give rise to the idea that there are universal moral laws given to us by universal reason, which resides in each of us. By virtue of this metaphorical conjunction, Strict Father family morality is seen as incorporated within every human being. What needs to be shown next is how Kant extends Strict Father family morality from the internal to the external, that is, how he makes it govern all moral relations among all human beings.

The Family of Man Metaphor

Kant's solution makes use the Family of Man metaphor, according to which all human beings belong to a single family and are all brothers and sisters. This metaphor entails that we all have a moral obligation to treat each other as we would family members, according to an ideal model of what a family is.

THE FAMILY OF MAN METAPHOR

Family	→	Humankind
Each Child	→	Each Human Being
Other Children	→	Every Other Human Being

This simple mapping has a number of important entailments:

Family Moral Relations	→	Universal Moral Relations
Family Moral Authority	→	Universal Moral Authority
Family Moral Laws	→	Universal Moral Laws
Family Moral Nurturance	→	Universal Moral Nurturance

In other words, this metaphor projects family moral structure onto a universal moral structure. For example, it is a consequence of this metaphor that just as each child in the family is subject to the same moral authority and moral laws, so each person in the world is subject to the same moral authority and moral laws. The obligation to nurture others in the family gets transformed into an obligation to nurture all humankind.

This metaphor, however, is very general. It does not say anything about what type of family humankind is to be. We, the authors, grew up with a version of the Family of Man metaphor in which the family was to be a Nurturant Parent family, not a Strict Father family. As a result we saw this metaphor as saying that we all have a primary obligation to reach out in empathy toward all human beings and to offer whatever nurturance is in our capacities. This is not Kant's version.

Kant had the Strict Father version of this metaphor. For Kant, the Family of Man is a Strict Father family. The universal moral laws are the precepts a strict father would give. As we shall see, every major tenet of his moral philosophy is a consequence of his Strict Father family morality.

Universal Morality as Strict Father Morality

Let us begin by fleshing out the Family of Man metaphor as Kant does, by imposing upon it a Strict Father interpretation, in which the family moral authority is the father, the father's commands are the family moral laws, and nurturance is the nurturance needed to become morally strong. To arrive at a

Kantian version of the Family of Man metaphor, we thus add the following constraints to the mapping given above:

Family Moral Authority = The Father
Father's Commands = Family Moral Laws
Nurturance = Nurturance To Be Morally Strong

Given this, the submapping

Family Moral Authority → Universal Moral Authority

becomes

Father   → Universal Moral Authority

Similarly, by such substitutions, we arrive at the new submappings:

Father's Commands   → Universal Moral Laws
Obedience To Father → Obedience To Universal Moral Laws

Putting all this together, the Kantian Strict Father version of the Family of Man metaphor goes like this:

THE STRICT FATHER FAMILY OF MAN METAPHOR

Strict Father Family		*Humankind*
Family	→	Humankind
Father	→	Universal Moral Authority
Each Child	→	Each Human Being
Other Children	→	Every Other Human Being
Father's Commands	→	Universal Moral Laws
Obedience To Father	→	Obedience To Universal Moral Laws
Family Moral Relations	→	Universal Moral Relations
Family Nurturance To Be Morally Strong	→	Universal Nurturance To Be Morally Strong

We are now one step away from being able to state the metaphoric structure of Kant's moral theory. What remains to be seen is the relationship between Kant's versions of the Family of Man metaphor and the Society of Mind metaphor. It is this relationship that is the basis for Kant's claim that morality is founded on universal human reason issuing absolute and universally valid moral commands.

In the Strict Father interpretation of the Society of Mind metaphor, Reason is metaphorically a Strict Father, and therefore the Moral Authority.

Father → Reason
Father = Moral Authority

Therefore,

Moral Authority = Reason

According to the Folk Theory of Essences, Reason is Universal Reason. Therefore,

Moral Authority = Universal Reason

This then, fills out the central metaphor by which Kant understands universal moral reason as a strict father who issues universal moral laws. Bringing all of this together, we get the following complex metaphor that defines Kantian moral theory:

THE UNIVERSAL MORALITY AS STRICT FATHER MORALITY METAPHOR

Strict Father Morality		_Universal Rational Morality_
Family	→	Humankind
Each Child	→	Each Human Being
Other Children	→	Every Other Human Being
Father	→	Universal Reason
Father's Moral Authority	→	Universal Moral Authority
Father's Commands	→	Universal Moral Laws
Obedience To Father	→	Obedience To Universal Moral Laws
Family Moral Relations	→	Universal Moral Relations
Family Nurturance To Be Morally Strong	→	Universal Nurturance To Be Morally Strong

Stated as such, this seems like a strange and arbitrary metaphor, an odd way to conceptualize universal rational morality. Yet this mapping is a product of completely sensible parts. Once we see what those parts are and how Kant assembled them, we can discern the internal logic of this complex metaphorical mapping.

At this point, we can see what the metaphorical logic is behind all of Kant's major moral doctrines. All of his major doctrines are inferences of the above metaphors.

Kantian Morality

In light of this mapping, we can now see that Kant understood his Judeo-Christian moral tradition according to a Protestant tradition that interpreted it as a form of Strict Father morality applied at large. According to the theological interpretation of this tradition, God the Father is the supreme moral authority. He issues absolute commands in the form of moral laws to which all of God's human creatures are subject. Morality consists in conforming our individual will to God's will, which is obedience to God. God punishes moral wrongdoing and rewards moral righteousness. Acting morally requires insight into what God commands us to do, plus developing the moral strength to do what is required by God. God's Reason establishes the ultimate moral ideal that defines our proper relations to other people and to ourselves. Any act that violates this ideal trespasses absolute moral boundaries established by God. We are commanded to love and nurture others as required to realize God's moral ideal, to help them achieve well-being, and to respect them as creatures of God.

Kant's ethics is a sophisticated form of the Strict Father morality that underlies this interpretation of the Judeo-Christian moral tradition. Kant rejects the idea that morality comes from God, but he accepts the other major claims of his moral tradition and tries to give a rational justification for them. In essence, what Kant does is to replace God's Reason with an equally transcendent Universal Reason possessed by all people. God's commandments, as moral laws, are thus transformed into the absolute moral laws issued by Universal Reason. Christianity's split of the self into soul and body is carried directly over into the Kantian picture as a split between our rational and bodily natures. The strength of will necessary to follow God's commandments translates as the strength of will necessary to overcome the passions of the body, to follow Reason, and thus to do one's ethical duty. Moral virtue is moral strength.

Given such a "replacement" of God by Reason, one might be tempted to try a very different analysis of Kant's moral theory than the one we just gave. One might try starting with a Strict Father version of Protestant Christianity. This would involve a metaphorical mapping from a Strict Father family onto Christian theology (A1, Lakoff 1996a, chap. 14). Then one might try a metaphorical mapping with Strict Father Christian theology as a source domain and Universal Rational Morality as a target domain. In such a mapping, God's Reason would map onto Universal Reason.

Such a metaphorical mapping captures important aspects of Kant's morality, but falls far short of a full explanation. We will discuss below why it fails to make sense of his categorical imperative and his Kingdom of Ends. In addition, such an analysis misses exactly what it is that makes Kant a key Enlightenment thinker. The analysis we have given uses central Enlightenment ideas—faculty psychology, the primacy of Reason, the notion of essence, and the idea of the Family of Man. The Enlightenment, after all, did not merely substitute Reason for God. Descartes' philosophy requires both Reason and God.

At this point we can give an overview of Kant's moral theory that shows exactly how each of his doctrines fits into the picture above. The above analysis explains how his moral doctrines make up a coherent whole, with no doctrines that are not motivated by one or more of four commonplace folk theories and metaphors.

Moral Authority

The metaphor of Moral Authority is central to Strict Father morality. It is also central to Kant's moral theory, which has three major doctrines concerning moral authority. First, the ultimate Moral Authority is Reason. Second, for Reason to be a moral authority it must be "pure," that is, free of any bodily taint. Third, Reason can be a moral authority only if it is universal.

To those raised in the Western tradition, these doctrines may seem so commonplace as to be self-evident. But why? They do not follow as tautologies from any principles of logic. Nor are they definitional; there are moral theories in which none of these is true. Why, exactly, did they seem self-evident to Kant? And why will they seem self-evident to so many readers?

The answer, we claim, is that all three doctrines are consequences of Strict Father morality as it functions both in the Strict Father version of our Judeo-Christian tradition and in Kant's conceptual system. First, the idea that moral authority is Reason comes out of faculty psychology, in which Reason functions as a Strict Father to tell Will how to act and to try to get Will to be as disciplined as possible. Second, the idea that a moral authority must be pure, uninfluenced by the passions of the body, is a hallmark of Strict Father morality. And third, the universality of Reason derives partly from the mapping of the Strict Father family onto the Family of Man in such a way that the strict father's commands, which must apply to all children, are mapped onto Universal Moral Laws that are binding on all people.

Kant's three basic doctrines on moral authority are summed up as follows:

> It is clear from the foregoing that all moral concepts have their seat and origin completely a priori in reason. . . . In this purity of their origin lies their very worthiness to serve us as supreme practical principles [universal moral laws]; and to the extent that something empirical [e.g., experience or feeling] is added to them, just so much is taken away from their genuine influence and from the absolute worth of the corresponding actions. (C2, Kant, *Grounding* 411; brackets added)

Moral Strength

The metaphor of Moral Strength is central to Strict Father morality, in which everything depends on the father's moral strength and his ability to develop it in his children. As we saw above, moral strength comes into Kantian moral theory via the role of Strict Father morality in faculty psychology. There Reason is the Strict Father who must develop the strength of Will, so that Will will be able to resist Passion. This use of moral strength makes it the fundamental virtue of Kantian moral theory.

If acting morally is acting from duty as specified by moral commandments, then doing one's duty requires great moral strength. Universal Reason (the Strict Father) tells Will how it ought to act in order to be moral. Unfortunately, there are strong forces acting against Reason to influence Will. As we saw above, those forces are both external (forces of evil) and internal (temptations of the flesh). There is a great battle going on within us for mastery of the Will:

> Man feels within himself a powerful counterweight to all the commands of duty, which are presented to him by reason as being so pre-eminently worthy of respect; this counterweight consists of his needs and inclinations, whose total satisfaction is summed up under the name of happiness (*Grounding* 405).

For Kant, duty is the requirement to act out of respect for moral law, where "respect" is one's "consciousness of the subordination of my will to a law without the mediation of other influence" (*Grounding* 402 note). So doing one's duty requires the strength of will to do what the moral law, as given by Reason, commands, without one's Will being "subordinated" to any forces of evil or temptation, no matter how strong they might be. For Will to do what Reason commands, it must be strong enough to fend off the assault of bodily passions, needs, and inclinations.

We can now see why Kant has the account of virtue that he has. Kant understands all virtue as the principal virtue of Strict Father morality, namely, Moral Strength. "Hence virtue is the moral strength of the will of a human being in obeying his duty" (C2, Kant, *Metaphysics* 405). Thus, virtue requires that man have "control over himself" (*Metaphysics* 408) and "not let himself be governed by his feelings and inclinations. . . . For unless reason takes the reins of government in its own hands, feelings and inclinations play the master over man" (*Metaphysics* 408).

Kant's Strict Father morality gives primacy to doing one's duty regardless of any forces influencing action in a contrary fashion. Consequently, Kant is left with a narrow and restrictive conception of virtue as no more than the moral strength to do one's duty. In classical moral theory, virtue is a state of character concerned with habits that allow one to choose wisely and well, typically with balance. But Kant's emphasis on the Moral Strength metaphor forces him to see virtue as operating primarily in the battle between the body and reason:

> Now, fortitude is the capacity and resolved purpose to resist a strong but unjust opponent; and with regard to the opponent of the moral dispositions within us, such fortitude is virtue. (*Metaphysics* 380)

> Virtue is the strength of a man's maxim in obeying his duty. All strength is known only by the obstacles it can overcome; and in the case of virtue the obstacles are the natural inclinations, which can come into conflict with moral purpose (*Metaphysics* 394)

Just as we would expect in any Strict Father morality, virtue as moral strength shows itself through self-discipline and self-constraint. In his *Lectures on Ethics*, therefore, Kant lists as one of our duties that of "self-mastery":

> Here is the rule: Seek to maintain self-mastery; thou wilt then be fit to perform thy self-regarding duties. There is in man a certain rabble of acts of sensibility which has to be vigilantly disciplined, and kept under strict rule, even to the point of applying force to make it submit to the ordinances of governance. This rabble does not naturally conform to the rule of the understanding, yet it is good only in so far as it does so conform. (C2, Kant, *Lectures* 138)

Kant follows out the implications of this metaphor of self-discipline, spelling out all of the forms of constraint, force, and vigilance that we must bring to bear on our sensuous nature. A moral person must "weaken the opposing forces," "divide them," "stamp out the tendency which arises from sensuous motive," and "discipline himself morally" (*Lectures* 139). If a person "surrenders authority over himself, his imagination has free play; he cannot discipline himself, but his imagination carries him away by the laws of association; he yields willingly to his senses, and, unable to curb them, he becomes their toy" (*Lectures* 140).

Kant goes to great lengths to describe the battle that rages between our bodily and rational natures. Our bodily needs and wants would reduce us to mere brutes, if they had their way, so that the moral will must develop remarkable strength to overcome this onslaught of temptation. Moral strength thus requires self-control, self-reliance, and self-discipline. These virtues, which are virtues only as a consequence of the metaphor of Moral Strength, are evident in Kant's powerful explanation of why various forms of servility are immoral.

> Do not become the vassals of men. Do not suffer your rights to be trampled underfoot by others with impunity. Incur no debts for which you cannot provide full security. Accept no favors which you might do without. Do not be parasites nor flatterers nor (what really differs from these only in degree) beggars. Therefore, be thrifty so that you may not become destitute. Complaining and whimpering, even merely crying out in bodily pain, are unworthy of you, and most of all when you are aware that you deserve pain. (*Metaphysics* 436)

Notice that what Kant is really describing here is the necessity, on his view, of not being dependent on others, insofar as that is possible. To be morally strong, one must be able to proceed on one's own, without help. Anything that puts you in a dependent relationship—such as incurring debt, becoming financially needy, becoming a dependent servant, and even letting bodily pain take control—is to be avoided whenever possible. Self-control, self-mastery, and full autonomy are the conditions for being able to act morally.

Moral Boundaries

Kantian morality, like all Strict Father morality, is a morality of constraint. In the location version of the Event-Structure metaphor, purposeful actions are understood metaphorically as self-propelled motions along paths toward destinations, or "ends." The term *end* for Kant is a purpose conceptualized via the Event-Structure metaphor as a destination. From the perspective of the Event-

Structure metaphor, morality for Kant is primarily a matter of determining what constraints there are on these metaphorically defined purposeful actions—what our destinations (or ends) should be, what means (or paths) we are permitted to take toward achieving those ends, and what forces affect our motion as we move metaphorically toward those ends.

Kant sees moral obligations as imposing forces that constrain us: We are "bound" by duty and we are morally "compelled" to act in certain prescribed ways. There are moral laws that are "binding" on all rational creatures.

Reason commands, dictates, and orders the will to choose in accordance with certain constraining moral principles and laws. Evils, both internal and external, are strong forces that would drive us off the straight path, overcome our reason and will, and make us slaves of our passions. They would force the will to act against reason. Therefore, the will needs a strong constraint to follow moral laws and stay on the path that leads to moral ends.

Moral Freedom

As we saw at the beginning of this chapter, the metaphor of Moral Freedom is one of those metaphors for morality given priority by the Strict Father model of the family. It is based on the Location Event-Structure metaphor, where purposeful action is self-propelled motion to a destination (an "end"), and freedom of action is unimpeded motion. In the metaphor, immoral actions are motions that interfere with others' reaching of their ends, that is, they keep others from acting to achieve their purposes.

The reason this is given priority by the Strict Father model of the family is that, in Strict Father morality, being self-reliant through being self-disciplined is a primary value, and the self-disciplined person cannot become self-reliant if people are interfering with the achieving his or her ends.

Moral Freedom, like the other metaphors that form the Strict Father complex, is a localized metaphor. It is not integrated into a consistent moral system. For example, it, in itself, says nothing about whether it is immoral to interfere with someone who has immoral ends.

Though Kant was using the Strict Father system with its isolated, localized metaphors, he was also building a systematic and consistent moral conceptual system. For Kant, the metaphor of Moral Freedom was not isolated and localized, but stood at the very center of his moral theory.

For Kant, the metaphor of Moral Freedom is intimately tied to the notion of a moral end: Choosing any end at all is, by definition, a matter of free will.

You haven't really "chosen" the end if you are forced to adopt it. The very possibility of choosing moral ends presupposes freedom to make the choice. As he says, "An end is an object of free choice" (*Metaphysics* 38). "Now I can indeed be forced by others to actions which are directed as means to end, but I can never be forced by others to have an end; I alone can make something an end for myself" (*Metaphysics* 381). Part of the *essence* of any moral end is that it has been freely chosen. Inhibiting freedom, therefore, is for Kant an interference with the possibility of choosing moral ends.

Since all moral ends issue from Reason, it follows that Reason must be free. If Reason were constrained by anything external to it, it could not choose freely and thus could not be the source of moral ends. Since Reason defines what we most essentially are, so the freedom to choose moral ends is part of our essence.

Moral Ends and Ends-in-Themselves

The notion of "ends-in-themselves" is notoriously the most esoteric and obscure of all Kant's ideas, and yet it lies at the very heart of his entire moral theory. For him, morality ultimately comes down to always treating others as "ends-in-themselves." What sense can we possibly make of an "end-in-itself"? Most of us think of ends as things we can achieve through our actions, as metaphorical destinations we are trying to reach, that is, as end points on a path of action. Ordinarily, we cannot conceive of an end with no means, of a metaphorical end point with no path that it lies at the end of. An "end-in-itself" is not one of our ordinary everyday concepts. And yet Kant takes it to be the very essence of morality. How can this be?

Moreover, Kant thinks that everyone has an absolutely binding universal moral duty to treat people as "ends-in-themselves." What, exactly is an "end-in-itself"? And how does this idea emerge out of the four commonplace metaphors and folk theories given above?

Consider ordinary ends, which we have no moral duty to pursue, ends we can choose freely whether to pursue or not. Often, these are ends we can achieve. For example, suppose you have it as your purpose to let some fresh air into your room. You open the window and your end is realized: The fresh air comes in. Your action of opening the window made your end real.

An end-in-itself, Kant says, is not like this. There is never a time at which it is unrealized. You do not and cannot bring it about by your actions. Kant calls this a "self-subsisting end" to capture the idea that it is prior to and independent of your desires and actions. This makes the concept seem mysterious. How can it be an "end" at all, if it has a prior existence independent of anything you do or even desire?

To answer such questions, we need to think first of what a moral end is. It is a morally permissible or obligatory purpose that we try to achieve through our actions. What defines what moral purposes are? Kant's answer is Universal Reason, which defines the category of universally obligatory or permissible moral purposes. By the Folk Theory of Essences, this category of moral purposes (that is, "ends") must have an essence—the essence of what makes something a moral end.

Here the logic of the Folk Theory of Essences enters. For something to exist "in itself" it must (1) not be caused by anything else and (2) must be caused only by itself. Essences, as defined by the folk theory, have just this strange property: They exist outside of time. They have always been there and always will be there. Nothing external causes them. But how can they "cause themselves"? The answer comes in two parts. First, since every member is in the category by virtue of having the properties of the essence, none of the category members would exist without it. Hence the essence is the causal source of all members of the category. Second, the essence is itself a category member, since it obviously has all the properties of the essence. Thus, all essences are self-causing and not caused by anything else, and so exist in themselves. (Kant uses the term *self-subsisting*.)

Since the essence of moral ends is an essence, it exists in itself. Moreover, the essence of moral ends is in the category of moral ends, so it is a moral end-in-itself. And since every moral end is an end, the essence of moral ends is an end-in-itself.

So far, so good. But Kant claims that *people* are ends-in-themselves. Given that the essence of moral ends is an end-in-itself, how does Kant get to the claim that *people*—all people—are ends-in-themselves? Notice that if all people have as part of their essence the essence of moral ends, then people become ends-in-themselves. We are now one short step away. For Kant, Universal Reason is the causal source of all moral ends. As such, it is the essence of all moral ends. Since all people have Universal Reason as part of their essence, they all have the essence of moral ends as part of their essence. Hence, all human beings are, by their rational nature, ends-in-themselves. And from that it follows that, because they are by nature rational beings, they are not means for any other end.

The logic of essences explains what might appear to be an anomaly in the notion that people can be ends-in-themselves. Notice that is it not people who are self-causing here. It is Universal Reason within people that is self-causing, since it is an essence.

This is the structure of the conceptual system that Kant took for granted, that was implicit in his thinking. His conclusion comes from three sources: (1) the Folk Theory of Essences, (2) the nature of moral ends, and (3) Universal Reason as a source of morality. The main element in all this structure is the Folk Theory of Essences, one of the four conceptual cornerstones of Kantian thought.

To show exactly how this logic works in detail, we have reconstructed the logical structure of this aspect of Kantian thought. We will be using the word *category* not in Kant's technical sense, but in the more commonplace philosophical sense. A category is a kind of thing. It is defined by a concept, and that concept characterizes the essence of the category. Here is the logic behind Kant's notion of an end-in-itself. What has to be shown here is (1) how an end could exist "in-itself" independent of the desires and purposes of any being and (2) why such an "end-in-self" is an end for everyone, when it need not be the desire or purpose of any particular person.

THE FOLK THEORY OF ESSENCES

An essence of a category is not caused by anything else.

Every essence is a member of the category it is an essence of.

An essence of a category is the causal source of all the members of the category; since it defines the category, the category and all its members would not exist without it.

Since an essence is in the category it is an essence of, and since it is the causal source of all members of the category, it is "self-causing."

Every essence exists "in itself" because it is self-causing and not caused by anything else.

The essence of a category exists as a member of that category "in itself."

Since categories and their essences are part of what defines Universal Reason, they are categories and essences *for everyone.*

MORAL ENDS

Moral ends form a category. Therefore, there is an essence of that category, the essence of moral ends.

As an essence, it exists in itself.

As a member of the category of moral ends, the essence of moral ends is a moral end-in-itself.

A moral end-in-itself is an end-in-itself.

The essence of moral ends is an end-in-itself.

Since an end-in-itself is the essence of the category of all ends, and since categories and their essences are the same for everyone, it follows that an end-in-itself is an end for everyone.

UNIVERSAL REASON AS THE SOURCE OF MORALITY

All moral ends follow from Universal Reason.

Therefore, Universal Reason is the causal source of all moral ends.

Therefore, Universal Reason is the essence of all moral ends.

Therefore, Universal Reason exists as an end-in-itself.

Universal Reason is the essence of our rational nature.

Therefore, rational nature exists as an end-in-itself.

All human beings have a rational nature.

Therefore, all human beings exist as ends-in-themselves.

Therefore, every human being is an end for everyone.

Therefore, no human being exists as a means to serve some other end.

In practical terms, here is what Kant means when he says that we should act only so as to respect rational nature as an end-in-itself. Suppose I were to treat you as a means to some end of mine. Under what conditions would I be treating you as an end-in-itself? Only if I engage your reason to determine whether my end is a moral end for you and to use your freedom to choose to be used as a means for my end. To do otherwise is to deny you the use of your freedom and reason and thereby deny you your status as an end-in-itself.

We are now in a position to understand what Kant means when he says, "Now I say that man, and in general every rational being, exists as an end in himself and not merely as a means to be arbitrarily used by this or that will" (*Grounding* 428).

He does not mean that no one can ever be used as a means. Suppose I hire you to paint my house, and you freely agree to do so. I am using you as a means to get my house painted, but Kant would find nothing immoral about this, since I am not violating your freedom. Though I use you as a means to one of my ends, I still recognize you as an end-in-itself. That is, since you are rational, you contain the essence of all moral ends. That includes freedom to choose your own ends, and I am not impinging on that. If, however, I were to put a gun to your head and make you paint my house, then I would be violating your ability to choose your own moral ends. I would be assaulting the locus of morality itself in you. That is what it means *not* to treat you as an "end-in-itself."

What Kant means by being an "end-in-itself" is this: Universal Reason is what allows each of us to give moral laws to ourselves and hence to set our moral ends. It is what gives us freedom—freedom to choose our own moral ends—and hence makes us morally independent. Kant calls such moral independence "autonomy."

We have suggested that the technical term *end-in-itself* refers to the essence of moral ends, which exists "in-itself." This essence defines all moral ends and is part of what we most essentially are because it is a consequence of our being rational. As free rational beings, we are ends-in-ourselves because we are the very condition of any moral action whatsoever. The fact that Reason resides in us means that *we* are what makes any moral action possible at all. To treat any human being as anything but an end-in-itself is to violate the very condition of morality.

Kant also claims that being an end-in-itself is the basis of all dignity. Here is the rationale. In Strict Father morality, it is independence (autonomy) that permits dignity. A person who is dependent, who is not self-reliant, has no dignity in that moral system. For Kant, it is the freedom that comes from being able to choose our own moral ends that gives us moral independence and, hence, dignity. As he says, "the dignity of humanity consists just in its capacity to legislate universal law" (*Grounding* 440).

We can now see what Kant means by the "Kingdom of Ends." It is the ideal state in which everyone acts morally. It is called the "Kingdom of Ends" for two reasons: In it everyone chooses only moral ends, and in it everyone treats everyone else as ends-in-themselves. In the Kingdom of Ends, therefore, the freedom of each person is maximized consistent with the freedom of every other person. In the Kingdom of Ends, everyone has dignity:

> Now morality is the condition under which alone a rational being can be an end-in-himself, for only thereby can he be a legislating member in the Kingdom of Ends. Hence, morality and humanity, insofar as it is capable of morality, alone have dignity. (*Grounding* 435)

Autonomy and Internal Evil

The metaphor of Moral Strength in the Strict Father model states that evil is a force in the world, both internal and external, and that one must be morally strong to stand up to it. If you are morally weak, you won't be able to stand up to evil, and so you will fall before it. In Strict Father morality, the body, as the seat of passion and desire, is a source of internal evil and so is a threat to moral action.

The application of Strict Father morality in the Society of Mind metaphor requires that Will be strong if it is to resist Passion and follow the dictates of Reason. In Kant's use of the Society of Mind metaphor, strength of Will is crucial for moral autonomy: You cannot give the law to yourself via your reason unless your will is strong enough to fend off internal evils, that is, bodily inclinations. We can see this in what Kant has to say about why it is immoral to let bodily passions overcome your rational capacities.

Consider, for example, Kant's account of our duties regarding "self-stupefaction through the immoderate use of food and drink" (*Metaphysics* 427) and "wanton self-abuse" (*Metaphysics* 424ff.). Are there things that I am not morally permitted to do to my own body, even if they do not harm others? What about drunkenness and gluttony? These are morally impermissible because they throw away our rational autonomy:

> When a man is drunk, he is simply like a beast, not to be treated as a human being; when he is gorged with food, he is temporarily incapacitated for activities which require adroitness and deliberation in the use of his powers. (*Metaphysics* 427)

The vices of drunkenness and gluttony make us unfit for rational deliberation and thereby diminish, or even discard temporarily, our autonomy as rational beings. When we do such things, we use ourselves for pleasure and escape alone.

A similar violation of autonomy occurs, according to Kant, whenever we use our bodies nonpurposively for sexual pleasure. Kant asserts that our sexual attributes are given to us for the natural end of procreation. The use of these attributes in any nonpurposive way is a violation of the moral order, understood metaphorically as a "natural order." Kant claims that "the end of nature in the cohabitation of the sexes is propagation, i.e., preservation of the race," (*Metaphysics* 426) and sex not directed toward this end is immoral. Kant attacks every conceivable form of sexual activity that cannot be directed toward procreation. He claims that any "unnatural" or "unpurposive" use of one's sexual attributes is immoral because "a man gives up his personality (throws it away) when he uses himself merely as a means for the gratification of animal drive" (*Metaphysics* 425). Kant even goes so far as to argue that such misuse of sexuality is far worse even than suicide, which is another form of using oneself merely as a means. Suicide requires courage to end one's misery, but "when one abandons himself entirely to an animal inclination, he makes himself an object of unnatural gratification, i.e., a loathsome thing, and thus deprives himself of all self-respect" (*Metaphysics* 425).

Given the vehemence of Kant's attack on unpurposive sex with another person where procreation is not possible, one can easily anticipate the scorn he heaps on masturbation, "when a man is stimulated not by an actual object but by imagining it, thus creating it himself unpurposively" (*Metaphysics* 425). In such awful cases, "fancy engenders a desire contrary to an end of nature" and it reduces one's own person to the status of a mere pleasure machine.

Moral Nurturance

In Strict Father morality, nurturance is subservient to moral strength. Nurturance is nurturance to be strong. Raising or teaching someone in such a way that they become morally weak is not nurturance. This is Kant's view. Nurturance serves a moral purpose. It is intended to help the child develop moral strength, learn what is right (universally), and be able to realize moral ends through self-discipline.

The primary duty of nurturance toward others is benevolence. Benevolence is a "practical love of all mankind" that is "the duty to make the ends of others (as long as they are not immoral) my own" (*Metaphysics* 450). Such benevolent concern for the well-being of others expresses itself as beneficence, that is, being "helpful to men in need according to one's means, for the sake of their happiness and without hoping for anything thereby" (*Metaphysics* 452). The question that must be answered in order to justify the duty of beneficence is why, beyond not harming another person, I should have a duty to make their (morally permissible) ends my ends. Why should the principle of respect for rational beings require anything more than leaving them alone (not interfering with their freedom, insofar as they act morally)?

Kant's answer to this question stems from his Strict Father morality. The point of helping others in need is that this makes it possible for them to act morally and to realize their moral ends. It is not appropriate to help others in a way that lets them remain morally weak and dependent. Rather, you are trying to help them develop moral strength and the ability to pursue ends that realize freedom and morality.

We have, for example, a duty to ourselves to develop our talents. Why? Because only if we develop our bodily and mental talents and abilities can we be morally strong beings capable of realizing moral ends. Kant explains:

> With regard to contingent (meritorious) duty to oneself, it is not enough that the action does not conflict with humanity in our own person as an end in itself; the action must also harmonize with this end. Now there are in humanity capacities for greater perfection which belong to the end that nature has in view as regards humanity in our own person. To neglect these capacities might perhaps be consistent with the maintenance of humanity as an end in itself, but would not be consistent with the advancement of this end. (*Grounding* 430)

In other words, self-fulfillment in itself is not a moral goal. Self-fulfillment is moral only when it makes you morally strong.

Our bodies and minds are not our own to dispose of as we please. We have a duty to be morally strong, to develop our moral capacities, and to seek moral perfection, since these are the very conditions for acting morally and being autonomous. Morality requires of us that we nurture ourselves, not merely out of self-interest, but even when it is difficult and painful to develop our talents and we would rather take the easier road. The end of human existence is morality—the autonomous, rational exercise of one's freedom in a way that treats all people as ends-in-themselves.

Self-nurturance is, then, the strengthening of your capacities—physical, mental, and moral—to enable yourself to pursue ends required by moral law. Nurturance serves moral strength, as required for the pursuit of moral perfection: "But as for what concerns perfection as a moral end, there is indeed . . . only one virtue (. . . moral strength of one's maxims). (*Metaphysics* 447)

The Categorical Imperative

The term *categorical* means "absolute." It contrasts with anything that is conditional, hypothetical, context-dependent, or contingent on personal desires. For Kant, a "hypothetical imperative" is a conditional requirement or command that depends on your purposes (i.e., your personal ends). By contrast, categorical imperatives place requirements on you regardless of what your personal ends might be.

Kant's concept of a categorical imperative comes directly out of Strict Father morality. The Strict Father (Universal Reason) issues certain commands, and the child (you) absolutely must follow them to the letter. Your needs are irrelevant. Your feelings are irrelevant. Your purposes are irrelevant. It is defined as being good for you:

There is one imperative which immediately commands a certain conduct without having as its condition any other purpose to be attained by it. This imperative is categorical. It is not concerned with the matter of the action and its intended result, but rather with the form of the action and the principle from which it follows. (*Grounding* 416)

Being moral is doing your duty. Doing your duty is acting out of respect for moral law and nothing else. Therefore, morality cannot be based on any feelings, needs, or purposes you might happen to have. Each version of the categorical imperative is a universal, unconditional, and absolutely binding moral law.

To get a sense of what Kant is proposing, here are paraphrases of his four versions, all of which he considers equivalent.

1. Act only according to that maxim by which you can at the same time will that it can become a universal law.
2. Act always so as to treat humanity (yourself or others) always as an end and never as a means only.
3. Act only according to those principles that, through universal reason, you give to yourself as universal moral laws.
4. Act so as to create a kingdom of ends.

The examples that Kant gives might sound on the surface as if they were part of a Nurturant Parent morality, or perhaps expressions of a principle of universal love, or just directives to be nice to people: Act as you think everyone should act; treat people as ends-in-themselves, not means; respect their freedom; be fair; be a moral idealist.

But it is easy to see that these are neither products of a Nurturant Parent morality, nor based on feelings of empathy, nor guidelines from Miss Manners. First, if you follow these prescriptions, your actions cannot be based on feelings such as love, or empathy, or friendship. These are ruled out of any moral considerations, because they are not based on Universal Reason.

Second, these are universal, absolutely binding moral laws that you have to obey. Being moral is obeying them for their own sake and for no other reason or motive. Whatever the effects of one's actions on others, it reduces morality to following the law only out of respect for law itself.

Third, the imperative to treat people always as ends-in-themselves, however noble as a principle of respect, is ultimately based on a principle of preserving individual freedom and is not essentially about nurturance, empathy, love, or kindness. Rather it is about freedom and independence, as construed within the Strict Father moral tradition. It is not that there is anything bad about freedom. Quite the contrary. It is to be cherished. But Kant's imperative always places freedom and independence first, giving it absolute priority over all other values in all circumstances. It *always* takes priority over love, community building, respect for nature, empathy, and so on. That is what makes it a Strict Father principle.

What This Means for Kantian Morality

So what? So Kantian morality *is* Strict Father morality. One imagines an orthodox Kantian saying, "Okay, so it is Strict Father morality, and rightly so, since that is the morality dictated by pure practical reason, subject to a few minor clarifications and revisions. You've simply found a clever way to describe the morality that issues from Universal Reason and that holds for all rational beings." There are several replies.

First, the cognitive analysis we've given explains what has hitherto resisted explanation, namely, how Kant's moral theory hangs together. It shows how it is a product of commonplace folk theories and metaphors of the Enlightenment. Moreover, it shows how Kant's logic follows from those folk theories and metaphors. Kant's doctrines do not come out of thin air; nor are they merely a random list. They are a product of one of the most systematic minds of all time, and we believe we have revealed a central part of the system. In addition, this analysis shows Kant to be using ordinary modes of reason—metaphors and folk theories common to his philosophical tradition—with extraordinary systematicity and originality.

Second, Kant's use of metaphoric reason shows that his moral theory does not emerge from "pure practical reason," which is supposedly literal and disembodied. But this, in itself, contradicts the very foundation of his moral theory. It is sobering to realize that Kant's moral theory is absolutely based on a view of concepts and reasoning that is inconsistent with empirical results in the cognitive sciences. Every aspect of second-generation cognitive science is at odds with the account of reason that Kant requires. What this means is that empirical results about the nature of mind can contradict philosophical theories of morality. Cognitive science presents us with an "is" that can contradict an "ought." When this happens, we maintain, we must opt for the most cognitively realistic position that is supported by the widest range of converging empirical evidence about the nature of mind.

This does not mean that Kant has nothing to teach us about morality. Far from it. One learns enormous amounts from reading Kant. But what we have learned about the mind from cognitive science does invalidate the central thrust of his theory that the foundations of morality lie in pure reason—something that does not and cannot exist. Kant's moral philosophy articulates key moral concepts, such as respect, freedom, autonomy, and moral law, from a Strict Father perspective, which has played a major role in the Protestant Christian tradition. We have a great deal to learn from his genius in systematically analyzing such concepts and their relations.

Finally, Kant's idea that the foundations of morality can have absolutely nothing to do with either human feeling or the fact that we have bodies is absurd. At best, it is a narrow and one-sided (the Strict Father side) attempt that captures only a small part of what goes into moral reasoning and the choice of moral ends. At its worst, it misses most of what is really important in our moral thinking. As Antonio and Hannah Damasio have demonstrated (B1, Damasio 1994), people with brain lesions that leave them reasoning without access to emotion simply cannot function in appropriate ways in a social environment. They certainly cannot function morally. This is an empirical result. The idea of a pure reason that can function in the moral domain independent of emotion is empirically untenable.

21

Analytic Philosophy

The dominant orientation in twentieth-century Anglo-American philosophy, known as "analytic philosophy," is defined principally by its preoccupation with language. The so-called linguistic turn that characterizes analytic philosophy is based on the belief that it is by analyzing language that we come to understand everything that supposedly matters to philosophy, such as concepts, meaning, reference, knowledge, truth, reason, and value. There is much to be said for this yoking of philosophy to linguistic analysis, for, as we have seen, language is one of our most important windows into the workings of the mind. It is not the *only* window, but it is the source of a vast majority of the evidence we currently have about cognition.

However, in twentieth-century analytic philosophy, this salutary concern with linguistic analysis came to be defined in a very narrow and unfortunate manner by philosophers influenced by the writings of Gottlob Frege. As a mathematician and logician, Frege was excessively concerned with justifying mathematics as universal and absolute, transcending all time, place, and culture. His attack on "psychologism," the view that mathematics is a result of the structure of the human mind rather than an objective mind-independent reality, led him to adopt a view of all meaning and thought as disembodied and formal. Frege mistakenly believed that the only way meanings could be shared and public was for them to be disembodied, abstract, yet objectively existing entities. He thought of this universal, objective realm as containing such entities as senses (meanings), propositions, numbers, functions, and other formal structures.

Frege contrasted this supposedly objective realm with the subjective realm of individual minds, which he saw as being populated with images, subjective

440

ideas, and feelings. Since he considered these aspects of imagination and feeling wholly subjective, he denied them any role in semantics, which was supposed to be about objective meanings and truth conditions. Consequently, meaning was thought to have nothing to do with our embodiment, our imagination, or our feelings.

Under Frege's influence, analytic philosophy—the philosophy whose central focus was language—defined itself as formal and logical analysis of allegedly universal, disembodied senses (meanings), propositions, and functions. Words had objective senses corresponding to them, and these senses picked out referents in the world. Meaning was therefore seen as an abstract relation between words and aspects of an objective, mind-independent world.

Given such a disembodied philosophy of language, it is not surprising that Anglo-American philosophy came to be concerned with the analysis of allegedly disembodied meanings, reference, truth conditions, and knowledge built up out of propositions. Nowhere in this picture of language is there any place for embodiment or imagination in conceptualization, reasoning, or knowledge. Because of his erroneous belief that nothing tied to the body or to imagination could generate shared meanings, Frege's philosophy of language could not account for an enormous part of natural language. As we shall see, this problem is shared by all approaches to analytic philosophy.

The Anatomy of Analytic Philosophy

As we saw in Chapter 12, analytic philosophy makes use of many of the entailments of our everyday metaphors for mind. It also makes use of views taken from Descartes. But analytic philosophy is an enormous enterprise, and analytic philosophers differ on which of these views they accept. We can make a bit more sense of the enterprise by making a few distinctions. First, we need to ask what characterizes the movement in general. Second, we need to distinguish between two forms of analytic philosophy, formalist philosophy (based on mathematical logic) and ordinary language philosophy. Last, we will take up the issue of meaning holism.

Our enterprise is not an exhaustive description of all forms of analytic philosophy, but a rough classification of the assumptions and motivations lying behind mainstream views in both the formalist and ordinary language versions of analytic philosophy.

Analytic Philosophy in General

As we saw in Chapter 12, the Thought As Language metaphor is central to the Anglo-American philosophical tradition. Its effect is to view the concepts expressed by language as linguistic symbols meaningless in themselves and requiring interpretation. What follows from that metaphor is that analysis of language is analysis of thought. Via this metaphor, linguistic analysis becomes conceptual analysis, which is the central tenet of analytic philosophy.

Now consider our practice of teaching children "the words for things," through which they learn "the meanings of words." This practice is accompanied by two common folk theories:

THE NAMING FOLK THEORY

Words pick out things in the world.

THE MEANING FOLK THEORY

Learning the meanings of words is learning to name things correctly.

When these folk theories are put together with the Thought as Language metaphor, according to which concepts are seen metaphorically as linguistic symbols (words), then concepts (represented by linguistic symbols) are seen as picking out things in the world and thus assigning meaning to words. This makes all meaning mind-independent, objective, and publicly accessible. Since the words of a language have an objective existence as symbols and are publicly accessible, and since entities in the world have a mind-independent, objective, and publicly accessible existence, it follows that meaning (the relation between the two) has a mind-independent, objective, and publicly accessible existence. These ideas are also central to analytic philosophy.

Of course, from a cognitive science perspective, the Naming and Meaning folk theories are oversimplifications at best and fallacies if you take them seriously. Words, that is, the phonological forms of lexical items, conventionally express concepts, which reside in human minds and which, as we have seen, get their meaning via their embodiment. Each of us, from childhood on, forms conceptual categories of embodied perceptions, actions, and other experiences. That is, we conceptualize the world through our embodied experiences and the shaping provided by the structures of our bodies and brains.

Meanings of concepts thus come through embodied experience. When an embodied concept expressed by a word accords with an embodied conceptualization of some object in the world, we speak of this situation as "the word naming the object." But when we speak this way, we are leaving out the roles of the mind, brain, and body, since we are not conscious of those roles. The sentence "Words pick out objects" is a manifestation in our conventional language of the Naming Folk Theory. It does not accurately represent the way that words have meanings and can be used to refer to objects in the world. It should be thought of as akin to expressions like "The sun rises," which is an expression of another scientifically false folk theory based on a common perception.

If you literally believe the Naming Folk Theory, you will believe that words pick out objects (irrespective of human bodies, brains, and minds). If you believe both folk theories, then you will believe that meanings are given by the way words pick out objects (again irrespective of human bodies, brains, and minds). If you also believe the Thought As Language metaphor, then it will seem natural that concepts are linguistic in nature, pick out things in the world, and get their meanings that way—objectively, without any significant role played by human bodies, brains, and minds.

The correspondence theory of truth follows immediately from these folk theories of language and the Thought As Language metaphor: If words get their meaning by picking out things in the world, then sentences express propositions about the world in itself and those propositions are true just in case the words fit the world. Because of this, analytic philosophy winds up with a truth-conditional theory of meaning: The meaning of a sentence is understood in terms of the conditions under which it is true. As a consequence, all meaning is literal, objective, and disembodied.

To sum up, mainstream analytic philosophy puts the Thought As Language metaphor together with the above folk theories to yield the following tenets:

A1. To analyze language is to analyze thought.
A2. Linguistic meaning is mind-independent, objective, and publicly accessible.
A3. The meaning of a linguistic expression is given by what it can correspond to in the world.
A4. The correspondence theory of truth: A sentence is true if the words fit the state of affairs in the world.

A5. All meaning is literal.
A6. Meaning is disembodied.

In addition, analytic philosophy inherits two tenets from Descartes:

A7. We can, just by thinking about our own ideas and the operations of our own minds, with care and rigor, come to understand the mind accurately and with absolute certainty.
A8. Since philosophical reflection is sufficient, no empirical study of language or thought is necessary. Only training in philosophical analysis via self-reflection is sufficient to answer philosophical questions, especially questions about the nature of meaning and truth. No empirical study is necessary; nor could it add anything.

Formalist Philosophy

The analytic tradition grew out of Frege's and Bertrand Russell's concern with the foundations of mathematics. In attempting to provide formal foundations for mathematics, they developed mathematical logic. They took logic and set theory to be the basis of all mathematics.

As a result, mathematical logic came to have a place at the center of analytic philosophy, and various rich philosophical traditions developed within the framework of a formalist philosophy, not just those of Frege and Russell, but those of Carnap and the Vienna Circle, Quine and the post-Quineans (e.g., Goodman, Davidson, and Putnam), and the possible-world semanticists like Kripke, Montague, and Lewis.

All of them shared not only the central tenets of analytic philosophy, but also additional views as well that defined the more specific enterprise of what we will call "formalist philosophy." The considerable differences among these thinkers are all set within a common heritage.

Formalist philosophy is a version of analytic philosophy distinguished by its acceptance of certain entailments of what we called in Chapter 12 the Thought As Mathematical Calculation metaphor. Formalist philosophy is founded on the notion that Thought Is Language, but it construes that language as the formal language of mathematics. Via these metaphors, all aspects of thought are understood in terms of mathematical symbolization, according to the following entailments of the Thought As Mathematical Calculation metaphor. Those entailments are:

F1. Just as in mathematics numbers can be accurately represented by sequences of written symbols, so concepts are seen as adequately represented by sequences of written symbols.
F2. Just as mathematical calculation is mechanical (i.e., algorithmic), so is reasoning.
F3. Just as there are systematic principles of mathematical calculation that work step-by-step, so there are systematic principles of reason that work step-by-step.
F4. Just as numbers and the principles of mathematical calculation are universal, so concepts and reason are universal.

These basic tenets of formalist philosophy follow from the Thought As Mathematical Calculation metaphor. The important details that became the heart of the formalist enterprise, as we shall see, came out of the development of mathematical logic. That development required three additional technical metaphors that are not versions of ordinary everyday metaphors. We will call them the Formal Language metaphor, the World As Set-Theoretical Structure metaphor, and the Formal Semantics metaphor.

The Formal Language Metaphor

The use of symbols in mathematics is made to fit with the use of symbols in language via what we have called the Formal Language metaphor (Chapter 12), in which certain systems of symbols are conceptualized in terms of written languages:

THE FORMAL LANGUAGE METAPHOR

Written Signs Of A Natural Language	$\rightarrow$	Abstract Formal Symbols
A Natural Language	$\rightarrow$	A Formal "Language"
Linguistic Elements	$\rightarrow$	Individual Symbols
Sentences	$\rightarrow$	Well-Formed Symbol Sequences
Syntax	$\rightarrow$	Principles For Combining Formal Symbols Or Transforming Sequences Into Other Sequences

The Formal Language metaphor allows for logic (the study of reasoning) to be cast technically in terms that are both mathematical and linguistic, as in tenets F5 and F6:

F5.   A formal "language" is a system of symbols in which individual symbols are conceptualized as individual linguistic elements and well-formed symbol sequences as sentences. Principles for combining symbols or transforming symbol sequences into other symbol sequences constitute the "syntax" of the formal "language."

F6.   A system of formal logic consists of a formal language, a set of axioms, which are sequences of symbols in that language taken as being true, and syntactic rules of transformation, which transform sentences taken as true into other sentences taken as true.

Set-Theoretical Models

In a formal language, the symbols are, in themselves, meaningless. A theory of meaning and truth along Fregean lines was developed. It assumed a particular metaphysics defining the notion of a world-state: The world at any moment consists of entities, with properties and relations holding between them. It further assumed that models of the world could be constructed using set theory. Each model contained a universe of entities, sets of entities, and sets of n-tuples of entities. The world was then conceptualized metaphorically as being a set-theoretical structure, according to the following metaphor:

THE WORLD IS A SET-THEORETICAL STRUCTURE

Abstract Mathematical		
Entities	$\rightarrow$	Real-World Entities
Sets	$\rightarrow$	One-Place Predicates
N-tuples	$\rightarrow$	N-Place Relations
Set Membership	$\rightarrow$	One-Place Predication
Membership In Set Of N-Tuples	$\rightarrow$	N-Place Predication

Note that conceptualizing predication in terms of set membership is a version of Aristotle's metaphor that Predication Is Category Membership: "That one term should be included in another, as in a whole, is the same as for the other to be predicated of all of the first" (C2, Aristotle, *Prior Analytics* 24b).

The Formal Semantics Metaphor

Once the world is conceptualized as a set-theoretical structure, the Fregean view of meaning can also be characterized using set theory via what we will call the Formal Semantics Metaphor. The "meaning" of an expression in a formal language is conceptualized as a relation between that expression and an element of a set-theoretical model. Truth in a world-state is conceptualized as satisfaction in a model of that world-state, where "satisfaction" is technically defined in terms of set membership.

THE FORMAL SEMANTICS METAPHOR

A Relation Between A Symbol And An Element Of A Model Of A World State	→	Meaning Of The Symbol
Set Membership In A Model	→	Truth Of A Predication

This apparatus, taken together, gives rise to a number of additional tenets of formal philosophy:

F7. The symbols of a formal language, in themselves, are meaningless. A formal language needs to be interpreted to become meaningful.

F8. A state of the world consists of a set of entities, properties, and relations between those entities.

F9. States of the world can be modeled using set-theoretical structures.

F10. Meaning is a relationship between symbols of a formal language and entities in a set-theoretical model.

F11. Truth is a correspondence between a symbol sequence indicating a predication in a formal language and a membership relation in a set-theoretical model.

During the mid-twentieth century, a mathematical theory of formal languages was developed by Emil Post that has been shown to be equivalent to proof theory in mathematical logic and to the theory of Turing machines (or recursive functions), which is taken as characterizing an algorithm. During this period, Alonzo Church formulated a claim that came to be accepted in formal philosophy:

F12. Church's Thesis: All functions we intuitively regard as computable are technically computable by a Turing machine—or an equivalent system, that is, a system of formal logic or a Post production system (C2, Kleene 1967, 240–241).

The point of Church's thesis was to claim that the previous intuitive notion of a precise procedure or formulation in mathematics could be formalized using the apparatus of Turing machines, or proof theory, or the theory of formal languages.

Church's thesis was about mathematics, but using the Thought As Mathematical Calculation metaphor, formal analytic philosophers metaphorically projected Church's thesis from mathematics to thought in general. A metaphorical version of Church's thesis became commonplace in formalized analytic philosophy during the 1950s. It can be stated as follows:

THE METAPHORICAL VERSION OF CHURCH'S THESIS

Any precisely formulated theory or collection of ideas can be stated in a formal system, that is, either a Turing machine, a system of formal logic, or a Post production system.

The metaphorical version of Church's thesis came to play the role of Russell's earlier claim that rigorous philosophy, mathematics, and science could only be carried on in mathematical logic. This led to a general view of scientific theories as systems of axioms in a mathematical logic. Recall that such symbols are meaningless in themselves, and therefore the axioms are meaningless as well. As with other symbolic expressions, such axioms are to be "given meaning," not through human understanding but by being mapped onto set-theoretical models that are assumed to be able to fit the world as it is.

F13. Scientific (or philosophical) theories are systems of axioms in a mathematical logic, where the symbols are meaningless and to be interpreted in terms of set-theoretical models.

Given this view, as Quine pointed out, the ontological claims of a philosophical or scientific theory are given by the entities in the set-theoretical models of world-states that the variables in such axioms can take as values. For example, given an axiom of the form "$(x) F(x) \supset G(x)$" [for all entities x, if predicate F is true of x, the G is true of x], the value of x will range over entities in the universe of the model. Those entities constitute the ontology of the theory, that is, they are the entities that the philosophical or scientific theory takes as "really existing." As Quine put it,

F14. To be is to be the value of a variable.

These are the general tenets that characterize formal philosophy. All of the particular versions start from here and elaborate further.

Ordinary Language Philosophy

Ordinary language philosophers rebelled against the claim that the analytic program could only be fulfilled within the framework of formal philosophy. Formal philosophers believed that ordinary language was too vague, ambiguous, and imprecise for doing precise conceptual analysis and that ordinary language had to first be translated into a formal language to which mathematical logic could apply. Russell even believed that ordinary language was too sloppy for doing philosophy, science, or mathematics and needed always to be translated into a formal language that was logically clear and precise.

Ordinary language philosophers argued instead that all the important conceptual distinctions that philosophy required could be made on the basis of ordinary language and that, moreover, mathematical logic was inadequate for characterizing the full richness of everyday language. Philosophers such as Strawson, Austin, and the later Wittgenstein sought to realize a nonmathematical general program of analytic philosophy by carefully attending to the nuances of ordinary language and its use in context.

Indeed, they argued against the adequacy of formal philosophy. Strawson argued that natural language contained presuppositions. These cannot be handled in two-valued classical formal logic of the sort used by Russell in his theory of descriptions, where every sentence is either absolutely true or absolutely false. Austin argued that speech acts like ordering and promising could not be either true or false and hence could not be dealt with by formal logic. To handle such cases, he proposed a theory of speech acts as pragmatic functions on basic propositions.

Paul Grice attempted to reconcile formal philosophy with ordinary language philosophy using a theory of what he called "conversational implicature." Gricean implicatures (informal inferences in conversational contexts) were intended to preserve Russell's version of the formal philosophical worldview for *semantics* while accounting for the ordinary language view of language *use*.

In all of these cases, there have been replies proposing how formal philosophy might be adjusted to deal with the Gricean phenomena. There have been presuppositional logics (e.g., C2, Van Fraassen 1968; A9, Gazdar 1979). Lakoff (C2, 1971) and Searle and Vanderveken (C2, 1985) have suggested how one might extend satisfaction in a model to handle speech-act phenomena. Gazdar even suggested how some Gricean implicatures might be handled in a formal logical system.

The later Wittgenstein argued, against formal logic, that categories such as *game* could not be characterized using necessary and sufficient conditions but were, instead, defined by *family resemblances*. His notion of a *language game* preserves the common Thought As Language metaphor, but challenges the notion that meaning can be given in terms of the objective world. A language game is a self-contained system of thought and action and is based on a "form of life" that can only be characterized in terms of what people do and think, how they live—and not in objective mind- and body-free terms.

W. B. Gallie (C2, 1956) argued that many concepts, such as art, democracy, and freedom, are "essentially contested"; that is, they necessarily mean different things to different people and so cannot be given definitions once and for all, as required by formal philosophy.

These ordinary language philosophers all made deep contributions to our understanding of language, and those contributions have found their way into empirical research in cognitive science and cognitive linguistics. However, ordinary language philosophy has for the most part still been handcuffed by certain founding assumptions of analytic philosophy in general as discussed above. Searle, for example, has given an extensive analysis of the structure of speech acts that attempts to preserve most of the major assumptions, A1 through A8 above, shared by the discipline. These limitations have prevented Searle and other contemporary analytic philosophers from gaining insight into the wide variety of syntactic, semantic, and pragmatic phenomena discussed in second-generation cognitive science and in cognitive linguistics.

Quine's Philosophy

One of the most popular doctrines in analytic philosophy is meaning holism, which arose from the philosophy of Willard Van Orman Quine. The doctrine arises within formalist philosophy, of which Quine, as a mathematical logician, is an advocate. To make sense of meaning holism, one must place it in the context of formal philosophy and take for granted tenets F1 through F14 above.

Quine has always been centrally concerned with ontological commitment. In the spirit of Occam's Razor, Quine has wanted to keep the "ontological furniture of the universe" to a minimum. Occam's Razor—keeping postulated entities to a minimum—is at the center of Quine's philosophy. Having physical objects in one's ontology, Quine reasonably argued, is unavoidable. Because scientific theories require the existence of natural kinds, that is, genera and species of plants and animals (e.g., a tiger), as well as types of substances (e.g., gold), kinds are also accepted by Quine as really existing entities. But he rebels against admitting into a philosophically respectable ontology anything that could be eliminated. That includes for him such things as essences (e.g., the properties that are necessary and sufficient for something to be a table) and arbitrary sets (e.g., the set consisting of your maternal grandmother, the number four, and Abraham Lincoln).

Like other formalist analytic philosophers, Quine takes the world as being made up, objectively, of entities, including the natural kinds. Quine saw especially clearly the relationship in formal analytic philosophy between Universal Reason (formal logic) and metaphysics. His motto, "To be is to be the value of a variable," states that relationship explicitly. Quine saw that, if you accept the tenets of formal analytic philosophy with formal logic as Universal Reason, then the choice of a logic *is* the choice of a metaphysics. The entities that you allow the variables in your logic to vary over *are* the entities whose existence you are committed to. For example, in a formula like "For all x, f(x)," the entities over which x can vary are the entities whose existence you are taking for granted. In other words, the choice of a logic is the choice of a form of Universal Reason, which is in turn a commitment to a particular structure of reality.

Quine then observes that the proper logic for philosophy must be first-order logic, not second-order logic. First-order logic contains variables over entities. Second-order logic contains variables over properties and relations. To see the difference, compare these sentences:

Rich people are selfish.
Fido has remarkable properties.

In formal logic, these sentences would be formalized as:

(For All x) If Person (x) and Rich (x), then Selfish (x).
For all x, if x is a person and x is rich, then x is selfish.

(For Some f) Property (f) and f (Fido) and Remarkable (f).
There are properties f that Fido has and that are remarkable.

The first sentence is "first-order," since *x* varies over entities, in particular, human beings. But the second sentence is "second-order," since *f* varies not over entities but over properties of entities. For example, *f* might vary over "rich" and "selfish."

Quine observed that to accept the use of second-order logic was to accept properties as being real things. But, Quine reasoned, if you are already committed to the existence of objects and the properties are inherent in the objects, it is an extra ontological commitment to consider the abstract properties as real things separate from the objects they are properties of. Moreover, since essences are conjunctions of properties, a commitment to the existence of properties as things is a commitment to the existence of essences as things. From this, we get two more of Quine's major philosophical commitments: first, that the correct logic is a first-order logic; and second, his nominalism, the claim that all that exists are the objects in the world (the things named by nouns). For Quine, this is simply a consequence of a commitment to Occam's Razor and to formalist philosophy.

Quine's imposition of Occam's Razor created a bifurcation in formalist philosophy. Quine wanted to keep his logic "extensionalist," that is, to limit it to first-order logic where the variables could take on as values only entities and natural kinds, but not properties or other nonobject entities. To grant ontological status to properties and nonobject entities, Quine thought, would be unscientific and would revert to a previous metaphysical tradition that postulated abstract entities immune to scientific verification or falsification, for example, essences.

Advocates of "intentional" logic had no such ontological scruples. They had no problem with the existence of Fregean senses, essences, or properties. They readily embraced second-order logic, which allowed them to quantify over such abstract entities. Higher-order logics also permitted the development of possible-world semantics, in the traditions of Montague and Kripke, in which there could exist possible as well as actual individuals. They saw intentional logics as more in the spirit of Frege. Some intentional logicians even developed Meinongian logics, in which impossible objects (like square circles) could exist in models. Impossible worlds and impossible objects were, for Quine, even more ontologically otiose than possible worlds: They were not even logically possible, much less real things.

Meaning Holism

The Quinean tradition thus requires limiting the logical "language" used by philosophers to first-order logic. The central portions of Quine's philosophy follow from an important technical result in first-order logic—the Löwenheim-Skolem theorem. Quine, in *Methods of Logic* (C2, 1959, 259), states the theorem as follows:

> *If a class of quantificational schemata is consistent, all its members come out true under some interpretation in the universe of positive integers.*

At the time it was proved, this theorem was shocking, because it contradicted the following widely held assumption. Suppose a mathematical logician had some ideas he wanted to express in the form of an axiom system using first-order logic, as Peano had in his axiomatization of arithmetic. It was assumed that the very structure of the axioms—their logical form—would limit the possible meaning of symbols in the axioms.

For example, suppose a logician tried to state, in axioms of first-order predicate calculus, a characterization of the real numbers. One might have expected that these "truths about the real numbers" could be satisfied only by the real numbers. Since Cantor was assumed to have proven that there are more real numbers than natural numbers (positive integers), it would follow that statements of truths about the real numbers, as opposed to the positive integers, could not be satisfied by the positive integers because there wouldn't be enough positive integers.

The shocking aspect of the theorem, as Quine says, is that

> the truths about real numbers can by a reinterpretation be carried over into truths about positive integers. This consequence has been viewed as paradoxical, in the light of Cantor's proof that the real numbers cannot be exhaustively correlated with integers. But the air of paradox is dispelled by this reflection: whatever disparities between real numbers and integers may be guaranteed in those original truths about real numbers, the guarantees are themselves revised in the reinterpretation.

In a word and in general, the force of the Löwenheim-Skolem theorem is that the narrowly logical structure of a theory—the structure reflected in quantification and truth functions, in abstraction from any special predicates—is insufficient to distinguish its objects from the positive integers. (C2, Quine 1959, 259–260)

So ends *Methods of Logic.*

From Quine's point of view, this mathematical theorem had far-reaching implications for formalist philosophy of language: The symbols of a formal language, in themselves, are meaningless. Technically, they can be assigned meanings only by linking them up with abstract formal entities in set-theoretical models. The symbols in the theorems of the formal system are countable, like the positive integers. It should not be surprising that one can always find a way to link up a countable number of abstract meaningless symbols in the formal language with a countable number of abstract meaningless entities in the models. For this reason, the axioms of a formal logic, being nothing but meaningless symbols, cannot be assumed to have any meaning at all until *all* the symbols are interpreted. This means that you can never have a partial axiomatic theory with a fixed interpretation.

Note that fixing the interpretation of some finite number of symbols is not enough to guarantee that even those symbols will have the interpretation you intend. The reason is that the interpretation of the remaining symbols can always be rigged so as to interact with, and thereby affect, the interpretation of the first bunch of symbols, since "interpretation" and "meaning" here mean nothing more than associating symbols via some mathematical function to entities in a model. The moral Quine draws is that the arbitrary symbols of a formal language can only be meaningfully interpreted in an ultimately fixed way *as a whole* all at once, not one or a number at a time. This is called *meaning holism,* another of the major pillars of Quine's philosophy.

One might think that such a result might be of interest to mathematical logicians but would hardly be earthshaking for philosophy, science, and other human enterprises. But formalist philosophy assumptions F1 through F14—which characterize a disembodied view of language, meaning, and thought—fatefully link all human enterprises using thought or language to mathematical logic. These assumptions, which define *human* language and all rational thought in terms of mathematical logic, make the consequences of the Löwenheim-Skolem theorem applicable to all human language and thought and therefore to virtually every human enterprise, beginning with philosophy.

The consequences are startling. They contradict much of previous analytic philosophy. First, they go against Frege's idea that a sense is a fixed interpretation in all situations in all languages for all time. In other words, they contradict Frege's central idea that words, via their objective senses, can pick out unique references to objects in the world. They contradict the idea that logic mirrors nature and that a correct logic is a correct guide to the structure of the world. They contradict Wittgenstein's early view that formal logic presents a picture of reality. But this is only the beginning. The consequences are even greater.

Consequence 1

Ontological Relativity: Philosophical ontologies are relativized to the way that reference is fixed **for an entire language.**

If "To be is to be the value of a variable" and if the values of variables cannot be fixed until the interpretation of the whole formal language is fixed, then a philosopher's ontology (what he counts as existing) is relative to the interpretation of his or her entire language (when the variables of his or her language are all assigned referents). This is startling to philosophers for its metaphysical implications: Words and even propositions cannot map one-by-one uniquely onto entities or states of affairs in the world.

Consequently, Quine concluded that there is no objectively correct way, determined by the world in itself, to uniquely specify referents for the symbols of a logical "language." For any set of symbols in a formal language purporting to be "about" the world, there can be no one objectively right way to "carve up the world" into entities corresponding to those symbols.

Consequence 2

There is no analytic-synthetic distinction.

An analytic sentence in a formal language is supposed to be true not by virtue of any fact about the world, but only by virtue of the meanings of the words in the sentence. Classical examples are cases like "Bachelors are unmarried" or "A triangle is a three-sided figure." Given our everyday understandings of the meanings of the words in these sentences, one might think that they might be true given the meanings of these words alone.

But suppose formal philosophy is taken for granted, with tenets F1 through F14 above assumed as part of an overall philosophical worldview. Then these sentences (appropriately "regimented" or placed in "logical form") are just meaningless symbols, for example, "(x) [B(x) ⊃ -M(x)]" and "(x) [T(x) ⊃ F(x)]." The predicates B, M, T, and F do not have any meanings in themselves, though we might want them to mean what we mean by *bachelor, married, triangle,* and *three-sided figure*. These formulas do not have any meaning until a meaning is assigned—and not just to them, but to *the entire formal language!* Thus, Quine points out, you cannot just fix the model-theoretic interpretations of B, M, T, and F alone and be sure that these formulas will be true by virtue of those interpretations alone. You have to fix the interpretations of every expression in the language in order to be sure. But then, it is not the meanings (model-theoretic interpretations) of just those terms that make the sentences true. It is the meanings of all the expressions of the formal language. Thus, no sentences can be true just by virtue of the meanings of the terms in those sentences alone. Hence, there are no analytic sentences.

This too might seem anything but earth-shattering. Why should anyone really care about philosophers' curiosities like analytic sentences? We can begin to see why people might care, however, when we look at consequence 3.

Consequence 3

No part of a scientific theory can be confirmed or disconfirmed; only the theory as a whole can be confirmed or disconfirmed.

In other words, individual theoretical sentences (or groups of them) can be confirmed or disconfirmed only by their role in the entire theory in which they are embedded. This is commonly called the Quine-Duhem thesis (Pierre Duhem, 1861–1916, was a French philosopher of science who held this view).

Within formalist philosophy, a scientific theory is taken to be a set of axioms in predicate calculus, together with the logical consequences of those axioms. If you are a formalist philosopher and assume the tenets of formalist philosophy, you will assume that a scientific theory is a set of axioms in a formal language. The Quine-Duhem thesis is then a consequence of the Löwenheim-Skolem theorem and meaning holism.

The reason is this: A part of a scientific theory is a finite subset of axioms. To confirm them is to discover that they are true. That presupposes that you know what they mean. But their meanings cannot be fixed; their interpretations could be changed by the interpretations of the additional axioms. But if the interpretations of these axioms is not fixed, then you don't know what the axioms mean and you certainly cannot be sure that they are true. Therefore, a part of a theory cannot be confirmed.

Of course, the same holds for falsification. If you cannot fix the meaning of only a subpart of a theory, you can't provide a counterexample to that subpart.

Auxiliary hypotheses (additional axioms) could always be added to change the meanings of the symbols in the earlier axioms and hence to avoid any putative counterexample. Only the complete theory as a whole can have counterexamples. In short, there can be no evidence for or against parts of a theory in isolation from the theory as a whole.

It must be borne in mind that this radical view, that a subpart of a theory cannot be falsified, depends upon assuming formalist philosophy and, with it, the axiomatic character of scientific theories. That is, it requires a disembodied view of language and thought. If thought is embodied and meaning is fixed through embodiment, then tenets F1 through F14 of formalist philosophy are false, the Löwenheim-Skolem theorem is irrelevant, and meaning holism cannot apply, since its presuppositions are false. Since second-generation cognitive science is in conflict with formalist philosophy, it is necessarily in conflict with meaning holism and the Quine-Duhem thesis.

Consequence 4

Translation is indeterminate.

Suppose again that one accepts formalist philosophy and meaning holism. What is to be meant by "translation"? Before one can have a "translation," one must have at least two meaningful "languages." According to formalist philosophy, each "language" is a formal language, consisting, in itself, of meaningless symbols. The symbols of the languages are different. In order to be made "meaningful," each formal "language" must be given an "interpretation," that is, an assignment of referents in a set-theoretical model of the world to the symbols of the language. Each formal language—each collection of meaningless symbols—has a different assignment of reference, that is, a different mathematical function pairing meaningless symbols to meaningless elements of a set-theoretical model.

According to formalist philosophy, a correct "translation" from formal language A to formal language B would be an assignment of each symbol of language A to a symbol of language B, so that the interpretation (that is, the reference assignment) of each symbol and each sentence of language A will have the same interpretation (same reference assignment) as the corresponding symbols and sentences of language B. Thus, all the true sentences of language A would be translated into corresponding true sentences of language B.

Could such a "translation" ever be determinate? That is, could you ever know if such a translation were correct? Quine points out that the answer is no. If you tried to do such a translation a little bit at a time, then you would never know if your translation so far was right, because a future interpretation (reference assignment) could always change a previous interpretation. The translation would have to be "all at once" in order to be sure that the languages "carved up the world" in the same way.

But even if the translation were all at once, you could never know for sure if you were right. To know for sure, there would have to be a determinate checking procedure, that is, an algorithm that could check if you got it right. But any such algorithm must run in time; it must have a starting point and must check bit by bit. However, at no point could the algorithm ever be sure that future interpretations would not change interpretations already checked. Since the number of sentences of a formal language are denumerably infinite, the algorithm would never finish checking in any finite time. Therefore, no algorithm could ever check that the "translation" was correct, and so "translation" is indeterminate.

This Quinean view of "translation" has been seen as applying to Kuhn's view that scientific theories are "incommensurable." If the language of one scientific theory cannot be "translated" into that of another, then theories must be incommensurable and cannot be seen as making comparable claims.

Quine's Technical Use of Words

Reading the work of Quine and those who have followed in his footsteps is a strange experience. Assumptions F1 through F14 of formalist philosophy are taken for granted as being true of actual human language and actual human thought. Words like *language, meaning, interpretation,* and *translation*—which are technical terms in formalist philosophy—are treated as though they were the normal English words, as though they could apply in their normal senses to a natural language such as English. Similarly, the set-theoretical models of world-states are treated as if they were the world itself, with the abstract mathematical entities of the models characterizing an objective reality. Quine's conclusions are commonly taken not as being merely about the formal systems of mathematical logicians, but about our ordinary language, thought, reality, and truth.

Quine and others are assuming the truth of many of the entailments of the metaphors and folk theories on which analytic and formalist philosophy are based. It is that metaphorical understanding of thought and language that makes thought and language appear disembodied and also makes much of formalist philosophy appear unproblematic. But if language and thought are embodied and if thought is metaphorical, then formalist philosophy, the entire structure on which meaning holism is based, goes up in smoke. It does not apply to real human thought and language.

Quine on Science

There is an important bifurcation in Quine's philosophy. The formalist philosophy and meaning holism we have described are widely shared with others in the formalist tradition. Given Quine's meaning holism, ontological relativity, and the Quine-Duhem thesis, one might think Quine would consider the development of an objective, sophisticated, believable science impossible. After all, meaning holism not only challenges the possibility of verificationism in the tradition of logical positivism but also appears to contradict Popper's falsificationism.

How does Quine get around his own meaning holism when it comes to science? He has to make a set of assumptions that go very much against the direction of his own meaning holism. First, he has to admit some sets into the "furniture of the universe." Natural kinds, for Quine, are sets that exist objectively in the world. "Kinds can be seen as sets, determined by their members. It is just that not all sets are kinds" (C2, Quine 1969, 118). He gives the example of fish (minus dolphins and whales) as being a natural kind.

Part of the job of science is to get the natural kinds right, to fit the variables in a formal scientific language to the natural kinds in the right way. The problem is not just the natural science at hand, for example, biology. The problem is how we can be sure that we've done the natural science right. Quine's second major assumption is that behaviorist psychology provides the appropriate tools for epistemology. This is what Quine calls "naturalized epistemology." "Epistemology, or something like it, simply falls into place as a chapter of psychology. It studies a natural phenomenon, viz., a physical human subject. . . . We can now make free use of empirical psychology" (C2, Quine 1969, 82–83). Behaviorist psychology is admissible because it is external, not internal—a matter of behavior, not mind.

To make this possible, Quine adds three more assumptions to these two. His third assumption is that the old observation sentences of logical positivism can be revised appropriately to eliminate the contribution of internal consciousness to "sense-data." "What to count as observation now can be settled in terms of the stimulation of sensory receptors, let consciousness fall where it may" (C2,

Quine 1969, 84). He sees the stimulation of sensory receptors as objective, external, independent of any interpretation, and hence not subject to the problems of the reliability of sense-data. "The observation sentence is the cornerstone of semantics . . . it is where meaning is firmest. Sentences higher up in theories have no empirical consequences they can call their own. . . . The observation sentence, situated at the sensory periphery of the body scientific, is the minimal verifiable aggregate; it has an empirical content all its own and wears it on its sleeve" (C2, Quine 1969, 89). As many critics have observed, this claim to the grounding function of observation sentences goes directly against his meaning holism!

Quine's fourth major assumption is actually a cluster of closely related assumptions all built upon the alleged existence of objective similarities between things in the world. He assumes that there is an objective similarity in the world between objects of the same kind, that human beings have innate abilities to notice similarities more or less accurately, that behaviorist psychology can discover these similarities in an objective way, and that scientific method can ultimately become sophisticated enough to figure out which of the perceived similarities is real.

> A man's judgments of similarity do and should depend on his theory, on his beliefs; but similarity itself, what the man's judgments purport to be judgments of, purports to be an objective relation in the world. It belongs in the subject matter not of our theory of theorizing about the world, but of our theory of the world itself. Such would be the acceptable and reputable sort of similarity concept. . . . If I say that there is an innate standard of similarity, I am making a condensed statement that can be interpreted, and truly interpreted, in behavioral terms. . . . Between an innate similarity notion or spacing of qualities and a scientifically sophisticated one, there are all gradations. Sciences, after all, differ from common sense only in degree of methodological sophistication. (C2, Quine 1969, 135, 123, 129)

Quine's fifth assumption—that science is correctly done a bit at a time—flies directly in the face of the Quine-Duhem thesis. Quine-Duhem says that theories cannot be confirmed a bit at a time, since reference cannot be fixed a bit at a time. But this, Quine assumes, can be overcome by behavioral psychology.

> [Objective similarity] does get defined in bits: bits suited to special branches of science. In this way, on many limited fronts, man continues to rise from savagery, sloughing off the muddy old notion of kind or similarity piecemeal, a vestige here and a vestige there. (C2, Quine 1969, 135)

These behaviorist assumptions about science are idiosyncratic to Quine. They are not widely accepted, while his holism, which comes directly from formalist philosophy and mathematical logic, has become a standard view in the formalist community.

Quine's program of "naturalized epistemology," it must be recalled, presupposes formalist philosophy and meaning holism, as well as these views on science. Quine's project in naturalized epistemology is to use behaviorist psychology together with formalist philosophy as a way to get at the true natural kinds that he attributes to the world. As a version of formalist philosophy and behaviorism, naturalized epistemology is as far away as one could imagine from the kind of empirically responsible philosophy we envision.

Rorty's Use of Meaning Holism

One of those who have accepted meaning holism but rejected Quine's view of science is Richard Rorty. Using a Quinean view of language, meaning holism, and ontological relativity, Rorty altogether gives up on any attempt to show that one can have certain foundational knowledge of the world. In doing so, he moves to a total relativity of meaning. For him there is no objective way to link the symbols of one's "language" to the entities of the objective world. Without a Quinean "naturalized epistemology" to ground meaning in some scientifically objective way, meaning becomes arbitrary, fixed by the accidents of history and always subject to reinterpretation at a later stage in history.

Rorty's move depends crucially on aspects of formal philosophy and on meaning holism. Accordingly, he must see mind and language as disembodied. Without an embodied notion of meaning that can allow meaning to be determined through bodily experience, his only choice is to completely accept relativity, utter historical contingency, and a coherence theory of truth.

Rorty has rejected Quine's program of "naturalizing epistemology," which, as Quine realized, would be needed to link a formal "language" to natural kinds in the world. The reason for the rejection is that behaviorist psychology, or any other empirical psychology, would also be subject to meaning holism. That is, it is impossible to fix a correct behaviorist psychology or any other psychology. Psychology therefore cannot do the job of repairing meaning holism for science. Rorty thus sees clearly that Quine's holism is incompatible with Quine's commitment to natural kinds and to a naturalized epistemology.

Rorty has realized that, without such a means of fixing meaning, Quinean formalist philosophy leads to an internal contradiction: It presupposes a correspondence theory of truth but, due to meaning holism, it leads to a coherence theory of truth. In the absence of an independent way of fixing meaning by actually studying the world and empirical human psychology, the classical task of epistemology—certain knowledge—dissolves. Rorty accepts this as an advance.

However one feels about Rorty's relativistic conclusion about knowledge (we reject it), Rorty has put his finger on something correct. The basis of all analytic and formalist philosophy, from Frege on, has been that human psychology is irrelevant to meaning and truth conditions. The very act of bringing into philosophy the empirical study of the human mind and placing it above a priori philosophy changes everything.

In keeping meaning holism, Rorty keeps the fundamental assumptions of analytic philosophy required for meaning holism to make sense. Thus he must reject the very possibility of any privileged role for the empirical study of the mind. Philosophy must remain a priori and above psychology for him.

Our view is, of course, in radical contrast to both Rorty's and Quine's, since it rejects the very foundations upon which the notion of meaning holism was founded. The embodiment of meaning, as empirically required by second-generation cognitive science, locates meaning in the body and in the unconscious conceptual system. This is inconsistent with the entire foundation of analytic philosophy, without which meaning holism makes no sense. Second-generation cognitive science directly contradicts Quine's rejection of anything like the cognitive unconscious and embodied thought, that is, any embodied characterization of meaning and inference.

What's Wrong with Analytic and Formalist Philosophy?

The results of second-generation cognitive science stand squarely opposed to the analytic and formalist philosophical traditions on precisely those issues that are the central themes of this book: (1) the embodiment of concepts and of mind in general; (2) the cognitive unconscious; (3) metaphorical thought; and (4) the dependence of philosophy on the empirical study of mind and language.

The contradictions could not be clearer. Take embodiment. The public nature of linguistic meaning and those aspects of meaning that are universal across cultures arise from the commonalities of our bodies and our bodily and social experience in the world. Frege was wrong. Psychology isn't purely subjective. From the commonalities of our visual systems and motor systems, universal features of spatial relations (image schemas) arise. From our common capacities for gestalt perception and motor programs, basic-level concepts arise. From the common color cones in our retinas and the commonalities of our neural architecture for color vision, the commonalities of color concepts arise. Our common capacity for metaphorical thought arises from the neural projections from the sensory and motor parts of our brain to higher cortical regions responsible for abstract thought. Whatever universals of metaphor there are arise because our experience in the world regularly makes certain conceptual domains coactive in our brains, allowing for the establishment of connections between them. The commonalities of our bodies, brains, minds, and experience makes much (though not all) of meaning public.

There is no Fregean abstract realm of disembodied senses and no mystical relations between such supposed senses and objects and categories in a supposed mind-independent world. Our brains and minds do not operate using abstract formal symbols that are given meaning by correlations to an allegedly mind-independent world that comes with categories and essences built in. The body and brain are where meanings arise in and through our interactions with the environment and other people.

It is not true that all thought is conscious and that we can know it completely through a priori philosophical reflection. Most of our thought is unconscious, and empirical investigation is necessary if we are ever to understand its nature.

Finally, the existence of metaphorical concepts and metaphorical thought does not gibe with the analytic and formalist worldviews, in which all concepts must be literal, defined by a purely objective relation between either Fregean senses or abstract symbols and a mind-independent world. Metaphors are products of body, brain, mind, and experience. They are pervasive in our everyday thought and in philosophy itself. They could only get their meaning through embodied experience.

If any of the metaphor analyses in Chapters 10 through 20 are correct, then the central theses of analytic and formalist philosophy cannot be, since analytic philosophy must necessarily deny the existence of conceptual metaphor.

Language and Poststructuralist Philosophy

Poststructuralist philosophy rests to a large extent on four claims about the nature of language:

1. The complete arbitrariness of the sign; that is, the utter arbitrariness of the pairing between signifiers (signs) and signifieds (concepts)
2. The locus of meaning in systems of binary oppositions among free-floating signifiers (*différance*)
3. The purely historical contingency of meaning
4. The strong relativity of concepts

Cognitive linguistics and other branches of cognitive science have shown all of these views about the nature of language to be empirically incorrect. Let us consider each of these claims in turn.

The Nonarbitrariness of the Sign

The doctrine of the arbitrariness of the sign rests on the false dichotomy between predictability and arbitrariness: Any form-meaning pairing that isn't predictable by general rule must be arbitrary. Most of language, however, is neither completely arbitrary nor completely predictable, but rather "motivated" to some degree. Let us take a very simple example of motivation—derivational morphology. The word *refrigerator* consists of the morphemes *re-frig-er-at-or*. Each morpheme has a meaning. If the meaning of the whole were simply a predictable function of the meaning of the parts, "refrigerator" would simply mean *something that makes things cold again*. That isn't exactly right; refrigerators are both more and less than that. But the meaning of the word is *not* arbitrary relative to the meaning of the morphemes. The meanings of the morphemes in each word motivate the meaning of the word as a whole. It would have been strange to call a computer a *refrigerator* and a refrigerator a *computer*. In general, most words with derivational morphology are cases in which the meaning of a word is motivated, but not predicted, by the meanings of its parts. The meanings of the parts, for example, *re-* and *frig-*, however, may be arbitrary. But given that arbitrariness, the meaning of "refrigerator" is motivated, not arbitrary. This is the normal situation. There does exist some arbitrariness. But given that, what we mostly find is not full arbitrariness, but motivation.

All conventional metaphorical expressions are cases of motivation. For example, words like *attraction, electricity,* and *magnetism* as used in discussion of a love relationship are not arbitrary. They are motivated by the meanings of the words in the source domain of physical force together with the general conceptual metaphor Love Is A Physical Force. All such metaphorical meanings of words are not arbitrary; they are motivated. They are not predictable, because one cannot predict that any given source-domain word will or will not be metaphorical. But given that a source-domain word does have a metaphorical sense, characterized in terms of a preexisting conceptual metaphor, that sense will not be arbitrary, but rather motivated. In general, central senses of words are arbitrary; noncentral senses are motivated but rarely predictable. Since there are many more noncentral senses than central senses of words, there is more motivation in a language than arbitrariness.

Iconicity is another form of nonarbitariness. For example, take a sentence like "John got out of bed and put on his shoes." This would normally be understood as indicating that John first got out of bed and then put on his shoes, not the reverse. In such sentences, the order of the clauses indicates the order of the actions.

Another iconic principle governs the order of adjectives in English: In a noun phrase, properties that are more inherent to the object designated by the head noun are closer to the head noun. Thus we can say *the beautiful big old red wooden house*, but not *the red wooden beautiful old big house* or *the wooden red old big beautiful house*, and so on. *Wooden* is closest to the head noun since it indicates what the house is made of, which is inherent to the house. *Beautiful* comes first because it is purely subjective. *Old* is neither inherent to the house (it doesn't start out old) nor purely subjective; instead, it is relative to some standard of age. Thus it comes closer to the head noun than *beautiful* and farther from it than *wooden*. *Red* is not completely inherent (you can repaint the house), but it is more inherent than *old*, which depends purely on the time of the utterance relative to the time at which the house was built. As a physical property *big* is less subjective than *beautiful*, but more subjective than *old*, since the standard of what counts as *big* is more subjective than the standard of what counts as *old*. (For a general discussion of iconicity, see A8, Haiman 1980; A3, Taub 1997.)

Différance

Consider the doctrine of *différance*, that the locus of meaning lies in systems of binary oppositions between free-floating signifiers. The idea here is that signs come in pairs *(a, b)*. Each sign is arbitrary, but each pair must be interpreted as opposites. So if *a* is interpreted as *male*, *b* must be interpreted as *female*; if *a* is *happy*, *b* is *unhappy*; and so on. Correspondingly, if *b* is interpreted as *male*, *a* must be interpreted as *female*; if *b* is *happy*, *a* is *unhappy*. A system of concepts is just a system of signs of this sort, a collection of systematic oppositions, but without any fixed meaning. What a sign "means" is the sum of the differences between it and other signs. Meaning is never given directly, but only via such oppositions.

Moreover, there is nothing about the world or people that fixes these interpretations. Given the assumption of the arbitrariness of the sign, each pair *(a, b)* can be interpreted as any opposition at all. Because of this, any given interpretation, say, *a* as *happy* and *b* as *unhappy*, could equally well be given an ironic reading with the interpretations reversed, with *a* as *unhappy* and *b* as *happy*. Ironic readings are thus natural, inherent in the process of interpretation itself.

Relativism

Relativism claims there are no semantic universals. Indeed, it claims that conceptual systems—systems of interpretation that vary from language to language—are incommensurable. Meaning is simply different in each language and culture, because the fitting of the signs of a language (signifiers) to things signified is arbitrary. Translation is therefore impossible. The traditional goals of cultural anthropology are also therefore impossible, since a person from one culture could not possibly understand the conceptual system of another culture. The best you can do is describe your own understanding of what happens in that culture.

Contingency

The historical contingency of meaning is a consequence of meaning incommensurability and the lack of universals. Meaning changes over time. What results are new and different conceptual systems with meanings that depend on historical circumstances. Understanding a previous period of the history of one's own culture is therefore as impossible as understanding another culture. Historians can at best interpret history from their own perspective, their own conceptual system.

Science therefore can have no privileged perspective. Since science must use a language and a system of concepts, the language of science is also an arbitrary imposition of a sign system on the world. There develop, naturally, distinct scientific theories that have different languages and are incommensurable. None is privileged, and there is no reason to accept one and reject another.

Ties to Rorty and Holism

It should be obvious that poststructuralist views bear resemblances to the Rortyan version of Quine's meaning holism. Though Quine, like the rest of analytic philosophy, is not bound by the doctrine of *différance,* the views on the arbitrariness of the sign, relativism, and contingency are parallel. Poststructuralism's arbitrary signs are akin to the arbitrary uninterpreted formal symbols in Quinean analytic philosophy. Also similar are the claims about relativism, the impossibility of translation, and the contingency of meaning.

The doctrine of free-floating signifiers constitutes a disembodied account of meaning, as in the case of Rorty and Quine. If meaning were embodied in our sense, then it would not be totally arbitrary. Moreover, the doctrine of the arbitrariness of the sign, as we have noted above, is at odds with our analysis of conceptual metaphors, which assign nonarbitrary meanings to signs. The poststructuralist theory of meaning is fundamentally at odds with virtually every finding of second-generation cognitive science.

But there is one place where the poststructuralist theory of meaning is confirmed by cognitive science, and it is an important place. One empirical finding of second-generation cognitive science is that meaning does change over time and differ across cultures in significant ways, but by no means totally. Universals and meanings are widespread across cultures, but there is also significant relativism.

In addition, the poststructuralist theory of meaning is at odds with the very existence of cognitive science, or any science. Since any science makes assumptions, those assumptions, it is claimed, invalidate any privileged status for the science. Evidence, it is claimed, has no privileged status. It is evidence only relative to those assumptions. Thus, the poststructuralists regard science as merely one arbitrary narrative among others, with no privileged status.

Convergent Methodologies

This view of science does not take into account either the nature of the assumptions behind or the use of convergent methodologies. It is the use of convergent evidence achieved via different methods that keeps science from being merely an arbitrary narrative. As we have seen, the assumptions of second-generation cognitive science are not assumptions about outcomes, but about method. Moreover, cognitive science uses not one methodology, but many different methodologies, each with different methodological assumptions. For example, evidence from cognitive psychology uses nearly a dozen different methodologies. The study of generalizations over inference patterns uses another methodology, as does the study of generalizations over polysemy, as does the study of generalizations over historical semantic changes. The more distinct methodologies with different assumptions there are that have to converge, the less likely it is that assumptions will predetermine the results. As convergent methodologies for accumulating evidence pile up, the probability that the "evidence" merely reflects assumptions gets vanishingly small.

An Embodied Philosophy of Language

The philosophy of language got off to a bad start with Frege and with the poststructuralist movements. The entire programs of both analytic and poststructuralist philosophy left out, and are fundamentally inconsistent with, everything that second-generation cognitive science has discovered about the mind, meaning, and language. Frege's overly narrow view of psychology led him to believe that the psychological was merely subjective and idiosyncratic and could never lead to anything public and universal. Frege's rabid antipsychologistic bent led him to deny any role in meaning for any aspect of the body or imagination. Frege missed the possibility that the body could ground meaning in an intersubjective way and that imaginative mechanisms like metaphor could preserve inference and thus be central to reason.

Where Frege sought absolute, timeless universals of meaning, the poststructuralists correctly perceived that conceptual systems have changed in important ways over time and vary in important ways across cultures. But they went to the opposite extreme, assuming that any account of meaning that was not timeless and universal had to be arbitrary and ever subject to change. They found in Saussurean linguistics as popularly portrayed a view of meaning that could fit that account. This too was a view that ignored the role of the embodiment of meaning. It also ignored the possibility that metaphors might also be grounded in the body and constrained by experience. Because they rejected science as merely an arbitrary narrative, they could not bring empirical studies of mind and language to bear critically on their own a priori philosophical assumptions.

In both cases, we see the incompatibility between a priori philosophical theorizing and empirical findings about the mind and language. If, on the contrary, we start with an empirically responsible philosophy—one that rests on the broadest convergent evidence—then the embodied and imaginative character of mind requires us to rethink the philosophy of language from the ground up.

22

Chomsky's Philosophy and Cognitive Linguistics

This book is primarily about the conflict between a priori philosophies and empirical findings in cognitive science. The conflict shows itself prominently in linguistics. Contemporary linguistics is a philosophically saturated discipline. Many of its founders and best-known practitioners have been trained in ordinary language philosophy, formalist philosophy, formal logic, or some combination of these. Many others, through their university training, have assimilated important philosophical assumptions from these and other traditions.

The assumptions made in those philosophies have found their way into contemporary linguistic theories. The result, we believe, has been a clash between empirical linguistics and a priori philosophical assumptions that, knowingly or not, have been adopted by certain linguistic theorists. It is absolutely vital for linguists to be aware of the effects philosophical theories have had on linguistics. Most important, we need to know whether philosophical assumptions are determining the results reported by linguists as empirical results. In addition, there is the fundamental question of whether it is possible to have a linguistic theory that is sufficiently free of substantial a priori assumptions that its conclusions are not determined in advance.

Perhaps the most philosophically sophisticated linguistic theorist of our time has been Noam Chomsky. He has written at length about his Cartesian inheritance and is well versed in the analytic tradition, especially the formalist tradition. Because of Chomsky's profound and far-reaching influence on contemporary linguistics, it is important to examine the philosophical assumptions underlying his theory of language. We will see first how his philosophical assumptions determine what he takes linguistics to be. Second, we will assess his theory in the light of empirical research on language and mind.

Chomskyan Linguistics

Chomsky's view of linguistics represents an amalgam of certain previous philosophical programs. Chomsky has blended parts of Cartesian philosophy with parts of formalist philosophy to form a philosophical worldview that has persisted throughout his career, despite extreme changes in his specific linguistic theories. His early transformational grammar was a reinterpretation of the linguistics of his teacher, Zellig Harris, and over the years he has incorporated additional elements of Harris's linguistics, as well ideas from Roman Jakobson, John R. Ross, James McCawley, Paul Postal, George Lakoff, and others with whom he has had fundamental disagreements.

In understanding Chomksy's linguistics, it is crucial to recognize that Chomsky's philosophical assumptions are paramount. They are taken for granted throughout his work and are not subject to question. Chomskyan linguistics is a philosophical project within a hybrid Cartesian-formalist philosophy. We can see this more clearly by looking at the details of what Chomsky adopted from earlier philosophy.

Chomsky's view of language is based on a Cartesian conception of the mind, discussed explicitly in his *Cartesian Linguistics* (C2, Chomsky 1966). Let us recall briefly some of the key components of Descartes' view of mind and reason that were appropriated and adapted by Chomsky.

1. *Separation of Mind and Body.* Descartes claimed that the mind—the seat of reason, thought, and language—is ontologically different in kind from the body. One need not, and should not, look to the body for an account of the autonomous workings of the mind.
2. *Transcendent Autonomous Reason.* Reason is a capacity of mind, not of the body. Reason is autonomous. It works by its own rules and principles, independent of anything bodily, such as feeling, emotion, imagination, perception, or motor capacities.
3. *Essences.* Every kind of thing contains an essence that makes it the kind of thing it is.
4. *Rationality Defines Human Nature.* There is a universal human nature, an essence shared by all and only human beings. What makes human

beings human—the only thing that makes them human and that defines their distinctive nature—is their capacity for rational thought and language (see Descartes' *Discourse on Method* I:6).

5. *Mathematics as Ideal Reason.* Descartes saw mathematics as the quintessential form of human reason. Correct human reason therefore had to have the same essential character as mathematical reason (see Descartes' *Rules for the Direction of the Mind*, Rule 4.)

6. *Reason as Formal.* The ability to reason is the ability to manipulate representations according to formal rules for structuring and relating these mental symbols. Logic is the core and essence of this rational capacity, and mathematics, Descartes argued, is the ideal version of thought, because it is the science of pure form.

7. *Thought as Language.* Descartes (in his letter to Mersenne; see Chapter 19) conceptualizes thought metaphorically as language, with complex ideas put together out of simple ones, as sentences are put together out of words. Universal reason thus makes possible a universal language, which would of course have a universal grammar. Language too would be mathematical and therefore purely formal.

8. *Innate Ideas.* Descartes argued that the mind must have implanted in it by God certain ideas, concepts, and formal rules that he thought could not have been acquired via experience (Letter to Mersenne, July 23, 1641). These a priori structures are just given to us by the nature of mind and reason, and so they are possessed by all rational creatures. Under the Thought As Language metaphor, language too would be innate.

9. *The Method of Introspection.* Just by reflecting on our own ideas and the operations of our own minds with care and rigor, we can come to understand the mind accurately and with absolute certainty. No empirical study is necessary.

From these elements of Cartesian philosophy, Chomsky appropriated the following view of language.

Chomsky's Cartesian View of Language

Reason, which makes us human beings, is for Descartes languagelike. Language takes on for Chomsky the role of reason in Descartes' philosophy; that is, language becomes the essence that defines what it is to be human. Language is mathematical in nature, and since mathematics is a matter of pure form, language, for Chomsky, is purely formal. Language is also universal and innate, an autonomous capacity of mind, independent of any connection to things in the external world. Language must also have an essence, something that makes language what it is and inheres in all language. That essence is called "universal grammar"; it is mathematical in character and a matter of pure form. Language does not arise from anything bodily. It can be studied adequately through introspective methods. Studying the brain and body can give us no additional insight into language.

These basic tenets of Chomsky's linguistics are taken directly from Descartes. The only major tenets of Descartes that Chomsky rejects are the existence of mental substance and the idea that reason/language is all conscious and that its workings are directly available to conscious reflection. Indeed, Chomsky deserves enormous credit for helping to bring into cognitive science the idea of the cognitive unconscious as it applies to grammar. It was largely through Chomsky's influence that first-generation cognitive scientists became aware of the enormous range of phenomena composing the cognitive unconscious.

Chomsky's Formalist View of Language

Chomsky's version of the Cartesian perspective already contains many aspects of formalist philosophy. First, there is the Thought As Language metaphor, in which reason is conceptualized as linguistic in nature. Second, there is the Thought As Mathematical Calculation metaphor. Jointly, these motivate the use of formal "languages" in formalist philosophy to characterize reasoning in terms of the manipulation of symbols, independently of what the symbols refer to.

Let us look in some detail at what Chomsky inherited from formalist philosophy.

> *Formal Languages:* A formal "language" (conceptualized via the Formal Language metaphor) is a system of symbols in which individual symbols are conceptualized as individual linguistic elements and well-formed symbol sequences as sentences. Principles for combining symbols or transforming symbol sequences into other symbol sequences constitute the "syntax" of the formal "language."

The principal technical idea in Chomsky's linguistic theory is that of a formal language, which was developed in mathematical logic. The mathematical theory of formal languages, developed by Emil Post, was the direct inspiration for Chomsky's formal theory of language and was the mathematical setting for that theory (C2, Rosenbloom 1950, 152–154).

> *Pure Meaningless Syntax*: The symbols of a formal language, in themselves, are meaningless. A formal language needs to be interpreted to become meaningful.

This is true by definition in formalist philosophy and in Post's theory of formal languages. Technically, the rules of a formal language "look at" only meaningless formal symbols in the language. By definition of what a "formal language" is, they cannot look at the only kind of "meaning" allowed in formalist philosophy, namely, relations between symbols and elements of set-theoretical models. Technically, Chomsky's theory requires that syntax be independent of semantics and that meaning cannot possibly enter into syntactic rules. This is not an empirical matter but a consequence of a priori philosophical assumptions. Whatever syntax is, from a formalist perspective it must be independent of meaning.

The study of *logical form* from the perspective of formalist philosophy is not part of semantics, but of syntax. Thus, Chomsky's theory permits the study of logical form as part of formal syntax. Technically, a logical form in itself is meaningless—just a bunch of symbols. In formalist philosophy, the semantic interpretation of the logical forms consists of connections between symbols and the world, or a model of it.

Before we turn to other ideas that Chomsky inherited from formalist philosophy, it is important to look at the basic metaphor Chomsky uses in defining his theory of grammar in *Syntactic Structures* (C2, Chomsky 1957), namely, that a natural language is a formal system (in the sense of Post).

CHOMSKY'S METAPHOR:
A NATURAL LANGUAGE IS A FORMAL LANGUAGE

A String Of Formal Symbols	→	A Sentence
A Set Of Such Strings	→	A Language
Rules For Generating Such A Set	→	A Grammar

A formal "language"—a purely mathematical entity—had already been conceptualized by logicians in terms of aspects of written natural languages, though it was technically characterized as a form of pure mathematics having nothing to do literally with real natural languages. This metaphorical conceptualization of systems of formal symbol strings as formal "languages" made Chomsky's metaphor appear natural to adherents of formalist philosophy. Indeed, Chomsky took it not as a metaphor for modeling natural language syntax, but as a truth.

In addition, Chomsky seems to have adopted the metaphorical version of Church's thesis as well as the formalist view that scientific theories are formalized axiom systems (as described in Chapter 21).

- The Metaphorical Version of Church's Thesis: Any precisely formulated theory or collection of ideas can be stated in a formal system, that is, either a Turing machine, a system of formal logic, or a Post production system.
- Scientific (or philosophical) theories are systems of axioms in a mathematical logic.

Chomsky thus inherited the idea that the only way to be precise and scientifically rigorous is to formulate one's theory in a formal system.

Chomsky also seems to have adopted the Quine-Duhem thesis to the effect that there can be no counterexamples to part of a theory, because there can always be other aspects of the theory added to get around them. Indeed, Chomsky has actively used a Quine-Duhem strategy, as we shall discuss momentarily.

Chomsky also inherited basic ideas from his teacher, Zellig Harris, and also from Roman Jakobson. From Harris he appropriated the idea of syntactic transformations and the idea of headed constructions (what is now called "X-bar theory"). From Jakobson, he took the idea of distinctive features.

Chomsky's theory of language thus comes in two parts. The first part is his a priori philosophical worldview, a blend of Cartesian and formalist philosophy. This is not subject to question or change. It defines a philosophical perspective that he calls "the generative enterprise." To engage in the enterprise is to accept the worldview. The second part is his specific linguistic theory at a given time, whose details have changed considerably several times over the years. While the linguistic theory is still under construction (as it will be as far into the future as we can see), it is not subject to counterexample, according to the Quine-Duhem thesis, since it is still partial. The generative enterprise, as Chomsky understands it, is a long-term philosophical project defined by an a priori philosophical worldview.

Chomsky's philosophical worldview constrains what "syntax" and "language" could possibly mean. (We will use quotation marks for words whose technical philosophical meanings in Chomskyan linguistics differ markedly from their ordinary uses.) Both his Cartesian and formalist perspectives require that "language" must be both mathematical and purely formal. Both require that it be autonomous, that is, that the "syntax" of a "language" be characterizable independent of meaning or of any other external input.

Chomsky's Cartesian philosophy requires that "language" be an autonomous faculty of mind. Its autonomy requires that "language" be independent of "external" aspects of body and brain. As an autonomous faculty in Chomsky's philosophy, "language" must be:

- independent of memory
- independent of attention
- independent of perception
- independent of motion and gesture
- independent of social interaction and culture
- independent of contextual knowledge
- independent of the needs of interpersonal communication

"Language" can only be a matter of pure form. "Syntax," what Chomsky takes to be the study of pure form, is therefore the essence of what constitutes "language."

Syntax, on this Chomskyan account, is thus *the* creative part of the human mind. It creates, *from nothing external to itself,* the structures of language upon which all human rationality is built. That is why it is an *autonomous* syntax.

What is the relationship between such a "language" and the human brain? "Syntax" is *instantiated* in the brain. But it must be instantiated there in such a way as to be causally independent of all nonlinguistic aspects of the brain. The brain is seen as having an autonomous "syntax module." To be autonomous, it cannot be affected causally by input from any "not purely syntactic" parts of the brain. The "syntax module" must therefore take no input that could have a causal effect from any parts of the brain concerned with perception, motor movement, attention, other kinds of memory, cultural knowledge, and so on.

The requirement that "syntax" be autonomous and causally self-sufficient, that it take no causally effective input external to itself, places a crucially important restriction on how "syntax" can be instantiated in the brain. A purely autonomous, exclusively generative "syntax" could only be instantiated as a neural module—either a localized module or a widely distributed subnetwork—*with no input!* Any input from anything outside of "syntax" itself would destroy its autonomy and its purely generative character. Just as a formal syntactic, purely generative system cannot be affected by input from anything outside itself, so a neural instantiation of such a purely generative formal syntactic system can have no neural inputs from any other (not purely syntactic) part of the brain or body.

For this reason, constraints on real-time linguistic processing cannot be part of "syntax." Real-time linguistic processing is subject to nonlinguistic constraints: limitations of memory and attention, constraints that the auditory system places on hearing, and motor constraints on pronunciation. These are constraints of the body, limitations of the brain, mouth, and ear. No such bodily constraints can enter into "pure syntax," since it is assumed to be an autonomous capacity of mind alone.

In addition, Chomsky's Cartesian philosophy requires that "language" define human nature, that it characterize what separates us from other animals. To do so, the capacity for language must be both universal and innate. If it were not universal, it would not characterize what makes us *all* human beings. If it were not innate, it would not be part of our essence. We must have an inborn universal "syntax" (called *universal grammar*) that characterizes humanly possible "languages" in a purely formal, autonomous, disembodied way. This universal "syntax," shared by all languages, defines the essence of what "language" is.

Chomsky's use of Cartesian essences is twofold here. The capacity for "language" defines the essence of human nature, and universal "syntax" defines the essence of language.

Moreover, "syntax" on this account could not have evolved through natural selection. Chomsky's Cartesian perspective rules out such a possibility. If "syntax" is to characterize the essence of human nature, if it is to define what *distinguishes* human beings from the apes, then it could not have been present in any form in the apes. If "syntax" is to define the essence of human nature, it must come all at once, by genetic mutation, not gradual selection. Chomskyan "syntax" is not something that can be shared in part with our simian ancestors. That is why Chomsky so adamantly opposes the attribution to animals of any language capacity at all. He thus views the study of animal communication as irrelevant to any study of the language capacity.

It is also a consequence of Chomsky's philosophical views that language cannot make use of any "lower capacities," any capacities like perception or movement that we share with other animals. Whereas most specifically human capacities build on evolutionarily prior capacities, Chomsky must deny that our language capacity has any such dependence.

These philosophical views restrict what Chomsky means by "language" and "syntax." Anything that does not fit these constraints cannot, according to this worldview, be "syntax" or part of "the language capacity." On the basis of these philosophical assumptions, Chomsky founds his linguistic theory.

In Chomsky's theory, "syntax" autonomously creates ("generates") the structures used in language. Of course, there is more to language than just "syntax." There are other components to a whole grammar, for example, semantic and phonological components that take structures created by the "syntax" as input and perform other operations on them. But it is "syntax" that characterizes the essence of "language," and so it must be autonomous and take no other input.

Because of its philosophical status, no empirical finding about natural languages could, in principle, affect this characterization of "syntax" or "language." Any putative finding suggesting that syntax is not autonomous cannot really be about "syntax" or "language" in Chomsky's sense, and so must be attributed to some other faculty or theoretical component. Chomsky's term *core grammar* applies to what is covered by his theory of "syntax." Anything outside of Chomskyan "syntax" is outside of "core grammar" and thus not part of what Chomsky's theory is about.

For example, suppose one found meaning or pragmatic context constraining the occurrence of a syntactic construction. This could not, according to Chomsky's philosophical worldview, be part of "syntax." It would have to be part of some other faculty or component dealing with semantics or pragmatics. Moreover, it would not be part of anything *essential* to "language." The parts of natural language that capture the *essence* of language are those parts characterizable by pure "syntax." Any aspects of real natural languages tainted by semantics or pragmatics, or processing or memory constraints—anything that smacks of the body, of communication between people, or of the nonmental physical world—is ruled out of the essence of "language," is not "core grammar," and is not a matter of "syntax."

This is not an arbitrary move, but a consequence of Chomsky's philosophical worldview. From Chomsky's perspective, science studies essences. Physics, for example, is concerned with the essence of matter, energy, force, space, and time in general. Linguistics, likewise, is concerned with the essence of "language," namely, pure "syntax." Phenomena outside of the essence of "language" are not worthy of being called "linguistics" and thus not "interesting" from the perspective of his linguistics. Because they do not contribute to a study of "syntax" in Chomsky's sense of the term, they do not contribute to an understanding of human nature, of what makes us quintessentially human. This includes the study of such allegedly "nonessential" aspects of language as semantics, pragmatics, discourse, linguistic processing, the neural instantiation of language, cultural linguistic differences, and animal communication. But "syntax," being the pure essence of language, is independent of these nonessential matters and is a higher scientific pursuit.

Chomsky thus accepts the Cartesian distinction between *grammaire genérale* (universal grammar) and language-specific grammar. Universal Grammar is the autonomous, universal, innate capacity of mind that characterizes the essence of human nature and distinguishes humans from apes. Universal Grammar is causally independent of the rest of the mind and body and spontaneously creates all "core" linguistic structures with no external input from anything else in the mind, brain, or external world. Language-specific grammar is much less important and interesting, just a collection of idiosyncrasies that particular languages happen to have, that are not universal, and that tell us nothing about the essence of human nature.

Chomsky's Philosophy and His Politics

There is an intimate link between the philosophy underlying Chomsky's linguistics and his political philosophy. The link comes through Chomsky's Cartesianism. As a Cartesian, Chomsky believes there is single universal human nature, that the mind is separate from and independent of the body, and that what makes us distinctively human is our mental capacities, not our bodies. Because of the independence of mind from body, we can think freely, free of any physical constraints. This gives us free will. Thus, by human nature, all people require maximum freedom. Being ruled by a government is thus inherently oppressive and an ideal political system is maximally anarchic.

Since what makes us human is our minds, not our bodies, and since the mind is separate from the body, what makes us essentially human is not material. Thus, it follows from this philosophical perspective that universal human nature does not include a need to acquire material possessions (beyond what is required to live). Capitalism is thus a perversion of universal human nature and nonstate socialism is in accord with universal human nature.

Putting these two views together, one arrives at the conclusion that the ideal form of government is a type of anarchistic socialism, which is why Chomsky favors anarcho-syndicalism. From this perspective, the major sins against universal human nature are oppression and greed on the part of capitalistic governments and large corporations. Much of Chomsky's political writings focus on uncovering instances of governmental oppression and corporate greed and on seeing world politics and history from this perspective.

In Chomsky's philosophy, rationality and freedom take center stage, while culture, aesthetics, and pleasure (e.g., religion, ritual and ritual objects, business and trade, music, art, poetry, and sensuality) play no essential role in universal human nature; for Chomsky, these things simply get in the way of proper politics and have nothing to do with reason and language. The same is true of one's bodily relation to the physical environment or to "lower" animals, which Chomsky, following Descartes, sees as devoid of language and reason and lacking in free will.

Our reason for including this excursion into Chomsky's political philosophy is simply to demonstrate the coherence of his overall philosophical views. His political philosophy derives from the same source as his linguistic philosophy. There is a reason why "language" for Chomsky does not include poetic language and why his "linguistic universals" do not include a consideration of the sensuality of language, of poetic universals and of the universal capacity to form imagery and metaphor and express them in language. It is also why one finds in his work no serious discussion of the role of culture in language.

In both Chomsky's linguistics and his politics, one finds the systematic Cartesian denial of the role of the body and of our animal nature in human nature.

Problems with Chomsky's Cartesian Linguistics

The philosophical assumptions behind Chomsky's linguistic theory are almost entirely inconsistent with empirical research on mind and language coming out of second-generation cognitive science. That research indicates that the syntax of a language is structured:

- not independently of meaning, but so as to express meaning
- not independently of communication, but in accordance with communicative strategies
- not independently of culture, but often in accord with the deepest aspects of culture
- not independently of the body, but arising from aspects of the sensorimotor system

There is a wide-ranging literature in cognitive, functional, and other types of linguistic research establishing this. What follows is an extremely brief account of some of the phenomena that have led many linguists to reject the Chomskyan paradigm.

Neuroscience and "Syntax"

From the perspective of neuroscience, Chomsky's idea of "syntax" is physically impossible. As we have seen, a completely autonomous Chomskyan "syntax" cannot take any causally effective input from outside the syntax itself. Such a "syntax" would have to be instantiated in the brain within a neural module, localized or distributed, *with no neural input* to the module. But this is physically impossible. There is no neural subnetwork in the brain that does not have neural input from other parts of the brain that do very different kinds of things. (For a critique from a neuroscientific perspective, see B1, Edelman 1992, 209ff.)

What Is Linguistics?

Chomsky's theory of language has also been criticized by rank-and-file linguists who are trying to provide complete, thorough, and detailed descriptions of languages from around the world as accurately as they can. Such linguists do not begin with Chomsky's philosophical worldview and the accompanying notions of "language" and "syntax."

The majority of linguists are concerned with describing particular languages in their entirety as well as looking for general properties. One of their criteria for what is meant by language includes everything involved in learning another language. For example, what does an English speaker have to learn to *fully master* Chinese or Navaho or Greenlandic Eskimo or Georgian?

This is the very opposite of what Chomsky is talking about. Chomsky is concerned with Universal Grammar, not language-specific grammar. Chomsky is looking for a purely syntactic essence, a set of parameters shared by all languages and known innately by all normal human beings. This turns out, even on Chomsky's assumptions, to be only a relatively small "core" of formal structures. It leaves out many of the features of most human languages, for example, evidential systems (A8, Chafe and Nichols 1986), classifier systems (A4, Craig 1986), politeness systems (A9, Brown and Levinson 1987), spatial relations systems (A8, Talmy 1983; A8, Levinson 1992-present), aspectual systems (D, Comrie 1976), and lexicalization systems (A8, Talmy 1985b).

Working linguists study *all* aspects of language, especially those that have to be learned. This includes the meanings of words and constructions, pragmatic, semantic, and discourse constraints on the use of constructions, classifier systems, evidential systems, spatial-relations systems, lexicalization systems, aspectual systems, processing differences, attention to differences in what meanings must be expressed syntactically, and a vast number of other things that fall outside of the Chomskyan characterization of "language" and "syntax." Chomsky's notion of "syntax" (or "core grammar") is so limited that it leaves out most of what you would have to learn in order to learn another language. Descriptive linguists require a theory that can deal with all those matters, which Chomsky's theory cannot.

From a Chomskyan perspective, this is not a criticism but a failure to understand what the scientific study of "language" is really about, namely, the autonomous mental faculty that defines the essence of human nature, that is, an autonomous and purely formal "syntax." Moreover, Chomsky appears to doubt that all that other stuff—meaning, pragmatics, discourse, cultural constraints, processing—can be precisely studied in a scientific manner at all. Here is where the metaphorical version of Church's thesis and Chomsky's view of science comes in. To study something scientifically is, for Chomsky, to study it using the methods of formalist philosophy. If it is not formalized using those methods, it is (by the metaphorical version of Church's thesis discussed in Chapter 21) not precisely formulated at all, that is, not rigorous, and therefore not scientific.

The Distributional Generalization Criterion

For the working linguist, what characterizes a suitably scientific study of syntax? A minimal requirement for scientific adequacy is a criterion that all sciences seem to share, namely, the statement of generalizations governing the occurrences of the phenomena in question. For example, laws of motion should state generalizations governing motion. Laws of electromagnetism should state generalizations governing electrical and magnetic phenomena. Similarly, the scientific study of syntax should look for generalizations governing the occurrence of syntactic elements. Traditional grammarians have a long list of what syntactic elements are. They include clause and phrase types, for example, as well as adverbial subordinate clauses, coordinate phrases and clauses, the ordering of syntactic categories (e.g., verbs, noun phrases, adverbs) within clauses, and much more. Syntax, on this view, is the study of generalizations over the distributions of occurrences of such syntactic elements.

This distributional generalization criterion is given highest priority by ordinary descriptive linguists as well as theoretical linguists concerned with full descriptive adequacy. For such linguists, getting the full description of particular languages right has priority over Chomsky's philosophy. The task of full linguistic description is both incompatible with Chomsky's philosophical assumptions and goes way beyond the Chomskyan notion of "core grammar" and its correlative notion of "syntax." *All* syntactic constructions in a language have to be characterized in full, whether Chomsky would consider them "core" or not. Generalizations must cover both "core" and "noncore" cases. On this non-Chomskyan view of linguistics, it then becomes an empirical question whether semantic and pragmatic considerations enter into such distributional generalizations. As we shall see below, they commonly do.

But from a Chomskyan perspective, "syntax" is constrained in such a way that semantic and pragmatic considerations in principle could not enter into "syntax." Chomsky's philosophy requires that his notion of "syntax" take precedence over the distributional generalization criterion. Any distributional generalizations over syntactic elements that require the inclusion of semantics or pragmatics in the statement of the generalization cannot be part of "syntax," since they are ruled out a priori by Chomsky's philosophical assumptions. The question here is which is to take precedence, the distributional generalization criterion or Chomsky's a priori philosophy.

Second-generation cognitive science is committed to looking at language from the broadest perspective. It includes, for starters, all those things you would have to learn if you were to learn a foreign language: the meanings, the pragmatics, the speech-act constructions, constraints on processing, and on and on. It includes mechanisms of grammatical change—called "grammaticalization"—which typically involve lexical items becoming part of the syntax. This phenomenon inherently crosses the line between the lexicon and syntax. Second-generation cognitive science is committed to studying much more of language than is Chomskyan linguistics. From such a perspective, Chomskyan linguistics studies only a tiny part of language.

Moreover, second-generation cognitive science gives priority to the distributional generalization criterion over any a priori philosophical assumptions. The result is a different view of what constitutes "syntax." Let us consider two well-known examples of what are syntactic phenomena on the basis of the distributional generalization criterion, but that, on Chomsky's philosophical grounds, lie outside of "core grammar" and thus cannot be part of Chomskyan "syntax."

So far we have discussed the general problems that Chomskyan philosophy faces in its attempt to define and restrict what counts as "linguistics." It is time to give some of the nitty-gritty details. We will discuss two cases out of hundreds in the literature that might be discussed. These cases reveal some of the problems that arise when you try to isolate a "pure syntax" independent of meaning, context, and communicative purposes. The examples are lengthy because they have to be to reveal all the issues involved.

"Main-Clause" Constructions in Subordinate Clauses

There are certain types of constructions that generative grammarians used to think occurred only in main clauses and that came to be called "main-clause constructions." Since they have peculiar syntaxes, these clause types attracted the attention of traditional syntacticians who have no Cartesian or formalist philosophical program determining their method. Here are some examples:

DEICTIC LOCATIVE CONSTRUCTIONS

Here comes the bus!
Syntax: A deictic locative (*here* or *there*) followed by an auxiliaryless verb (*come, go,* or *be*) followed by the subject.

INVERTED EXCLAMATIONS

Boy! Is he ever tall!
Syntax: Optional *Boy! Man! God! Christ!* . . . followed by a question construction containing a nontemporal use of *ever.*

WH-EXCLAMATIONS

What a fool he is! What fun we had! What idiots we were!
Syntax: An initial predicate noun phrase beginning with *What* followed by subject and inflected *be.*

RHETORICAL QUESTIONS

Who on earth can stop Jordan?
Syntax: Wh-question with negative polarity postnominal modifier (*on earth, in the world,* etc.)

REVERSAL TAGS

It's raining, isn't it? It isn't raining, is it?

Chomskyan linguists of the late 1960s and early 1970s were interested in these constructions, because their apparent restriction to main clauses would seem to be a purely syntactic constraint. However, Lakoff (A4, 1987, case study 3) pointed out that such cases could occur in certain adverbial subordinate clauses, but not others. In short, these constructions were not limited to main-clause position.

According to the notion of "syntax" used by ordinary working grammarians—the distributional generalizations criterion—occurrence or nonoccurrence in subordinate clauses is a syntactic matter. Thus the occurrence of constructions with unusual syntax in main or subordinate positions is a matter of syntax.

Compare *if*-clauses and *because*-clauses. Both take clauses with ordinary clausal syntax:

> I'm leaving *because my bus is coming.*
> I'm leaving *if my bus is coming.*

But of these, only final *because*-clauses take the unusual clausal syntax:

> I'm leaving *because here comes my bus.*
> *I'm leaving *if here comes my bus.*

In general, all the above so-called main-clause constructions occur in final *because*-clauses, but not *if*-clauses. This is a syntactic phenomenon that calls for an explanation. Here is another example:

> The Bulls are going to win *because no one can stop Jordan.*
> The Bulls are going to win *if no one can stop Jordan.*

> The Bulls are going to win *because who on earth can stop Jordan?!*
> *The Bulls are going to win *if who on earth can stop Jordan?!*

Lakoff found in addition that other clause types work like *because*-clauses, namely, *although-, except-, since-,* and *but*-clauses. He also found that not all normally main-clause constructions occur in *because*-type subordinate clauses. For instance, imperative and simple question clauses do not.

> You're upset *because I'm ordering you to go home.* (indicative syntax)

> *You're upset *because go home!* (imperative syntax: no subject or auxiliary)

> I'm curious *because I want to know who stole the money.* (indicative syntax)
> *I'm curious *because who stole the money?* (interrogative syntax)

This raises a question about what, precisely, governs the distribution of such syntactic elements: Exactly which so-called main-clause constructions can occur in exactly which subordinate clauses? That is, one would like to find a generalization of the form: Constructions of type A can occur in subordinate clauses of type B under conditions C. If there is such a generalization, is it a "purely syntactic" one?

As Lakoff (A4, 1987) observes, there is such a generalization. First, the so-called main-clause constructions that can occur in subordinate clauses are those that convey statement speech acts, either directly or indirectly. All the constructions listed above have that property. In fact, even interrogative constructions can occur in these adverbial clauses if they are rhetorical questions conveying a negative statement. For example, "Who wants to watch a really dull movie?" can be a rhetorical question conveying "No one wants to watch a really dull movie." Both the negative statement and the interrogative construction can occur in *because*-clauses, but only when the interrogative construction is a rhetorical question conveying a negative statement.

> I'm leaving *because no one wants to watch a really dull movie.*
> I'm leaving *because who wants to watch a really dull movie?!*

Thus, we can see that in a generalization of the form "Constructions of type A can occur in subordinate clauses of type B under conditions C," type A consists of constructions that convey statements (directly or indirectly). That is, the generalization is about speech acts, including indirect speech acts. *It is a pragmatic generalization that unites the syntactic constructions!*

What about type B clauses? What distinguishes the permissible clauses introduced by *because, although, except, since,* and *but* from the impermissible clauses introduced by *when, where, while, as,* and so on? In clauses of the form "*X* because/since *Y*," *Y* is a reason for *X*. In the relevant clauses of the form "*X* although/ except/ but *Y*," *Y* is a reason for *not X*. For example, "I would stop but who's tired?" conveys "I would stop but I'm not tired," in which not being tired is a reason for not stopping. In short, type B clauses are those that express a reason for or against the main clause. That is, *the generalization is semantic in nature,* namely, expressing a reason.

As for condition C, it is straightforward. The content of the statement conveyed by the type A syntactic construction must be identical to the reason expressed in the type B subordinate clause. Thus, the general principle is as follows:

> Syntactic constructions of type A can occur in final-position adverbial subordinate clauses of type B under condition C.
> A: Conveying a statement of proposition P (directly or indirectly) in context.
> B: Expressing a reason R for or against the content of the main clause.
> C: P = R.

The general principle governing this phenomenon contains conditions that are syntactic (syntactic constructions occurring in final-position adverbial subordinate clauses), semantic (the clauses express reasons), and pragmatic (the syntactic constructions convey statement speech acts).

In short, a prima facie syntactic phenomenon, namely, which syntactic constructions occur in which final-position adverbial subordinate clauses, is governed by semantic and pragmatic conditions. (For greater detail, see A4, Lakoff 1987.) If generalizations governing the distribution of syntactic constructions constituted "syntax," then this phenomenon would be a counterexample to Chomsky's claim that "syntax" is autonomous.

What Counts as Grammar?

Is this phenomenon a counterexample to the Chomskyan claim that "syntax" is independent of semantics and pragmatics? What Fillmore has called "an ordinary working grammarian" without a presupposed philosophical worldview might think so, on the basis of the distributional generalization criterion. "Syntax" for the ordinary grammarian has to do with distributional generalizations over such traditionally syntactic elements as syntactic constructions (like those given), adverbial subordinate clauses, and final position.

But recall that in Chomskyan linguistics, the presupposed Cartesian-formalist philosophy defines "syntax" in a very different, philosophy-driven way. By definition, any phenomenon that is governed by semantic and pragmatic con-

ditions cannot be a matter of "syntax." Therefore, the phenomenon characterized by the given general principle cannot be considered part of "syntax." Is it
a problem for Chomskyan theory that the traditional grammarian's prima facie
notion of syntax does not accord with "syntax" as given by Chomsky's philosophy? Not at all.

In Chomsky's theory, terms like *syntax, grammatical, language,* and *core
grammar* are all technical terms in the theory and their meaning is subject to
meaning holism and the Quine-Duhem thesis. The traditional grammarian's
notion of "syntax" as defined by the distributional generalization criterion
plays no technical role in Chomsky's theory.

According to meaning holism and the Quine-Duhem thesis, theoretical terms
like *grammatical, core grammar, syntax,* and *language* do not have fixed referents until the theory as a whole is completed. Thus, it is not an externally
given, empirical fact that a given sentence is or is not "grammatical" and
"core." At one stage of theory development a given sentence might be considered "grammatical" and "core" (if it can be handled as part of "syntax") and
at a later state a theorist might consider the same sentence "ungrammatical" or
"noncore" if purely formal syntactic rules cannot account for it as a well-
formed sentence of the language. Accordingly, Chomsky has long distinguished
between sentences that speakers find "acceptable" and those that are "grammatical." Acceptability or unacceptability is, for Chomsky, a judgment speakers happen to give for reasons of performance. *Grammaticality,* however, is a
theory-internal technical term. A "grammatical" sentence is what a "grammar" in the theory generates and hence what the philosophical subject matter
called "syntax" is to account for.

This technical use of the term *grammatical* is sometimes confusing to readers. If you are not yourself a Chomskyan linguist, the terms *grammatical* and
ungrammatical will have an ordinary nontechnical (not theory-internal) meaning for you. Ungrammatical sentences will be ill-formed sentences of your language, sentences that violate something in the grammar or lexicon of the
language you speak. "Grammaticality" and "ungrammaticality" in this sense
are thus judgments that a speaker makes and therefore define in large measure
the empirical subject matter of linguistics. The issue, in a real sense, is what the
term *grammatical* is to mean.

Take, for example, the sentences:

1. *I'm leaving because here comes the bus.*
2. **I'm leaving if here comes the bus.*

We, the authors, consider the starred sentence ungrammatical and the unstarred sentence grammatical. However, from a Chomskyan perspective, our judgment would be one of "acceptability," but not necessarily "grammaticality."

As a result, there are various ways in which a Chomskyan theory could accommodate the general principle concerning the occurrence of main-clause phenomena in subordinate clauses—assuming for argument's sake that the generalization given above is correct. Here are three possibilities within a Chomskyan perspective in which the starred sentences above indicate unacceptability, not ungrammaticality.

Possibility 1: Both sentences 1 and 2 would be "grammatical." The "syntax" would allow type A constructions in all subordinate clauses. The general principle would be part of the pragmatic component, not the "syntax." As such it would mark sentences like (2) as pragmatically unacceptable, with (1) as pragmatically acceptable.

Possibility 2: Both sentences 1 and 2 would be "ungrammatical." The syntax would allow type A constructions only in main clauses, which are syntactically defined. However, the general principle given above could be considered a principle of performance whereby the "ungrammatical" sentence 1 could be used as an acceptable sentence under the conditions given.

Possibility 3: Both sentences 1 and 2 are not part of "core grammar" and so are not part of the subject matter of the theory of Universal Grammar. They are part of language-specific grammar, not Universal Grammar and so are not of scientific interest.

In short, meaning holism applied to the technical terms *grammaticality* and *core grammar* insulates the theory from any such putative counterexamples. What makes this possible is that there is no theory-external constraint in Chomsky's philosophy on what the crucial terms *syntax, grammatical, language,* and *core grammar* are to mean. What counts as "syntax" is strictly defined by the philosophy. The other notions can, via the Quine-Duhem thesis and meaning holism, change what they refer to in order to accommodate the philosophy's a priori account of "syntax."

Let us now turn to a second well-known case that makes the same point.

The Coordinate Structure Constraint

In classical Chomskyan transformational grammar, certain constructions are formed from simple clauses by "movement." For example, in "Who was John hitting?" the relative clause would be based on the simple clause *John was hitting someone* and would be formed by changing *someone* to *who*, moving *who* from the position after *hitting* to the beginning of the clause, and moving *was* from the position after the subject *John* to the position before the subject. The original position of an element moved to the beginning or end of a clause is traditionally marked with an underlined blank space, as in "Who was John hitting ___?"

In addition, certain sentences are "coordinate" and contain "conjuncts." For example, "John ate a hamburger and Bill drank a beer" has two conjoined clauses, *John ate a hamburger* and *Bill drank a beer.* In "John [[ate a hamburger] and [drank a beer]]" there is a coordinate structure, indicated by outer brackets, containing two conjoined verb phrases, *ate a hamburger* and *drank a beer,* placed in inner brackets.

Ross (D, 1967) discovered what he called the coordinate structure constraint. It was framed within Chomksy's transformational grammar in terms of a constraint on the "movement" of a constituent.

> *The Coordinate Structure Constraint:* No constituent can be moved out of a coordinate structure unless it is moved out of all conjuncts.

As stated, this is a purely syntactic constraint. It mentions coordinate (or conjoined) constituents (which are purely syntactic) and movement rules, which in Chomskyan theory are purely syntactic. No discussion of meaning appears in the constraint.

The following sentence is grammatical, as the constraint predicts, since movement occurs out of both conjuncts:

> What did [[John eat __] and [Bill drink __]]?

On the other hand, when movement occurs out of only one conjunct, the sentence is ill-formed:

> *What did [[John eat hamburgers] and [Bill drink __]]?
> *What did [[John eat __] and [Bill drank milk]]?

This phenomenon has been found in a wide variety of languages and has been postulated as a universal. Ross stated it as a purely syntactic constraint, and Chomsky has used that statement to support his argument for innate syntactic universals.

Chomsky has argued the following: Children throughout the world have acquired the same purely syntactic constraint. They did so without having been corrected or told that the ill-formed sentences were ill-formed. Therefore, there is no basis on which they could have learned the constraint. If they have it and couldn't have learned it, it must be innate. Here, then, is a putative example of an innate, purely syntactic universal.

Shortly after Ross discovered the constraint in 1966, he discovered a potential counterexample:

What did John [[go to the store] and [buy __]]?

Here there is movement out of only the second conjunct. This violates the constraint.

Lakoff at the time argued that the *and* in this sentence meant *to* and that the conjoined structure wasn't a "true conjunct." But Lakoff's argument depended on looking at the meaning of the *and,* not just at the syntax, and so did not remove the counterexample to the constraint as being syntactic.

Despite this, the potential counterexample was ignored for nearly two decades, until Goldsmith (D, 1985) and Lakoff (A8, 1986) and two of Goldsmith's students, William Eilfort and Peter Farley, discovered additional counterexamples. In these, there is movement from the first, but not the second, of two VP (verb phrase) conjuncts:

Goldsmith: How much can you [[drink __] and [still stay sober]]?
Farley: That's the stuff that the guys in the Caucasus [[drink __] and [live to be a hundred]].
Eilfort: That's the kind of firecracker that I [[set off __] and [scared the neighbors]].

At this point, well-formed sentences had been found that violated the constraint for two clauses in all the possible ways: movement from the second but not the first conjunct, and movement from the first but not the second conjunct.

Lakoff (A8, 1986) went further. He showed that counterexamples also occurred in multiple-conjunct sentences. Here are some examples. Second, third and fifth conjuncts:

What did he [[go to the store,] [buy __,] [load __ in his car,] [drive home,] [and unload __]]?

First and third conjuncts:

> How many courses can you [[take __ for credit,] [still remain sane,] and [get all A's in __]]?

Second of three conjuncts:

> Sam is not the sort of guy who you can just [[sit there,] [listen to __,] and [stay calm]].

First, third, and fifth of six conjuncts:

> This is the kind of brandy that you can [[sip __ after dinner,] [watch TV for a while,] [sip some more of __,] [work a bit,] [finish __ off,] [go to bed,] and [still feel fine in the morning]].

Second, fourth, and fifth conjuncts:

> I [[went to the toy store,] [bought __,] [came home,] [wrapped __ up,] and [put __ under the Christmas tree]] one of the nicest little laser death-ray kits I've ever seen.

The conclusion is clear: From the point of view of pure syntax, there is no co-ordinate structure constraint at all for VP conjuncts. Elements can be moved from any combination of conjuncts at all.

Lakoff then went on to offer a semantic description of the conditions for moving conjuncts, since not just any movement works.

Condition 1: The classical cases of all-or-nothing movement are instances of semantic parallelism, where each conjunct is an instance of a general semantic category. For example,

> What did [[John eat __] and [[Bill drink __]]?

is a case where both conjuncts involve consuming food. Note the bizarreness of

> *What did [[John eat __] and [Bill tune __]]?

This would make sense only if, say, John were eating musical instruments or Bill were tuning food. Only such weird contexts would make the conjuncts parallel, both cases of either acting on musical instruments or acting on food.

Condition 2: The multiple-conjunct cases that violate all-or-nothing movement are natural sequences of events, in which VPs with no movement either set a scene or change a scene, but are not part of the sequence. For example, consider

What did John [[go to the store,] [buy __,] [put __ in his car,] [drive home,] and [unload __]]?

This portrays a natural sequence of events as characterized by frame semantics. Going to the store and driving home, which have no movement, are changes of scene. The other predicates are all predicated of the same topic, indicated by *what.*

Condition 3: The two-conjunct cases that violate all-or-nothing movement are also kinds of natural sequences of events—causal sequences (causing, enabling) or their negations (not preventing). The causal relations in the cases cited are: Conjunct 1 does not prevent Conjunct 2: Drinking that much does not prevent staying sober.

How much can you [[drink __] and [still stay sober]]?

Conjunct 1 enables Conjunct 2: Drinking that stuff enables them to live to be a hundred.

That's the stuff that the guys in the Caucasus [[drink __] and [live to be a hundred]].

Conjunct 1 causes Conjunct 2: Setting off that kind of firecracker caused the neighbors to be scared.

That's the kind of firecracker that I [[set off __] and [scared the neighbors]].

Lakoff's conclusion was that semantics—frame semantics (in Fillmore's sense [A6, 1982b, 1985]), not model-theoretic semantics—governs movement out of coordinate verb phrases. Natural sequences of events are characterized relative to such Fillmorean semantic frames. There is no purely syntactic coordinate structure constraint. The all-or-nothing cases are cases with a different semantics—semantically parallel structure.

The Status of Counterexamples

Lakoff (A8, 1986) took this as one of hundreds of cases in which semantics governs syntax—prima facie counterexamples throughout the linguistics literature to Chomsky's claim that syntax is autonomous and independent of semantics. However, Lakoff was using the traditional grammarian's notion of syntax—the distributional generalization criterion—not Chomskyan philosophy's notion of "syntax."

From the perspective of Chomsky's philosophy, "syntax" must be autonomous. Otherwise, it isn't "syntax." A Chomskyan would have no trouble at all adjusting these cases to Chomsky's philosophy while keeping the coordinate structure constraint as purely syntactic. Here are two ways to do it:

POSSIBILITY 1

A. Keep the coordinate structure constraint in Universal Grammar (pure syntax).

B. The so-called counterexamples violate the constraint and so are "ungrammatical." However, the principle given above is a principle of performance that renders such "ungrammatical" sentences acceptable.

POSSIBILITY 2

A. Keep the coordinate structure constraint in Universal Grammar (pure syntax).

B. The so-called counterexamples are not in "core grammar" and are part of the subject matter of language-specific grammar, not Universal Grammar. They are therefore outside the subject matter of the theory and not appropriate for true scientific study, which studies only the essence of language.

Of course, there is nothing in Chomsky's philosophical view of "syntax" that requires that the coordinate structure constraint be part of syntax. As a result, Chomskyan linguistics, faced with data like that just presented, could accept the conclusion that there is no purely syntactic coordinate structure constraint. This would not lead to the conclusion that syntax makes use of semantics and pragmatics. Instead, there is a third possibility:

POSSIBILITY 3

A. Accept the account of coordinate structures given above, in which there is no purely syntactic coordinate structure constraint, but rather a semantically and pragmatically governed coordinate structure constraint.
B. Redefine "grammatical" so that violations of the coordinate structure constraint are not "ungrammatical," but rather pragmatically unacceptable. That is, "What did John eat hot dogs and Bill drink?" would now be called "grammatical," but pragmatically unacceptable.
C. The semantically and pragmatically governed coordinate structure constraint would now not be part of "syntax" at all. "Syntax" would still be autonomous and pure.

In all three possibilities, the Quine-Duhem thesis has entered. The notion "grammaticality" is embedded in the theory and tied to the concept of "syntax." "Grammaticality" is not independently determinable outside of the theory. Speakers' judgments of which sentences are acceptable say nothing about grammaticality per se. This is to be expected from the Quine-Duhem thesis.

As long as Chomsky's philosophical assumptions are preserved, the idea that "syntax" is semantics- and pragmatics-free is immune to counterevidence. It is protected by meaning holism and the Quine-Duhem thesis.

To sum up, within Chomsky's philosophy, there is no possible counterevidence to such basic claims as what constitutes "syntax," what is "core grammar," and whether "syntax" is autonomous. Those claims are part of a philosophically based research program and are not the kind of thing that could be subject to counterexamples, given the philosophy. This raises an interesting question. Can one find evidence against Chomsky's philosophy itself?

What's Wrong with the Philosophy Chomsky Appropriated for His Linguistics?

The empirical findings of second-generation cognitive science are at odds with Chomsky's philosophical worldview on virtually every point. Indeed, Chomsky's philosophy inherits what is wrong with both Cartesian and formalist philosophy.

The contradictions should be familiar by now. Cartesian philosophy is inconsistent with findings about the embodiment of mind. The mind is embodied, not disembodied. Concepts are embodied. Concepts get their meaning through the brain and body and through embodied experience. They are not part of a disembodied innate faculty of pure mind. Our spatial-relations concepts, upon which much of our most basic forms of reason depend, arise through the structure of the brain, its topographic maps and other physical structures (B2, Regier 1996) . Our aspectual concepts characterizing the way we structure events arise through motor control (B2, Narayanan 1997a, b). And most of our abstract concepts structured by metaphor—time, causation, even our concepts of mind itself—are grounded in bodily experience. Abstract reason is simply not an autonomous, body-free faculty of mind.

The embodiment of concepts also contradicts the formalist philosophy that underlies Chomsky's linguistic theory. Because concepts are embodied and not just symbols, thought is not "linguistic," not just a matter of symbol manipulation. Concepts are not adequately representable by meaningless symbols. The Thought As Language metaphor does not reflect scientific truth. Meaning arises through the body and brain, not via the disembodied connection of symbols with the world (or a set-theoretical model of it). The embodiment of concepts plus the existence of conceptual metaphor are inconsistent with the classical correspondence theory of truth. Since meaning holism and the Quine-Duhem thesis depend on all the apparatus of formalist philosophy, they too become invalid.

Syntax cannot be autonomous, that is, affected by no nonsyntactic input. If it were, it would have to be instantiated in the brain in an autonomous fashion—in a module or distributed subnetwork with no input! Any effect by a nonsyntactic input would destroy autonomy. But there is no part of the brain, no module or subnetwork of neurons, that has no neural input! That is a physical impossibility.

The Intrusion of Philosophy into Science

This book is centrally concerned with the relationship between science and philosophy. It claims that empirical scientific results, especially converging results about the mind and language that have been arrived at using multiple methods, take precedence over a priori philosophical theories. Indeed, we are claiming that second-generation cognitive science requires a new approach to philosophy, an embodied philosophy that will be consistent with its findings about the embodiment of mind, the cognitive unconscious, and metaphorical thought.

The study we have just presented of Chomskyan linguistics is hardly an accidental one. The aspects of Cartesian and formalist philosophy assumed in Chomskyan linguistics are very much like those assumed in first-generation cognitive science. The overlap isn't total, but it is significant.

Philosophy enters into science at many junctures. The job of the cognitive science of philosophy is to point out philosophy when it sees it, analyze the conceptual structure of the philosophy, and note its consequences. Chomskyan linguistics is a perfect case of a priori philosophy predetermining specific scientific results. Cognitive linguistics, which we will discuss shortly, emerged hand in hand with second-generation cognitive science and over the years has become a part of it. It is not subject to the intrusion of Cartesian and formalist philosophy. It has its own philosophical presuppositions, of course, but these are methodological, having to do with the demand for generalizations, converging evidence, and cognitive reality. They do not specify in advance what empirical investigation must discover, beyond the constraint that it will take the form of empirical generalizations.

Second-generation cognitive science, and cognitive linguistics with it, did not begin with a full-blown inherited philosophy. It was free to discover empirically whether concepts are embodied, whether there is metaphorical thought, and whether syntax is or is not independent of semantics. Its findings about embodiment and metaphorical thought contradict established philosophy, but they were not initially assumed. It looks for converging evidence. It is free to study the mind and language in all their manifestations.

Cognitive Linguistics

Cognitive linguistics is a linguistic theory that seeks to use the discoveries of second-generation cognitive science to explain as much of language as possible. As such, it accepts the results of second-generation cognitive science and does not inherit the assumptions of any full-blown philosophical theory. Its assumptions are methodological assumptions: that the proper methods are to search for the most comprehensive generalizations, to be responsible to the broadest range of converging evidence, and to adapt linguistic theory to empirical discoveries about the mind and the brain. Since the late 1970s, cognitive linguistics has developed into a field of its own with an extensive literature. (For an introduction to that literature, see the References, Sections A1–A9.) The range and depth of the literature is so great that we cannot survey most of the results. All we can do is provide a sense of the main results as they bear on philosophical issues.

Cognitive Semantics

Cognitive semantics studies human conceptual systems, meaning, and inference. In short, it studies human reason. We have already discussed much of the research on cognitive semantics. The most basic results are these.

- Concepts arise from, and are understood through, the body, the brain, and experience in the world. Concepts get their meaning through embodiment, especially via perceptual and motor capacities. Directly embodied concepts include basic-level concepts, spatial-relations concepts, bodily action concepts (e.g., hand movement), aspect (that is, the general structure of actions and events), color, and others.

- Concepts crucially make use of imaginative aspects of mind: frames, metaphor, metonymy, prototypes, radial categories, mental spaces, and conceptual blending. Abstract concepts arise via metaphorical projections from more directly embodied concepts (e.g., perceptual and motor concepts). As we have seen, there is an extremely extensive system of conceptual metaphor that characterizes abstract concepts in terms of concepts that are more directly embodied. The metaphor system is not arbitrary, but is also grounded in experience.

Such embodied mechanisms of conceptualization and thought are hidden from our consciousness, but they structure our experience and are constitutive of what we do consciously experience.

Cognitive Grammar: Grammar as Symbolization

In a cognitive grammar, there can be no autonomous syntax since there can be no input-free module or subnetwork in the brain. Moreover, by studying generalizations over distributions of syntactic elements, it has been found empirically that those generalizations in hundreds of cases in English alone require reference to semantics, pragmatics, and discourse function. The cases cited in the previous section concerning adverbial clauses and coordinate structures are typical.

In sorting out these generalizations, cognitive linguists such as Langacker, Lakoff, and Fauconnier have been led to the conclusion that there are no autonomous syntactic primitives at all. Syntax is real enough, but it is neither autonomous nor constituted by meaningless, uninterpreted symbols. Rather, it is the study of symbolization—the pairing of meaning with linguistic expressions, that is, with phonological forms and categories of phonological forms. Each symbolization relation is bipolar: It links a conceptual pole with an expression pole. At each conceptual pole is a category of concepts; at each expression pole is a category of phonological forms. For example, take the central sense of English *on,* as in "The cup is *on* the table." The conceptual pole would be a complex image schema made up of the primitives Contact, Support, and Above. The phonological pole would consist of the phonemic representation /an/. The grammatical category of prepositions, as discussed below, is a radial category with a center defined by the pairing of spatial-relations image schemas with phonological forms. This view of syntax, with much more complex examples, has been worked out meticulously in Ronald Langacker's classic two-volume work, *Foundations of Cognitive Grammar* (A8, 1986, 1991); a shorter discussion appears in his *Concept, Image, and Symbol* (A8, 1990).

From a neural perspective, symbolization is just a way of discussing neural connectivity. *The grammar of a language consists of the highly structured neural connections linking the conceptual and expressive (phonological) aspects of the brain.* This includes grammatical categories, grammatical structures, and lexical items. Since both semantics and phonology are grounded in the sensorimotor system, such a view of grammar makes good sense from the neural perspective. Far from being autonomous, grammar links these bodily-grounded systems. The terms *input* and *output* would be misleading here, since connectivity flows in both directions between these systems, which are independently grounded in the body.

The conceptual pole includes cognitive mechanisms involved in the processing of conceptual content, for example, what is remembered, old and new conceptual information, shifts of attention from one conceptual entity to another, viewpoints taken on situations, and the conceptual structure of discourse. The subfield of functional grammar, which we see as part of the cognitive linguistics enterprise, is concerned with the way conceptual cognitive functions enter into the structure of language via symbolization relations. Cognitive functions like these have been found to govern important aspects of such grammatical phenomena as anaphoric relations, the ordering of syntactic elements, and grammatical constructions.

In sum, a grammar consists of such symbolization relations. Syntax consists of higher-order categories of phonological forms that are at the expression pole of symbolization relations. Symbolization relations are connections linking two bodily-grounded systems, the conceptual and phonological systems. Syntax is not autonomous, but exists only by virtue of a system of (conceptual-phonological) symbolization relations.

Thus, a grammar is not an abstract formal system, but a neural system. The properties of grammars are properties of humanly embodied neural systems, not of abstract formal systems.

The Lexicon

In the simplest cases, lexical items are pairings of phonological forms with individual concepts. But such simple cases are rare exceptions. Polysemy is the norm. Most words have a number of systematically related meanings. Many cases of polysemy (by no means all) are sanctioned by conceptual metaphors—cross-domain mappings in the conceptual system. Thus, a word like *come* with a central sense concerning motion in space has additional senses, defined by metaphor, in the domain of time. Indeed, each of the basic time metaphors yields a separate extended sense of *come*. In a sentence like "Christmas is coming," the Moving Time metaphor extends *come* to the time domain. In "We're coming up on Christmas," the Moving Observer metaphor extends *come* to the time domain in a different way. The word *come* is thus paired not with just one concept, but with a radial category of concepts that has a central member and extensions, many of which are metaphorical. Most lexical items are polysemous, with their polysemy defined by systematic conceptual relations such as metaphor and metonymy.

Though the central sense may be arbitrarily paired with a phonological form, the extended senses are paired with that form because the central sense is. Given the arbitrary pairing of *come* with its spatial-motion sense, it is not arbitrary that *come* is paired with its temporal senses. They are, rather, motivated by the independently existing time metaphors. Since most words are multiply polysemous and have motivated noncentral senses, most pairings between phonological forms and meanings are motivated.

Semantic and Syntactic Categories

What is a noun? We all learned in high school that a noun is the name of a person, place, or thing, that is, a bounded physical entity. That's not a bad place to start. The names of persons, places, and things—bounded physical entities—are certainly the best examples of nouns. Of course, there many more kinds of nouns than that.

Before we look at other kinds of nouns, let us consider what "name of" means in this case. The *name-of* relation is the relation between something conceptual and something phonological, like the relationship between the concept of a chair and the phonological form *chair*. A chair is a thing and *chair* as a noun is the name of that thing. From a neural perspective, the *name-of* relation is one of activation. When we hear and understand language, the phonological form activates the concept; in speaking, the concept activates the phonological form. Particular cases of naming are conceptual-phonological pairings. The word *name* designates the phonological pole of such a pairing.

Because we are neural beings, we categorize. Because neural systems optimize, we extend categories radially, adding minimal extensions to the central category structures that we already have. Because children's earliest categories are perceptual-motor categories, we all have a central category of bounded physical objects that is extended as we grow older. Neural optimization extends the central subcategory of bounded physical objects to a radial category on the basis of existing conceptual metaphors and other neurally based cognitive mechanisms. The result is a radial category centered around bounded physical objects (persons, places, and things) and extended from this simple center in many ways. Conceptual metaphor extends persons, places, and things to metaphorical persons, places, and things of the kind we discussed in the chapters in Part II: states (metaphorical locations), activities (metaphorical objects, locations, or paths), ideas (metaphorical objects or locations), institutions (metaphorical persons), and other metaphorically comprehended abstract concepts. There are, of course, other systematic cognitive extensions of the center, for example, to pluralities and masses. The resultant category of bounded entities (physical and metaphorical) is what Langacker (A8, 1986, chap. 5) calls the conceptual category of Things. These category extensions vary, sometimes very considerably, from language to language.

Nouns name Things. That is, each particular Thing is expressed (via the naming relation) by a phonological form. The general conceptual category of Things induces a corresponding category consisting of the phonological forms naming those Things. That category of phonological forms is the category of Nouns.

The Noun category is, therefore, at the phonological pole of the conceptual Thing category. The relation that links the Thing category to the Noun category is called the Noun-relation. The Noun-relation is a category consisting of naming-relations between particular things and particular phonological forms. In short, Nouns symbolize Things.

However, the category of those phonological forms that happen to be Nouns does not exist independently of the concepts that those phonological forms symbolize. A phonological form is a Noun by virtue of the kind of concept it symbolizes.

The category of Things is a radial category. It has a universal center, namely, persons, places, and things. But this center can be extended differently in different languages. Therefore the concepts named by the Nouns of a language are not all universal.

Similar accounts can be given for verbs (with actions at the center of the conceptual category), adjectives (with properties at the center of the conceptual category), and prepositions (with spatial relations at the center of the conceptual category). All of these central senses will have various types of extensions to noncentral senses, often via metaphor.

Consider a simple example. What is called *syntactic tense* is a linguistic form. For example, in English *-ed* is the past-tense form. But tense is not merely an arbitrary sign. It has something to do with time. In particular, tense markers express how a proposition is located in time relative to the speech act (or another reference point in a discourse). For example, in "John worked hard yesterday," the tense marker is *-ed*, which expresses the temporal location of the working as past relative to the time of utterance. In other words, a linguistic form is a tense by virtue of what it means.

Correspondingly, a tensed clause is a sequence of linguistic forms—not just any sequence of signs, but a sequence of phonological forms that expresses a proposition that is located in time relative to the speech act (or a temporal reference point). Again, a sequence of phonological forms is a tensed clause by virtue of what it means. *Tensed clause* is a syntactic category, but one that has no existence as a category independent of what the forms mean. In short, syntactic categories—that is, categories of phonological sequences—are always induced by what those phonological forms express conceptually.

A hierarchical conceptual structure containing propositions located in time thus induces a corresponding hierarchical syntactic structure containing tensed clauses. Consider the sentence "John believes that Harry left." The conceptual content of *John believes __* expresses a temporally located proposition with a blank for the content of the belief to be filled in. The proposition expressed by *Harry left* fills it in. Here the proposition that *Harry left* is embedded within the more complex proposition that *John believes that Harry left.* The propositional structure induces the corresponding hierarchical clause structure at the expression pole where the clause *Harry left* is inside the larger clause *John believes that Harry left.*

So far we have seen two aspects of syntactic structure, both induced by conceptual structure and the symbolization relation. First, there are syntactic categories like Noun, Verb, Adjective, Preposition, and Clause. Second, there is the hierarchical structure of tensed clauses. This is, of course, just the bare bones of grammar.

Grammatical Constructions: More Than the Sum of the Parts

In cognitive linguistics, a grammar consists of grammatical categories (of the sort we have been discussing) and of grammatical constructions. A construction is a pairing of a complex conceptual structure with a means of expressing that conceptual structure—typically by word order or markings of some sort. Constructions include, at their conceptual pole, constraints on cognitive functions such as given versus new information and attentional focus. Each construction states constraints on how complex content is expressed phonologically in the given language. Neurally, constructions consist of complexes of neural connections between conceptual and phonological categories.

Each grammatical construction indicates (1) how the meanings of the parts of the construction are related to the meaning of the whole construction; (2) how the conceptual combination is expressed in linguistic form (e.g., by linear order or by morphological marking); and (3) what additional meaning or cognitive function is expressed by virtue of (1) and (2).

Part (3) is especially important. Grammatical constructions have their own conceptual content. Consider a classical example like "Harry sneezed the tissue off the desk." *Sneeze* is basically an intransitive verb, as in "Harry sneezed." To sneeze is to forcefully and suddenly expel air through the nose. By itself, it does not take a direct object or a directional adverb. But when *sneeze* is placed in the forced-motion construction with an immediately following direct object *(the tissue)* and directional adverb *(off the desk),* the result is more than just the sum of the meanings of *sneeze, the tissue,* and *off the desk.* To *sneeze the tissue off the desk* means *to exert force on the tissue by means of sneezing with the result that the tissue moves off the desk.* Much of this meaning is contributed by the construction itself, not just by the linguistic expressions in it.

This is called the caused-motion construction, and it is discussed in detail in Goldberg's *Constructions* (A8, Goldberg 1995).

Constructional Polysemy

A word, as we have seen, can have many systematically related concepts that form a radial category at its semantic pole. Similarly, a grammatical construction can be polysemous, with the polysemy also expressed by a radial category of systematically related concepts at its semantic pole. For example, the central meaning of the caused-motion construction can be extended to noncentral cases by the Event-Structure metaphor, in which the exertion of force is mapped onto causation and the location is mapped onto a state. Thus, "Bill talked Harriet into a state of bliss" is a variant of the caused-motion construction, in which Bill caused Harriet to change to a blissful state by talking. The Actions Are Locations variant of event structure also extends the central meaning of this construction. Thus, the sentence "Bill talked Harriet out of leaving" means that, by talking to her, Bill caused Harriet to change so as not to leave. Here once more, force is mapped onto causation, motion onto change, performing an action onto being in a location, and not performing the action onto being out of the location. Goldberg discusses constructional polysemy at length (A8, Goldberg 1995).

The Embodiment of Grammatical Constructions

Grammatical constructions are not arbitrary ways of putting meaningless forms together. Instead, they are means of expressing fundamental human experiences—embodied experiences. Dan Slobin has argued that children learn grammatical constructions as expressions of their most basic prelinguistic early experiences. This suggests that, given a radial category of senses for a construction, the central sense should express experiences common in early childhood. The study of grammatical constructions seems to bear this out. For example, the central meaning of the caused-motion construction is something physical and prelinguistic that we all learn to do as young children, namely, to exert bodily force on something resulting in its motion.

Children very early learn a version of the deictic *there*-construction, as in "Da ball" meaning *There's the ball*. The function of the deictic *there*-construction is to point out to someone something that is at a location in your field of vision at the time of speaking. (For a thorough discussion of how the meaning of the deictic *there*-construction induces the syntax of the construction, see A4, Lakoff 1987, case study 3.)

Compositionality of Constructions

Constructions state generalizations. They characterize how grammatical forms are used to express specific conceptual content and cognitive functions. Each grammatical construction can be seen as a condition governing how complex concepts are expressed in a language. Constructions compose (i.e., fit together) by superimposition. They fit together when their joint conditions are met.

One kind of grammatical construction is dedicated to stating the constraints under which other grammatical constructions can fit together. A good example of this is the principle we discussed above governing when otherwise "main-clause" constructions can be embedded in subordinate clauses. Here is the general principle as stated above:

Syntactic constructions of type A can occur in final-position adverbial sub-ordinate clauses of type B under condition C.
A: Conveying a statement of proposition P (directly or indirectly) in context.
B: Expressing a reason R for or against the content of the main clause.
C: P = R.

This principle is in the form of a grammatical construction stating conditions under which a certain kind of content can be expressed in a certain form.

A in the construction is a condition on conceptual content. The speech act in context conveys a statement of a proposition P. "Syntactic constructions of type A" picks out the expression pole corresponding to the conceptual pole meeting condition A. Condition A defines a conceptual category that can be expressed by many different constructions (e.g., deictic *there*-constructions, rhetorical questions).

A subordinate clause is a clause inside a larger clause. In an adverbial subordinate clause, the proposition expressed by the clause fills a semantic role of a two-place predicate expressed by the subordinating conjunctions *because*, *if*, *although*, and so on, which express conditions under which things happen. Among the conditions are locations (*where*-clauses), times (*when*-clauses), reasons (*because*-clauses), and hypothetical conditions (*if*-clauses).

B in the construction is also a condition on conceptual content: The type of subordinate clause expresses a reason for or against the content of the main clause.

C in the construction expresses an identity of the reason and the proposition conveyed by the statement.

"Final position" is a constraint on linguistic form. It tells where in the sentence the clause in question can occur relative to other elements of the sentence, namely, after them.

Colorless Green Ideas

In cognitive linguistics, the grammatical structure of a sentence is given by constructions. Each construction has a semantic pole with a hierarchical semantic structure and a phonological pole with a hierarchical expressive structure (of categories of phonological forms). Above the level of the individual concept, the conceptual categories are general: property, thing, process, manner, and so on. Thus, a sentence like "Shameless wild women live happily" has a conceptual analysis at one level like Property-Property-Entities-Process-Manner, where the phonological pole of the sentence indicates the order in which the general semantic categories occur in the English sentence. Thus, English permits high-level semantic sequences of words in the order Property-Property-Entities-Process-Manner. This is a complex syntactic structure pairing concepts and constraints in phonological order.

Now consider a sequence of English words such as "Colorless green ideas sleep furiously." Individually, each word fits a general conceptual category: *colorless* and *green* are properties, *ideas* are entities, *sleep* is a process, and *furiously* is a manner. Thus, "Colorless green ideas sleep furiously" fits the syntactic (i.e., conceptual-phonological) sequence Property-Property-Entities-Verb-Manner. Although the sentence as a whole is meaningless except in highly contrived situations and poetic conceits, the word sequence does fit a syntactic (i.e., conceptual-phonological) structure of English. For the reverse sequence, "Furiously sleep ideas green colorless," there is no such higher-level conceptual-phonological structure in English that fits those words in that order.

This example is given, of course, because it was Chomsky's original example arguing that there is an autonomous syntactic structure in English. Cognitive grammar accounts for such cases better than Chomsky's account did, since Chomsky's theory of autonomous, semantics-free syntax did not account for the fact that "Colorless green ideas sleep furiously" does fit a permissible pairing of higher-level semantic concepts expressed in the given order. That is, *green* is understood as a property modifying *ideas*, which are understood as entities. *Furiously* is understood as a manner that modifies *sleep*, which is understood as a process. There is a partial semantics that is understood when those words with their meanings are put together in that order. The theory of cognitive linguistics accounts for that naturally.

The Language Capacity and Linguistic Universals

In cognitive linguistics, the human language capacity is seen as something radically different than it is in Chomskyan linguistics.

First, it is seen fundamentally as a neural capacity, the capacity to neurally link parts of the brain concerned with concepts and cognitive functions (attention, memory, information flow) with other parts of the brain concerned with expression—phonological forms, signs in signed languages, and so on. In short, grammar is the capacity to symbolize concepts. The constraints on grammars are neural, embodied constraints, not merely abstract formal constraints. Categorization tends to be radial and graded. Contextual constraints are natural.

Second, the structure of language is inherently embodied. Both basic grammatical categories and the very structure imposed by constructions derive from the structure of our embodied experience.

Third, syntactic categories are induced by conceptual categories. Conceptual structure arises from our embodied nature. There is no autonomous syntax completely free of meaning and cognition.

Fourth, grammatical constructions are pairings of complex conceptual categories and cognitive functions with their means of expression.

Fifth, the language capacity is the total capacity to express concepts and cognitive functions. Thus, the range of concepts that can be expressed in any language is part of the human language capacity. Whatever means of expression there is in any language is part of the human language capacity. Where Chomskyan linguistic theory narrows the language capacity to what is true of all languages, cognitive linguistics considers the language capacity in the broadest terms as what is involved in any part of any language.

Sixth, grammatical universals are universals concerning the pairing of form and content; they are not universals of form alone (whatever that could mean). Moreover, there is more to language and to linguistic universals than grammar. Linguistic universals include conceptual universals (e.g., primitive spatial relations, universal conceptual metaphors), universals of cognitive function, and universals of iconicity.

Innateness

The traditional innate-versus-learned dichotomy is simply an inaccurate way of characterizing human development, including linguistic development. Part of the idea of innateness is that we are born with certain capacities, which we keep. They are part of our genetic heritage, which we always have with us. Moreover, innateness is usually tied to the notion of essence. What we essentially are is what we were born with. What we learn later is incidental, not necessarily part of who we are.

This picture makes no sense whatever from a neural perspective. We are born with a vast number of neural connections, a great many of which die off within the first few years of life, depending on which are used and which are not. Moreover, new connections grow, again depending in part on the connections used. From this it is clear that much of what is given at birth is not present five years later. But what is given at birth is supposed to be innate and thus something that cannot be lost. The neural facts don't fit the philosophical theory of innateness.

Moreover, the connections present at birth are too dense to perform normal adult human functions. Development requires that connections must die off. That means that learning requires a loss of what we were born with. But in the classic picture, learning just adds to what we were born with. Neurally, the classic picture doesn't work.

In addition, we have the capacity to grow new connections depending on the connections already in use. Is this an innate capacity or not? The fact that we can all do this seems to make it innate. But where it happens depends on experience, which is not innate. This capacity seems to be both innate and not innate. Again, the dichotomy doesn't hold.

In short, the innate-versus-learned dichotomy makes very little sense given what we have learned about human brains.

Much of language from a neural and cognitive perspective makes use of capacities that are not purely linguistic. Since our conceptual system to a large extent grows out of our sensorimotor system, one must decide how much of our sensorimotor capacities are innate. Since our motor capacities develop in the womb, it is not clear how relevant it is just what we do and don't have at birth.

Genetics doesn't help here much, since genes code multiple functions. Moreover, genes do not even come close to fully determining the details of neural connectivity at birth.

Since there is no autonomous syntax, the issue of the innateness of autonomous syntax does not arise. If you do not believe in Cartesian philosophy, if you do not accept the disembodied mind and essences, then the issue of innate ideas loses its philosophical significance.

What is innate about language is commonly equated with what is universal about language. But we have seen that much that is universal about language concerns universals of common experiences, which occur after birth. Those universals are due, not just to what we are born with, but also to universals of experience that depend on common environmental factors. They include universals of the conceptual poles of grammatical constructions, universals of spatial relations, and universals of metaphor.

In summary, cognitive linguistics recognizes neural reality and does not adhere to the innate-versus-learned dichotomy.

Some Philosophical Implications of Cognitive Linguistics

Cognitive linguistics is not founded on an a priori philosophical worldview beyond the basic methodological assumptions outlined above. But given the use it makes of second-generation cognitive science and the contribution it makes to our understanding of concepts, reason, and language, it has significant philosophical implications. It both provides a basis for a strong critique of traditional philosophical views and leads to what we have called an experientialist view of philosophy.

Experientialist Philosophy

Each of the hidden neural and cognitive mechanisms we have mentioned helps make up not only our conceptual systems but our very experience.

- We experience objects as colored in themselves, even though it is now known that they are not. The neural system responsible for the internal structure of our color categories also creates for us the experience of color.
- We experience space as structured by image schemas (as having bounded regions, paths, centers and peripheries, objects with fronts and backs, regions above, below, and beside things). Yet we now know that space in itself has no such structure. The topographic maps of the visual

field, the orientation-sensitive cells, and other highly structured neural systems in our brains not only create image-schematic concepts for us but also create the experience of space as structured according those image schemas.

- We experience time in terms of motion and resources, even though neither of those is inherent in time itself. Our metaphors for conceptualizing time in terms of motion not only create a way to comprehend and reason about time in terms of motion but also lead us to experience time as flowing by or ourselves as moving with respect to time.

- We experience the imbalance of an unrighted wrong. Yet the notion of Justice as Balance is not part of an objective universe. The Moral Accounting metaphor not only provides us a way to conceptualize justice in terms of balance but permits us to experience unrighted wrongs as imbalance and the righting of wrongs as recovery of balance.

Our experience of the world is not separate from our conceptualization of the world. Indeed, in many cases (by no means all!), the same hidden mechanisms that characterize our unconscious system of concepts also play a central role in creating our experience. This does not mean that all experience is conceptual (far from it!); nor does it mean that all concepts are created by hidden mechanisms that shape experience. However, there is an extensive and important overlap between those mechanisms that shape our concepts and those that shape our experience.

There is an extremely important consequence of this. For the most part, it is our hidden conceptual mechanisms, including image schemas, metaphors, and other embodied imaginative structures, that make it possible for us to experience things the way we do. In other words, our cognitive unconscious plays a central role not only in conceptualization but in creating our world as we experience it. It was an important empirical discovery that this is true, and it is an equally important area for future research to discover just how extensive this phenomenon is.

Common Sense

We have evolved so that the hidden mechanisms of meaning produce a global experience for us that allows us to function well in the world. Our preponderance of commonplace basic experiences—with basic-level objects, basic spatial relations, basic colors, and basic actions—leads us to the commonsense theory of meaning and truth, that the world really, objectively is as we experience it and conceptualize it to be. As we have seen, the commonsense theory works very well in ordinary simple cases precisely because of the nature of our embodiment and our imaginative capacities. It fails in cases where there are conflicting conceptualizations or worldviews, and such cases are quite common.

Because the mechanisms of conceptualization are hidden from us, those mechanisms are not included in our commonplace understanding of truth. But truth for a language user, in fact, is relative to our hidden mechanisms of embodied understanding.

Embodied Truth

A person takes a sentence as "true" of a situation if what he or she understands the sentence as expressing accords with what he or she understands the situation to be.

What the classical correspondence theory of truth misses is the role of human beings in producing the human notion of truth. Truth doesn't exist without (1) beings with minds who conceptualize situations and (2) a language conventionally used by those beings to express conceptualizations of situations. Those conceptualizations required to produce the very notion of truth are themselves produced by the hidden mechanisms of mind. To understand truth for a language user, one must make those mechanisms of conceptualization visible. That is one of the central enterprises of cognitive science and cognitive linguistics.

This becomes especially clear in the case of metaphorical thought. The embodied correspondence theory of truth for language users allows us to understand what we ordinarily mean by truth in cases where metaphorical thought or a particular framing is used to conceptualize a situation. As we saw, when we conceptualize time as a resource—and live by this metaphor—then we experience time as limited resource that can be wasted or saved or squandered or used wisely. If we conceptualize a situation in terms of Time As A Resource, then it might be true that I *wasted* a lot of your time or that you *squander* your time, even though time independent of the metaphor is not in itself a resource. If we extended the metaphor to include stealing time, then it might *become true*—that is, accepted as true by people in our culture—that most workers steal 2.2 hours per week from their employers.

The embodied correspondence theory of truth characterizes what language users normally take truth *for them* to be. In itself, it does not characterize a scientific account of truth. For such an account, we must return to the theory of embodied scientific realism in Chapter 7. Suppose, for example, that an astrologer provides an explanation of an event in astrological terms. Relative to the astrologer's metaphorical worldview, he or she would take that explanation as true. The embodied correspondence theory of truth accounts for the astrologer's understanding of what is *true for that astrologer*. But embodied scientific realism contradicts astrological truth, since it requires, at the very least, broad convergent evidence and predictability. Astrology fails to meet that standard of truth. Cognitive semantics can thus make sense both of truth for a person relative to a worldview and truth relative to reasonable scientific standards.

Worldview

A worldview is a consistent constellation of concepts, especially metaphorical concepts, over one or more conceptual domains. Thus, one can have, for example, philosophical, moral, and political worldviews. Worldviews govern how one understands the world and therefore deeply influence how one acts. Multiple worldviews are commonplace, and people commonly shift back and forth between them. Cultures differ considerably in worldview. Within cognitive linguistics, the study of worldview is an enterprise of considerable importance.

The entities and actions that are characterized by our conceptual systems, including our systems of metaphor, characterize our ordinary metaphysics—what we take as existing (a subject-self distinction, causal paths, essences, mental vision, moral contagion, wasted time, and so on). Our everyday metaphysics is not fanciful. It gets us through our everyday lives. Nonetheless that metaphysics is constituted by metaphor and other embodied conceptual structures.

What Can an Empirically Responsible Philosophy of Language Do?

In summary, there is a mature and extensively worked out empirical theory of language that accords with second-generation cognitive science and is not skewed by the kind of philosophical assumptions that shape formalist theories in the Chomskyan tradition. Practitioners of cognitive linguistics see this empirical theory as better stating generalizations over the full range of linguistic phenomena, as better fitting convergent evidence, and as in far better accord with nonphilosophically driven results in cognitive science.

As one would expect, cognitive linguistics, since it is consistent with and extends second-generation cognitive science, is not in accord with analytic philosophy, in either the formalist or ordinary language versions. Nor is it in accord with poststructuralist philosophy or with Chomsky's mix of Cartesian and formalist philosophy. Cognitive linguists see this as an advantage, a freeing of their science from a priori philosophy that restricts and distorts the study of language, while allowing their science to fit important results in cognitive science that other theories cannot fit—results about spatial relations, metaphor, metonymy, framing, blending, classifiers, aspectual systems, polysemy, radial categories, mental-space phenomena, grammaticalization, iconicity, and so on. Indeed, one reason cognitive linguists have written on these topics is that they lie outside the purview of philosophically constrained theories, yet constitute an overwhelming proportion of linguistic phenomena. Linguistics is the arena in which one can most clearly see the constraining effects of a priori philosophical worldviews.

We have been arguing for an experientially responsible philosophy, one that incorporates results concerning the embodiment of mind, the cognitive unconscious, and metaphorical thought. Cognitive linguistics, which incorporates such results, provides an empirically responsible linguistic theory that could be the basis for an empirically responsible philosophy of language. But what would such a philosophy of language do, given the job already done by cognitive science and linguistics?

One extremely important function for such a philosophy of language would be the sort of work done in this book, namely, a cognitively responsible analysis of important concepts and an application of such analysis to philosophically important texts and areas of culture.

Given that our language never just fits the world, that it always incorporates an embodied understanding, it becomes the job of the philosophy of language to characterize that embodied understanding accurately and to point out its consequences. Under such a reconceptualization, the philosophy of language, using cognitive linguistics, becomes applicable to every human endeavor. Its job is to reveal the cognitive unconscious in an empirically responsible way and to show why such revelations matter. It is a job of urgent and extraordinary importance in many areas of life—in morality, politics, economics, education, interpersonal relations, religion, and throughout our culture.

23

The Theory of
Rational Action

The dominant view throughout the history of Western philosophy is that there is an essence that makes us human beings and that that essence is rationality. Reason has traditionally been defined as our human capacity to think logically, to set ends for ourselves, and to deliberate about the best means for achieving those ends. Reason is understood throughout the tradition as a conscious process that operates by universal principles.

More specifically, the classic view of rationality that we have inherited is defined by the following assumptions:

1. Rational thought is literal.
2. Rational thought is logical (in the technical sense defined by formal logic).
3. Rational thought is conscious.
4. Rational thought is transcendent, that is, disembodied.
5. Rational thought is dispassionate.

In addition, there is the traditional distinction between theoretical and practical reason. Theoretical reason is contemplative. It aims at describing and explaining phenomena and is therefore a matter of justified beliefs. Practical reason, on the other hand, aims at satisfying desire through action, and so it employs the results of theoretical reason to determine the best way to act so as to satisfy desire. It is the source of principles governing both means-end reasoning and moral behavior. Despite occasional challenges (e.g., from John Dewey), this distinction has dominated philosophy.

Throughout this book we have argued that everyday human reason does not fit this classical view of rationality at all. Most of ordinary human thought—thought carried out by real "rational animals"—is metaphoric, and hence not literal. It uses not only metaphor but also framing, metonymy, and prototype-based inferences. Hence it is not "logical" in the technical sense defined by the field of formal logic. It is largely unconscious. It is not transcendent, but fundamentally embodied. Basic inference forms arise partly from the spatial logic characterized by image schemas, which in turn are characterized in terms of the peculiarities of the structures of human brains and bodies. The same is true of aspectual reasoning—reasoning about the way we structure events, which appears to arise out of our systems of motor control. Metaphorical thought, which constitutes an overwhelming proportion of our abstract reasoning, is shaped by our bodily interactions in the world.

Moreover, we now know from Antonio Damasio's *Descartes' Error* (B1, 1994) that assumption 5 is false. As Damasio shows from studies in neuroscience, those who have lost the capacity to be emotionally engaged in their lives cannot reason appropriately about social and moral issues. That is, they cannot choose appropriate ends for themselves and cannot carry out the means to those ends. Emotional engagement, Damasio argues, is an absolutely necessary component of means-end rationality. In short, the basic tenets of classical rationality as described above are undermined by the embodied and imaginative character of human reason.

Does this mean that most human beings, most "rational animals," are irrational in their everyday thought? Does it mean that philosophers too can never be rational? Not at all. We have argued throughout Part III that the best of philosophical reasoning has always used and must use exactly the same bodily-based and imaginative resources as our everyday human reason. If tenets 1 through 4 of the classical view of rationality are accepted, then Plato, Aristotle, Descartes, Kant, and all Anglo-American analytic philosophy would have to be branded as irrational. That is hardly a conclusion that most philosophers would like to accept; nor is it one that we embrace. The problem lies with the "rational ideal." It is empirically incorrect. Tenets 1 through 5 do not really characterize what makes us rational animals.

The Theory of Rational Action

Why does it matter what philosophers have thought about rationality? Who cares whether real human thought fits the classical philosophical view of what is "rational"? Suppose we are right that second-generation cognitive science has dispelled the classical view of what makes us rational animals. What difference does it make?

The classical view of rationality has been enshrined in what has come to be called the *rational-actor model* or *theory of rational action*. It is a mathematical theory that treats "rational choice" as being literal, logical, disembodied, dispassionate, and consciously calculable. The theory of rational action plays a major role in contemporary economics and international relations theory and is coming to play an ever larger role in human affairs. A version of game theory is commonly used to characterize in a precise mathematical way what "rational choice" is.

Research by George Lakoff and Robert Powell (personal communication) shows that the theory of rational choice has a metaphorical structure and that metaphorical thought plays a crucial role in its application in any context. The point of this analysis is to challenge once again the classical philosophical view of rationality. Our larger purpose is to dispel the view that the rational-actor model simply describes the world as it is, that it naturally governs all practical reasoning and social action.

As we shall argue, the rational-actor model is, instead, a human imposition, an attempt to use a certain mathematics and at least three layers of metaphor to model very specific, narrowly defined, highly idealized situations—idealizations that are not and cannot be defined by the model itself or by any other mathematical means. Any application of rational-choice theory outside such narrow and metaphorically defined situations is irrational from a larger human perspective. Since rational-choice theory itself cannot define the situations in which it can be applied, its application is a matter of human judgment. To make such judgments using as much information as possible, one must be aware of how metaphorical thought is used in any such application of the theory. Only in this way can we approach the question of where the model is useful and where it might be harmful.

Just Mathematics, Not an Inherent Aspect of the World

The mathematical theory of games, as used in rational-choice theory, is just abstract mathematics. The question that Lakoff and Powell asked was why one particular form of mathematics and not some other should be considered as defining "rational action." To answer the question, they first had to isolate the pure mathematics of game theory from the *interpretation* of that mathematics as being about rational action. This was not a trivial enterprise, since most game theory texts teach the mathematics together with the usual interpretation of the mathematics. Here is what Powell and Lakoff discovered when they separated out the mathematics from its interpretation.

The theory of strategic action and rational choice can be seen as having three parts: a formal mathematical structure plus two layers of interpretive mappings. The formal mathematical structure is just mathematics. Technically, it is a version of formal language theory, with some probability theory added. By itself, the mathematical structure can say nothing at all about the nature of rationality. It is the two layers of metaphorical interpretation that allow the mathematics to be understood as a theory of strategic action and rational choice.

The Metaphorical Structure of
Game-Theoretical Models of Rational Action

The mathematical theory of rational action, which is a form of game theory, is all too often taken as literally defining the essence of all means-end rationality. Moreover, this essence of rationality is seen as being purely mathematical and therefore not subject to question.

What follows is a technical analysis of what the pure mathematics is and how that mathematics must be interpreted by at least three layers of metaphors before it can be applied to any subject matter such as economics or international relations. In fact, two layers of metaphor are absolutely required before the mathematics can be considered as having anything whatever to do with rational action. Without those layers, the mathematics consists merely of symbols and formal relationships, which in themselves say nothing about rational choice.

The question immediately arises: Do all these layers of metaphor matter? The answer will be a clear yes. What is called the "mathematics of rational choice" actually builds in a hidden moral worldview and some rather bizarre and not very "rational" assumptions. These are not in the mathematics itself, which is just mathematics, but in the largely hidden interpretation of the mathematics.

Mathematizing Means-End Rationality Through Metaphor

The basic idea of means-end rationality is simple: You have desires you want to fulfill and purposes you want to achieve. There are things you want not to happen. How, using Reason, can you choose to act most efficiently and effectively to maximize achieving your desires and purposes while minimizing unwanted outcomes?

The assumption usually made is that reason is Universal Reason, universally valid rational principles that govern how the world works. Mathematics is taken as embodying just such principles. The question then is how to mathematize means-end rationality. The assumption is that this can be done literally, that there is a mathematics that literally fits means-end rationality.

How can you mathematize the choice of means for maximizing the achievement of purposes and minimizing unwanted outcomes? As we saw in Chapter 11, achieving purposes is most commonly conceptualized as reaching a destination, via the Event-Structure metaphor. In that metaphor, actors are conceptualized as travelers and courses of action as paths that lead to destinations. An action is motion along a path. The state resulting from an action is a location. The choice among actions is the choice among paths. The Event-Structure metaphor has the effect of spatializing action to achieve a purpose as motion to reach a destination. This is the first metaphorical step in the mathematization of achieving purposes.

To mathematize desirable and undesirable outcomes, some arithmetic is necessary. What is desirable and undesirable must become numbers. A common metaphor links desire and numbers, the metaphor we discussed in Chapter 14, Well-Being Is Wealth, in which an increase in well-being is seen as a gain and a decrease in well-being as a loss. This metaphor turns desirability—what is good for you—into accounting.

By forming a conceptual blend of these two metaphors, we reach the first step toward a mathematization of means-end rationality. The achievement of a result is reaching a destination: If the result is desirable, you get money at the destination (a *payoff*); if it is undesirable, money is taken from you (a *loss* or *cost*). The more desirable the result, the more money you get. The less desirable the result, the more money you lose.

The state in which you are making a choice is a location you are in. The possible courses of action open to you are possible paths that lead to destinations.

A choice among courses of action is a choice among paths. The "rational" choice is the one that will allow you to get the most money or lose the least.

This is one level of metaphor. It allows us to spatialize courses of action and the achievement of results. And it allows us to conceptualize the relative desirability of those results in terms of accounting. But these ideas are still not mathematicized. The next step is to take our spatialization in terms of locations and paths to other locations and visualize it metaphorically as a "tree," with the initial location as the "root," the trunk and branches as the paths, and the branching points as intersections of paths—places where one must make a decision as to which way to go.

The trees are further visualized schematically in terms of points (or *nodes*) and directed lines (or one-way arrows) connecting the points. The root is visualized as a point with arrows coming out of it, but none going into it. The tips of the branches or ends of the paths are visualized as points with arrows going to them, but no arrows emanating from them. This is the shape of a very primitive "decision tree."

We now need to turn all this into mathematics. That is, we need metaphors to conceptualize trees (or branching paths) in which you get payoffs or losses at the ends of branches in terms of some well-known mathematics. The goal is to be able to compute the "best" course of action, the one where you come out with the highest number at the end.

These metaphors and a description of the mathematics are described below. They are set off from the text so that those who want to examine them closely can do so, while those who prefer to skip over the details can.

The Nature of the Mathematicization

The theory of rational action requires a mathematics that can be used to metaphorically conceptualize (1) a branching tree structure, (2) the gain or loss of money at the tips of the branches, and (3) the totality of possibilities for gain or loss.

Branching tree structures and possible constraints on them are formalizable in the mathematics of formal languages, in which the mathematical elements are meaningless symbols and the structural relations among the symbols are stated in terms of logical axioms or "production rules," rules for manipulating the symbols. In the mathematics described in the box below, symbols are given for such entities as:

The branching "nodes," that is, the locations at which a decision has to be made as to which path to take next (metaphorically, the decision points)

The traveler or travelers (metaphorically, the actor or actors)

The paths from location to location (metaphorically, the courses of action)

The motion(s) of the traveler(s) (metaphorically, the actions)

The final locations (metaphorically, the resulting states)

The starting location (metaphorically, the initial state)

A sequence of movements (metaphorically, a history of actions)

In the mathematics of labeled tree structures, these symbols are called *nodes, lines,* and *relations.*

For example, a branching node (a location) is a symbol, N. A traveler is a symbol, P. "Symbol P labels symbol N" is conceptualized as "P bears the relation L to N" and symbolized as "L(P, N)," which metaphorically stands for "Traveler P is at location N," which in turn metaphorically stands for Actor P is in state N. Here you can begin to see the layers of intervening metaphor between "rational action" and the mathematical symbolization.

Similarly, the payoffs and losses are symbolized metaphorically as positive and negative numbers. The possible combinations are constrained by the logic of probability theory. The overall mathematics is the theory of formal languages and the theory of probability.

Even More Metaphor

The theory of rational action gets interesting and very complex with more than one actor. Here are some of the possible complications:

- A "gain" for one actor may be a "loss" for another.
- A given result may be more important for one actor than for another.
- The actors may know different things.
- There may be some randomness involved; the results may not be just a matter of choice.
- The actors each have "rational" strategies for how best to respond to the actions of others.
- There may or may not be a way to maximize everybody's payoffs.

There are additional metaphors for mathematizing these concepts:

- The relative importance of an outcome is conceptualized metaphorically as a number, with greater importance being represented as a higher number.
- A "knowledge set" (a set of states in which you have the same knowledge) is conceptualized metaphorically as a set of states with the same immediate choices for action; that is, as a set of locations with the same choice of paths leading to the next location. The assumption here is that, since knowledge necessarily informs rational decisions, states in which you have the same rational choices are states in which you have the same knowledge.
- Randomness is conceptualized metaphorically as an action on the part of Nature, which is conceptualized metaphorically as an actor and hence as a traveler; it is symbolized by a special symbol, P_0, since there is only one Nature. Randomness is further metaphorized as probability, constrained by the logic of probability theory.

Even probability theory has a metaphoric structure.

- The probability that a single event will happen in the future is conceptualized metaphorically as the distribution of previous events of a similar kind in the past. This requires a categorization of events as "similar."
- The occurrence of an event is conceptualized metaphorically in terms of set theory as a set of world-states in which the event occurs. This allows the Boolean logic of sets to apply.
- A *conjunction* of events is the set *intersection* of the sets of world-states in which those events occur. A *disjunction* of events is the set *union* of the set of world-states in which the events occur. The *negation* of an event is the set *complement* of the set of world-states in which the event occurs.
- Boolean logic is then metaphorically conceptualized in terms of the arithmetic of the numbers between zero and one. Set intersection is metaphorized as multiplication, set union as addition, the totality of possibilities as 1, and complete impossibility as 0. This was George Boole's grand metaphor for mapping arithmetic onto classical propositional logic.

Putting all these metaphors together, we get the metaphorical arithmetization of probability. Here is an example of how all these metaphors fit together to give an arithmetic of "probability":

The probability of the conjunction of two events

= the number assigned to the intersection of their world-states

= the product of multiplying the numbers assigned to each of their world-states

= the product of multiplying the numbers assigned to each event as being its "probability"

= the product of multiplying the numbers assigned to each event by taking the percentage of the occurrence of previous "similar" events in previous "similar" situations, given some understanding of what "similar" means.

Equilibrium

Next we come to the crucial notions of a strategy and an equilibrium. A *strategy* intuitively is a decision as to what courses of action to take under foreseeable circumstances. This has to be rendered metaphorically into mathematics of the sort we are discussing. Strategies are metaphorically conceptualized as mathematical functions. The output of the function is a choice of a course of action, metaphorically, a choice of path at a location in a tree. The input to the function is a "knowledge set," a set of states in which the actor appears to have the same options. The mathematical function can be seen metaphorically as telling you what choice to make in each case, given the available knowledge.

Next, we have to understand a "best reply." Pick a particular actor. Look at the "strategies" for all the other actors. The "best reply" for that actor is that strategy that maximizes that actor's payoff, given the strategies of all the other actors.

Finally, there is the notion of a *Nash equilibrium*, named for its inventor, John Nash, who won the Nobel Prize for developing the concept and its mathematics (not given here). The Nash equilibrium is the set of strategies, such that each strategy is the best reply for all the actors. That is, it is the overall set of strategies that will allow all to maximize their payoffs.

These are the basic ideas of the theory of rational action. There are an enormous number of variations on them. Each variation has somewhat different concepts, different assumptions, and different mathematical details. Our purpose in providing this general form is not to give the mathematics of any particular model, but to give you an idea of what goes into such a model from the viewpoint of its implicit, largely unconscious, metaphorical structure.

The point of the analysis is to show that the mathematics alone, with no metaphorical interpretation, says nothing whatever about rational choice. Moreover, even with the metaphorical interpretation, the model cannot be applied without an artificially constructed version of a situation to apply it to. That constructed situation consists of "stylized facts," which are themselves arrived at using complex forms of cognition, including implicit moral choices. Without such stylized facts, the rational-actor model cannot be put to use. Therefore, the rational-actor model, even with its layers of metaphor, cannot characterize rational action in any inherent way independently of the cognitive and ethical enterprise of stylizing facts.

In the following box, there is a symbolic representation of certain aspects of the formal mathematics. The reader uninterested in these details should skip to the end of the box.

The Formal Mathematical Structure

Game theory is a mathematical structure (technically, a formal language) consisting of the following sets of symbols under the following constraints:

N: A set of symbols, N_i.
T: A set of symbols, T_i.
L: A set of two-place relations, L_i.

These are to be understood as the nonterminal nodes N (the ones not at the ends of paths), the terminal nodes T (the ones at the ends of paths), and the lines L connecting the nodes in a decision tree. They fit the usual formal syntax and axioms for a tree whose branches are unordered. In each tree there is a unique root, N_0.

There are other sets of symbols to be interpreted as "labels" for each of the nodes and lines. Each node and line in a tree is labeled by elements of the following kinds:

P: A set of labels, P_j, for members of N (the nonterminal nodes).
M: A set of labels, M_j, for members of L (the lines).
V: A set of sequences of (positive and negative) numbers $(V_1, \ldots, V_n)$ that label members of T (the terminal nodes).

(continues)

The P's stand for actors or "players," the M's for movements or actions, and the V's for numbers indicating payoff values. But in themselves, these are just a bunch of meaningless symbols structured by axioms, from which one can deduce theorems, which are other bunches of meaningless symbols. That is all the mathematics is without an interpretation. We have given these mathematical symbols in their abstract form to dramatize the fact that mathematics, by itself, can tell us nothing whatever about rational choice. To apply mathematics to rational choice, interpretive metaphorical mappings are needed.

The First Layer of Interpretative Mappings

This metaphorical mapping interprets the symbols of the formal language in spatial terms.

THE BRANCHING PATHS METAPHOR

Each N_i	$\rightarrow$	A Location (A Bounded Region In Space)
Each L_i	$\rightarrow$	A Path Between Two Locations
Each P_i	$\rightarrow$	A Traveler In A Location
Each M_i	$\rightarrow$	A Self-propelled Movement By The Traveler Along The Path
Each T_i	$\rightarrow$	A Destination (A Terminal Location)

Given all this, one can define a "branching" B_i from a nonterminal node N_i as the set of all L_j's, such that, for some X, $L_j (N_i, X)$. The B's are interpreted by the following interpretive mapping.

Each B_i	$\rightarrow$	A Choice Of Paths Available To The Traveler

Now one can define H: The set of sequences of line labels from the root to the terminal nodes. An interpretive mapping interprets H:

Each H_i	$\rightarrow$	A Sequence Of Paths Taken By The Traveler

THE PAYOFF METAPHOR

Each V_i	$\rightarrow$	A "Payoff" (an amount given to, or taken from the *i*th traveler at each final destination)
Positive Numbers	$\rightarrow$	Gains
Negative Numbers	$\rightarrow$	Losses

(continues)

The resulting interpretation of the mathematical symbols is that there is a set of locations and paths linking them. Travelers start at one initial location and choose paths along which they move to new locations. When they get to a destination, each traveler receives some money. So far, this says nothing whatever about the general notion of rational choice. What is needed is a another layer of interpretive metaphorical mappings.

The Second Layer of Interpretive Mappings

Most of the work in this layer of mappings is done by the Event-Structure metaphor (see Chapter 11).

THE EVENT-STRUCTURE METAPHOR

Each Location (N_i)	→	A State
Each Path (L_i)	→	A Course Of Action
Each Traveler	→	An Actor
Each Movement	→	An Action
Each Destination (T_i)	→	A Resultant State
Each Choice Of Paths (B_i)	→	A Choice Among Courses Of Action
Each Movement History (H_i)	→	A History Of Actions

In addition, one other well-known conceptual metaphor (see Chapter 14) is needed: Well-Being Is Wealth, in which increases in well-being are seen as "gains" and decreases in well-being as "losses."

WELL-BEING IS WEALTH

Each Payoff	→	A Degree Of Well-Being
Positive Numbers	→	"Gains" In Well-Being From A History Of Actions
Negative Numbers	→	"Losses" In Well-Being From A History Of Actions

Here is the result of the second interpretive mapping: There are a number of states and means to get to other states. Actors start in a given initial state and choose actions by means of which they arrive at new states. When the actors reach the state resulting from all their actions, each actor has a resulting degree of well-being.

Given this, rational choice is defined as follows:

A choice of a course of actions is "rational" if it results in a maximization of well-being.

(continues)

In short, rational action is the maximization of causal profit. Only with these two layers of metaphorical interpretation can the abstract mathematics say anything at all about rationality.

Extensions Forming a Theory of Strategic Action

At this point, one can add definitions so that one can use the mathematics to prove theorems. The theorems are understood as being about rational choice only by virtue of the layers of metaphor interpreting the mathematics. Here is how the basic structure given above can be extended. The extension is discussed in the text at length.

- Define I_i: A set of sets of nodes, such that each node has the same set of B_j's (branches) available to the traveler (actor) at that point and the actor does not have information to distinguish among the nodes.
- Interpretation: $I_i \rightarrow$ a set of historical situations that are indistinguishable to a given actor, P_i. That is, the set of paths leading to the immediately following nodes look the same.

I is called an "information set." It is a set of states in which a given actor has the same information about his next possible action.

- Extension: Add symbol P_0 to the set P.
- Extension: Add PR: The set of numbers between 0 and 1.
- Extension: Add the Constraint: If a node is labeled P_0, then its branches are labeled PR_i, where the sum of all the $PR_i = 1$.
- Interpretation: Each $PR_i \rightarrow$ A Probability
- Interpretation: $P_0 \rightarrow$ Nature (conceptualized metaphorically as an actor)
- Define a *strategy*, S, as a mapping for each player that specifies an action at each of its information sets.
- Define $S(P_i, I_j)$ = The set of all M_k such that M_k is an action available to P_i at I_j.
- Define $S_{-i} = \{S_1, S_2, \ldots, S_{i-1}, S_{i+1}, \ldots, S_n\}$, that is, the set of the strategies of the actors other than P_i.
- Define a *best reply* for actor$_i$ to a set of strategies S_{-i} as a strategy for actor P_i that maximizes his payoff assuming that everyone else plays according to S_{-i}.
- Define a *Nash Equilibrium* as a set of strategies S such that S_i is a best reply to S_{-i} for all actors.
- Define the *equilibrium path* as the path through the tree traced when each actor follows its equilibrium strategy.

If you take this model as literal truth, you will believe that actors (i.e., consumers, firms, governments) really act according to the model, that is, that *real histories will always be equilibrium paths*. If you do not take the model as literal truth, you can still find it useful for thinking about situations.

Since the model includes two parts, the mathematics and the metaphorical interpretation of the mathematics, it is extremely important to tell them apart and to know what the metaphors used in the model entail.

Metaphorical Entailments

We are now in a position to see why it matters that rational-choice theory is metaphorical. The mathematics, combined with the metaphors, yields a set of metaphorical entailments:

- Results of courses of action can always be ranked preferentially.
- Preference is transitive.
- Actors are unitary, distinct, and volitional (in full control of their choices).
- A history can be broken down into a discrete sequence of actions.
- There is a final resultant state in a history.
- At each point in a history, future courses of action are uncertain, but there is a well-defined set of possibilities, each with a distinct probability of occurrence.
- The probability of courses of action at one point in history is independent of all previous occurrences. (This can be changed in alternative versions.)
- If two subgames at different points in a history are identical, then their historical differences don't matter.
- The model is literal. Within the model there are no alternative interpretations of actions.
- There is no "cost" to using this mathematical model.

All these metaphorical entailments have doubtful validity at best in the real world. There is no end to history. There are always multiple ways to interpret a state or action. Preference is commonly not transitive; nor are clear preferences always assignable. Actors are commonly not in full conscious control of their choices.

The limitations of rational-choice theory are well known to practitioners. There are, nonetheless, a small number of true believers who take rational-choice theory as literally true, rather than as a metaphorical application of precise mathematics that may be useful in limited situations.

Kahneman-Tversky "Irrationality"

Kahneman and Tversky and their coworkers (A10), in a long sequence of brilliant experiments, have shown that most people are "not rational," that is, they do not reason in everyday life in accord with the laws of probability and the rational model. From the perspective of second-generation cognitive science, what Kahneman, Tversky, and their coworkers have actually shown is not that people are irrational, but rather that most people reason using frames and prototypes and hence do not reason literally and "logically," in the technical sense of either formal or probabilistic "logic." In other words, human reason is far richer than the rational-actor model and probability theory recognize. Metaphorical, frame-based, and prototype reasoning are cognitive mechanisms that have developed in the course of human evolution to allow us to function as well as possible in everyday life. It would be truly irrational *not* to use the cognitive mechanisms that, in general, allow us to function as well as possible overall. What Kahneman and Tversky have really demonstrated, with important evidence, is that people really do reason using metaphors, frames, and prototypes.

Many of the classical Kahneman-Tversky experiments show that most subjects in certain reasoning tasks ignore or are oblivious to the laws of probabilistic logic. According to probability theory, the probability of any event A is always greater than or equal to the probability of event A conjoined with event B. For example, in rolling a die, you are more likely to role a six than a six followed by a two. Or in a more mundane case, it is more likely that you will get a check in the mail than that you will both get a check in the mail and get wet in a rainstorm. Kahneman and Tversky concoct cases in which subjects use conceptual framing and prototype-based reasoning that violate this basic law of probability.

Here is an example. The probability that an earthquake in California will cause a flood in which a thousand people drown is necessarily lower than the probability of a flood causing a thousand fatalities anywhere in North America. Yet, Kahneman and Tversky found that subjects will judge the deadly flood triggered by a California earthquake more likely than the fatal flood alone. They correctly see that the source of this result is that people reason using cognitive models that they take as prototypical.

Presumably, most of their subjects had a mental model in which earthquakes occur with reasonably high probability in California, a model in which earthquakes cause tidal waves, and a model in which sudden tidal waves cause a large number of deaths. When these cognitive models are evoked by the question and linked together, they imply that the probability of a California earthquake triggering the highly fatal flood is higher than any other well-known scenario for such a highly fatal flood, most of which have many fewer than a thousand fatalities because of excellent warning systems for most large floods. This form of reasoning gives the wrong answer to the Kahneman-Tversky question, but is the reasoning "irrational"?

This is the stuff of everyday common sense. It is the way we ordinarily use our cognitive models to contextualize and make sense of situations. Moreover, this is the type of reasoning that we mostly need. If we did not reason automatically and unconsciously using prototypes and conceptual frames in such a contextualized manner, we would probably not survive.

But back to the question. This type of reasoning gave the wrong answer on the Kahneman-Tversky query. Is it "irrational"?

First, we mostly don't have a choice, since reasoning using cognitive models is usually part of the operation of the cognitive unconscious. Moreover, most of the time such forms of reasoning are useful and not misleading. Being able to reason automatically and unconsciously using cognitive models as prototypes has considerable survival value and works almost all the time, hundreds of times a day. It would be very strange to say that our most productive, efficient, and effective form of reasoning, the reasoning necessary for survival in the everyday world, is irrational. It is only in those limited contexts in which the classical view of rationality is appropriate that such reasoning appears to be "irrational." It is not irrational at all; it is simply contextually inappropriate. Different contexts call for different forms of reason, and there are contexts in which conscious probabilistic reasoning *is* appropriate.

But however you want to use the word *irrational,* Kahneman and Tversky are right. People do not normally reason using probability theory and the rational-actor model in their everyday lives. And it is a good thing they do not. These forms of reasoning are not appropriate to most situations that we find ourselves in every day. And most of the time we couldn't use them if we tried, because most of our reason is unconscious, while those forms of reason are conscious and so have only a very limited range of real use.

Modeling Real Situations Using Rational-Choice Models

Suppose we want to apply a mathematical model of "rational choice" together with its two layers of metaphor to a real world situation. What is required?

The real world situation must be understood in some conventional or stylized way. A "stylized situation" is a situation conceptualized in just the right way so that the rational-actor model can map onto it. In other words, the situation as stylized must have what the rational-choice models have: distinct actors, initial states, actions, states resulting from the actions, final states, and quantifiable (and therefore comparable) degrees of well-being for each actor that results from that actor's course of action. Some stylized situations must have a distinct beginning and end; others need not, depending on the details of the model to be used.

If an actual situation does not have properties like those specified by a particular model, rational-choice theorists have the following modeling options: (1) They may try to adjust the present version of rational-choice theory to make it appropriate to the actual situation. (2) They may suggest a "restylization" of the facts and/or a new model. Or (3) they may declare the situation intractable from the perspective of the present theory of rational choice.

In addition, the mathematical model requires the use of additional metaphors. Here are some common examples:

For international relations: A State Is A Person (A Rational Actor)
For economics: A Firm Is A Person (A Rational Actor), or A Consumer Is
    A Rational Actor

If the mathematical model with its metaphorical interpretation "fits" the situation as stylized, then we call it a "model" *of* the situation. The hard work here is done by the "stylization" of the situation.

The Necessity of Metaphor in Using
Rational-Choice Theory in Modeling

The mathematical model discussed above becomes a model of "rational action" in a given stylized situation only by virtue of all the metaphors stated above. This fact must not be overlooked. The mathematics by itself is just abstract mathematics: formal language theory plus probability theory. The mathematics is not part of the objective world. To apply the mathematics one must add three layers of metaphor and then "stylize"—that is, reconceptualize—the situation so that highly metaphorized mathematical models can "fit."

What is missing from this analysis is a serious study of the "stylized facts." In such cases, a single unified perspective on a situation is chosen and most of the situation is left out as irrelevant. Whereas the mathematics of models of rational action has been studied in great detail, there is comparatively little, if any, study from a cognitive perspective of the cognitive mechanisms that are used in coming up with stylized facts.

The Issue of Aptness

We are not saying that the use of the rational-actor model can never be valid because it is metaphorical and because the facts it is to fit must be stylized. There may be situations in which the metaphors are apt and stylization of the facts is apt. But the rational-actor theory in itself cannot distinguish when this is so. This absolutely crucial aspect of any "rational" application of the model is necessarily outside of the model itself.

The fact that the model cannot tell you when it is rational to use the model is not new or surprising in any way. Anyone actually engaged in modeling real situations using rational choice theory knows this very well.

The Construction of "Rational" Realities

As we pointed out in our discussion of time (Chapter 10), time is often conceptualized as a moneylike resource that can be wasted, spent wisely, squandered, and budgeted. Time in itself is not moneylike or even resourcelike. But it can be conceptualized that way via metaphor. Moreover, we can construct, and have constructed, institutions in which that metaphor is made true, for example, businesses in which employees are paid by the hour or some other period of time.

Just as institutions have been constructed according to the Time Is Money metaphor, so institutions have been constructed according to the rational-actor model. Contemporary economic markets are such institutions. In markets, the Well-Being Is Wealth metaphor is taken as a truth: It is just assumed that maximizing well-being for firms really is maximizing wealth. "Rational action" for a firm in a market is sometimes defined as nothing more than acting so as to maximize wealth, that is, to maximize profits and minimize costs and losses. In such cases, corporations are the "rational actors." Contemporary "free" markets are institutions that have been constructed to fit various models of "rational action." And the metaphors defining those models have been made real in the construction and maintenance of markets.

One might think that free markets are just part of nature and that rational-actor models just model the way firms and consumers behave in a state of nature. That is simply not true. Contemporary markets are carefully crafted, legislated, and monitored so that various kinds of rational-choice models can be used effectively, so that business can be made "rational" and kept that way. In short, the rational-choice model is not just *de*scriptive of natural behavior; rather, it has been made *pre*scriptive, with markets tailored so that such models can be most effectively used. As a result, many corporations make a great deal of money using models of rational choice. The market is structured and maintained so that this remains possible.

Maximizing "rationality" in economics has a beneficial side: It maximizes rational control and minimizes the economic effects of "irrationalities" such as natural disasters, natural business cycles, unscrupulous individuals, and corruption. It is for this reason that markets are structured so that such models can be applied.

Is there anything wrong with this?

Many things. Take the environment for starters. When markets are structured primarily by models of "rational action," the environment can only be seen in such models as a resource. Intrinsic environmental values are not modeled. Permanent environmental destruction, bit by bit, is not a loss in the model. Who would it be a loss to? Nature is not a rational actor. Monetary profits are not "profits" that go to nature—to plants and animals and ecosystems. The loss of the natural world is not a "loss" to any corporation—to any rational actor in such a model, much less to them all. The only way it can be conceived as a "loss" to a corporation is if its destruction reduces the available resources for future profit-making actions by a corporation. Even worse, money spent to clean up pollution is added to the gross domestic product and to the profits of the corporations doing the cleanup. Pollution then becomes a source of economic benefit—a good!

Or take the issues of bioregionalism and ecological diversity. When nature is seen as a resource for corporate profits or for maximizing the gross domestic profit of a nation, the result is often monoculture. Whole countries have been turned over mostly to single crops—say, bananas, or pineapples, or peanuts—in the name of maximizing profit. What is lost is ecological diversity and ways of life based on that diversity. When well-being literally is seen as wealth, then other forms of human well-being suffer. Bioregions and cultures are destroyed.

Current "free trade" policies are, to a large extent, an attempt to change more of the world to fit some version of the rational-actor model for business, in which well-being *is* seen as literal wealth for corporations or nations. From the perspective of corporate and overall national wealth, this may be "rational," a matter of extending rational control over the vagaries of economics. From an ecological and cultural perspective, it is profoundly irrational, that is, destructive of other vital forms of well-being—the long-term well-being of the natural world, of indigenous forms of cultural life, and of values crucial to the human spirit.

To bring an area of life into accord with "rational choice" is to force life into the mold of a specific complex of metaphors—for better or worse, all too often for the worse. An example is the trend to conceptualize education metaphorically as a business, or through privatization to make education a business run by considerations of "rational choice." In this metaphor, students are consumers, their education is a product, and teachers are labor resources. Knowledge then becomes a commodity, a thing with market value that can be passed from teacher to student. Test scores measure the quality of the product. Better schools are the ones with higher overall test scores. Productivity is the measure of test scores per dollar spent. Rational-choice theory imposes a cost-benefit analysis in which productivity is to be maximized. Consumers should be getting the "best education" for their dollar.

This metaphor stresses efficiency and product quality above all else. In doing so, it hides the realities of education. Education is not a thing; it's an activity. Knowledge is not literally transmitted from teacher to student, and education is not merely the acquisition of particular bits of knowledge. Through education, students who work at it become something different. It is what they become that is important. This metaphor ignores the student's role, as well as the role of the student's upbringing and the culture at large. It ignores the nurturing role of educators, which often can only be very labor-intensive. And it ignores the overall social necessity for an ongoing, maintained class of education professionals who are appropriately reimbursed for the immense amount they contribute to society.

Another example is health maintenance organizations, which measure the productivity of doctors in terms of number of patients seen per unit time and in which doctors are also a commodity in a labor market. Both are cases of rational-choice metaphors made real at a cost, a cost that is not in the rational-choice models themselves. What tends to get lost is education, health, professionalism, dedication, community, and human dignity.

Morality and "Rational" Choice

It is important to recall that such cases do not reveal that there is anything wrong with rational-choice models in themselves. When used as *de*scriptive models, they can be of use when the models fit the reality of human well-being or perhaps even ecological or cultural well-being. Such models do, of course, have limitations, the greatest limitation being that all of the manifold forms of well-being must be conceptualized simplistically as if well-being were wealth, as if manifold forms of well-being could be reduced to comparable numbers and subject to maximization. Such models reduce multiple values to single values. Moreover, they eliminate intrinsic value and include only comparative values, subject to maximization. They leave no room for something of incomparable intrinsic worth, something that stands outside the whole framework of market value.

When might a rational-choice model accurately model a situation? First, the model must be used *descriptively*, not *prescriptively*—to describe the world, not to change it. Second, the situation must have a single form of well-being that can be accurately modeled by numbers. Third, the values of the situation must be comparative, not intrinsic. In such cases, rational-choice models can be accurate and useful descriptive tools.

But the cases we have just been discussing are not like this at all. They are cases in which (1) there are intrinsic values; (2) there are multiple values that cannot be reduced to single numbers; and (3) the models are being used to change the world, not just to describe it. In such cases, the change is defined in terms of the values used in the models, for example, maximizing literal profits.

The choices of what such values should be are moral choices, not "rational" (i.e., interest-maximizing) choices. In short, any use of a rational-choice model to change the world, to make it more "rational," is a moral choice. *Any* use!

The prescriptive use of rational-choice models is never purely *objective* and never independent of choices made within some moral system. As we have seen, our moral systems are themselves metaphorical. Metaphorical moral systems are at work in any use of rational-choice models to change the world on a significant scale.

The Case of Foreign Policy and War

Game theory and models of rational choice have had a profound effect on our lives in ways that we are mostly unaware of. For example, they been used since the early days of the cold war as ways of making foreign policy and war more "rational." In order to use rational-actor models for foreign policy, nations must be conceptualized metaphorically as people with interests—national interests. What is used is a Nation As Person metaphor. It is in the interest of a person to be healthy and strong. In the Nation As Person metaphor, health for a person maps onto overall economic health for a nation, and strength for a person maps onto military strength. Maximizing the national interest, according to this metaphorical logic, is maximizing the nation's overall wealth relative to other nations and its military strength.

What the Nation As Person metaphor hides are the real people and all the forms of well-being they individually require. The metaphor also hides all ecological values that do not translate into wealth and military strength.

For the purpose of foreign policy, the nation is conceptualized as a person in a world community, a community of nations. There are neighbor nations, friendly nations, hostile nations, client states, and competitor states. Some nations are seen as mature, grown-up states able to take care of themselves. Maturity for a nation is industrialization. The nonindustrialized nations are seen as immature or "backward," unable to take care of themselves in a world economy. They are "developing" nations.

The goal of U.S. foreign policy is to maximize our national interest, that is, our national wealth and military strength. To make foreign policy rational, models of rational choice have been introduced. To make the world more rational, to bring more of world politics and the world economy under rational control, it has been regarded as important to spread the application of rational-choice models as far as possible by measures such as "free trade" treaties and by maximizing political stability.

The application of the rational-actor model to foreign policy is thus not merely an attempt to describe the world but to change it to conform to rational-actor models. After all, models of rational actors cannot apply generally unless as many actors as possible are using the same models of rationality. To this end, the United States has been training foreign policy scholars and military and economic leaders from around the world in the use of such models. International economic institutions such as the World Bank also make use of such models.

What is hidden in the international use of models in which the rational actors are nations or corporations? The answer again is the multiple forms of well-being required by individual people, indigenous cultures, and the environment.

As a case in point, consider the Gulf War. Rational-actor models were constructed and consulted (it is not publicly known how seriously) in the decision to invade Kuwait and Iraq. The congressional debate prior to the war was framed in terms of the question asked in all rational-actor models: Do the "gains" outweigh the "losses"? Is it "worth it" to go to war?

The principal nations involved, the United States and Iraq, were conceptualized in such models as rational actors, with the national interests seen in terms of economic health and military strength of the nations involved. In the congressional debate, the following were the main issues considered as possible gains or losses for the United States: The continued availability of plentiful, cheap Middle East oil; the balance of power in the region; political prestige at home and abroad; the lives and health of U.S. soldiers; and the cost of armaments used and of logistical support.

The lives and health of Iraqi civilians, both during and as a later result of the war, were not factored in as possible "costs" or "losses" to the United States; that is, they were left out of the stylized facts debated in Congress. Not to include the lives of noncombatant civilians in Iraq was a moral decision (implicit or explicit) that shaped the "stylization" of the situation.

A year after the war, the CIA estimated that more than one million Iraqis, mostly women, children, and the elderly, had died directly from the war or indirectly due to the destruction of the infrastructure of the country—sewage processing plants, hospitals, electric generators, and the like. Massive destruction occurred to the ecology of the Gulf. U.S. troops allowed Saddam Hussein's elite troops to escape and did not unseat Saddam Hussein. Indeed, his brutal dictatorship has become even more brutal and more entrenched.

Yet the United States has considered the war a "success"; the *New York Times* even considered it a "bargain," since the "losses" were so small. The million Iraqi lives lost did not count as "losses" or "costs" in such tallies. We still have access to cheap oil. The region is still stable; with Saddam Hussein in control, Iraq has not fragmented. The Kurds have not formed their own country and threatened Turkey. The Shiites, the majority in the country, have not gone their own way and supported Iran. Saddam Hussein's continuing rule, though far from optimal for the United States, has maintained regional stability and hence rationality in the foreign policy domain. From the perspective of the rational-actor model, the war *was* a success.

Whether or not a particular model of rational action in that situation was actually used in the decision to go to war, the debate in Congress and the decision by the administration was structured by the style of thinking that characterizes such models; that is, the situation was stylized in terms of gains and losses of our assets or potential assets. The lives and suffering of the enemy nation's innocent civilians and the ecology of the country do not count as our assets and so cannot count as our losses (except perhaps for propaganda purposes by the enemy). Using such a style of thought has moral implications. Our decision to go to war, or at least the military strategy used during the war, might have been different had those lives been counted as *our* losses.

Rational-actor models in themselves are, of course, morally blameless. They are just models: mathematics plus metaphors. The way that they and the stylizations of situations are used is another matter. That is where human judgment and morality enter in. Such judgment and morality are not part of "rationality" as defined by models of rational action, since they are not in the models themselves.

But the use of such models to change the world is not morally neutral. Rational-actor models are metaphorical human constructions humanly imposed. They are not a feature of the world in itself. They can be used insightfully or not, fruitfully or not, morally or not. How we choose to use them is not a "rational choice" as defined in the models. Not to understand this is truly irrational.

Morality and Rationality

Most people most of the time do not reason according to the rational-actor model, nor even according to the traditional philosophical ideal of rationality as literal, formal, conscious, disembodied, and unemotional. Real human reason is embodied, mostly imaginative and metaphorical, largely unconscious, and emotionally engaged. It is often about human well-being and about ends determined by human well-being. Since morality concerns well-being and since our conceptions of morality arise from our modes of well-being, morality enters into human reason most of the time. It not only affects the choice of ends, but also the kinds of reasoning done in achieving those ends. Rationality almost always has a major moral dimension. The idea that human rationality is purely mechanical, disengaged, and separable from moral issues is a myth, a myth that is harmful when we live our lives according to it.

The Autonomous Rational Self

The traditional notion of rationality, together with Kant's idea of autonomy, gave rise to the view of human beings as autonomous rational actors, with complete freedom of the will and a transcendent rationality that allows them to think anything at all and to freely choose their purposes and beliefs. This view is false.

As Foucault has pointed out throughout his works, and as we saw in our discussion of morality (Chapter 14) and Kantian ethics (Chapter 20), we are greatly constrained in the way we can think. The cognitive unconscious is a principal locus of power in the Foucaultian sense, power over how we can think and how we can conceive of the world. Our unconscious conceptual systems, which structure the cognitive unconscious, limit how we can think and guarantee that we could not possibly have the kind of autonomy that Kant ascribed to us.

Moreover, the ways in which our rationality is embodied makes anything like full autonomy impossible. There are two reasons. First, many of our concepts arise from built-in constraints on the body, for example, spatial-relations concepts. Second, as we learn our concepts, they become parts of our bodies. Learned concepts are embodied via permanent or very long-term changes in our synapses. Much of our conceptual system, so deeply embodied, cannot become unlearned or overridden, at least not by some act of will and almost never quickly and easily.

Does this mean that we are forever enslaved to our unconscious conceptual systems? To some extent, yes. But to an important extent, no. We will always think in terms of containment, paths, the Event-Structure metaphor, and many other concepts that are so strongly and deeply embodied in our brains that we will always be using them.

But we also have considerable cognitive flexibility, which provides for a limited but crucial freedom of conceptualization. Because we have multiple metaphors for our most important concepts, those metaphors can sometimes be reprioritized. It may be possible to learn to use certain metaphors rather than others and to learn new metaphors. Occasionally we become aware of some of our metaphors and their connections to each other, which may generate new ways of understanding. Because complex concepts and worldviews consist of basic concepts and metaphors bound together in complexes, it may be possible to learn new complexes. And because we are conscious beings capable of reflection, we may be able to learn to monitor the use of our cognitive unconscious, provided that we learn how it operates.

Cognitive science has something of enormous importance to contribute to human freedom: the ability to learn what our unconscious conceptual systems are like and how our cognitive unconscious functions. If we do not realize that most of our thought is unconscious and that we think metaphorically, we will indeed be slaves to the cognitive unconscious. Paradoxically, the assumption that we have a radically autonomous rationality as traditionally conceived *ac-*

tually limits our rational autonomy. It condemns us to cognitive slavery—to an unaware and uncritical dependence on our unconscious metaphors. To maximize what conceptual freedom we can have, we must be able to see through and move beyond philosophies that deny the existence of an embodied cognitive unconscious that governs most of our mental lives.

24

How Philosophical
Theories Work

Each of us goes through life armed with philosophical views about all manner of things: morality, politics, God, knowledge, human nature, the meaning of life, and a vast array of other important life issues. Most of these views we inherit from our culture. We are seldom, if ever, conscious of what such philosophical views are, we find it difficult to articulate them explicitly, and we tend to be unaware of all their implications for our lives.

That is where philosophical theories can be helpful. Philosophical theories are our conscious, systematic attempts to develop coherent, rational views about our world and our place in it. They help us understand our experience, and they also make it possible for us to reflect critically on our views and to see where and how they ought to be changed.

Since its emergence as a distinctive mode of thinking among the pre-Socratics, philosophy, as conceived in the Western world, has perennially viewed itself as the ultimate form of rational thought. Philosophy has tended to see itself as the final arbiter of what counts as understanding, knowledge, and rational inquiry. At the heart of this conception is a view of reason as being capable of reflecting directly on its own operations. If there really did exist such a transcendent, universal, fully conscious, self-critical reason, and if philosophical methodology did give us direct access to it, then philosophy would be the highest form of a priori reasoning about the nature and limits of human cognition and experience. This is precisely what Kant thought philosophy was.

But this is not what philosophy is. We have seen some of the evidence from the cognitive science of the embodied mind showing that reason does not have such a disembodied, transcendent, fully conscious character. This, in turn, shows why philosophy is not pure reason reflecting on itself. The existence of the cognitive unconscious at the heart of our thinking and reasoning undermines any view of reason as transparent and directly self-reflective, as well as any aprioristic view of philosophy. The cognitive sciences reveal that reason is embodied and that it cannot know itself directly.

Therefore, for reason to know itself, and for philosophy to become sufficiently self-critical, it must at the very least make use of empirical methods from the cognitive sciences that allow us to explore the workings of the cognitive unconscious. Once philosophy becomes empirically responsible in this way, surprising things begin to happen. As we saw in Part II, many of our most basic philosophical concepts (e.g., time, causation, self, mind, morality) look very, very different from traditional philosophical analyses. They are not literal and they don't have classical category structure. Instead, they are defined by multiple metaphors and are tied to structures of our embodied experience.

Philosophy itself also turns out to be very different from what we thought before. Instead of being the activity of a pure reason, it is the activity of an embodied reason. It operates through the cognitive unconscious and thus makes use of all of the imaginative resources of the cognitive unconscious. It is grounded in and constrained by structures that depend on the nature of our bodies and the environments we live in.

The analyses we have given in this part of the book take various aspects of historically important philosophical theories as themselves objects of study, using the methods of the cognitive science of the embodied mind. These exercises in the cognitive science of philosophy are obviously very incomplete and narrow in scope. We have in no way been claiming to have explained a philosophy or to have reduced it to a handful of metaphors. Rather, our analyses of the metaphors underlying various aspects of several important philosophical theories are meant to show how cognitive science can help us understand a good deal of what makes a philosophy tick. That is, it helps us understand how the parts of a philosophy hang together, how our conceptualization and reasoning are constrained by the metaphors, and how we can evaluate some of the merits and faults of a philosophy. We believe that analyses of the sort provided in this part of the book (under the label "the cognitive science of philosophy") yield the following insights about the nature of philosophical theories.

Philosophy Rests on Shared Conceptual Metaphors

Philosophers use the same cognitive resources that everyone else does when they think and reason. They operate with the same general metaphors and metonymies that define our various folk theories, that populate the cognitive unconscious, and that are the shared property of whole cultures and traditions. We have seen that philosophers employ a relatively small number of conceptual metaphors that form the core of their central doctrines in fields ranging from metaphysics and epistemology to ethics and political theory. It is these metaphors, taken for granted throughout the body of a philosopher's work, that make the philosophy a unified theory and not a mere laundry list of concepts and claims. Such core metaphorical mappings define the inference patterns common throughout the philosopher's reasoning and reveal the generalizations that link a philosopher's key doctrines.

Whenever a philosophical theory seems intuitive to us, it is primarily because it is based on metaphors that are deeply embedded in our cognitive unconscious and are widely shared within a culture. A theory will resonate for us just insofar as it orchestrates many of the conceptual metaphors that make up our everyday folk theories. Nobody would understand Kant's moral theory at all if it didn't make use, albeit creatively, of the same metaphors that underlie our cultural models of morality.

Metaphysics as Metaphor

From Thales to Heraclitus, Plato to Aristotle, Descartes to Kant, Russell to Quine, it is the core metaphors at the heart of each philosopher's thought that define its metaphysics. Each of those source-to-target mappings project the ontology of a given source domain to form the ontology of the relevant target domain. For example, since the Pythagoreans took the Being Is Number metaphor as foundational, they projected the ontology of mathematical objects onto Being in general. They thought of numbers as chunks of space with distinctive shapes, and so they saw the world as made up of concatenations of such shapes. Or, when Descartes appropriated the Understanding Is Seeing metaphor, he thereby accepted an ontology of the mental realm that required mental counterparts to visible objects, people who see, natural light sources, and so forth. His metaphysics of mind is populated with metaphorical counterparts to these entities, and he reasons about them using patterns of inference imported from the domain of vision to the domain of mind and thought.

Metaphorical metaphysics of this sort is not some quaint product of antiquated and naive philosophical views. Rather, it is a characteristic of *all* philosophies, because it is a characteristic of all human thought. Thus, as we saw in our account of mind, most of analytic philosophy is defined by an interweaving of several metaphors (Thought Is Language, Thinking Is Mathematical Calculation, The Mind Is A Machine, etc.) that are shared within our culture. Any contemporary philosophical view that employs the Thinking Is Mathematical Calculation metaphor thereby appropriates its distinct ontology, in which all thoughts must be unitary entities, just as numbers are conceived to be. Just as much for Quine as for Thales, metaphysics stems from metaphors.

Philosophical Innovation

There is nothing deflationary about this view of the metaphoric nature of philosophical theories. Showing that philosophies are built up from metaphors, metonymies, and image schemas does not diminish their importance. On the contrary, it reveals just how marvelous such philosophical systems really are. Philosophers are not simply logic-choppers who fine-tune what their culture already knows in its bones. Instead, they are the poets of systematic thought. Philosophy at its best is creative and synthetic. It helps us put our world together in a way that makes sense to us and that helps us deal with the problems that confront us in our lives. When philosophers do this well they are using our ordinary conceptual resources in very extraordinary ways. They see ways of putting ideas together to reveal new systematic connections between different aspects of our experience. They sometimes give us the means for criticizing even our most deeply rooted concepts. They show us ways to extend our metaphors and other imaginative structures to deal with newly emerging situations and problems. Thus, Kant, almost single-handedly, generated the notion of moral autonomy (and its metaphors) that has become a defining feature of the modern view of moral responsibility.

Constrained Philosophical Imagination

To set out the defining metaphors of a philosophy is not necessarily to critique it. Many philosophers with a traditional view of language and meaning mistakenly believe that discovering the metaphorical underpinnings of a theory somehow undermines it. This is just false. It is based on an objectivist, literalist view of language and mind that does not, and cannot, recognize the existence of conceptual metaphor.

Two decades ago, Paul DeMan (C2, 1978) attracted considerable philosophical attention with his analyses of metaphors lying at the heart of philosophical theories. This was thought to be disturbing, because it denied the literalist view of concepts and meaning, and also because DeMan held a view of metaphor as unstable and indeterminate in meaning. DeMan's analyses of Locke's view of mind and language and Kant's treatment of judgment do, indeed, challenge the literalist view. However, DeMan was mistaken in claiming that metaphor is destabilizing and indeterminate. He expresses this view of the indeterminacy and unreliability of metaphor in his remarks on metaphors in Kant's philosophy:

> The considerations about the possible danger of uncontrolled metaphors, focused on the cognate figures of support, ground, and so forth, reawaken the hidden uncertainty about the rigor of a distinction that does not hold if the language in which it is stated reintroduces the elements of indetermination its sets out to eliminate. (C2, DeMan 1978, 27)

DeMan is wrong to claim that such metaphors destabilize philosophical theories. We have seen how conceptual metaphors ground abstract concepts through cross-domain mappings using aspects of our embodied experience and how they establish the inferential structures within philosophies. As our analyses show, conceptual metaphors are anything but loci of indeterminateness and uncertainty. Metaphors are the very means by which we can understand abstract domains and extend our knowledge into new areas. Metaphor, like any other embodied, imaginative structure, is not a philosophical liability. Rather, it is a remarkable gift—a tool for understanding things in a way that is tied to our embodied, lived experience. Identifying philosophers' metaphors does not belittle them. Instead, it helps us understand the power of philosophical theory to make sense of our lives. The extended analyses of philosophers' conceptual metaphors that we have given in this book show that it is the metaphors that unify their theories and give them the explanatory power they have. There is no philosophy without metaphor.

Only two things are denied by the presence of conceptual metaphor in philosophy: (1) There is no philosophy built up solely from literal concepts that could map directly onto the mind-independent world. (2) There is no transcendent, disembodied, literal reason that is fully accessible to consciousness. Neither of these things is necessary in order to do philosophy. On the contrary, a belief in them is an obstacle to cognitively realistic, empirically responsible philosophical views that have a bearing on our lives.

How Philosophy Is Changed

What difference does any of this cognitive analysis make for our understanding of philosophy? Plenty. Let us consider some examples of how we might think differently about a particular philosophical view once we have studied it from the perspective of second-generation cognitive science.

We saw how several of Aristotle's most famous doctrines are the consequence of his weaving together of conceptual metaphors. Take, for instance, his fateful view of logic as purely formal. This view emerges in the following way. Predications Are Categories. That is, to predicate an attribute of a thing is to place it within a category. Categories are understood metaphorically as abstract containers. Syllogisms, as forms of deductive reasoning, work via a container logic (e.g., *A* is in *B,* and *B* is in *C,* so *A* is in *C*). We saw also that Aristotle's founding metaphor was Ideas Are Essences. To conceptualize a thing is to categorize it, which is to state its essence, the defining attributes that make it the kind of thing it is. For Aristotle, then, the essences of things in the world, since they are what constitute ideas, can actually be *in* the mind. And for the essence to be in the mind, it cannot be the substance or matter of a thing; rather, it must be its *form:* Essences Are Forms. So, if our ideas are the forms of things, and we reason with the forms of things, then logic is purely formal, abstracting away from any content.

Seeing these tight connections among the metaphors explains for us the logic of Aristotle's arguments and shows us why he has the doctrines he has. Once we see this, we see also that there is no absolute necessity about this particular view of things. It is a view based on a metaphorical logic that uses one particular set of conceptual metaphors. However, there are other possible metaphors for understanding logic and reasoning in ways inconsistent with the metaphors Aristotle used to characterize "logic."

The fact that the same patterns of inference occur with different content was taken by Aristotle to be empirical verification of his view that logic is a matter of form. From the perspective of cognitive semantics, of course, there is a very different explanation for the same empirical observations. Via the metaphor that Categories Are Containers (that is, bounded regions in space), the logic of containers is mapped onto all categories conceptualized in the cognitive unconscious in terms of containment, that is, of bounded regions of space. *Modus ponens* and *modus tollens* are examples of the logic of containment:

Embodied *Modus Ponens:* If Container A is inside Container B and X is inside Container A, then X is inside Container B.
Embodied *Modus Tollens:* If Container A is inside Container B and X is outside Container B, then X is outside Container A.

Here X is either another container or a specific entity. Apply the Categories Are Containers metaphor, and we get the equivalent of the Aristotelian principles.

The general applicability of these principles to any such categories, regardless of the specific content of the categories, is an instance of embodied content: the concept of containment and the Categories Are Containers metaphor. Symbolic logic is disembodied and therefore an inaccurate, misleading way to characterize such embodied principles of human logic. Symbolic logic involves the manipulation of meaningless symbols and therefore misses the embodied character of these forms of human reason.

Once we use the tools and methods of second-generation cognitive science to understand Aristotle's logic, we may need to rethink his logic and see another, more cognitively realistic, view of logic. The alternative to formal, disembodied reason is an embodied, imaginative reason.

The same situation holds for Descartes's conception of mind and his idea of self-reflection. That conception is built on the Understanding Is Seeing metaphor, with all of its many submappings: Ideas Are Objects, Reason Is Light, Knowers Are Seers, Intelligence Is Visual Acuity, and so on. Somehow Reason is supposed to shine its light on its own internal operations, even as they are occurring. In this way Self-Knowledge Is Self-Reflection.

Some of these metaphors may be apt or not in various contexts. However, cognitive science suggests that the particular metaphor of self-reflection is cognitively unrealistic. It ignores the pervasive and indispensable workings of the cognitive unconscious. Thus, we have strong reasons for questioning the adequacy of the entire system of metaphors that jointly give rise to the notion of direct self-reflection.

Finally, recall the set of metaphors that together make up the Language of Thought metaphor that underlies so much of contemporary analytic philosophy. We saw that these metaphors, such as Thought Is Language, Thinking Is Mathematical Calculation, Ideas Are Objects, and The Mind Is A Machine (nowadays, a computer), are all deeply embedded in our cultural folk models of mind, thought, and language. They are then brought together in a unique way by contemporary analytic philosophers to form the Language of Thought metaphor, in which, for example, ideas are symbols (of a language of thought) that get their meaning via the Fregean metaphor that such symbols correspond to things in the world.

We cannot emphasize strongly enough just how pervasive and widely influential such metaphors have been in defining the goals and methods of analytic philosophy. Because of this, it is worth reviewing once more those entailments of our everyday metaphors for the mind that appear prominently in one or another version of analytic philosophy.

THOUGHT AS LANGUAGE

Thought has the properties of a language.
Thought is external and public.
The structure of thought is accurately representable as a linear sequence of written symbols of the sort that constitute a written language.
Every thought is expressible in language.

THE MIND AS BODY SYSTEM

Thoughts have a public, objective existence independent of any thinker.
Thoughts correspond to things in the world.

THOUGHT AS MOTION

Rational thought is direct, deliberate, and step-by-step.

THOUGHT AS OBJECT MANIPULATION

Thinking is the manipulation of mental objects.
Thoughts are objective. Hence, everyone can have the same thoughts; that is, thought is universal.
Communicating is sending ideas to other people via language.
Thoughts have a structure, just as objects do.
The structure of thoughts can be uniquely and correctly analyzed, just as the structure of an object can.

THOUGHT AS MATHEMATICAL CALCULATION

Just as numbers can be accurately represented by sequences of written symbols, so thoughts can be adequately represented by sequences of written symbols.

Just as mathematical calculation is mechanical (i.e., algorithmic), so rational thought is.

Just as there are systematic universal principles of mathematical calculation that work step-by-step, so there are systematic universal principles of reason that work step-by-step.

Just as numbers and mathematics are universal, so thoughts and reason are universal.

The Mind As Machine

Each complex thought has a structure imposed by mechanically putting together simple thoughts in a regular, describable, step-by-step fashion.

In addition, much of analytic philosophy has also inherited some important metaphorical entailments from Aristotle. Although analytic philosophy eschews Aristotle's central metaphor, Ideas Are Essences, it accepts many of the entailments of that metaphor within the Aristotelian worldview. Here are some examples:

First, concepts (what the mind "grasps") are defined by inherent characteristics of things in the world.

Second, there is Aristotle's definition of definition: "A definition is a phrase signifying a thing's essence" (*Topics* 102a). That is, a definition is a set of necessary and sufficient conditions for something to be the kind of thing it is. Thus, a definition of a concept is a collection of necessary and sufficient conditions through which we can grasp the inherent characteristics of kinds of things in the world.

In other words, all concepts are literal, defined directly in terms of the features of kinds of things in the world. The meaning of concepts is therefore literal, defined in terms of the properties of kinds of things in the world in itself.

A third entailment of Aristotle's metaphors is Logic Is Formal, as discussed previously.

A fourth is Aristotle's theory of metaphor: Metaphor is not conceptual, since it is not literal; it is therefore a matter of the use of words. Moreover, it is not the proper use of words (which is literal), and so it is an improper use of words, appropriate to rhetoric and poetry rather than ordinary speech. Metaphors, to be comprehensible at all, must be based on similarity (the principal relationship between concepts). Moreover, the similarity must be an objective feature of the external world, since concepts are defined in terms of such features.

These entailments of Aristotle's metaphors are widely accepted without question by a great many Anglo-American philosophers. The same is true of many of the entailments of Descartes' metaphors:

Thought is essentially disembodied, and all thought is conscious.

We can, just by thinking about our own ideas and the operations of our own minds, with care and rigor, come to understand the mind accurately and with absolute certainty.

Nothing about the body, neither imagination, nor emotion, nor perception, nor any detail of the biological nature of the body, need be known in order to understand the nature of the mind.

Finally, certain of Kant's principal moral theses, which are entailments of his metaphors, have also been widely adopted within Anglo-American philosophy. They are:

There are universal moral laws.

We can know these universal moral laws through reason alone, reflecting on itself.

Therefore, no empirical facts can have any bearing on what we *ought* to do. ("You can't get an *ought* from an *is*.")

Universal Reason is what gives us freedom—freedom to choose our own moral ends—and hence makes us morally independent, that is "autonomous."

Though Anglo-American philosophers are by no means all Kantians, these entailments of Kant's moral metaphors are commonplace in much of Anglo-American moral theory.

It is sobering to realize that students studying Anglo-American philosophy are taught all (or at least most) of these metaphorical entailments as truths. They collectively define the core of the Anglo-American philosophical worldview. Yet there is nothing sacred or absolute about these metaphorical entailments. As it happens, all of them are at odds with the view of mind and language emerging from second-generation cognitive science. These are metaphorical entailments that ignore the embodiment of our concepts and reasoning. They ignore the cognitive unconscious that operates via conceptual metaphors, metonymies, and image schemas. Even though they have been massively influential in determining the course of much contemporary analytic philosophy, it may be time to give them up in favor of more cognitively realistic conceptions of the mind, language, and morality.

The cognitive science of philosophy thus does not just describe how philosophies work. It does that, and that is important work. But it also frequently gives us a basis for evaluation and criticism of philosophies. It allows us to bring our empirical understanding of the mind into the study of philosophies old and new.

Part IV

Embodied Philosophy

25
Philosophy in the Flesh

We are philosophical animals. We are the only animals we know of who can ask, and sometimes even explain, why things happen they way they do. We are the only animals who ponder the meaning of their existence and who worry constantly about love, sex, work, death, and morality. And we appear to be the only animals who can reflect critically on their lives in order to make changes in how they behave.

Philosophy matters to us, therefore, primarily because it helps us to make sense of our lives and to live better lives. A worthwhile philosophy will be one that gives us deep insight into who we are, how we experience our world, and how we ought to live.

At the heart of our quest for meaning is our need to know ourselves—who we are, how our mind works, what we can and cannot change, and what is right and wrong. It is here that cognitive science plays its crucial role in helping philosophy realize its full importance and usefulness. It does this by giving us knowledge about such things as concepts, language, reason, and feeling. Since everything we think and say and do depends on the workings of our embodied minds, cognitive science is one of our most profound resources for self-knowledge. That is the guiding ideal of this book.

Empirically Responsible Philosophy

The question is clear: Do you choose empirical responsibility or a priori philosophical assumptions? Most of what you believe about philosophy and much of what you believe about life will depend on your answer.

We have been arguing for an empirically responsible philosophy—a philosophy informed by an ongoing critical engagement with the best empirical science available. We are promoting a *dialogue* between philosophy and cognitive science. Ideally, they should co-evolve and mutually enrich each other.

Philosophical sophistication is necessary if we are to keep science honest. Science cannot maintain a self-critical stance without a serious familiarity with philosophy and alternative philosophies. Scientists need to be aware of how hidden a priori philosophical assumptions can determine their scientific results. This is an important lesson to be drawn from the history of first-generation cognitive science, where we saw how much analytic philosophy intruded into the initial conception of what cognitive science was to be.

On the other hand, philosophy, if it is to be responsible, cannot simply spin out theories of mind, language, and other aspects of human life without seriously encountering and understanding the massive body of relevant ongoing scientific research. Otherwise, philosophy is just storytelling, a fabrication of narratives ungrounded in the realities of human embodiment and cognition. If we are to know ourselves, philosophy needs to maintain an ongoing dialogue with the sciences of mind.

Why Empirical Responsibility Matters in Philosophy

Empirical responsibility in philosophy is important because it makes better self-understanding possible. It gives us deeper insight into who we are and what it means to be human. The shift from the disembodied mind to the embodied mind is dramatic. This book gives some sense of just how dramatic. But it is only a beginning. To get an idea of the sweeping implications of the use of cognitive science in philosophy, let us consider three final topics: what a person is, what evolution is, and what it means for spirituality to be embodied.

What a Person Is

We began this book with the claim that we, as members of our culture, have inherited significantly false philosophical views of what a person is. These are not merely the views of professional philosophers. They are widespread views that have influenced every aspect of our lives from morality to politics, to religion, medicine, economics, education, and on and on. They are so commonplace that we barely notice how they influence our lives.

At this point in the book we can now give a detailed account of what the traditional Western concept of the person is and what we think it should be replaced with. Putting these two views side-by-side is revelatory. We're not who we thought we were. What we do is not what we thought we were doing.

Here is the comparison. It is well worth pondering what your life would be like if your understanding of what you are were to change in this way.

The Traditional Western Conception of the Person

Disembodied Reason

- The Objective World: The world has a unique category structure independent of the minds, bodies, or brains of human beings.
- Universal Reason: There is a Universal Reason that characterizes the rational structure of the world. It uses universal concepts that characterize the objective categories of the world. Both concepts and reason are independent of the minds, bodies, and brains of human beings.
- Disembodied Human Reason: Human reason is the capacity of the human mind to use some portion of Universal Reason. Reasoning may be performed by the human brain, but its structure is defined by Universal Reason, independent of human bodies or brains. Human reason is therefore disembodied reason.
- Objective Knowledge: We can have knowledge of the world via the use of Universal Reason and universal concepts.
- Human Nature: The essence of human beings, that which separates us from the animals, is the ability to use Universal Reason.
- Faculty Psychology: Since human reason is disembodied, it is separate from and independent of all bodily capacities: perception, bodily movement, feelings, emotions, and so on.

Literal Reason

- Literal Concepts: Objective knowledge and objective truth require that universal concepts characterize the objective features of the world. Such concepts must be literal, that is, capable of directly fitting the features of the world.
- Monolithic Conceptual Systems: Any conceptual system capable of literally fitting the world must be univocal and self-consistent.
- Universal Means-End Rationality: Universal Reason provides a way to calculate how to maximize our self-interest in purely literal terms. Thus, people have the capacity to be self-interest maximizers.

Radical Freedom

- Free Will: Will is the application of reason to action. Because human reason is disembodied—that is, free of the constraints of the body—will is radically free. Thus, will can override the bodily influence of desires, feelings, and emotions.
- Conscious Reason: Reason is conscious. If it were not, unconscious reason would determine our actions and will would not be wholly free.

Objective Morality

- Objective Universal Morality: Morality is objective; there is an absolute right and wrong for any given situation.
- Rational Universal Morality: Morality is also rational. It is a system of universal principles (moral laws) that arise either from a universal notion of what is "good" or from Universal Reason itself.

This conception of the person is assumed in much of Western religion. In the Judeo-Christian tradition, Universal Reason is God's Reason, which human beings have the capacity to partake of. The locus of consciousness and reason is identified with the soul. Since the soul is separate from the body and not subject to physical constraints, it is seen as being able to live on after the death of the body. God gives moral commandments. They are rational because they derive from God's Reason. Since human beings partake of God's Reason, they can grasp these moral laws. Since people have radically free will, they can choose whether or not to follow moral laws. A strong will is necessary to overcome any temptations to violate moral laws.

This view of the person also lies behind the traditional European distinction between the natural sciences and the humanities. What is subject to physical law can be studied scientifically—the physical world, including biology. But, being radically free and not subject to laws of physical causation, the mind is seen as not amenable to scientific study. A different, "interpretive" methodology is supposedly required for the human sciences. For this reason, cognitive science has not been taken seriously within traditional humanistic fields of study.

The traditional Western view of the person is, as we have seen, at odds *on every point* with the fundamental results from neuroscience and cognitive science that we have been discussing. An actual human being has neither a separation of mind and body, nor Universal Reason, nor an exclusively literal conceptual system, nor a monolithic consistent worldview, nor radical freedom.

The Conception of an Embodied Person

Since Socrates, the fundamental invocation of philosophy has been to "know thyself." To know ourselves individually, we must know what we are like as human beings. Here is the new view of the person emerging from all the results that we have discussed.

Embodied Reason

- Embodied Concepts: Our conceptual system is grounded in, neurally makes use of, and is crucially shaped by our perceptual and motor systems.
- Conceptualization Only Through the Body: We can only form concepts through the body. Therefore, every understanding that we can have of the world, ourselves, and others can only be framed in terms of concepts shaped by our bodies
- Basic-Level Concepts: These concepts use our perceptual, imaging, and motor systems to characterize our optimal functioning in everyday life. This is the level at which we are maximally in touch with the reality of our environments.
- Embodied Reason: Major forms of rational inference are instances of sensorimotor inference.
- Embodied Truth and Knowledge: Because our ideas are framed in terms of our unconscious embodied conceptual systems, truth and knowledge depend on embodied understanding.
- Embodied Mind: Because concepts and reason both derive from, and make use of, the sensorimotor system, the mind is not separate from or independent of the body. Therefore, classical faculty psychology is incorrect.

Metaphoric Reason

- Primary Metaphor: Subjective experiences and judgments correlate in our everyday functioning with sensorimotor experiences so regularly that they become neurally linked. Primary metaphor is the activation of those neural connections, allowing sensorimotor inference to structure the conceptualization of subjective experience and judgments.

- Metaphorical Reasoning: Conceptual metaphors permit the use of sensorimotor inference for abstract conceptualization and reason. This is the mechanism by which abstract reason is embodied.
- Abstract Reason: By allowing us to project beyond our basic-level experience, conceptual metaphor makes possible science, philosophy, and all other forms of abstract theoretical reasoning.
- Conceptual Pluralism: Because conceptual metaphors, prototypes, and so on structure abstract concepts in multiple ways, we have a conceptual system that is pluralistic, with a great many mutually inconsistent structurings of abstract concepts.
- No Universal Means-End Rationality: Because we think using multiple metaphors and prototypes, there is, in most cases, no clear and unequivocal "self-interest" for a person that can be maximized. Thus, there is no objective Universal Means-End Rationality that can always calculate how to maximize that which typically does not take a clear form—one's supposedly objective "self-interest." Thus, people cannot be self-interest maximizers.

Limited Freedom

- Unconscious Reason: Most of our thought is below the level of consciousness.
- Automatic Conceptualization: Because our conceptual systems are instantiated neurally in our brains in relatively fixed ways, and because most thought is automatic and unconscious, we do not, for the most part, have control over how we conceptualize situations and reason about them.
- The Difficulty of Conceptual Change: Because our conceptual systems are mostly unconscious and neurally fixed, conceptual change is at best slow and difficult. We cannot freely change our conceptual systems by fiat.
- Embodied Will: Since reason is embodied, and since will is reason applied to action, our will cannot transcend the constraints of the body.

Embodied Morality

- No "Higher" Morality: Our concepts of what is moral, like all our other concepts, originate from the specific nature of human embodied experience. Our conceptions of morality cannot be objective or derive from a "higher source."

- Metaphoric Morality: Moral concepts are mostly metaphorical, based ultimately on our experience of well-being and family.
- The Pluralism of Human Moral Systems: Because each person's conceptual system contains a multiplicity of moral metaphors, some of which are mutually inconsistent, we each have within us a moral pluralism.

Human Nature Beyond Essentialism

- Human Nature Without Essence: Cognitive science, neuroscience, and biology are actively engaged in characterizing the nature of human beings. Their characterizations of human nature do not rely on the classical theory of essences. Human nature is conceptualized rather in terms of variation, change, and evolution, not in terms merely of a fixed list of central features. It is part of our nature to vary and change.

What Evolution Isn't

There is a common folk theory of evolution, that evolution is a competitive struggle to survive and reproduce. This folk theory has normative implications: Competitive struggle to survive and reproduce is natural. Moreover, it is good, because it got us where we are.

This folk theory is everywhere in our culture. It is used metaphorically to justify forms of free-market economics, educational reform, the basis for legal judgments, and the conduct of international relations. What the folk theory comes down to in all of its metaphorical applications is that it is natural to pursue one's self-interest competitively and to fail to do so is irrational. The normative implication is that the social order, in every domain, is naturally and optimally governed by principles of competitive self-interest and that anything that interferes with that is unnatural and immoral.

We have two substantive things to say about this folk theory and its applications. (1) The view of self-interest that emerges from it is empirically incorrect. (2) The view of evolution encapsulated in the folk theory is based on an inaccurate metaphor for what evolution is. Both the folk theory and its application are therefore thoroughly misguided.

Selfishness Versus Altruism

It is not surprising that the issue of altruism has taken center stage in many contemporary attempts to understand ethics in the context of evolution. Altruism becomes a "problem"—indeed, *the* problem—of moral theory, given the traditional notion of a person as it has developed since the Utilitarians and Darwin. The problem is this: Why should anyone act altruistically when people are by nature "rational," where "rationality" is taken to be the maximizing of one's self-interest? Why should altruism ever override selfishness?

There is a long history in the Western tradition of seeing rationality in terms of the effective pursuit of self-interest. It goes at least as far back as Epicurus' focus on pleasure and pain as the sources of all action: The rational person seeks pleasure and avoids pain. Well-being is the maximization of the pleasure available to us.

Enlightenment psychology for the most part saw human beings as motivated principally by the desire to maximize satisfactions. Enlightenment economic theory assumed this psychological model. It went on to define means-end rationality as the efficient calculation of means to well-defined, quantifiable ends. Given that rationality was taken as *the* defining characteristic of human nature, it came to be seen as *natural* for human beings to use their reason to maximize their perceived self-interest. Utilitarianism assumed this view of human nature and sought a utopian moral system based on it. In this "ideal" social system, each individual would have the maximum freedom to pursue his or her own self-interest consistent with others having the same freedom.

Darwinian evolutionary theory was widely interpreted—or rather misinterpreted—to fit the view of ourselves as self-interest maximizers by nature. Evolutionary theory is, in itself, an account of the survival of species in terms of adaptation to ecological niches. Darwinian adaptation was misleadingly metaphorized by others in terms of "competition," a competitive struggle for scarce resources in which only the strong and cunning emerge victorious, garnering the goods necessary for life and happiness. The evolutionary "success" of human beings in this "competition" was then attributed in social Darwinism to human rationality: Those who maximize their self-interest best win the competitive struggle.

The combined legacy of utilitarianism and social Darwinism is a deeply entrenched view of human rationality as the maximization of self-interest. The mathematical version of this is the rational-actor model, discussed in Chapter 23. As we saw, the rational-actor model requires the use of at least three levels of metaphor. It is not the literal mathematization either of some alleged "rational structure of the world" or of human rationality. Indeed, as we have seen, real human rationality makes use of unconscious conceptual framings, prototypes, metaphors, and so on.

The very notion of a well-defined, global, and consistent "self-interest" for any human being over any significant length of time makes no sense. It is ruled out by the following considerations:

1. Most of our reasoning is unconscious, so most determination of self-interest in our everyday lives is not done at the level of conscious choice.
2. Our unconscious conceptual systems make use of multiple metaphors and prototypes, especially in the area of metaphors for what is right and what is good and ought to be pursued. Thus, in most cases there is not a univocal, self-consistent notion of "the good" or of the "best outcome."
3. Since our unconscious reasoning about what is a "best outcome" often conflicts with our conscious determination of the "best outcome," there is no single unitary consistent locus of "self-interest."

In short, the nature of human conceptual systems makes it impossible for us to be objective maximizers of a univocal, consistent self-interest.

We can now see that the moral problem of the apparent conflict between selfishness (or "rationality") and altruism is ill-defined, because the notion of rationality is empirically incorrect: We are not and cannot be rational self-interest maximizers in the traditional sense. Moreover, the notion of altruism is also ill-defined. As we saw in Chapter 14, moral systems are defined relative to idealized family models (e.g., the Strict Father and Nurturant Parent models). What counts as "altruistic" is very different in different family-based moralities.

To see why, let us consider the simplest of examples: donating money and time to a political cause that you believe is morally right. According to a politically conservative version of Strict Father morality, banning abortion would be a good example of such a "morally right" cause, while a politically liberal version of Nurturant Parent morality would maintain that guaranteeing a woman's right to choose is the "morally right" thing to do. From either point of view, donating money and time is regarded as a purely altruistic act done for the sake of what is morally right, for the welfare of others and of society as a whole.

What this shows is that, even to understand what we take altruism to be in a given case, one must look to the family-based moral frameworks that structure the cognitive unconscious. The study of those moral frameworks and their consequences, which relies on techniques coming from cognitive science, is central to what a humanly adequate future moral theory must focus on.

Evolution and Family-Based Moral Theory

In evolutionary theory, survival is keyed to the ability to fit ecological niches. Fitting a niche and hence surviving can occur for many reasons: coloration that hides one from predators, a large number of offspring, the availability of food, and so on.

Unfortunately, evolutionary biology has acquired in the popular mind a Strict Father interpretation, in which the survival of those that fit their niche becomes, metaphorically, Evolution Is The Survival Of The Best Competitor. This Strict Father metaphor for evolution has then been applied metaphorically from the evolution of species to natural changes of all sorts that have nothing literally to do with the science of evolutionary biology. The metaphor is Natural Change Is Evolution.

To these metaphors has been added a crucial folk theory, the Folk Theory of the Best Result: Evolution produces the best result. The reasoning is based on the metaphor of the Moral Order (see Chapter 14), in which human beings are ranked as "better" than animals, plants, and other aspects of nature. If evolution produced us, then evolution must create "improvements." Nothing of this sort is part of literal evolutionary theory.

The two metaphors Natural Change Is Evolution and Evolution Is Survival Of The Best Competitor, together with the Folk Theory of the Best Result, have combined to yield the composite metaphor Natural Change Is Survival Of The Best Competitor, which produces the best result. This composite metaphor, arising from Strict Father morality, has been used to argue, whenever change is needed, for the introduction of an artificially constructed form of "evolution"—the imposition by law of market-driven competition. There are two issues here. First, the idea that the market functions like evolution is metaphorical, based on the metaphor Evolution Is Survival Of The Best Competitor. Second, real evolution is a natural process and does not have to be constructed by law and government enforcement. It is thus a bit ironic that this argument is made on the grounds that such change is "evolutionary," hence natural and productive of the best result.

An example of this form of argumentation is the argument for the privatization of public schools. Suppose that, through legislation (an artificial means) and through a government-run school voucher program (an artificially created market), public schools are privatized. "Natural evolution" will then take place: Schools will have to compete, only the best competitors will survive, and those schools that cannot compete will cease to exist. The surviving schools, by the Folk Theory of the Best Result, will be the best schools. It is an argument entirely based on two metaphors and a folk theory, all of which derive from Strict Father morality.

Many people do not notice that Evolution Is Survival Of The Best Competitor is, indeed, a metaphor, much less a Strict Father metaphor. One way to reveal its metaphorical character is to contrast it with a metaphor for evolution that takes the perspective of Nurturant Parent morality: Evolution Is The Survival Of The Best Nurtured. Here "best nurtured" is taken to include both literal nurturing by parents and others and metaphorical nurturing by nature itself. Where fitting an ecological niche is being metaphorized as winning a competition in one case, it is metaphorized as being cared for by nature in the other. Both are metaphors for evolution, but they have very different entailments, especially when combined with the metaphor Natural Change Is Evolution and the folk theory that evolution yields the best result. Putting these together yields a very different composite metaphor for natural change, namely, Natural Change Is The Survival Of The Best Nurtured, which produces the best result.

Applied to the issue of whether public schools should be privatized, this metaphor would entail that they should not be. Rather, public schools need to be "better nurtured," that is, given the resources they need to improve: better-trained and better-paid teachers, smaller classes, better facilities, programs for involving parents, community involvement, and so on.

The point is that both conceptions of evolution—as survival of the best nurtured and as survival of the best competitors—are metaphors. They are not literally about evolution, but arise instead from moral theories. Natural Change Is Evolution is not a literal truth. It too is a metaphor. Not all natural changes work by the mechanisms of evolutionary biology, which literally concerns biological species, reproduction, and ecological niches in the physical environment only. Moreover, it is not a truth that evolution produces the "best" result; it is only a folk theory and one with no basis in fact.

Once again the tools of cognitive science allow us to see what we otherwise might not see. In these examples, it was the cognitive unconscious and metaphorical thought that were illuminating. Let us turn for our last example to a case in which embodiment helps us to comprehend what is often presented to us as incomprehensible and disembodied: spiritual experience.

The Embodied Mind and Spiritual Life

Your body is not, and could not be, a mere vessel for a disembodied mind. The concept of a mind separate from the body is a metaphorical concept. It can be a consequence, as it was for Descartes, of the Knowing Is Seeing metaphor, which in turn arises from the embodied experience since birth of gaining knowledge through vision. The concept of a disembodied mind is also a natural concomitant of the metaphorical distinction between Subject and Self.

It is crucial to understand, at this point, exactly why the Subject is independent of the Self in these metaphors. Recall from Chapter 13 the three most fundamental forms of experience from which the primary Subject-Self metaphors arise:

1. The correlation between body control and the control of physical objects.
2. The correlation between being in one's normal surroundings and being able to readily control the physical objects in one's surroundings.
3. The correlation between how those around us evaluate our actions and the actions of others and how we evaluate our own actions.

Each experience is an embodied experience. In each case, there is a Person that is the source-domain model for the Subject. In (1), it is the Person manipulating the physical objects; in (2), it is the Person located in familiar surroundings; and in (3), it is the Person evaluating the actions of other people.

In each primary metaphor, that Person, who has an independent existence, maps onto the Subject. Because the general Subject-Self metaphor arises from these primary experiences, and because in each case the Person that maps onto the Subject has an independent existence, so the Subject must have an existence independent from the Self.

Moreover, because these kinds of experiences are part of everyone's daily life throughout the world, the corresponding primary metaphors, wherever they arise, will take a form in which the Subject has an existence independent of the Self. In short, our very concept of a *disembodied* mind arises from *embodied* experiences that every one of us has throughout our life.

Furthermore, these metaphors express a common phenomenological experience we all have: In virtually all of our acts of perception, the bodily organs of perception (eyes, ears, nose, tongue, skin) are not what we are attending to. For example, when we walk down the street and look at a house, we are normally not attending to our eyes, much less to the visual system of our brains. The fact that what we attend to is rarely what we perceive with gives the illusion that mental acts occur independent of the unnoticed body (C2, Leder 1990).

In summary, we all have a metaphor system that conceptualizes our minds as disembodied. We all have constant phenomenological experience that reinforces the illusion of a disembodied Subject. Yet, cognitive science shows that our minds are not, and cannot be, disembodied. Moreover, as we have just seen, cognitive science explains why we *think* that our minds are disembodied.

What difference does it make if there is no such thing as a disembodied mind? Why does it matter if reasoning with a rich conceptual system like ours requires having bodies and brains basically like ours? Of what consequence is it that our metaphorical Subject-Self split is only metaphorical, that there is no Subject independent of the body that can leave the Self or float from Self to Self? And why is it important to understand where such metaphors come from, how they emerge from embodied experience, and why they arise spontaneously around the world?

All this matters vitally in the realm of spiritual and religious life. What we have called variously the Subject or the disembodied mind is called in various religious traditions the Soul or Spirit. In spiritual traditions around the world, the Soul is conceptualized as the locus of consciousness, subjective experience, moral judgment, reason, will, and, most important, one's essence, that which makes a person who he or she is.

One might imagine a spiritual tradition in which such a Soul is fundamentally embodied—shaped in important ways by the body, located forever as part of the body, and dependent for its ongoing existence on the body. The results about the mind discussed throughout this book in no way rule out the existence of *that* kind of Soul, an embodied Soul.

But that is not the way the Soul or Spirit is conceptualized in a great many of the world's spiritual traditions, and as we have just seen, there is a good scientific reason why. The universal embodied experiences that give rise to the metaphors of Subject and Self produce in our cognitive unconscious a concept of a Subject as an independent entity in no way dependent for its existence on the body. Because of those universal embodied experiences, this idea has arisen in many places spontaneously around the world.

And yet, as commonplace and "natural" as this concept is, no such disembodied mind can exist. Whether you call it mind or Soul, anything that both thinks and is free-floating is a myth. It cannot exist.

Requiring the mind and Soul to be embodied is no small matter. It contradicts those parts of religious traditions around the world based on reincarnation and the transmigration of souls, as well as those in which it is believed that the Soul can leave the body in sleep or in trance. It is not consistent with those traditions that teach that one can achieve, and should aspire to achieve, a state of pure consciousness separate from the body.

It is also at odds with one of two traditions in Christianity, what Marcus Borg has called the "monarchial" model, one in which God is distant. He contrasts this with what he calls "spirit" models, in which God is immanent (E, Borg 1997, chap. 3). The tradition of the distant, monarchial God is centered around the idea that we are essentially disembodied Souls not of this world, that we are inhabiting our bodies only during an earthly sojourn, and that our ultimate purpose is to "dwell with God" elsewhere, in heaven, not on earth. In the distant-God Christian tradition, morality is tied to the disembodiment of Soul. Christians are supposed to live a holy life focused on transcending all the things of this world—bodily desire, material possessions, fame, worldly success, and long life. This disembodied, otherworldly conception of spirituality and transcendence downplays one's relation to the world, the natural environment, and all other aspects of embodied existence. Christians are commanded to act morally toward others and to be good stewards of the earth, because that is what God requires for their salvation, so that they may go to heaven and be united with God in the realm beyond this earthly world. Thus that form of Christianity ties morality, and the reason to be moral, to the disembodiment of mind and Soul.

Spiritual Experience Is Embodied

If there is no disembodied mind or Soul, then what is the locus of the real spiritual experience that people have in cultures around the world? This experience can only be embodied. It must be a consequence of what is happening in our bodies and brains. Exactly how the body and brain give rise to spiritual experience is an empirical question for cognitive science and one well beyond the scope of this book.

What we can begin to address, however, is a much more limited question, though an important one. The concept of spirituality in our culture has been defined mostly in terms of disembodiment and transcendence of this world. What is needed is an alternative conception of embodied spirituality that at least begins to do justice to what people experience.

What embodied sense can be made of *transcendence?* How are we to understand our sense of being part of *a larger all-encompassing whole,* of *ecstatic* participation—with awe and respect—within that whole, and of the *moral engagement* within such experience? Where is the *mystery* to be found in a spiritual experience that is embodied? And what is *revelation* there? Finally, what does the concept of *God* become in an embodied spirituality?

Intimations of the Spiritual in the Cognitive

The embodied mind is part of the living body and is dependent on the body for its existence. The properties of mind are not purely mental: They are shaped in crucial ways by the body and brain and how the body can function in everyday life. The embodied mind is thus very much of this world. Our flesh is inseparable from what Merleau-Ponty called the "flesh of the world" and what David Abram (E, Abram 1996) refers to as "the more-than-human world." Our body is intimately tied to what we walk on, sit on, touch, taste, smell, see, breathe, and move within. Our corporeality is part of the corporeality of the world.

The mind is not merely corporeal but also passionate, desiring, and social. It has a culture and cannot exist culture-free. It has a history, it has developed and grown, and it can grow further. It has an unconscious aspect, hidden from our direct view and knowable only indirectly. Its conscious aspect characterizes what we take ourselves as being. Its conceptual system is limited; there is much that it cannot even conceptualize, much less understand. But its conceptual system is expandable: It can form revelatory new understandings.

A major function of the embodied mind is empathic. From birth we have the capacity to imitate others, to vividly imagine being another person, doing what that person does, experiencing what that person experiences. The capacity for imaginative projection is a vital cognitive faculty. Experientially, it is a form of "transcendence." Through it, one can experience something akin to "getting out of our bodies"—yet it is very much a bodily capacity.

Recall (from Chapter 3) that, in dreaming, the high-level motor programs of our brains can be active and connected to our visual systems while their input to our muscles is inhibited. In preparing to imitate, we empathically imagine ourselves in the body of another, cognitively simulating the movements of the other. That cognitive simulation, when "vivid," is the actual activation of motor programs with input to the muscles inhibited, which results in the "feel" of movement without moving. The experience of such a "feel" is a form of empathic projection. There is nothing mystical about it. It is what we do when we imitate. Yet this most common of experiences is a form of "transcendence," a form of *being in the other*.

Imaginative empathic projection is a major part of what has always been called spiritual experience. Meditative traditions have, for millennia, developed techniques for cultivating it. Focus of attention and empathic projection are familiar cognitive capacities that, with training, can enhance our sense of being present in the world.

Empathic projection is, within Nurturant Parent morality, also the major capacity to be developed in the child. Empathy—the focused, imaginative experience of the other—is the precondition for nurturant morality. Empathy links moral values to spiritual experience.

Empathic projection is possible not only with other people, but with animals. Pet owners commonly identify with the experience of their pets. Ethologists, as part of their close observation of animals, develop the capacity to feel in their bodies what the animals they are observing are doing. Part of the awe we feel observing animals in the wild—the trotting coyote, the soaring eagle, the playful porpoise, the aggressive jay, and the magnificent diving whale—is that we too feel some of the sense of trotting and soaring, playing and diving. Shamans in aboriginal cultures around the world observe animals closely by empathically "becoming" the animals, and ritual practices in a wide range of aboriginal religions employ the movements of animals to achieve an ecstatic experience, an experience of being in the body of a very different kind of being.

Empathic Projection and Immanence

The environment is not an "other" to us. It is not a collection of things that we encounter. Rather, it is part of our being. It is the locus of our existence and identity. We cannot and do not exist apart from it. It is through empathic projection that we come to know our environment, understand how we are part of it and how it is part of us. This is the bodily mechanism by which we can participate in nature, not just as hikers or climbers or swimmers, but as part of nature itself, part of a larger, all-encompassing whole. A mindful embodied spirituality is thus an ecological spirituality.

An embodied spirituality requires an aesthetic attitude to the world that is central to self-nurturance, to the nurturance of others, and to the nurturance of the world itself. Embodied spirituality requires an understanding that nature is not inanimate and less than human, but animated and more than human. It requires pleasure, joy in the bodily connection with earth and air, sea and sky, plants and animals—and the recognition that they are all more than human, more than any human beings could ever achieve. Embodied spirituality is more than spiritual *experience*. It is an ethical relationship to the physical world (E, Abram 1996; Spretnak 1991, 1997).

Such an empathic connection with the more than human world is seen in many religious traditions as an encounter with the divine in all things. In theology, this is technically called *panentheism*. Here is Marcus Borg's description (E, Borg 1997):

Panentheism as a way of thinking about God affirms both the transcendence of God and the immanence of God. For panentheism, God is not a being "out there." The Greek roots of the word point to its meaning: *pan* means "everything," *en* means "in," and *theos* means "God." God is more than everything (and thus transcendent), yet everything is in God (hence God is immanent). For panentheism, God is "right here," even as God is also more than "right here."

In the Jewish mystical tradition, the Kabbalah (E, Matt 1995, 24) views God in the same way:

Do not say, "This is a stone and not God." God forbid! Rather all existence is God, and the stone is a thing pervaded by divinity.

Here is a metaphor for God in which empathic projection onto anything or anyone is contact with God.

This is an embodied spirituality based in empathy with all things. The primacy of empathy is at the center of Nurturant Parent morality. It is this empathetic dimension of spiritual experience that links the spiritual to the moral via nurturance—to the responsibility to care for that with which we empathize. It is thus an activist moral attitude not just toward individuals, but toward society and the world.

But empathic connection to the world is only one dimension of spirituality that the body makes possible. It is the body that makes spiritual experience passionate, that brings to it intense desire and pleasure, pain, delight, and remorse. Without all these things, spirituality is bland. In the world's spiritual traditions, sex and art and music and dance and the taste of food have been for millennia forms of spiritual experience just as much as ritual practice, meditation, and prayer.

The mechanism by which spirituality becomes passionate is metaphor. An ineffable God requires metaphor not only to be imagined but to be approached, exhorted, evaded, confronted, struggled with, and loved. Through metaphor, the vividness, intensity, and meaningfulness of ordinary experience becomes the basis of a passionate spirituality. An ineffable God becomes vital through metaphor: The Supreme Being. The Prime Mover. The Creator. The Almighty.

The Father. The King of Kings. Shepherd. Potter. Lawgiver. Judge. Mother. Lover. Breath.

The vehicle by which we are moved in passionate spirituality is metaphor. The mechanism of such metaphor is bodily. It is a neural mechanism that recruits our abilities to perceive, to move, to feel, and to envision in the service not only of theoretical and philosophical thought, but of spiritual experience.

*   *   *

Cognitive science, the science of the mind and the brain, has in its brief existence been enormously fruitful. It has given us a way to know ourselves better, to see how our physical being—flesh, blood, and sinew, hormone, cell, and synapse—and all things we encounter daily in the world make us who we are.

This is philosophy in the flesh.

Appendix :
The Neural Theory of
Language Paradigm

Three Models of the Embodiment of
Mind and Language

How can brains function as minds? Our brains consist of enormously complex and highly structured networks of neurons. How do the particular, intricate neural structures that human brains have yield the full range of human concepts? Exactly what kind of neural structures characterize what kinds of concepts and why? How do the neural systems in human brains learn the specific kinds of concepts they learn and the language that expresses those concepts?

These are the questions that have been taken up in the Neural Theory of Language (NTL) research group at the International Computer Science Institute (ICSI) at Berkeley, a collaboration since the late 1980s between Jerome Feldman, George Lakoff, and their students. At the heart of the modeling effort is Feldman's idea of structured connectionism (B2, 1982, 1985, 1988), which can be used to model highly specific brain structures. The central enterprise of the group is to provide neural models of embodied cognition, especially the acquisition and use of language and thought as described in cognitive linguistics. The group has also included Lokendra Shastri, whose theory of neural binding is currently used in much of the group's modeling efforts (B2, Shastri and Ajjanagadde 1993; Shastri 1996). Structured connectionism, from this perspective, becomes the central link between language and thought, on the one hand, and the highly specific neural structures of the brain, on the other, since it can simultaneously model neural computation and the forms of computation required by language and thought.

The group's research consists of the neural modeling of tasks involving the learning and use of human concepts and language. Within such an enterprise, one can discover with great precision how specific neural structures of the kind found in the brain can learn the specific kinds of concepts that are central to human language.

The job of this kind of neural modeling is to understand *how* a network of neurons doing neural computation can perform such a task, not *where* in the brain it is performed. The where-question is the leading research question for neuroscience; the how-question requires neural modeling research. Both questions must ultimately be asked, and their answers must mesh.

In order to separate the questions and ask them precisely, the neural modeling project at ICSI has constructed the following research paradigm.

The NTL Research Paradigm

Since there is an enormous gap between physical brain structures and the level of human concepts and language, the NTL group has developed a paradigm for ultimately bridging that gap in a small set of precise steps, using research methodologies already in place within cognitive science. The job of the paradigm is to provide an overall unified methodology for cognitive science.

The NTL paradigm comes in two parts. First, there is a common paradigm shared widely throughout virtually all of contemporary cognitive science, in which there is a description of high-level cognition at the top level, a description of the relevant neurobiology at the bottom level, and an intermediate level of neural computation relating these. The job of the neurocomputational level is both (a) to model the workings of the neural system described at the neurobiological level and (b) by virtue of modeling the neural system, to show via methods of neural computation how the cognitive effects at the top level are achieved by the neurobiology at the bottom level.

THE COMMON PARADIGM

Top Level: Cognitive
Middle Level: Neurocomputational
Bottom Level: Neurobiological

Going directly from the level of cognition to the level of neurobiology is a giant step. The strategy of the NTL group at Berkeley is to break the giant step of neural computation down into three levels so that the task becomes more manageable. The result is a five-level paradigm developed by Feldman and his students David Bailey and Srini Narayanan:

THE NTL PARADIGM

Level 1: Cognitive Science and Cognitive Linguistics
Level 2: Neurally Reducible Conventional Computational Models

Level 3: Structured Connectionist Models
Level 4: Computational Neuroscience
Level 5: Neuroscience

Consider, for example, the link between Level 4 and Level 5. Computational neuroscience models the brain as if it were "circuitry," with axons and dendrites seen as "connections" and with activation and inhibition as positive and negative numerical values. Neural cell bodies are conceptualized as "units" that can do basic numerical computations (adding, multiplying, etc.). Synapses are seen as points of contact between connections and units. Chemical action at the synapses determines a "synaptic weight"—a multiplicative factor. Learning is modeled as change in these synaptic weights. Neural "firing" is modeled in terms of a "threshold," a number indicating the amount of charge required for the "neural unit" to fire. The computations are all numerical.

Since brain circuitry is enormously complicated, structured connectionist models seek a simplified representation of such circuitry, in which equivalent computations are carried out by neural circuitry of minimal complexity. Thus, the link between Level 3 and Level 4 is one of simplification: the minimal structured circuitry that will do the same job as models of the actual "brain circuitry."

Sometimes the relation between the level of analysis in cognitive science and cognitive linguistics and the structured connectionist level can be given directly, as in the model by Regier discussed below, in which Level 2 is skipped over since it is unnecessary.

But even the simplest neural models that carry out a complex task can be so complicated that an intermediate level of representation is helpful in doing the modeling. In the NTL paradigm, conventional computational systems from computer science are sought out that have the basic properties of neural systems: parallel operation, distributed control (no internal clock or centralized controller), ability to react quickly and effectively to changing contexts, resource dependency, and so on. Modified Petri nets, for example, have these properties. Such conventional computational mechanisms are only used when it is known how to map them directly to structured connectionist neural models, and it is common for novel forms of conventional-style models to be constructed along these guidelines. Thus, only neurally reducible computational models are used in the modeling. Very few conventional computational systems satisfy all the requirements.

The circle from Level 5 back to Level 1 is closed by neuroscience research, which provides direct evidence linking cognitive science and linguistics to the actual brain. The links from Levels 1 to 2 to 3 to 4 to 5 concern computation. They are attempts to answer the how-question, how networks of neurons can, via neural computation, characterize thought and language. The direct link from Level 1 to Level 5 is an attempt to answer the where-question: Where in the brain are the computations performed?

No ontological relevance is ascribed to the level of conventional computations. It is there purely for convenience, to make the job of neural modeling easier. The neural computations attributed to the brain are characterized on Level 3, the structured connectionist level. These must be equivalent to the neural computations carried out by the more complex brain circuitry on Level 4, which more directly models the details of the physical brain.

Three Models

Real neural networks in the brain, by their nature, do things. The purpose of neural modeling is to find out how they can do what they do. That means that neural modeling research is the study of tasks that neural systems carry out.

The NTL group has, so far, undertaken three major neural modeling tasks:

1. The Spatial-Relations Learning Task
2. The Verbs of Hand Motion Learning Task
3. The Motor Control and Abstract Aspectual Reasoning Task

In each case, it has been shown that neural structures modeling aspects of the perceptual and motor systems *can* carry out the given task for concepts, and that, so far as anyone can tell thus far, those perceptual and motor models *are required* to carry out the task. In each case, the modeling was carried out in sufficient detail to prove the point.

To get some appreciation for what these modeling experiments have achieved, here is a brief account of each task.

The Spatial-Relations Learning Task

A few kinds of simple figures (squares, circles, triangles) are presented in various spatial relations, both static and moving (*in, on, through, above*, and so on) on a simple computer model of a retina (n x m pixels). One figure is chosen as Landmark and the other as Trajector (e.g., if the circle is under the square, the square is Landmark and the circle is Trajector.)

Native speakers of various languages (e.g., Russian, Bengali, Chinese, English) give the spatial-relations term for the relation (e.g., *under* in a scene in which the circle is under the square). All this is the input to a computational neural model. The job of the model is to learn the spatial-relations system of the language and the spatial-relations terms so that the neural system can correctly give the right names for new spatial configurations presented on the computer screen. Part of the challenge is to show how the learning can take place with no negative evidence, that is, without being told what answers are wrong.

This task was undertaken by Terry Regier in his dissertation and his results are published in *The Human Semantic Potential* (B2, Regier 1996). Regier approached the task starting with the basic results described above about spatial relations from the field of cognitive linguistics.

Regier's Model

Regier first tried standard PDP (parallel distributed processing) connectionist models whose only significant neural structure consists of a fully connected input layer of neurons, an output layer, and one or more "hidden layers" in between trained by association. PDP networks learn via statistical associations. Regier was not able to get any pure PDP models to learn static spatial relations such as In or Above. Such models appear to have technical limitations that make it impossible to learn such tasks.

Regier then asked whether a hybrid model would work. The hybrid consisted of two parts: (1) A structured connectionist model of particular neural structures of the sort that have been found in the visual system of the brain. Its job was to learn the spatial features (e.g., containment, contact, aboveness) relevant for characterizing the structure of spatial relations and for making minimal distinctions among them. (2) A PDP connectionist model for learning via back propagation. Given the features computed by (1), its job was to learn how these features were associated in lexical items.

Regier found that the structured part of the hybrid model could pick out the relevant features, while the PDP associative part of the model could learn how they come together in particular words. Such a hybrid model could both represent the structure of a significant range of spatial-relations concepts for a significantly wide range of languages and learn the terms for those concepts in the languages tested.

Indeed, Regier's model could *learn without negative evidence.* Conventional PDP connectionist models must be trained on both positive examples of a categorization to be learned and negative examples that do not fit the categories. But human beings learn without negative examples. Regier's learning model duplicated the human feat, the first model to do so. Its accuracy with no negative evidence is extraordinarily high (in the 0.999 range most of the time). Moreover, it accurately displayed prototype effects (degrees to which the spatial-relations term was appropriate) without being trained on prototypes.

What allowed the system to learn and represent spatial relations was his modeling of neural structures of the sort found in the brain's visual system. Regier's overall model made use of models of the following neural structures:

- Topographic maps of the visual field: layers of cells in the visual cortex organized so that cells near each other in the layer respond to stimuli near each other in the retina. Such maps preserve nearness: Nearness of stimuli in the input results in nearness of cell activation in the output.
- Orientation-sensitive cells: cells that respond maximally to lines at a given orientation.

- Center-surround receptive fields: Certain cells respond maximally when they take inputs of a certain kind from a central area and of another kind from a surrounding area, for example, a green center and a red surround.
- Neural gating: an architecture allowing cell A to fire if it gets activation from cell B, but only if it also gets activation from cell C.
- Filling-in: an architecture allowing the flow of activation in a map from outside to inside, known to exist from the work of Ramachandran and Gregory (B1, 1991).

(For the full details of the use of how Regier's neural model makes use of results from neuroscience, perceptual psychology, and psychophysics, see Regier B2, 1996).

The key to what allowed Regier to make his model work is the following idea. As we mentioned above, Len Talmy (A8, 1983) had shown that certain elementary spatial-relations concepts are topological in nature. For example, the container schema remains a container no mater how much you bend its boundary or how big or small you make it. Regier modeled topological spatial relations using three kinds of structures working together: (1) topographic maps of the visual field, which occur in our visual systems; (2) center-surround receptive fields; and (3) a filling-in mechanism of the sort found by Ramachandran and Gregory (B1, 1991). Talmy had also observed that certain spatial-relations primitives are orientational in nature, for example, front and back, up and down. Regier conjectured that orientational spatial primitives could be modeled by orientation-sensitive cells. Regier's conjectures allowed him to construct a model that worked for a limited but important range of cases (two dimensions, not three; no force dynamics).

The limitations of Regier's results are important to bear in mind. He took models of parts of the perceptual system to characterize spatial relations. But there is no reason to believe that the brain does it in just the way specified in his model. Indeed, there are good reasons to believe it does not. He did show that such neural apparatus from the visual system *can* characterize the elementary spatial-relations concepts sufficiently well so that novel cases can be learned for a wide range of cases in rather different languages. This is no mean feat, but it is only suggestive.

What Regier's model does is give us our first glimpse of how the structure of the brain, together with our bodily capacities for perception and movement *might* create concepts and forms of reason. It is a kind of existence proof. In the model, linguistic and conceptual *categories* are created using *perceptual* apparatus from the visual system, and language is accurately learned for those concepts. Regier's results suggest that the absolute distinction in faculty psychology between the *perceptual* and the *conceptual* is illusory. In Regier's model, linguistic and conceptual *categories* that are about space are created using the plausible neural mechanisms of spatial *perception*.

The word *create* here is all-important. Conceptual categories of spatial relations are created as the result of the structure of our brains plus our experience of our bodies and how they function in space and how things are named in our language. Conceptual categories of spatial relations are not "things" that exist in the world independently of living beings and just happen to be instantiated in human brains. Spatial conceptual categories come into being because of the bodies and brains and spatial experiences we have.

Regier's results are limited, but the very fact that *perceptual* mechanisms can serve *conceptual* functions is suggestive. Consider the so-called abstract uses of spatial-relations terms, as in "I'm *in* a depression," "Prices went *up*," and "He's *beside* himself." As we saw (Chapters 4 and 5), these and other nonspatial uses of spatial-relations concepts are given their nonspatial meanings systematically by conceptual metaphors that preserve their spatial logics. The point is that even nonspatial metaphorical uses of spatial concepts in reasoning use the same mechanisms of *perceptual inference* as are used by the spatial concepts.

In such cases, it is not far-fetched to assume that the neural mechanism permitting this is like the one used in dreaming, in which we use the visual system of our brains to "see" images based on input not coming from our retinas. Similarly, perceptual mechanisms, freed by neural inhibition from their purely perceptual functions, could be used to perform abstract reason.

Again, we see the importance of the idea of embodied realism. Spatial-relations concepts arise from our bodies and brains together with visual input and are not objectively existing entities in space. But we have evolved so that we can use them to impose a conceptual structure on the world that allows us to function well in the kind of space we live in. When we see bees *in* the garden, there is no physical container walling in the bees. But the mental container we impose on the garden allows us to walk around the space "containing" the bees. The embodied realism of our conceptual system allows us to function well, day in and day out. We take our spatial relations for granted because they work for us. But it is mistaken to think that they are just objectively given features of the external world. Instead, we make the best of what our brain, especially our visual system, affords us.

Concepts of Bodily Movement

Do bodily movement and motor control enter into the definition of concepts? Since we have concepts of bodily movements, it would be awfully strange if the range of possibilities for bodily movements did not enter into the range of possibilities for concepts of bodily movement in the world's languages.

Consider for a moment some of the verbs used for hand motion in English:

seize, snatch, grab, grasp, pick up, hold, grip, clutch, put, place, lay, drop, release, pull, push, shove, yank, slide, flick, tug, nudge, lift, raise, lower, lob, toss, fling, tap, rap, slap, press, poke, punch, rub, shake, pry, turn, flip over, rotate, spin, twirl, squeeze, pinch, twist, bounce, stroke, wave, caress, stack, salute, and many, many more . . .

The conceptual system of English must be capable of making all the conceptual distinctions among these verbs.

But that's only English. Other languages make distinctions that English doesn't. Moreover, each language has its own unique collection of linguistic gaps that reflect conceptual differences in the concepts named. Here a few examples:

In Tamil, *thallu* and *ilu* correspond to English *push* and *pull*, except that they connote a sudden action as opposed to a smooth continuous force. The latter reading can be obtained by adding a directional suffix, but there is no way to indicate smooth pushing or pulling in an arbitrary direction.

In Farsi, *zadan* refers to a large number of object manipulations involving quick motions. The prototypical *zadan* is a hitting action, though it can also mean *to snatch* (*ghaap zadan*) or to strum a guitar or play any other musical instrument.

In Cantonese, /mīt/ covers both pinching and tearing. It connotes forceful manipulation by two fingers, yet is also acceptable for tearing larger items when two full grasps are used.

It should be clear that what the body can do enters into, and helps to define, the *conceptual* range covered by each verb in each language.

The Learning Task for Verbs of Hand Motion

The task is to create a computational model of a neural network that can learn the verbs used for hand motion in an arbitrary language. Given a hand action, the network should be able to conceptualize the action correctly so as to name it correctly. In addition, given a name of an action, the network should be able to give the right instructions to a model of an arm so that the arm can perform the action correctly.

This task was carried out by David Bailey in his Berkeley dissertation (B2, Bailey 1997). To start, Bailey needed a computerized model of a body, with all the indicated muscles and joints. Luckily, one was available: Jack, at the University of Pennsylvania. To carry out the arm movements correctly, a collection of motor synergies would have to be assumed, taken from the literature on motor control. Synergies are low-level, self-contained motor actions like pivoting the wrist, tightening a grip, releasing a grip, or extending the index finger. In addition, Bailey would need a mechanism for executing motor schemas coordinating all the synergies in the appropriate sequence in real time and being able to adjust to conditions in the world.

Working with Srini Narayanan, Bailey accomplished this using an appropriately adapted version of Petri nets, a well-understood computational mechanism from computer science. To get a neural model, Bailey and Narayanan showed how the modified Petri nets could be mapped onto a structured connectionist network.

Computational to Connectionist Mappings

A modified Petri net is a computational mechanism structured as follows:

- There are "transitions" that "do things" in real time, that is, they activate other processes in an appropriate sequence with concurrence of processes.
- Before and after each transition is a "place," a state of part of the system.
- Transitions only occur if they have sufficient "resources." A unit of such a resource is called a "token." Each transition requires a certain number of tokens in order to "fire." That number is called a "threshold."
- The firing of transitions is thus not governed by a clock or a central controller. A transition fires whenever it has the appropriate resources, that is, whenever enough tokens of a "resource" are in the "places" just before the transition.
- The firing of the transition removes a certain number of tokens (that number is called "weight 1") from the places before the transition and puts a certain number (called "weight 2") of tokens in the places after the transition.
- It takes an amount of time for a transition to fire. This is called a "delay."
- In a modified Petri net, a number of tokens in a given place for long enough can "decay," that is, there can be a lessening of the number of tokens in a place if they stay there long enough.
- Weight 1, the threshold weight, can decrease over time.
- The transitions can branch out from a place and converge to a place from various other places, using up and producing resources as it progresses.
- It takes input from other networks and gives output to them.
- Because transitions fire whenever they have enough tokens in the input place, Petri nets are asyncronous—instead of firing at given time, they fire whenever their needs are satisfied. Thus they are also controlled locally.

A Petri net is, therefore, a network with branchings and convergences possible at each "place." Processing is the "firing" of transitions that "start" at one place and "move" to another place, using up resources and producing more resources as the processes take place. Each Petri net can take input from and give output to other nets.

Petri nets of the type used in NTL modeling can be mapped onto structured connectionist models. Here are the correspondences used to guide the mapping. It is by means of such a mapping that a well-understood computational mechanism can be reduced to a neural model.

The Reduction Of Petri Nets To Structured Connectionist Models

Transitions	$\rightarrow$	Small Clusters Of Neurons
Places	$\rightarrow$	Synapses
Token Count	$\rightarrow$	Level Of Activation
Weight $_1$	$\rightarrow$	Threshhold
Weight $_2$	$\rightarrow$	Synaptic Weight (Gain)
Duration Of Transition	$\rightarrow$	Time Neuron Takes To Fire
Token Decay	$\rightarrow$	Activation Decay
Change In Weight $_1$	$\rightarrow$	Change In Threshold

Petri nets model just one aspect of neural behavior, namely, neural control systems operating in real time. These are called *executing schemas,* or *X-schemas* for short. To model the learning of verbs of hand movement, more is needed than Petri nets. Bailey used two other computational mechanisms: (1) attribute-value structures to model the relationship between parameters and their values for low-level motor synergies, and for the feature-structure of verbs; and (2) Steve Omohundro's Model Merging Algorithm (B2, 1992) to model connectionist recruitment learning at the computational level. These two conventional computational devices, taken together with Petri nets, support the full task at the computational level. They can then be mapped directly onto the structured connectionist level, where all those computational tasks are carried out by a single complex neural structure. By virtue of the computational-to-neural-level mappings, the exact computational functions carried out by the complex network at the neural level can be understood.

The mechanisms at the computational level (Level 2) form a bridge between Levels 1 and 3, since they do two modeling jobs at once: the modeling of cognitive and linguistic behavior at Level 1 and the modeling of computational function by the neural structures at Level 3.

The Bailey Model

Bailey's idea was to perform the learning task in the following manner. He began with videos of Jack performing hand movements and then had informants label the movements with verbs. Bailey then constructed a learning mechanism, including all three kinds of computational devices, that would learn the verbs from the actions, with the computer acting as the agent carrying out the actions. Bailey's model contrasts with Regier's, in which the computer was acting as observer. The system learned the verbs so that it could both (1) recognize an action and name it correctly and (2) perform the correct action, given the verb.

The resulting system works quite well. On the sample of 18 verbs from English used, the recognition rate was 78 percent and the command-obeying rate was 81 percent. The model has also been tried on Farsi, Russian, and Hebrew data.

Implications

Here is the philosophical significance of Bailey's system: The system matches words directly with motor schemas in the form of neural networks capable of giving the appropriate signals to motor synergies that can move the body, in this case the arm. The "conceptual structure" is the system for controlling the body. Yet the system does learn the correct distinctions among the words for hand movements quite reliably. In short, the fundamental *conceptual roles* for making the right *linguistic distinctions* among the verbs are played by *features of the motor system.*

It is worth stopping for a moment to consider how strange this will sound to any scholar trained within the tradition of faculty psychology, in which anything conceptual is simply a different kind of thing than anything physical like a motor schema or a motor synergy. Any talk of features of the motor system doing the job of conceptual structures will sound like a category mistake—even for verbs whose subject matter is bodily movement. Of course, scholars trained in that tradition tend not to study verbs characterizing bodily actions.

What needs to be borne in mind is that a motor schema is not a subcortical motor synergy, but rather a cortical structure that is connected to subcortical synergies. It is a highly structured neural network that characterizes the overall structure of a bodily movement, linking together all the parts of the movement and all the right values of parameters like force and direction.

Concepts of bodily movement expressed in natural language often include information about more than just movement—goals, social factors, and so on. But suppose we separate off such additional factors for the moment and concentrate only on the part of the concept that is about bodily movement. From the perspective of neural computation—the perspective of the information contained and represented at the neural level—exactly what is the difference between the neural characterization of (1) a concept characterizing the overall organization and performance of a complex movement made up of simpler movements and (2) a higher-order motor schema that can actually organize and perform such a complex movement by activating simpler movements? From the perspective of information processing at the neural level, there is no difference at all! That is, at the neural level, the conceptual structure of a body-movement concept would be doing the same kind of job as the neural structure actually carrying out the movement. It should not be that surprising in such cases that the neural conceptual structure characterizing complex bodily movements should look like the neural schema capable of controlling and carrying out the same complex bodily movements.

Suppose we took the core meaning of a motor concept to *be* the motor control schema (or schemas) used to perform the movement. Could the neural motor schemas actually do the work of the neural conceptual schemas? There are at least two types of semantic work that have to be done. The first semantic job is that all the concepts expressed by the verbs of bodily movement in the given language have to be distinguished from one another. Bailey's verb-learning task is a test of that capacity. The fact that his system works to learn the verbs correctly indicates that the neural system can carry out the semantic job of making conceptual distinctions.

The second semantic job that has to be done by concepts is inferential in nature: The conceptual schemas for such verbs must be able to carry out all the relevant inferences. As we shall see shortly in our discussion of Narayanan's research, the "logical," that is, inferential properties of the verbs can be characterized by the same neural schemas that can control body movement. The actual control of movement, on the one hand, and the inferential structures about the control of movement, on the other, contain the same information. At the level of neural computation, the same information is characterized in performance as in reasoning about performance.

This is a startling result. From the perspective of faculty psychology, where mind and body are just different kinds of things, it should be impossible. Yet, from the perspective of neural modeling, it is the only thing that makes sense. From that perspective, the information characterizing each detail of a high-level motor action is the same as the information needed to activate the performance of each detail of a higher-level motor action.

The Operation of Motor Schemas Without Muscle Movement

Bailey implicitly assumes that the meanings of verbs of bodily movement are the very motor schemas for carrying out those movements. Bailey is obviously not claiming that the meaning of the verb *grasp* is the actual bodily movement of grasping, that every time you think of the meaning of *grasp* you physically move your hand in a grasping motion.

Rather, his model reflects the implicit assumption that the system of motor schemas and parameter values for motor synergies in the brain can operate without the muscles of the body actually moving. This happens all the time during sleep when we dream. In our dreams we experience our bodies moving. Studies of the brain during dreaming show that the parts of the brain dedicated to motor schemas are active during dreaming, even though our bodies are neurally inhibited from moving during dreaming (B1, Hobson 1994). Since the motor schemas can operate while activation to the muscles is inhibited, it is possible that such inhibition of the neural links from the motor schemas to the muscles occurs when we imagine moving or when we reason about moving without doing it. Understanding the meaning of grasping and reasoning about grasping may activate the motor schema in the brain for grasping even though our muscles are not engaged. Thus, motor schemas that can, and do, control the body could in principle function in reasoning when they are not functioning to control the body.

Movements and Descriptions of Movements

The motor schemas in Bailey's system will carry out "movements" of Jack's virtual body. But concepts for such movements must also be able to fit movements by other people that are being seen. How could a motor schema for performing an action also function to recognize someone else's action?

Consider for a moment what happens when we are imitating someone else's movements. We are able to coordinate how we see someone else's movements with movements of our own. There must, in other words, be neural coordination between our visual system and our motor system.

To see that this is possible, try this experiment: Construct a mental image of someone carrying out some basic movement. Now physically carry out the movement in your image. In general we can do such things. Mental images are computed neurally in the brain's visual system. Here too there must be neural coordination between the visual system and the motor system. That is, there must be neural mechanisms for coordinating neural motor schemas with what is seen or imagined in the visual system.

Now close your eyes, move your hand so as to reach out and lift a cup, and as you do it, form a mental image of yourself doing it. Being able to do this again requires coordination between the motor system and the visual system.

Given such coordinating neural links between the visual and motor systems, it is possible for a motor schema to function as a pattern recognition device for what you see or imagine someone else doing. Similarly, it can be used to generate a mental image of what someone is doing. This means that a motor schema is capable of carrying out a recognition function as well as a control function. Nigel Goddard (B2, 1992) has built a connectionist system that recognizes actions in terms of such schemas.

Bailey's system only shows that the right distinctions among the verbs of hand motion can be made in a learning task on the basis of motor schemas. Bailey did not show that those motor schemas can actually carry out abstract conceptual inferences. Such a demonstration is absolutely necessary if a motor schema is even to be considered a candidate for the meaning of a verb of bodily movement. That demonstration is carried out in Srini Narayanan's companion dissertation (B2, 1997a).

How Motor Control Projects to the General Logic of Events and Actions

Narayanan, in working with Bailey on characterizing motor-control schemas, made an interesting discovery that, in retrospect, should have been obvious. He discovered that all high-level motor-control schemas (above the level of the motor synergy) have the same basic control system structure:

- Getting into a state of readiness
- The initial state
- The starting process
- The main process (either instantaneous or prolonged)
- An option to stop
- An option to resume
- An option to iterate or continue the main process
- A check to see if a goal has been met
- The finishing process
- The final state

First, you have to reach a state of readiness (e.g., you may have to reorient your body, stop doing something else, or rest for a moment). Next, you have to do whatever is involved in starting the process (e.g., to lift a cup, you first have to reach for it and grasp it). Then you begin the main process; while doing it, you have an option to stop, and if you do so, you may or may not resume. For example, you might be lifting the cup. You can then repeat the main process or continue it. You can then check to see if you achieved a purpose you have previously set. Finally, you can do whatever it takes to finish the process, and then you are in a final state.

These are the phases of just about any bodily movement, and Narayanan has worked out the details of how any motor schema can be modeled in such a way as to make all this explicit, in the form of Petri nets that are reducible to structured connectionist neural networks. He has also shown how such structures can bind to low-level motor synergies, perform motor control, embed one control schema within another, and connect different types together in a full range of combinations to produce complex motor schemas from simple ones.

To a linguist, this general control structure is thoroughly familiar. It defines the general structure of actions and events, what linguists call "aspect" (see D, Comrie 1976; Vendler 1967; Dowty 1979; Moens and Steedman 1988; and A8, Langacker 1986). Each action verb has an inherent aspect. For example, *tap* is inherently iterative: the central process is normally repeated (unless otherwise specified); *pick up* has a purpose and a final state; *run* has no inherent final state; *slip* is nonvolitional; *walk* is durative (it takes an extent of time); and *leave* is nondurative (it is instantaneous).

Each language has lexical, grammatical, or morphological means of affecting the aspect of a verb. In English, the word *begin* focuses on the starting phase of an action or event. A form of *have* plus the verb plus the suffix *-ed*, as in "Sam *has* pick*ed* up the cup," indicates completion. A form of *be* plus the verb plus the suffix *-ing* indicates the action is in process, as in "Sam *is* pick*ing* up the cup."

In addition, there is a logic of aspect. A classic example is the so-called perfective paradox: "John is walking" entails that John has walked, while "John is walking to the store" does not entail that John has walked to the store. This problem, and just about all other classical logical and conceptual problems of aspect, turns out to be handled naturally by Narayanan's general schema for motor control. The general control schema can also operate independently of motor control and be used to structure other processes, such as the recognition of an action or the formulation of a complex plan.

Narayanan then set out to demonstrate that the same general control schema that can control, say, hand motions can do abstract reasoning about the structure of events, that is, about aspect. To do so, he picked the following task.

The Motor Control/Abstract Reasoning Task

Pick an abstract nonphysical domain of discourse that is commonly conceptualized and talked about in terms of bodily movements. Narayanan chose international economics.

By searching the Internet, find news stories on international economics in serious publications (The *Economist,* The *Wall Street Journal,* The *New York Times)* that use bodily movements to metaphorically discuss international economics. Examples: "France Falls into a Recession," "Germany Pulls It Out," "India Releases Stranglehold on Business."

Work out the metaphors of bodily movements used to reason and talk about economics.

Construct a neural theory of metaphor that will handle these cases and work generally.

Show how the same neural control system that is capable of performing motor control and does so in Bailey's thesis will carry out the appropriate logical inferences for such news stories.

Narayanan carried out this task (B2, 1997) at the level of neurally reducible computational systems, systems with the basic properties of neural systems that, by known methods, can be readily mapped onto structured connectionist neural models. I will refer to Narayanan's model as "neural," even though, technically, is only directly mappable by known methods onto a neural model.

Take, for example, "France fell into a recession," "Germany pulled it out." The schemas for falling into a hole and pulling someone out of a hole are applied in a model of physical falling and pulling someone out of a hole. These schemas model the physical processes in ways that could, in principle, guide high-level motor control.

Metaphors such as Action Is Motion, A Recession Is A Hole, and More Is Up are modeled by neural connections linking the physical and economic domains.

The physical language in the news story activates a mental simulation of physical action, using neural control structures (with muscle control assumed to be inhibited). The results of the physical simulation are then projected back via metaphorical connections to the domain of economics, constituting inferences about economics made by means of motor-control simulations.

The philosophically important point is that abstract reasoning about economics can be done by the same structured neural network that has the capacity to control high-level motor schemas. The reasoning about economics is clearly part of the human rational capacity. The motor control is part of the capacity for bodily movement.

Narayanan's result does not *prove* that such abstract reasoning about economics using physical metaphors is actually done via our system of motor control. It is, however, another existence proof. Our neural capacities for motor control *can* be used to carry out abstract reasoning. The same neural circuitry that can move the body can be used to reason with.

References

This is a topic-oriented list of references. It includes both works cited and other works that are of either an introductory or a supplementary nature. It is intended to allow the reader entry to the literature, rather than to be exhaustive.

References in the text refer to the letters and numbers that structure this list. The list of categories appears first, then the category-by-category references.

Reference List Organization

The Reference List

A. *Cognitive Science and Cognitive Linguistics*

The Baumgartner-Payr and Solso-Massaro books provide some sense of the range of questions that cognitive scientists consider. Gardner is a history of first-generation cognitive science.

Baumgartner, P., and S. Payr. 1995. *Speaking Minds: Interviews with Twenty Eminent Cognitive Scientists.* Princeton: Princeton University Press.

Gardner, H. 1985. *The Mind's New Science: A History of the Cognitive Revolution.* New York: Basic Books.

Solso, R. L., and D. W. Massaro. 1995. *The Science of the Mind: 2001 and Beyond.* New York: Oxford University Press.

A1. Metaphor Theory

The most popular introduction to the field is the Lakoff-Johnson book. Lakoff 1993 is the most recent general survey of the field prior to the present book. Johnson 1981 is a survey of previous approaches to the study of metaphor. *Cognitive Linguistics* is a general journal devoted not only to metaphor research but to the whole gamut of cognitive approaches to linguistics.

Brugman, C. 1985. The Use of Body-Part Terms as Locatives in Chalcatongo Mixtec. In *Report No. 4 of the Survey of California and Other Indian Languages,* 235–290. University of California, Berkeley.

Cienki, A. Straight. 1997. An Image-Schema and Its Metaphorical Extensions. Manuscript, Emory University.

Fernandez-Duque, D., and M. Johnson. Forthcoming. Attention Metaphors: How Metaphors Guide the Cognitive Psychology of Attention. *Cognitive Science.*

Grady, J. 1997. Foundations of Meaning: Primary Metaphors and Primary Scenes. Ph.D. dissertation, University of California, Berkeley.

———. 1998. The Conduit Metaphor Revisited: A Reassessment of Metaphors for Communication. In J. P. Koenig, ed., *Discourse and Cognition: Bridging the Gap.* Stanford: CSLI/Cambridge.

———. Forthcoming. THEORIES ARE BUILDINGS Revisited. *Cognitive Linguistics.*

Grady, J., and C. Johnson. Forthcoming. Converging Evidence for the Notion of *Subscene.* In J. Moxley and M. Juge, eds., *Proceedings of the Twenty-Third Annual Meeting of the Berkeley Linguistics Society.* Berkeley: Berkeley Linguistics Society.

Grady, J., S. Taub, and P. Morgan. 1996. Primitive and Compound Metaphors. In A. Goldberg, ed., *Conceptual Structure, Discourse, and Language.* Stanford: CSLI/Cambridge.

Johnson, C. 1997a. The Acquisition of the "What's X Doing Y?" Construction. In E. Hughes, M. Hughes, and A. Greenhill, eds., *Proceedings of the Twenty-First Annual*

Boston University Conference on Language Development 2: 343–353. Somerville, Mass.: Cascadilla Press.

———. 1997b. Metaphor vs. Conflation in the Acquisition of Polysemy: The Case of SEE. In M. K. Hiraga, C. Sinha, and S. Wilcox, eds., *Cultural, Typological and Psychological Issues in Cognitive Linguistics*. Current Issues in Linguistic Theory 152. Amsterdam: John Benjamins.

———. 1997c. Learnability in the Acquisition of Multiple Senses: SOURCE Reconsidered. In J. Moxley, J. Juge, and M. Juge, eds., *Proceedings of the Twenty-Second Annual Meeting of the Berkeley Linguistics Society*. Berkeley: Berkeley Linguistics Society.

Johnson, M. 1987. *The Body in the Mind: The Bodily Basis of Meaning, Imagination, and Reason*. Chicago: University of Chicago Press.

———. 1993. Conceptual Metaphor and Embodied Structures of Meaning. *Philosophical Psychology* 6, no. 4: 413–422.

———. 1997. Embodied Meaning and Cognitive Science. In D. Levin, ed., *Language Beyond Postmodernism: Saying and Thinking in Gendlin's Philosophy*, 148–175. Evanston, Ill.: Northwestern University Press.

Johnson, M., ed. 1981. *Philosophical Perspectives on Metaphor*. Minneapolis: University of Minnesota Press.

Klingebiel, C. 1990. The Bottom Line in Moral Accounting. Manuscript, University of California, Berkeley.

Kövecses, Z. 1986. *Metaphors of Anger, Pride, and Love: A Lexical Approach to the Structure of Concepts*. Philadelphia: John Benjamins.

———. 1988. *The Language of Love: The Semantics of Passion in Conversational English*. Lewisburg, Penn.: Bucknell University Press.

———. 1990. *Emotion Concepts*. New York: Springer-Verlag.

Lakoff, A., and M. Becker. 1992. Me, Myself and I. Manuscript. University of California, Berkeley.

Lakoff, G. 1990. Metaphor and War: The Metaphor System Used to Justify War in the Gulf. Distributed by electronic mail, December 1990. Reprinted in Harry Kreisler, ed., *Confrontation in the Gulf: University of California Professors Talk About the War* (Berkeley: Institute of International Studies, 1992). Also in Brien Hallet, ed., *Engulfed in War: Just War and the Persian Gulf* (Honolulu: Matsunaga Institute for Peace, 1991). Also in *Journal of Urban and Cultural Studies* 2, no. 1: 1991. Also in *Vietnam Generation Newsletter* 3, no. 2 (November 1991). Also in *The East Bay Express,* February 1991.

———. 1993. The Contemporary Theory of Metaphor. In A. Ortony, ed., *Metaphor and Thought*. 2d ed., 202–251. Cambridge: Cambridge University Press.

———. 1995. Metaphor, Morality, and Politics, or Why Conservatives Have Left Liberals in the Dust. *Social Research* 62, no. 2: 177–214.

———. 1996a. *Moral Politics: What Conservatives Know That Liberals Don't.* Chicago: University of Chicago Press.

———. 1996b. The Metaphor System for Morality. In A. Goldberg, ed., *Conceptual Structure, Discourse, and Language.* Stanford: CSLI/Cambridge.

———. 1997. How Unconscious Metaphorical Thought Shapes Dreams. In D. J. Stein, ed., *Cognitive Science and Psychoanalysis.* New York: American Psychoanalytic Association.

Lakoff, G., and M. Johnson. 1980. *Metaphors We Live By.* Chicago and London: University of Chicago Press.

Lakoff, G., and R. Núñez. 1997. The Metaphorical Structure of Mathematics: Sketching Out Cognitive Foundations for a Mind-Based Mathematics. In L. English, ed., *Mathematical Reasoning: Analogies, Metaphors, and Images.* Hillsdale, N.J.: Erlbaum.

Lakoff, G., and M. Turner. 1989. *More Than Cool Reason: A Field Guide to Poetic Metaphor.* Chicago and London: University of Chicago Press.

Núñez, R., V. Neumann, and M. Mamani. 1997. Los mapeos conceptuales de la concepción del tiempo en la lengua Aymara del norte de Chile [Conceptual mappings in the understanding of time in northern Chile's Aymara]. *Boletin de Educación de la Universidad Católica del Norte* 28:47–55.

Reddy, M. 1979. The Conduit Metaphor. In A. Ortony, ed., *Metaphor and Thought,* 284–324. Cambridge: Cambridge University Press.

Sweetser, E. 1990. *From Etymology to Pragmatics: Metaphorical and Cultural Aspects of Semantic Structure.* Cambridge: Cambridge University Press.

———. Forthcoming. Our Father, Our King: What Makes a Good Metaphor for God.

Taub, S. 1990. Moral Accounting. Manuscript, University of California, Berkeley.

Turner, M. 1987. *Death Is the Mother of Beauty.* Chicago: University of Chicago Press.

———. 1991. *Reading Minds: The Study of English in the Age of Cognitive Science.* Princeton: Princeton University Press.

Winter, S. 1989. Transcendental Nonsense, Metaphoric Reasoning and the Cognitive Stakes for Law. *University of Pennsylvania Law Review* 137.

———. Forthcoming. *A Clearing in the Forest.* Chicago: University of Chicago Press.

A2. Experimental Studies in Metaphor

The journal *Metaphor and Symbol* is devoted primarily to empirical psychological research on metaphor. Gibbs 1994 is an excellent overview of that research.

Albritton, D. 1992. The Use of Metaphor to Structure Text Representations: Evidence for Metaphor-Based Schemas. Ph.D. dissertation, Yale University.

Boroditsky, L. 1997. Evidence for Metaphoric Representation: Perspective in Space and Time. In M. G. Shafto and P. Langley, eds., *Proceedings of the Nineteenth Annual Conference of the Cognitive Science Society*. Mahwah, N.J.: Erlbaum.

Gentner, D., and D. R. Gentner. 1982. Flowing Waters or Teeming Crowds: Mental Models of Electricity. In D. Gentner and A. L. Stevens, eds., *Mental Models*. Hillsdale, N. J.: Erlbaum.

Gentner, D., and J. Grudin. 1985. The Evolution of Mental Metaphors in Psychology: A Ninety-Year Retrospective. *American Psychologist* 40: 181–192.

Gibbs, R. 1994. *The Poetics of Mind: Figurative Thought, Language, and Understanding*. Cambridge: Cambridge University Press.

Gibbs, R., and J. O'Brien. 1990. Idioms and Mental Imagery: The Metaphorical Motivation for Idiomatic Meaning. *Cognition* 36: 35–68.

Kemper, S. 1989. Priming the Comprehension of Metaphors. *Metaphor and Symbolic Activity* 4: 1–18.

Nayak, N., and R. W. Gibbs. 1990. Conceptual Knowledge in Idiom Interpretation. *Journal of Experimental Psychology: General* 116: 315–330.

A3. Metaphor in Gesture and American Sign Language

McNeill is a classic book on the nature of spontaneous gesture and the first study of metaphoric gesture. Taub and Wilcox are the major sources on metaphor in American Sign Language.

Cienki, A. 1998. Metaphoric Gestures and Some of the Verbal Metaphorical Expression. In J. P. Koenig, ed., *Discourse and Cognition: Bridging the Gap*. Stanford: CSLI/Cambridge.

McNeill, D. 1992. *Hand and Mind: What Gestures Reveal About Thought*. Chicago: University of Chicago Press.

Taub, S. 1997. Language in the Body: Iconicity and Metaphor in American Sign Language. Ph.D. dissertation, University of California, Berkeley.

Wilbur, R. B. 1987. *American Sign Language: Linguistic and Applied Dimensions*. Boston: Little, Brown.

Wilbur, R. B., M. E. Bernstein, and R. Kantor. The Semantic Domain of Classifiers in American Sign Language. *Sign Language Studies* 46: 1–38.

Wilcox, P. 1993. Metaphorical Mapping in American Sign Language. Ph.D. dissertation, University of New Mexico.

A4. Categorization

Lakoff 1987 is a survey of categorization research up to the mid-1980s. The papers by Rosch are the classics in prototype theory.

Barsalou, L. W. 1983. Ad-Hoc Categories. *Memory and Cognition* 11: 211–227.

———. 1984. Determination of Graded Structures in Categories. Technical report, Psychology Department, Emory University.

Berlin, B., D. Breedlove, and P. Raven. 1974. *Principles of Tzeltal Plant Classification.* New York: Academic Press.

Craig, C., ed. 1986. *Categorization and Noun Classification.* Philadelphia: John Benjamins.

Hunn, Eugene. 1977. *Tzeltal Folk Zoology: The Classification of Discontinuities in Nature.* New York: Academic Press.

Kay, P. 1983. Linguistic Competence and Folk Theories of Language: Two English Hedges. In *Proceedings of the Ninth Annual Meeting of the Berkeley Linguistics Society,* 128–137. Berkeley: Berkeley Linguistics Society.

Lakoff, G. 1972. Hedges: A Study in Meaning Criteria and the Logic of Fuzzy Concepts. In *Papers from the Eighth Regional Meeting, Chicago Linguistic Society,* 183–228. Chicago: Chicago Linguistic Society. Reprinted in *Journal of Philosophical Logic* 2 (1973): 458–508.

———. 1987. *Women, Fire, and Dangerous Things: What Categories Reveal About the Mind.* Chicago and London: University of Chicago Press.

McNeill, D., and P. Freiberger. 1993. *Fuzzy Logic.* New York: Simon & Schuster.

Mervis, C. 1984. Early Lexical Development: The Contributions of Mother and Child. In C. Sophian, ed., *Origins of Cognitive Skills.* Hillsdale, N.J.: Erlbaum.

———. 1986. Child-Basic Object Categories and Early Lexical Development. In U. Neisser, ed., *Concepts and Conceptual Development: Ecological and Intellectual Factors in Categorization,* 201–233. Cambridge: Cambridge University Press.

Mervis, C., and E. Rosch. 1981. Categorization of Natural Objects. *Annual Review of Psychology* 32: 89–115.

Rosch, E. (E. Heider). 1973. Natural Categories. *Cognitive Psychology* 4: 328–350.

———. 1975a. Cognitive Reference Points. *Cognitive Psychology* 7: 532–547.

———. 1975b. Cognitive Representations of Semantic Categories. *Journal of Experimental Psychology: General* 104: 192–233.

———. 1977. Human Categorization. In N. Warren, ed., *Studies in Cross-Cultural Psychology.* London: Academic Press.

———. 1978. Principles of Categorization. In E. Rosch and B. B. Lloyd, *Cognition and Categorization,* 27–48. Hillsdale, N.J.: Erlbaum.

———. 1981. Prototype Classification and Logical Classification: The Two Systems. In E. Scholnick, ed., *New Trends in Cognitive Representation: Challenges to Piaget's Theory,* 73–86. Hillsdale, N.J.: Erlbaum.

Rosch, E., and B. B. Lloyd. 1978. *Cognition and Categorization.* Hillsdale, N.J.: Erlbaum.

Rosch, E., C. Mervis, W. Gray, D. Johnson, and P. Boyes-Braem. 1976. Basic Objects in Natural Categories. *Cognitive Psychology* 8: 382–439.

Schwartz, A. 1992. *Contested Concepts in Cognitive Social Science.* Honors thesis, University of California, Berkeley.

Smith, E. E., and D. L. Medin. 1981. *Categories and Concepts.* Cambridge, Mass.: Harvard University Press.

Taylor, J. 1989. *Linguistic Categorization: Prototypes in Linguistic Theory.* Oxford: Clarendon Press.

Tversky, B., and K. Hemenway. 1984. Object, Parts, and Categories. *Journal of Experimental Psychology: General* 113: 169–193.

Wittgenstein, L. 1953. *Philosophical Investigations.* New York: Macmillan.

Zadeh, L. 1965. Fuzzy Sets. *Information and Control* 8: 338–353.

A5. Color

Berlin and Kay is the classic study of basic color-term universals. Kay and McDaniel brought together cognitive science and color research in neuroscience. Hilbert, Hardin, and Thompson are the best overall works on philosophy and color.

Berlin, B., and P. Kay. 1969. *Basic Color Terms: Their Universality and Evolution.* Berkeley and Los Angeles: University of California Press.

DeValois, R. L., and K. DeValois. 1975. Neural Coding of Color. In E. C. Careterette and M. P. Friedman, eds., *Handbook of Perception,* vol. 5, *Seeing.* New York: Academic Press.

DeValois, R. L., and G. H. Jacobs. 1968. Primate Color Vision. *Science* 162: 533–540.

Hardin, C. L. 1988. *Color for Philosophers: Unweaving the Rainbow.* Cambridge, Mass.: Hackett.

Hilbert, D. R. 1987. *Color and Color Perception: A Study in Anthropocentric Realism.* Stanford University: Center for the Study of Language and Information.

———. 1992. Comparative Color Vision and the Objectivity of Color (Open Peer Commentary on E. Thompson, et al., Ways of Coloring: Comparative Color Vision as a Case Study for Cognitive Science). *Behavioral and Brain Sciences* 15: 38–39.

Kay, P., and C. McDaniel. 1978. The Linguistic Significance of the Meanings of Basic Color Terms. *Language* 54: 610–646.

Thompson, E. 1995. *Colour Vision: A Study in Cognitive Science and the Philosophy of Perception.* London and New York: Routledge.

A6. Framing

Fillmore is the major source for empirical linguistic research. Schank and Abelson started the major artificial intelligence approach. Holland and Quinn introduced the techniques to anthropology.

Fillmore, C. 1975. An Alternative to Checklist Theories of Meaning. In *Proceedings of the First Annual Meeting of the Berkeley Linguistics Society,* 123–131. Berkeley: Berkeley Linguistics Society.

———. 1976. Topics in Lexical Semantics. In P. Cole, ed., *Current Issues in Linguistic Theory,* 76–138. Bloomington: Indiana University Press.

———. 1978. The Organization of Semantic Information in the Lexicon. In *Papers from the Parasession on the Lexicon*, 1–11. Chicago: Chicago Linguistic Society.

———. 1982a. Towards a Descriptive Framework for Spatial Deixis. In R. J. Jarvella and W. Klein, *Speech, Place, and Action*, 31–59. London: Wiley.

———. 1982b. Frame Semantics. In Linguistic Society of Korea, ed., *Linguistics in the Morning Calm*, 111–138. Seoul: Hanshin.

———. 1985. Frames and the Semantics of Understanding. *Quaderni di Semantica* 6: 222–253.

———. 1997. *Lectures On Deixis*. Stanford: CSLI/Cambridge.

Holland, D. C., and N. Quinn, eds. 1987. *Cultural Models in Language and Thought*. Cambridge: Cambridge University Press.

Schank, R. C., and R. P. Abelson. 1977. *Scripts, Plans, Goals, and Understanding*. Hillsdale, N.J.: Erlbaum.

A7. Mental Spaces and Conceptual Blending

Coulson, S. 1997. Semantic Leaps: The Role of Frame-Shifting and Conceptual Blending in Meaning Construction. Ph.D. dissertation, University of California, San Diego.

Fauconnier, G. 1985. *Mental Spaces: Aspects of Meaning Construction in Natural Language*. Cambridge, Mass.: MIT Press.

———. 1997. *Mappings in Thought and Language*. Cambridge: Cambridge University Press.

Fauconnier, G., and E. Sweetser, eds. *Spaces, Worlds, and Grammar*. Chicago: University of Chicago Press.

Fauconnier, G., and M. Turner. 1994. Conceptual Projection and Middle Spaces. Department of Cognitive Science Technical Report 9401. University of California, San Diego.

———.1996. Blending as a Central Process of Grammar. In A. Goldberg, ed., *Conceptual Structure, Discourse, and Language*. Stanford: CSLI/Cambridge.

———. 1998. Principles of Conceptual Integration. In J.-P. Koenig, ed., *Discourse and Cognition: Bridging the Gap*. Stanford: CSLI/Cambridge.

———. Forthcoming. Conceptual Integration Networks. *Cognitive Science*.

———. Forthcoming. *Making Sense*.

Freeman, M. 1997. Grounded Spaces: Deictic-Self Anaphors in the Poetry of Emily Dickinson. *Language and Literature* 6:1, 7–28.

Grush, R., and N. Mandelblit. 1998. Blending in Language, Conceptual Structure, and the Cerebral Cortex. In J.-P. Koenig, ed., *Discourse and Cognition: Bridging the Gap*. Stanford: CSLI/Cambridge.

Mandelblit, N. 1995. Beyond Lexical Semantics: Mapping and Blending of Conceptual and Linguistic Structures in Machine Translation. In *Proceedings of the Fourth International Conference on the Cognitive Science of Natural Language Processing*. Dublin.

———. 1997. Grammatical Blending: Creative and Schematic Aspects in Sentence Processing and Translation. Ph.D. dissertation, University of California, San Diego.

Oakley, T. 1995. Ghost-Brother. In Presence: The Conceptual Basis of Rhetorical Effect. Ph.D. dissertation, University of Maryland.

Recanati, F. 1995. Le present ipistolaire: Une perspective cognitive. *L'information grammaticale* 66 (juin): 38–45.

Robert, A. 1998. Blending in the Interpretation of Mathematical Proofs. In J.-P. Koenig, ed., *Discourse and Cognition: Bridging the Gap*. Stanford: CSLI/Cambridge.

Sweetser, E. 1997. Mental Spaces and Cognitive Linguistics: A Cognitively Realistic Approach to Compositionality. Fifth International Cognitive Linguistics Conference.

Turner, M. 1995. *The Literary Mind*. New York: Oxford University Press.

———. 1996. Conceptual Blending and Counterfactual Argument in the Social and Behavioral Sciences. In P. Tetlock and A. Belkin, eds., *Counterfactual Thought Experiments in World Politics*. Princeton: Princeton University Press.

———. Forthcoming. Backstage Cognition in Reason and Choice. In A. Lupia, M. McCubbins, and S. Popkin, eds., *Elements of Political Reasoning*.

Turner, M., and G. Fauconnier. 1995. Conceptual Integration and Formal Expression. *Metaphor and Symbolic Activity* 10:3, 183–203.

———. 1998. Conceptual Integration in Counterfactuals. In J.-P. Koenig, ed., *Discourse and Cognition: Bridging the Gap*. Stanford: CSLI/Cambridge.

A8. Cognitive Grammar and Image Schemas

Barsalou, L. 1997. Perceptual Symbol Systems. Manuscript, Psychology Department, Emory University.

Brugman, C. 1981. *Story of* Over: *Polysemy, Semantics, and the Structure of the Lexicon*. New York and London: Garland.

Casad, E. 1982. *Cora Locationals and Structured Imagery*. Ph.D. dissertation, University of California, San Diego.

Casad, E., and R. W. Langacker. 1985. "Inside" and "Outside" in Cora Grammar. *International Journal of American Linguistics* 51: 247–281.

Chafe, W., and J. Nichols, eds. 1986. *Evidentiality: The Linguistic Coding of Epistemology*. Norwood, N.J.: Ablex.

Goldberg, A. 1995. *Constructions: A Construction Grammar Approach to Argument Structure*. Chicago and London: University of Chicago Press.

Haiman, J. 1980. The Iconicity of Grammar: Isomorphism and Motivation. *Language* 56, no. 3: 515–540.

Johnson, M. 1989. Image-Schematic Bases of Meaning. *RSSI (Recherches Semiotique, Semiotic Inquiry)* 9, nos. 1–3: 109–118.

———. 1991. The Imaginative Basis of Meaning and Cognition. In S. Kuchler and W. Melion, eds., *Images of Memory: On Remembering and Representation*, 74–86. Washington, D.C.: Smithsonian Institution Press.

Lakoff, G. 1986. Frame Semantic Control of the Coordinate Structure Constraint, *Proceedings of the Twenty-First Annual Meeting of the Chicago Linguistic Society*. Chicago: Chicago Linguistic Society.

———. 1987. There-Constructions. In G. Lakoff, *Women, Fire, and Dangerous Things*, Chicago: University of Chicago Press.

———. 1996. Reflections on Metaphor and Grammar. In S. Thompson and M. Shibatani, eds., *Festschrift for Charles Fillmore*. Philadelphia: Benjamins.

Langacker, R. 1983. Remarks on English Aspect. In P. Hopper, ed., *Tense and Aspect: Between Semantics and Pragmatics*, 265–304. Amsterdam: John Benjamins.

———. 1986, 1991. *Foundations of Cognitive Grammar*. 2 vols. Stanford: Stanford University Press.

———. 1990. *Concept, Image, and Symbol: The Cognitive Basis of Grammar*. Berlin and New York: Mouton de Gruyter.

Levinson, S., ed. 1992–present. *Working Papers from the Cognitive Anthropology Research Group*. Nijmegen, The Netherlands: Max Planck Institute.

Lindner, S. 1981. *A Lexico-Semantic Analysis of Verb-Particle Constructions With Up and Out*. Ph.D. dissertation, University of California, San Diego.

Slobin, D. 1970. Universals of Grammatical Development in Children. In G. B. Flores d'Arcais and W. J. M. Levelt, eds., *Advances in Psycholinguistics: Research Papers Presented at the Bressanone Conference on Psycholinguistics, . . . 1969*. Amsterdam: North-Holland.

———. 1985. Cross-linguistic Evidence for the Language-Making Capacity. In D. Slobin, ed., *A Cross-Linguistic Study of Language Acquisition*, vol. 2, *Theoretical Issues*. Hillsdale, N.J.: Erlbaum.

Talmy, L. 1983. How Language Structures Space. In H. L. Pick and L. P. Acredolo, eds., *Spatial Orientation: Theory, Research, and Application*. New York: Plenum Press.

———. 1985a. Force Dynamics in Language and Thought. In *Papers from the Parasession on Causatives and Agentivity*. Chicago: Chicago Linguistic Society.

———. 1985b. Lexicalization Patterns: Semantic Structure in Lexical Forms. In T. Shopen, ed., *Language Typology and Syntactic Description*, vol. 3. Cambridge and New York: Cambridge University Press.

A9. Discourse and Pragmatics

Green and Levinson are excellent introductory pragmatics texts. Schiffren and the Brown-Yule book provide excellent ways into the discourse literature.

Brown, G., and G. Yule. 1983. *Discourse Analysis*. Cambridge: Cambridge University Press.

Brown, P., and S. C. Levinson. 1987. *Politeness: Some Universals in Language Usage*. Cambridge: Cambridge University Press.

Gazdar, G. 1979. *Pragmatics: Implicature, Presupposition and Logical Form*. New York: Academic Press.

Goffman, E. 1981. *Forms of Talk*. Oxford: Blackwell.

Gordon, D., and G. Lakoff. 1975. Conversational Postulates. In P. Cole and J. L. Morgan, eds., *Syntax and Semantics*, vol. 3, *Speech Acts*, 83–106. New York: Academic Press.

Green, G. 1989. *Pragmatics and Natural Language Understanding*. Hillsdale, N.J.: Erlbaum.

Grice, P. 1989. *Studies in the Way of Words*. Cambridge, Mass., and London: Harvard University Press.

Gumperz, J. J. 1982a. *Discourse Strategies*. Cambridge: Cambridge University Press.

———. 1982b. *Language and Social Identity*. Cambridge: Cambridge University Press.

Hall, E. T. 1976. *Beyond Culture*. New York: Doubleday, Anchor. Reprint, 1981.

Keenan, E. O. 1976. The Universality of Conversational Implicature. *Language in Society* 5:67–80.

Lakoff, R. 1973. The Logic of Politeness, or Minding your P's and Q's. In *Papers from the Ninth Regional Meeting of the Chicago Linguistic Society*, 292–305. Chicago: Chicago Linguistic Society.

Levinson, S. C. 1983. *Pragmatics*. Cambridge: Cambridge University Press.

Saville-Troike, M. 1989. *The Ethnography of Communication: An Introduction*. 2d ed. Oxford: Blackwell.

Schiffrin, D. 1994. *Approaches to Discourse Analysis*. Oxford, Eng., and Cambridge, Mass.: Blackwell.

Scollon, R., and S. W. Scollon. 1995. *Intercultural Communication: A Discourse Approach*. Oxford, Eng., and Cambridge, Mass.: Blackwell.

Stubbs, M. 1983. *Discourse Analysis: The Sociolinguistic Analysis of Natural Language*. Chicago: University of Chicago Press; Oxford: Blackwell.

Tannen, D. 1986. *That's Not What I Meant!: How Conversational Style Makes or Breaks Your Relations with Others*. New York: Morrow.

———. 1991. *You Just Don't Understand: Women and Men in Conversation*. New York: Ballantine.

Tannen, D., ed. 1993. *Framing in Discourse*. New York and Oxford: Oxford University Press.

van Dijk, T. 1985. *Handbook of Discourse Analysis*. 4 vols. New York and London: Academic Press.

Weiser, A. 1974. Deliberate Ambiguity. *Papers from the Tenth Regional Meeting of the Chicago Linguistic Society*, 723–731. Chicago: Chicago Linguistic Society.

———. 1975. How Not to Answer a Question: Purposive Devices in Conversational Strategy. *Papers from the Eleventh Regional Meeting of the Chicago Linguistic Society*, 649–660. Chicago: Chicago Linguistic Society.

A10. Decision Theory: The Heuristics and Biases Approach
These are sample papers from a huge literature. They are chosen largely because they demonstrate framing and prototype effects.

Kahneman, D., and A. Tversky. 1983. Can Irrationality Be Intelligently Discussed? *Behavioral and Brain Sciences* 6: 509–510.

———. 1984. Choices, Values, and Frames. *American Psychologist* 39: 341–350.

Tversky, A., and D. Kahneman. 1974. Judgment Under Uncertainty: Heuristics and Biases. *Science* 185: 1124–1131.

———. 1981. The Framing of Decisions and the Psychology of Choice. *Science* 211: 453–458.

———. 1988. Rational Choice and the Framing of Decisions. In D. E. Bell, H. Raiffa, and A. Tversky, eds., *Decision Making: Descriptive, Normative, and Prescriptive Interactions*, 167–192. Cambridge: Cambridge University Press.

B. Neuroscience and Neural Modeling

B1. Basic Neuroscience
Churchland is a wonderful introduction to neuroscience for philosophers. Edelman not only surveys his own views on neuroscience but also gives an account of how cognitive linguistics meshes with neuroscience research. Damasio reviews evidence showing that emotion is crucial to rationality.

Churchland, P. S. 1986. *Neurophilosophy: Toward a Unified Science of the Mind/Brain.* Cambridge, Mass.: Bradford/MIT Press.

Churchland, P. S., and T. Sejnowski. 1992. *The Computational Brain.* Cambridge, Mass.: Bradford/MIT Press.

Crick, F. 1994. *The Astonishing Hypothesis: The Scientific Search for the Soul.* New York: Scribner.

Damasio, A. 1994. *Descartes' Error: Emotion, Reason, and the Human Brain.* New York: Grosset/Putnam.

Dehaene, S. 1997. *The Number Sense.* Oxford and New York: Oxford University Press.

Edelman, G. 1992. *Bright Air, Brilliant Fire: On the Matter of Mind.* New York: Basic Books.

Hobson, J. A. 1994. *The Chemistry of Conscious States.* Boston: Little, Brown.

Hubel, D. 1988. *Eye, Brain, and Vision.* New York: Scientific American Library.

Hubel, D., and T. Wiesel. 1959. Receptive Fields of Single Neurons in the Cat's Visual Cortex. *Journal of Physiology* 148: 574–591.

———. 1977. Functional Architecture of the Macaque Monkey Visual Cortex. *Proceedings of the Royal Society of London, Series B,* 198: 1–59.

Jeannerod, M. 1997. *The Cognitive Neuroscience of Action.* Oxford: Blackwell.

Ramachandran, V. S., and R. L. Gregory. 1991. Perceptual Filling-in of Artificially Induced Scotomas in Human Vision. *Nature* 350: 699–702.

Zeki, S. 1993. *A Vision of the Brain*. Oxford: Blackwell.

B2. Structured Connectionist Modeling

Ajjanagadde, V., and L. Shastri. 1991. Rules and Variables in Neural Nets, *Neural Computation* 3: 121–134.

Bailey, D. 1997. *A Computational Model of Embodiment in the Acquisition of Action Verbs*. Ph.D. dissertation, Computer Science Division, EECS Department, University of California, Berkeley.

Bailey, D., J. Feldman, S. Narayanan, and G. Lakoff. 1997. Modeling Embodied Lexical Development. In M. G. Shafto and P. Langley, eds., *Proceedings of the Nineteenth Annual Conference of the Cognitive Science Society*. Mahwah, N.J.: Erlbaum.

Ballard, D. 1997. *An Introduction to Natural Computation*. Cambridge, Mass.: Bradford/MIT Press.

Feldman, J. 1982. Dynamic Connections in Neural Networks. *Biological Cybernetics* 46: 27–39.

———. 1985. Four Frames Suffice: A Provisional Model of Vision and Space. *Behavioral and Brain Sciences* 8: 265–289.

———. 1988. Computational Constraints on Higher Neural Representations. In E. Schwartz, ed., *Proceedings of the System Development Foundation Symposium on Computational Neuroscience*, Cambridge, Mass.: Bradford/MIT Press.

Feldman, J., and D. H. Ballard. 1982. Connectionist Models and Their Properties. *Cognitive Science* 6: 205–254.

Feldman, J., and L. Shastri. 1985. Evidential Reasoning in Semantic Networks. *Proceedings of the Ninth International Joint Conference on Artificial Intelligence*. 465–474.

Feldman, J., G. Lakoff, A. Stolcke, and S. Weber. 1990. Miniature Language Acquisition: A Touchstone for Cognitive Science. *Proceedings of the Twelfth Annual Conference of the Cognitive Science Society*, Cambridge, Mass.: MIT Press. 686–693.

Goddard, N. 1992. The Perception of Articulated Motion: Recognizing Moving Light Displays. Ph.D. dissertation, University of Rochester.

Hummel, J., and I. Biederman. 1990. *Dynamic Binding in a Neural Network for Shape Recognition*. Technical Report 90–95. Department of Psychology, University of Minnesota.

Lange, T. E., and M. G. Dyer. 1989. High-Level Inferencing in a Connectionist Network. *Connection Science* 1, no. 2: 181–217.

Narayanan, S. 1997a. Embodiment in Language Understanding: Sensory-Motor Representations for Metaphoric Reasoning About Event Descriptions. Ph.D. Dissertation, Department of Computer Science, University of California, Berkeley.

———. 1997b. Talking the Talk Is Like Walking the Walk: A Computational Model of Verbal Aspect. In M. G. Shafto and P. Langley, eds., *Proceedings of the Nineteenth Annual Conference of the Cognitive Science Society*. Mahwah, N.J.: Erlbaum.

Nenov, V. 1991. Perceptually Grounded Language Acquisition: A Neural/Procedural Hybrid Model. Ph.D. dissertation, University of California, Los Angeles.

Omohundro, S. 1992. *Best-First Model Merging for Dynamic Learning and Recognition*. Technical Report TR–92–004. International Computer Science Institute, Berkeley.

Regier, T. 1995. A Model of the Human Capacity for Categorizing Spatial Relations. *Cognitive Linguistics* 6–1: 63–88.

———. 1996. *The Human Semantic Potential: Spatial Language and Constrained Connectionism*. Cambridge, Mass.: MIT Press.

Shastri, L. 1988a. A Connectionist Approach to Knowledge Representation and Limited Inference. *Cognitive Science* 12, no. 3: 331–392.

———. 1988b. *Semantic Networks: An Evidential Formalization and Its Connectionist Realization*. Los Altos, Calif.: Morgan Kaufmann; London: Pitman.

———. 1995. Structured Connectionist Models. In M. Arbib, ed., *The Handbook of Brain Theory and Neural Networks*, 949–952. Cambridge, Mass.: MIT Press.

———. 1996. Temporal Synchrony, Dynamic Bindings, and SHRUTI, A Representational but Non-Classical Model of Reflexive Reasoning. *Behavioral and Brain Sciences* 19, no. 2: 331–337.

———. 1997a. A Model of Rapid Memory Formation in the Hippocampal System. In M. G. Shafto and P. Langley, eds., *Proceedings of the Nineteenth Annual Conference of the Cognitive Science Society*. Mahwah, N.J.: Erlbaum.

———. 1997b. *Rapid Learning of Binding-Match and Binding-Error Detector Circuits via Long-Term Potentiation*. Technical Report TR–97–037. International Computer Science Institute, Berkeley.

———. 1997c. Recent Advances in SHRUTI. In F. Maire, R. Hayward, and J. Diederich, eds., *Connectionist Systems for Knowledge Representation and Deduction*. Neurocomputing Research Center, Queensland University of Technology.

Shastri, L., and V. Ajjanagadde. 1993. From Simple Associations to Systematic Reasoning. *Behavioral and Brain Sciences* 16:3, 417–494.

Shastri, L., and J. A. Feldman. 1986. Neural Nets, Routines and Semantic Networks. In N. E. Sharkey, ed., *Advances in Cognitive Science*. Chichester, Eng.: Ellis Horwood Publishers, 158–203.

Weber, S. 1989. A Structured Connectionist Approach to Direct Inferences and Figurative Adjective-Noun Combinations. Ph.D. dissertation, University of Rochester.

C. Philosophy

C1. Cognitive Science and Moral Philosophy

Classical moral philosophy assumed that the empirical study of the mind could not affect moral issues. These five works challenge that assumption.

Churchland, P. M. 1995. *The Engine of Reason, the Seat of the Soul: A Philosophical Journey into the Brain.* Cambridge, Mass.: MIT Press.

Flanagan, O. 1991. *Varieties of Moral Personality: Ethics and Psychological Realism.* Cambridge, Mass.: Harvard University Press.

Johnson, M. 1993. *Moral Imagination: Implications of Cognitive Science for Ethics.* Chicago: University of Chicago Press.

———. 1996. How Moral Psychology Changes Moral Philosophy. In L. May, A. Clarke, and M. Friedman, eds., *Mind and Morals: Essays on Ethics and Cognitive Science*, 45–68. Cambridge, Mass.: MIT Press.

———. 1998. Cognitive Science and Morality. In W. Bechtel and G. Graham, eds., *A Companion to Cognitive Science.* Oxford: Blackwell.

C2. Philosophical Sources

Aristotle. *The Basic Works of Aristotle.* Edited by R. McKeon. New York: Random House, 1941.

Austin, J. L. 1975. *How to Do Things with Words.* Cambridge Mass.: Harvard University Press.

Bermudez, J., A. Marcel, and N. Eilan, eds. 1995. *The Body and the Self.* Cambridge, Mass: MIT Press.

Block, N. 1980. What Is Functionalism? In N. Block, ed., *Readings in Philosophy of Psychology*, vol. 1, 171–184. Cambridge, Mass.: Harvard University Press.

Chomsky, N. 1957. *Syntactic Structures.* The Hague: Mouton.

———. 1965. *Aspects of the Theory of Syntax.* Cambridge, Mass.: MIT Press.

———. 1966. *Cartesian Linguistics.* New York: Harper & Row.

———. 1975. *Reflections on Language.* New York: Pantheon.

———. 1981. *Lectures on Government and Binding.* Cambridge, Mass.: MIT Press.

———. 1986. *Knowledge of Language: Its Nature, Origin, and Use.* New York: Praeger.

———. 1991. Linguistics and Adjacent Fields: A Personal View. In A. Kasher, ed., *The Chomskyan Turn*, 3–25. New York: Blackwell.

———. 1995. *The Minimalist Program.* Cambridge, Mass.: MIT Press.

Davidson, D. 1978. What Metaphors Mean. *Critical Inquiry* 5, no. 1: 31–47.

DeMan, P. 1978. The Epistemology of Metaphor. *Critical Inquiry* 5, no. 1: 13–30.

Dennett, D. 1991. *Consciousness Explained.* Boston: Little, Brown.

Descartes, R. [1628] 1970. *Rules for the Direction of the Understanding.* In E. S. Haldane and G. R. T. Ross, eds., *The Philosophical Works of Descartes.* 2 vols. Reprint, Cambridge: Cambridge University Press.

———. [1637] 1970. *Discourse on Method.* In E. S. Haldane and G. R. T. Ross (eds.), *The Philosophical Works of Descartes.* 2 vols. Reprint, Cambridge: Cambridge University Press.

———. [1641] 1970. *Meditations on First Philosophy.* In E. S. Haldane and G. R. T. Ross, eds., *The Philosophical Works of Descartes.* 2 vols. Reprint, Cambridge: Cambridge University Press.

———. [1644] 1970. *Principles of Philosophy*. In E. S. Haldane and G. R. T. Ross, eds., *The Philosophical Works of Descartes*. 2 vols. Reprint, Cambridge: Cambridge University Press.

Dewey, J. 1922. *Human Nature and Conduct: An Introduction to Social Psychology*. New York: Holt.

———. 1925. *Experience and Nature*. Chicago: Open Court.

Dummett, M. 1993. *Origins of Analytical Philosophy*. Cambridge, Mass.: Harvard University Press.

Flanagan, O. 1992. *Consciousness Reconsidered*. Cambridge, Mass.: MIT Press.

Fodor, J. 1975. *The Language of Thought*. New York: Crowell.

———. 1981. *Representations*. Cambridge, Mass.: MIT Press.

———. 1987. *Psychosemantics: The Problem of Meaning in the Philosophy of Mind*. Cambridge, Mass.: MIT Press.

Frege, G. 1892. On Sense and Reference. In P. Geach and M. Black, eds., *Translations from the Writings of Gottlob Frege*. 2d ed., 56–78. Oxford: Blackwell, 1966.

Gallagher, S. 1995. Body Schema and Intentionality. In J. Bermudez, A. Marcel, and N. Eilan, eds., *The Body and the Self*, 225–244. Cambridge, Mass.: MIT Press.

Gallie, W. B. 1956. Essentially Contested Concepts. *Proceedings of the Aristotelian Society* 31: 167–198.

Grice, H. P. 1975. Logic and Conversation. In P. Cole and J. Morgan, eds., *Syntax and Semantics*, vol. 3, *Speech Acts*, 41–58. New York: Academic Press.

———. 1989. *Studies in the Way of Words*. Cambridge, Mass.: Harvard University Press.

Hart, H. L. A., and A. M. Honoré. 1958. *Causation in the Law*. Cambridge, Mass.: Harvard University Press.

Haugeland, J. 1985. *Artificial Intelligence: The Very Idea*. Cambridge, Mass., and London: MIT Press.

Hobbes, T. 1651. *Leviathan*. London: Dent & Sons.

Johnson, M. 1991. Knowing Through the Body. *Philosophical Psychology* 4, no. 1: 3–18.

———. 1992. Philosophical Implications of Cognitive Semantics. *Cognitive Linguistics* 3, no. 4: 345–366.

Kant, I. [ca. 1780] 1980. *Lectures on Ethics*. Translated by L. Infield. Indianapolis, Ind.: Hackett.

———. [1785] 1983. *Grounding for the Metaphysics of Morals*. Translated by J. Ellington. In *[Kant's] Ethical Philosophy*. Indianapolis, Ind.: Hackett.

———. [1797] 1983. *Metaphysics of Morals*. Translated by J. Ellington. In *[Kant's] Ethical Philosophy*. Indianapolis, Ind.: Hackett.

Kleene, S. C. 1967. *Mathematical Logic*. New York: Wiley.

Lakoff, G. 1971. Linguistics and Natural Logic. *Synthese*. Reprinted in Davidson and Harman, eds., *Semantics in Natural Language*. New York: Reidel, 1972.

Leder, D. 1990. *The Absent Body*. Chicago: University of Chicago Press.

Le Poidevin, R., and M. MacBeath, eds. 1993. *The Philosophy of Time*. Oxford: Oxford University Press.

Locke, J. [1690; 5th ed. 1700] 1964. *An Essay Concerning Human Understanding*. Edited by J. Yolton. London: Dutton.

McCauley, R. 1986. Intertheoretic Relations and the Future of Psychology. *Philosophy of Science* 53: 179–199.

———. 1996. Explanatory Pluralism and the Co-Evolution of Theories in Science. In R. McCauley, ed., *The Churchlands and Their Critics*, 17–47. Oxford: Blackwell.

Merleau-Ponty, M. 1962. *Phenomenology of Perception*. Translated by C. Smith. London: Routledge and Kegan Paul.

Montague, R. 1974. *Formal Philosophy*. New Haven, Conn.: Yale University Press.

Neisser, U., ed. 1993. *The Perceived Self: Ecological and Interpersonal Sources of Self-Knowledge*. Cambridge: Cambridge University Press.

Plato. *The Collected Dialogues of Plato*. Edited by E. Hamilton and H. Cairns. Princeton: Princeton University Press. 1961.

Prior, A. N. 1993. Changes in Events and Changes in Things. In R. Le Poidevin and M. MacBeath, eds., *The Philosophy of Time*, 35–46. Oxford: Oxford University Press.

Putnam, H. 1981. *Reason, Truth, and History*. Cambridge: Cambridge University Press.

———. 1988. *Representation and Reality*. Cambridge, Mass.: Bradford/MIT Press.

Quine, W. V. O. 1959. *Methods of Logic*. Rev. ed. New York: Holt, Rinehart, and Winston.

———. 1960. *Word and Object*. Cambridge, Mass.: MIT Press.

———. 1969. *Ontological Relativity and Other Essays*. New York and London: Columbia University Press.

Rorty, R. 1979. *Philosophy and the Mirror of Nature*. Princeton: Princeton University Press.

———. 1989. *Contingency, Irony, and Solidarity*. Cambridge: Cambridge University Press.

Rosenbloom, P. 1950. *The Elements of Mathematical Logic*. New York: Dover.

Russell, B. 1903. *Principles of Mathematics*. 2d ed. New York: Norton.

Searle, J. 1969. *Speech Acts*. Cambridge: Cambridge University Press.

———. 1990. Artificial Intelligence: A Debate. *Scientific American* 262 (1): 25–37.

———. 1995. *The Construction of Social Reality*. New York: Free Press.

Searle, J., and D. Vanderveken. 1985. *Foundations of Illocutionary Logic*. Cambridge and New York: Cambridge University Press.

Sheets-Johnstone, M. 1990. *The Roots of Thinking*. Philadelphia: Temple University Press.

Van Fraassen, B. C. 1968. Presupposition, Implication, and Self-Reference. *Journal of Philosophy* 65: 136–152.

Varela, F., E. Thompson, and E. Rosch. 1991. *The Embodied Mind: Cognitive Science and Human Experience*. Cambridge, Mass.: MIT Press.

Wheelwright, P., ed. 1966. *The Presocratics*. New York: Odyssey Press.

Wittgenstein, L. 1953. *Philosophical Investigations*. New York: Macmillan.

D. Other Linguistics

Comrie, B. 1976. *Aspect*. Cambridge Textbooks in Linguistics. Cambridge: Cambridge University Press.

Dowty, D. 1979. *Word Meaning and Montague Grammar*. New York: Synthese Language Library, Reidel.

Goldsmith, J. 1985. A Principled Exception to the Coordinate Structure Constraint. In *Papers from the Twenty-First Regional Meeting, Chicago Linguistic Society, Part I*. Chicago: Chicago Linguistic Society.

MacWhinney, B. 1995. The CHILDES Project: Tools for Analyzing Talk. Hillsdale, N.J.: Erlbaum.

Malotki, E. 1983. *Hopi Time*. Berlin: Mouton.

Moens, M., and M. Steedman. 1988. Temporal Ontology and Temporal Reference. In *Proceedings of the Association for Computational Linguistics—88*, vol. 4, no. 2 (June): 15–29

Ross, J. R. 1967. *Constraints on Variables in Syntax*. Ph.D. dissertation, Massachusetts Institute of Technology. Published as *Infinite Syntax* (Norwood, N.J.: Ablex, 1985).

Vendler, Z. 1967. *Linguistics in Philosophy*. Ithaca, N.Y.: Cornell University Press.

Whorf, B. L. 1956. *Language, Thought, and Reality: Selected Writings of Benjamin Lee Whorf*. Edited by J. B. Carroll. Cambridge, Mass.: MIT Press.

E. Miscellaneous

Abram, D. 1996. *The Spell of the Sensuous: Perception and Language in a More-Than-Human World*. New York: Pantheon.

Borg, M. J. 1997. *The God We Never Knew: Beyond Dogmatic Religion to a More Authentic Contemporary Faith*. San Francisco: Harper San Francisco.

Gilligan, C. 1982. *In a Different Voice: Psychological Theory and Women's Development*. Cambridge: Cambridge University Press.

Hawking, S. 1988. *A Brief History of Time*. New York: Bantam.

Matt, D. C. 1995. *The Essential Kabbalah: The Heart of Jewish Mysticism*. San Francisco: Harper San Francisco.

Minsky, M. 1986. *The Society of Mind*. New York: Simon & Schuster.

Spretnak, C. 1991. *States of Grace: The Recovery of Meaning in the Postmodern Age*. San Francisco: Harper San Francisco.

———. 1997. *The Resurgence of the Real: Body, Nature, and Place in a Hypermodern World*. Reading, Mass.: Addison-Wesley.

Index